T0396622

Poorly
Soluble Drugs

Pan Stanford Series on Pharmaceutical Analysis

Gregory K. Webster
Chief Editor

J. Derek Jackson and Robert G. Bell
Editors

Titles in the Series

Published

Vol. 1

Poorly Soluble Drugs: Solution and Drug Release

Gregory K. Webster, J. Derek Jackson, and Robert G. Bell, eds.

2017

978-981-4745-45-1 (Hardcover)
978-981-4745-46-8 (eBook)

Forthcoming

Vol. 2

Chromatographic Method Development

Gregory K. Webster and Laila Kott, eds.

2018

Pan Stanford Series on Pharmaceutical Analysis
Volume 1

Poorly Soluble Drugs

Dissolution and Drug Release

edited by

Gregory K. Webster
J. Derek Jackson
Robert G. Bell

PAN STANFORD PUBLISHING

Published by

Pan Stanford Publishing Pte. Ltd.
Penthouse Level, Suntec Tower 3
8 Temasek Boulevard
Singapore 038988

Email: editorial@panstanford.com
Web: www.panstanford.com

British Library Cataloguing-in-Publication Data
A catalogue record for this book is available from the British Library.

Poorly Soluble Drugs: Dissolution and Drug Release

Copyright © 2017 Pan Stanford Publishing Pte. Ltd.

ISBN 978-981-4745-45-1 (Hardcover)
ISBN 978-981-4745-46-8 (eBook)

Printed in the USA

Contents

Foreword

Roughly twenty years ago, it became clear that drug product development had entered a new era of difficulty with the increased throughput of therapeutically effective but poorly soluble drug candidates. Gone are the days when all drug candidates were rapidly dissolving and absorbing drugs that were relatively easy to formulate and even easier to test. Now the development of an in vitro method for poorly soluble drugs is not boring. Rather, it's a stimulating endeavor.

High-throughput screening has contributed to the invention and discovery of many new poorly soluble molecules. This book, *Poorly Soluble Drugs: Dissolution and Drug Release*, is most timely as the authors are up to the challenge of sharing the knowledge and tools to tackle the in vitro testing and manufacturing of these products. This work is unique in that it has provided the linkage between testing and formulating the products with equal importance given to each aspect.

In vitro testing of poorly soluble drug products is especially challenging and important since dissolution is the rate-limiting step to drug absorption and exposure. The path forward is clear: methods must be able to take advantage of this characteristic by providing meaningful elucidation of the release rate or, in some cases, the actual release mechanism, and hence giving critical clinically relevant information.

Poorly soluble drugs require special attention during formulation and manufacturing to enhance the effectiveness of the drug through methods as simple as reducing particle size to the much more complex areas of formulation manipulation and engineering technology to increase in vivo concentrations and adsorption.

The practical matter is that the demands from regulators, the globalization of pharmaceuticals, and the competitive arena of market share drive the need to quickly educate and strengthen the knowledge of scientists working on these products. This book is quite essential to this effort.

The development of clinically relevant dissolution methods for drug products with limited water solubility has been a challenge for scientists in the drug industry as well as the regulatory agencies. The trend has started with the powerful tools available through quality by design (QbD) to create a clinically relevant dissolution test. Designing robust dosage forms of poorly soluble actives employs a thorough understanding of the components, matrix, and variability, thus following QbD concepts. This book gives a thorough investigation of the role of QbD with poorly soluble dosage forms, including design of experiments (DOE).

Development scientists are tasked with making these compounds soluble in a medium that is foreign to the poorly soluble drug but is necessary for oral drug formulation absorption. Aqueous solubility is the primary gauge of the success or failure of a drug and drug product. Solubility and dissolution performance in the gastrointestinal tract are critical for the bioavailability, and hence efficacy, of the product.

There are some emerging topics that are starting to acquire additional in-depth understanding—in particular, topics such as sink versus non-sink conditions in the dissolution method, contribution of solid-state properties, the chemistry of surfactants, in silico modeling, dose dumping, and capsule properties. The chapters in this book give these and other new topics well-referenced and refreshingly up-to-date attention. The work in this book bridges with established art and then builds links to, in some cases, entirely new directions.

The authors are from industry and academia, giving a well-rounded approach to this unique topic that has not been treated in book form to date. The subject is treated well beyond current guidances and USP chapters, a step much further than the status quo. I know the authors personally or by reputation, and they are experts in their areas. Many have a long history of direct involvement with the in vitro release test from the simpler testing equipment and methods to more complex and in some cases closer to the in vivo condition.

In vitro testing shows that the product is dissolved and therefore available for absorption and therapeutic effect thus linking what occurs in the patient's body to the efficacy of the product. The FDA and USP have emphasized the dissolution test for this reason as a proof that a commercial product on the market for many years will still be efficacious if it passes that test developed with the biobatch formulation. Hence the push to improve and make more robust the dissolution methods to link to in vivo performance. A way to forecast the in vivo performance is by making the dissolution test conditions as close to in vivo conditions as is possible. Approaches to assist the analyst in developing a sensitive method to characterize the release rate are explored thoroughly in this book along with the topics of in vitro and in vivo correlations and relationships.

Historically, a defining moment for poorly soluble drugs is the Biopharmaceutics Classification System, where the poorly soluble drug was described and characterized with some clarity. At that time, it became apparent that biowaivers for poorly soluble dosage forms were in most part unobtainable. With the exception of in vitro and in vivo correlations, clinical studies seemed to always be necessary, and little has changed over the years in this regard. The book offers insight into the development of predictive dissolution methods. Furthermore, knowing that poorly soluble drugs are uniquely sensitive to the testing environment (e.g. equipment design, vibration and de-aeration) is helpful when interpreting dissolution results.

Formal education of the industry analyst may not be provided for this topic. Because developing methodology for poorly soluble drugs demands more resources and research, this work will be helpful to the analyst to work more efficiently and solve problems more rapidly with this new knowledge in hand.

I commend the authors for their very considerable effort in bringing out this valuable publication.

Vivian Gray
Managing Director
Dissolution Technologies, Inc.
Hockessin, DE, USA
October 2015

Chapter 1

The Modern Pharmaceutical Development Challenge: BCS Class II and IV Drugs

Gregory K. Webster,[a] Robert G. Bell,[b] and J. Derek Jackson[c]

[a] *AbbVie Inc., Global Research and Development, 1 N. Waukegan Rd., North Chicago, IL 60064, USA*
[b] *Drug & Biotechnology Development, LLC, 406 South Arcturas Avenue, Suite 5, Clearwater, FL 33765, USA*
[c] *Flexion Therapeutics, Inc., 10 Mall Road, Suite 301, Burlington, MA 01803, USA*
gregory.webster@abbvie.com, rgb@drugbiodev.com, djackson@flexiontherapeutics.com

1.1 Introduction

Since the 1960s, pharmaceutical companies have been charged with monitoring the characteristics of their oral dosage forms dissolving in controlled media. Early dissolution testing focused on the quality control of the dosage form manufacturing. Dissolution testing provided unique capabilities in monitoring integrated production parameters that affect the dissolution rate: tablet hardness, excipient control, particle size, etc. The technique became required for the routine testing of oral dosage forms worldwide.

Poorly Soluble Drugs: Dissolution and Drug Release
Edited by Gregory K. Webster, J. Derek Jackson, and Robert G. Bell
Copyright © 2017 Pan Stanford Publishing Pte. Ltd.
ISBN 978-981-4745-45-1 (Hardcover), 978-981-4745-46-8 (eBook)
www.panstanford.com

The early history of dissolution testing is well documented in the literature.[1-3]

The additional advantage of early dissolution testing was the relative simplicity of the active pharmaceutical ingredients (API) and their associated dosage forms. These soluble drugs were readily bioavailable as they tended to be both highly soluble and highly permeable in the gastrointestinal tract. Thus, for highly soluble and permeable drugs, dissolution assessment was adequate to ensure clinical performance through simple standardized solutions such as water and acidic media. Discrimination focused primarily on the disintegration and dispersion from solid oral dosage forms with the active ingredients behaving predictably. Under these conditions, the dissolution test was effective in predicting performance in human clinical trials and was denoted as a clinically relevant method. With further evolution in drug formulation technologies, dissolution testing harmonized on the four basic apparatus still in use today and commonly referred to by their United States Pharmacopeia (USP) designations of apparatus 1, apparatus 2, apparatus 3, and apparatus 4.[4] Other USP apparatus designations, such as apparatus 7, tend to be for non-oral dosage forms.

Today, dissolution testing has broadened its applications and scope in moving forward from a simple quality control test to use in predicting in vivo/in vitro correlations (IVIVC) for soluble and permeable dosage forms and clinical relevance for many others. The simple formulations found in the developing years of dissolution testing have given way to more complex technologies to advance the bioavailability of less soluble molecules. The goal of this book is to move past the existing dissolution texts referenced earlier, which primarily denotes dissolution testing for soluble drugs, and to focus on the issues of dissolution testing current with the molecules in development today. This text builds upon the solid foundation of the earlier works[1-3] to the current application of dissolution and drug release technologies with an emphasis on poorly soluble drugs.

1.2 Changing Drug Emphasis

The dissolution platforms of USP apparatus 1–4 were key technologies in facilitating soluble drugs to market. The techniques, instruments, and the simple buffer systems used ideally characterized the dosage forms being developed. However, as the active pharmaceutical ingredients (API) became increasingly more complex (in terms of solubility and permeation in the gastrointestinal tract), these simple drug release mechanisms did not correlate. The Biopharmaceutics Classification System was developed by Amidon et al.[6] and published as a guidance by the U.S. Food and Drug Administration (FDA) for predicting the intestinal drug absorption of oral dosage forms.[7] The BCS system has become the gold standard to categorize and estimate oral drug absorption based on the drug's solubility and intestinal permeability characteristics.

1.2.1 BCS Classification System

The BCS categorization of drugs is based on the premise that as a drug dissolves, this concentration is available to move across the membrane and correlate to intestinal absorption. Gastric solubility is established through in vitro chemical testing at various conditions and pH's representative of the human gastrointestinal tract. The permeability of the drug is based on initial lipophilicity testing and further studied in animal models, tissue studies, cultured epithelial cells such as Caco-2 testing, and ultimately in humans (mass balance, absolute bioavailability intestinal perfusion testing, etc.).

With the drug solubility and permeability established, the BCS system segregates drugs into four classes, as illustrated in Table 1.1.

Per the FDA Guidance, the target drug is deemed "highly soluble" when the highest dose strength is soluble in <250 mL of aqueous media over a pH range of 1 to 7.5. The drug is deemed "highly permeable" if the extent of absorption in humans is determined to

Table 1.1   BCS drug classification

	High solubility	Low solubility
High permeability	Class 1 High solubility High permeability	Class 2 Low solubility High permeability
Low permeability	Class 3 High solubility Low permeability	Class 4 Low solubility Low permeability

be >90% of an administered dose. In addition, a drug formulation is deemed "rapidly dissolving" when > 85% of the labeled amount of drug substance dissolves within 30 minutes using USP apparatus 1 or 2 and a volume of ≤500 mL of buffered media.

Knowing the drug's BSC category allows the pharmaceutical scientist to evaluate the rate-limiting step in the absorption of the drug. For class 1 drugs the high solubility and high permeability of the API indicate that the absorption in the gastrointestinal tract should be dissolution rate limited. For class 2 drugs, because the drug is less soluble but still very permeable, this class of drugs should also be dissolution rate limited. With the solubility being high in class 3 but permeability of the drug low, the API in this class becomes absorption rate limited. For this class of drugs, dissolution is seldom clinically relevant. The FDA has recently issued a guideline denoting that, under special cases, a class 3 drug can be addressed with class 1 specifications for biowaiver studies.[8] However, this guidance does not propose that class 3 drugs can readily achieve IVIVC via traditional dissolution testing. IVIVC for class 1 and 3 drug products is not likely unless drug dissolution is significantly slowed due to formulation (e.g., MR formulation) or the compound is borderline BCS 1 with respect to solubility. This is acknowledged in the FDA Guidance for Industry for Dissolution Testing of IR solid oral dosage forms.[9] Class 4 drugs rely upon transporters and other biological means to transport across the membrane. Dissolution may be able to characterize this transport, but as with BCS class 3 drugs, dissolution is challenged to characterize the absorption of these drugs.

1.2.2 Poorly Soluble Drugs

A main focus of this book is to focus on the dissolution and drug release of BCS class 2 drugs. These drugs are permeable but with limited solubility. As of 2006, BCS class 2 drugs made up approximately a third of the global pharmaceutical market.[10] As will be illustrated in Chapter 5, formulation technologies have gone a long way to increasing the bioavailability of these molecules. Typically, these formulations are often working with amorphous or nanoparticle material technologies.

1.3 The Dissolution Market

Dissolution is a significant technology found in the laboratories of every major pharmaceutical business concern with oral dosage forms and predominately uses chromatography and spectroscopy for final quantitative analysis. As such, perhaps it is time that dissolution testing warrants a chapter in undergraduate instrumental analysis textbooks. In 2009, approximately 3400 instruments were sold with a market expecting to grow at 8% annually.[11] Today this market is valued at over $150 million and tied directly to the pharmaceuticals market. The pharmaceutical industry accounts for approximately 75% of these sales. The remaining 25% is split between contract research organizations, biotech, academia and agriculture.

As seen in Fig. 1.1, quality control testing dominates the demand for dissolution testing.

As Fig. 1.2 shows, in 2009, the largest dissolution vendor was Varian/Agilent. The market was diversified with several vendors; however, even with this diversity, the technique still today revolves around the standard USP designations of apparatus 1 and 2, with less significant portion of the market operating apparatus 3 and 4.

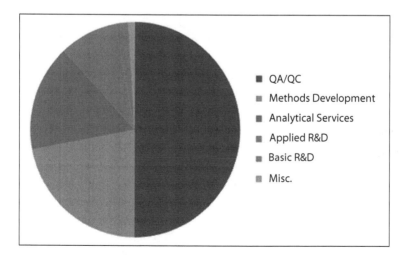

- QA/QC
- Methods Development
- Analytical Services
- Applied R&D
- Basic R&D
- Misc.

Figure 1.1   Dissolution testing by function in 2009 (data from Ref. 11).

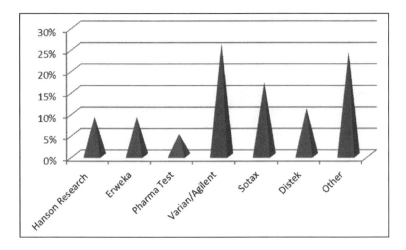

Figure 1.2 Vendor diversity in 2009 (data from Ref. 11).

However, as will be seen in Chapters 8 and 9, these later instrument technologies may play a more significant role in the dissolution and drug release testing in the years to come.

1.4 Dissolution and Drug Release in the Pharmaceutical Industry

The goal of this text is for leading scientists in the field to discuss the current applications of dissolution and drug release to poorly soluble molecules. Each chapter provides a current view from leading practitioners of the material presented. While much of the instrumentation of dissolution and drug release has not changed dramatically from the traditional USP apparatus, the approaches and applications have moved forward with the changing molecules being brought to market. This book builds upon the technique from its initial and early developments. Details on the development and theory of dissolution and drug release can be found in the literature.[1-3]

1.4.1 Solubility Determinations for Pharmaceutical API

Aqueous solubility for the active pharmaceutical ingredient (API) in the final formulation is a significant factor that influences the pharmacokinetic profile of a drug. There are various analytical methodologies and computational models for the prediction of solubility of the API. The solubility in the body can often be different from that determined in common in vitro buffer systems that are typically used for pharmaceutical quality control processes. The quality of computational models is also affected by the accuracy of the experimental solubility data. High-throughput discovery processes have driven the development of high throughput screening processes for measurement of physicochemical determinations including lipohilicity, pK_a and solubility, but the accuracy of these is often compromised by the requirement for speed and the use of non-biorelevant media in the determination. Kofi Asare-Addo and Barbara R. Conway from the University of Huddersfield will discuss the nuances regarding the solubility determinations for APIs.

1.4.2 Use of Surfactants in Dissolution Testing

Before any relevant dissolution or drug release testing can occur in formulated product, the analytical target must be dissolved in the test matrix. In dissolution testing, this is often achieved by adding surfactants to the dissolution media. The primary goal for the surfactant is to keep the target analyte in solution after it dissolves. Surfactants also aid in material wetting by reducing the surface tension in the media. However, ideally the surfactant should not alter the mechanism of release by accelerating the dissolving or erosion of the drug formulation itself. Amit Gupta of Zydus reviews the use of surfactants in dissolution media to expand this technique to poorly soluble drugs as well.

1.4.3 Intrinsic Dissolution Evaluation of Poorly Soluble Drugs

Michele Georges Issa and Humberto Gomes Ferraz of the University of San Paulo discuss the aspects involved in the intrinsic dissolution assay and its use in the evaluation of poorly soluble drugs. The most commonly used apparatuses and the calculation of the intrinsic dissolution rate (IDR) are discussed, as well as some relevant examples of applications and a brief explanation of the variables of the method and the use of experimental designs. Applying IDR throughout all stages of a medicinal compound's development—from the synthesis of a new chemical entity up to the quality assurance of the API—development scientists can better understand key parameters in selecting lead candidate parameters for optimum bioavailability using relatively small quantities of material.

1.4.4 Strategies for Oral Delivery of Poorly Soluble Drugs

The advances in combinatorial chemistry and high-throughput screening in the past couple of decades have enabled discovery of new chemical entities (NCEs) for a variety of complex and diverse biological targets. Many biological targets are highly lipophilic or hydrophobic in nature and there has been a significant increase in the number of NCEs with poor aqueous solubility. While oral delivery continues to be the most commonly used route of administration for NCEs, poor aqueous solubility can result in significant development challenges such as incomplete absorption, highly variable bioavailability, and highly variable pharmacokinetic profiles in preclinical species and humans. A number of conventional and enabling formulation technologies are now available to address poor solubility and the resulting poor biopharmaceutical performance of NCEs. However, a systematic evaluation and proactive selection of the optimal formulation technology is typically not available during discovery and a number of promising NCEs are terminated because of poor biopharmaceutics and lack of adequate exposure for preclinical safety assessment. The goal of this chapter is to describe a systematic approach for generating a cross-functional package of data comprising physicochemical, biopharmaceutical, ADME, PK/PD, and delivery technology evaluation to enable oral delivery of poorly water-soluble NCEs. Akash Jain, Dev Prasad, and Sudhakar Garad discuss the general concepts of formulating poorly soluble molecules into viable dosage forms from pre-clinical through commercial development.

1.4.5 A Staged Approach to Pharmaceutical Dissolution Testing

The application of dissolution methodologies develops as the target molecule goes from discovery through the various stages in pharmaceutical development. As more information becomes known about the drug and its chosen dosage form(s), the expectation of the dissolution method performance increases as well. By late Phase 2b/3 of clinical development, the dissolution method should be in its finalized form. The results of the clinical tests and manufacturing history are reviewed to propose relevant specifications for commercially marketed material. Gregory Webster, Paul Curry, and Xi Shao of AbbVie review the development of the dissolution method as the drug traverses the stages of pharmaceutical development.

1.4.6 Development and Application of in vitro Two-Phase Dissolution Method for Poorly Water-Soluble Drugs

A two-phase dissolution test, containing both aqueous and organic phases, is designed to be more physiologically relevant than conventional single-phase dissolution tests with the involvement of an "absorptive environment." This in vitro test simulates both kinetic processes of drug dissolution and partitioning into the intestinal membrane. Case studies reported in this chapter illustrate that this two-phase dissolution test enables a better opportunity for establishing in vitro and in vivo relationship (IVIVR) for several drugs, including immediate and extended dosage forms, compared with conventional single-phase dissolution methods. In particular, the two-phase dissolution test may be extremely useful in assessing the metastable supersaturated state of BCS II drugs, including the duration and degree of supersaturation, effect of polymeric precipitation inhibitors, and an overall effect upon drug partition. In addition, theoretical models of the two-phase dissolution systems with emphasis on drug partition kinetics are discussed. Ping Gao, Yi Shi, and Jon Miller of AbbVie review the application of biphasic dissolution to poorly soluble drugs.

1.4.7 Use of Apparatus 3 in Dissolution Testing of Poorly Soluble Drug Formulations

Brian Crist of Agilent Technologies discusses the use of apparatus 3 in the dissolution of poorly soluble drug formulations. The USP apparatus 3 reciprocating cylinder was originally designed for testing extended release products and exposing drug formulation to pharmacokinetic and mechanical properties similar to those found in the GI tract. The instrument has proven useful for poorly soluble compounds, chewable formulation, immediate release, delayed release, and numerous modified-release products. At the time of this writing, adaptions to the reciprocating cylinder to contain microspheres within a dialysis membrane are under evaluation so this may provide additional benefits to biorelevant drug release profiling micro- and nanoparticles. Because of the apparatus's ability to characterize the release profile of early drug formulation candidates, the apparatus has the potential to provide knowledge-based assessments of formulations developed through Quality by Design (QbD) and further characterization of the products design space for post-approval manufacturing.

1.4.8 Use of Apparatus 4 in Dissolution Testing of Poorly Soluble Drug Formulations

Geoffrey N. Grove of Sotax along with Rajan Jog and Diane Burgess of the University of Connecticut provide a brief history of the development of the USP 4 technique, pharmacopeial considerations, and regulatory considerations to help in laying the groundwork for when to choose USP 4. It also covers an overview of a variety of different system configurations and flow cell selection and design covering both compendial and non-compendial design choices. It further presents method parameters for considerations such as pump selection and flow rates and highlights choices which may be pertinent to working with poorly soluble drugs.

The chapter also offers a comprehensive review of tested drugs, including article titles, instrumentation, and system design notes. Experimental conditions including media types, filter types, flow rates, and various other parameters are summarized with emphasis placed on the relevance of the results for choosing USP 4. It is the authors' hope that this review of the current literature will serve as a useful reference and decision-making tool.

1.4.9 Dissolution of Nanoparticle Drug Formulations

John Bullock of Shire LLC discusses some of the many challenges associated with the dissolution of nanoparticle drug formulations. Numerous types of nanoparticle-based drug formulation technologies have been developed over the last couple of decades in order to provide a range of different enhanced drug delivery properties for therapeutic and diagnostic agents. In the field of oral drug delivery, the most common nanoparticle structures can be categorized as either pure drug nanoparticles consisting of essentially 100% drug or an assortment of different nano-sized structures in which the active drug is encapsulated or dispersed in a solid, semi-solid, or liquid state within a formulation matrix. Regardless of which of these two categories of drug nanoparticles are used in an oral dosage form, the fundamental challenges encountered and approaches pursued for developing suitable techniques to measure drug release are generally similar for both pure drug nanoparticles and matrix-type nanoparticle formulations. Two of the more distinguishing characteristics of dissolution methods for these nanoparticle formulations include the challenges encountered in appropriately sampling and processing dissolution samples containing small nano-size drug particles and the potential for much faster dissolution rates compared to formulations manufactured with conventional-size drug particles. This review starts with a brief discussion of the theoretical underpinnings controlling solubility and dissolution of drug particles in general as well as certain considerations more important for nano-size drug particles, with an emphasis on pure drug nanoparticles. Following this discussion, practical guidelines to lead the practitioner in developing a suitable method are provided that include considerations for the dissolution media used, instrumentation considerations, and sampling and processing approaches that have proven successful for a number of different types of nanoparticle formulations. In conjunction with the discussion of instrumentation considerations, a variety of in situ analysis techniques that have recently emerged and demonstrated advantages for dissolution testing of rapidly dissolving nanoparticle formulations are reviewed, and some alternative in vitro release techniques that have been described for evaluating the performance of nanoparticle drug formulations are examined.

1.4.10 Dissolution of Lipid-Based Drug Formulations

Stephen Caffiero from Boehringer Ingelheim describes the general nature of lipid-based drug formulations and their dissolution behavior. The availability of the drug for in vivo absorption can be enhanced by presentation of the drug as dissolved within a colloidal dispersion. The industry trend of an increasing number of poorly soluble drugs formulated in lipophilic matrices has resulted in a collaboration of expert groups at United States Pharmacopeia (USP) authoring a general chapter with specific focus on dissolution testing of these formulations (general chapter USP <1094>Liquid-Filled Capsules—Dissolution Testing and Related Quality Attributes), which will be discussed, as well as relevant topics such as a capsule rupture test, emulsion droplet size determinations, and a case study.

1.4.11 Dissolution of Stabilized Amorphous Drug Formulations

Justin Hughey of Banner Life Sciences discusses the critical role of dissolution testing in characterizing stabilized amorphous drug formulations (SADFs). Amorphous systems continue to be an important and growing technology in the oral delivery of BCS class 2 drugs. The use of high-energy forms of a drug substance, such as amorphous forms, coupled with precipitation inhibiting excipients has proven a valuable strategy in solubility and bioavailability enhancement. This chapter explains how the interplay of chemistry, physical form, functional excipients, supersaturation, and thermodynamics come together in dissolution testing to provide insight into the solution-mediated mass transfer and phase transition phenomena associated with SADFs.

The underlying considerations in drug delivery system design, dissolution method parameters, media selection, and the nuances of sink and non-sink conditions are discussed with illustrative case studies. These studies clearly demonstrate the power of dissolution and precipitation testing in understanding and advancing this important drug delivery approach.

1.4.12 Dissolution of Pharmaceutical Suspensions

A suspension consists of insoluble solid particles dispersed in a liquid medium. The most common reason to develop a suspension dosage form is limited aqueous solubility of the active pharmaceutical ingredient (API) at the dosage required. Another common reason to use suspensions is that they typically offer improved chemical stability compared to solutions. Suspensions also offer advantages in taste masking and a more convenient dosage form for certain patients (e.g., pediatrics). In this chapter, Beverly Nickerson, Michele Xuemei Guo, Kenneth Norris, and Ling Zhang of Pfizer focus on dissolution and drug release of suspensions in dosage forms that are dosed by the oral route of administration and include oral suspensions, suspensions for reconstitution, and suspensions in capsules.

1.4.13 Biorelevant Dissolution

Mark McAllister and Irena Tomaszewska of Pfizer discuss testing poorly soluble drugs in biorelevant dissolution conditions. In general, the approaches adopted to improve the physiological relevance of a dissolution test can be broadly categorized into one of two groups. The first group encompasses mechanistic approaches in which a single aspect of the dissolution process is controlled to study the impact of a physiological variable such as media or hydrodynamics. The second group includes methods and equipment designed to simulate multiple aspects of the gastrointestinal tract and deliver a holistic simulation of luminal conditions encompassing fluids, digestion, transit, and absorption. It is recognized that for particular aspects of a dosage form or API dissolution that a mechanistic (reductionist) approach may allow a detailed assessment of individual phenomena such as supersaturation, precipitation, and re-dissolution. In contrast, a holistic simulation approach considers the summation or net effects of multiple processes such as digestion, transit, and absorption on dosage form performance, for example when assessing complex food effects. This chapter describes the development of biorelevant dissolution testing for both mechanistic and holistic approaches and assesses the biological relevance of modifications made to media (composition and volume), hydrodynamics, and integration of an absorptive component within the dissolution test.

1.4.14 Clinically Relevant Dissolution for Low-Solubility Immediate-Release Products

Paul Dickinson, Talia Flanagan, David Holt, and Paul Stott of AstraZeneca discuss clinically relevant dissolution for low-solubility immediate-release products. A structured approach to development that endeavors to identify and evaluate risk to clinical performance and, where appropriate, test the impact of these risks in vivo is presented. Each compound should be considered on its own merits but the application of a proposed 5 step approach will ensure that all important factors are considered and will ultimately lead to the establishment of a robust control strategy. The benefits of such an approach include enhanced security of product supply, the ability to optimize the manufacturing process and demonstrate the (lack of) impact of any proposed change, and an improved assurance of the clinical quality of product supplied to patients. Dissolution testing and its associated specification is expected to serve a number of purposes as the demands of discriminatory power and demonstration of complete release can be at conflict. Further work in our scientific understanding and regulatory harmonization are required if we are to realize the full benefit of a move towards more clinically relevant dissolution specifications. The authors are firmly of the belief that the development of increased knowledge and understanding proposed under the auspices of QbD are predicated on an insight into and control of in vivo performance.

1.4.15 Method Validation and QbD for Dissolution Testing of BCS Class II/IV Products

Alger Salt of GlaxoSmithKline discusses the QbD approach to method development and validation for dissolution testing. QbD principles provide a structured approach to the development and delivery of products, technologies, and processes. Historically, QbD has been directed at manufacturing processes and product attributes. QbD principles are now being applied to analytical methods. Method transfers are part of a successful product lifecycle because work initiated in R&D is ultimately moved into the QC laboratories at manufacturing sites.

Prior to embracing the QbD principles analysts learned (and many times re-learned) that problems with analytical methods were often discovered during method transfers. This is not a good time to discover such problems, because it is usually difficult to make changes to the method and fixing the problems can be frustrating and expensive.

The four stages of the QbD approach are listed below. Definitions follow for each, but the control definition stage is the focal point in this chapter and is widely considered to be the primary component of the QbD approach.

1. Design intent
2. Design selection
3. Control definition
4. Control verification

1.4.16 Regulatory Considerations in Drug Release Testing of BCS Class II/IV Products

Robert Bell of Drug & Biotechnology Development and Laila Kott of Takeda Pharmaceuticals summarize the current guidances and how they are related to compounds of low solubility. We review a history and a progression of the instrumentation, as well as the evolution of the guidance documents. A summary of three classification systems that are used to describe all types of compounds is presented. The classifications systems include the 1995 biopharmaceutics classification system (BCS) for drug products, the biopharmaceutics drug disposition classification system (BDDCS) published in 2005, and the developability classification system (DCS) presented for the first time in 2010.

A discussion on the applicability of using biorelevant media for formulation development and the possibility of its use for quality control (QC) testing is presented. The guidance, as it stands, on this topic is discussed and dissected. Included is an outline on the setting of dissolution specifications as they relate to current guidance and industry trends. Finally, there is a summary of the current key guidance documents and emerging regulatory topics.

1.4.17 Dissolution of Capsule-Based Formulations

The availability of a compound formulated in a liquid-filled capsule for absorption depends on the initial dissolution and rupture of the capsule shell and subsequent release and dissolution of its fill contents in the GIT fluids. These two processes need to be monitored at the time of release and during the shelf life of a capsule product. Liquid-filled capsules pose unique challenges during the development and application of dissolution methods because of the complex nature of the shell and fill materials. The shell material is prone to changes in its mechanical properties or cross-linking of gelatin, which results in changes in its solubility. The fill material, on the other hand, may exhibit changes in particle size distribution or the polymorphic nature of the suspended material in the suspension fill or crystallization of a solubilized compound from a solution fill. In the latter case, the crystallization of the solubilized compound can occur either in the capsule dosage form or when the fill material encounters in vitro and in vivo aqueous fluids.

Dissolution testing is a highly valuable tool to characterize liquid-filled capsule products in vitro and is used routinely (a) to assess batch-to-batch quality, (b) to monitor changes in the quality of a product during its shelf life, (c) to assess product sameness after scale-up and post-approval changes (SUPAC), (d) to comply with biowaiver requirement for a lower strength of a product, and (e) to comply with biowaiver requirement for a product intended for local action in the GIT. In addition, a dissolution method designed to produce in vitro/in vivo correlations (IVIVC) or in vitro/in vivo relationship (IVIVR) can be used to predict potential bioequivalency or bioinequivalency between products. The intent of this chapter by Rampurna Gullapalli from Dart Neurosience LLC is to provide an in-depth discussion on the factors affecting in vitro and in vivo dissolution of liquid-filled capsule products, development of dissolution methods for their routine QC testing, and modifications to these methods to produce potential IVIVC and IVIVR.

1.4.18 Emerging and Non-compendial Drug Release Techniques

Dissolution testing of conventional and non-conventional dosage forms was introduced in the 1960s, and now serves as an essential compendial test, which is used by all pharmacopoeias for evaluating drug release. Apart from the standardized tests, there are several non-compendial methods that are qualified and validated for use in drug release testing.

This chapter is divided into three sections. The first section briefly discusses existing compendial methods along with any non-compendial modifications. The second section describes non-compendial apparatus for different types of dosage forms. The final section explains the various detection techniques such as UV imaging with Raman spectroscopy, FTIR-ATR, and fiber optics. Namita Tipnis and Diane Burgess of the University of Connecticut provide their review of emerging technologies on the horizon for dissolution and drug release testing in pharmaceutical laboratories.

References

1. Banakar, U. V. *Pharmaceutical Dissolution Testing.* CRC Press: Boca Raton, FL, 1991.
2. Hanson, R., Gray, V. *Handbook of Dissolution Testing*, 3rd ed. Dissolution Technologies, Inc.: Hockessin, DE, 2004.
3. Dressman, J. J., Kramer, J. *Pharmaceutical Dissolution Testing.* CRC Press: Boca Raton, FL, 2005.

4. The United States Pharmacopeia and The National Formulary (USP–NF), Chapter <711>. U.S. Pharmacopeial Convention, Rockville, MD, 2015.

5. Kirchhoefer, R., Peeters, R. Dissolution. In *Analytical Chemistry in a GMP Environment: A Practical Guide*; Miller, J.M., Crowther, J.B., eds. Wiley-Interscience: Hoboken, NJ, 2000.

6. Amidon, G. L., Lennernäs, H., Shah, V. P., Crison, J. R. A theoretical basis for a biopharmaceutics drug classification: the correlation of in vitro drug product dissolution and in vivo bioavailability, *Pharm. Res.*, 1995, 12, 413–420.

7. US Food and Drug Administration. The Biopharmaceutics Classification System (BCS) Guidance. 2009. http://www.fda.gov/AboutFDA/Centers Offices/OfficeofMedicalProductsandTobacco/CDER/ucm128219.htm

8. US Food and Drug Administration. Waiver of in vivo Bioavailability and Bioequivalence Studies for Immediate-Release Solid Oral Dosage Forms Based on a Biopharmaceutics Classification System. 2015. http://www.fda.gov/downloads/Drugs/GuidanceComplianceRegulatory Information/Guidances/ucm070246.pdf

9. US Food and Drug Administration. Dissolution Testing of Immediate Release Solid Oral Dosage Forms. 1997. http://www.fda.gov/downloads/drugs/guidancecomplianceregulatoryinformation/guidances/ucm070237.pdf

10. Takagi, T., Ramachandran, C., Bermejo, M., Yamashita, S., Yu, L. X. Amidon, G.L. A provisional biopharmaceutical classification of the top 200 oral drug products in the United States, Great Britain, Spain, and Japan, *Mol. Pharm.*, 2006, 3(6), 631–643.

11. Global Assessment Report, 11th ed. Strategic Directors International: Los Angeles, 2010, 434–438.

Chapter 2

Solubility Determinations for Pharmaceutical API

Kofi Asare-Addo and Barbara R. Conway

Department of Pharmacy, University of Huddersfield, Queensgate, Huddersfield HD1 3DH, UK

k.asare-addo@hud.ac.uk, b.r.conway@hud.ac.uk

2.1 Introduction

The complex and time-consuming process of drug development, from discovery of a new chemical entity (NCE) to the authorization of marketing for a new drug, can span a period of 12–20 years.[1] Undesirable physicochemical attributes are a major cause of attrition in the drug development process. In a typical physicochemical screening process (i.e., pK_a, solubility, permeability, stability, and lipophilicity), poor solubility is a key factor limiting successful development.[1,2] Compounds with insufficient solubility are more likely to fail during discovery and development as inadequate solubility not only impacts other property assays, masking additional undesirable properties, but also influences both pharmacokinetic and pharmacodynamic characteristics of the

Poorly Soluble Drugs: Dissolution and Drug Release
Edited by Gregory K. Webster, J. Derek Jackson, and Robert G. Bell
Copyright © 2017 Pan Stanford Publishing Pte. Ltd.
ISBN 978-981-4745-45-1 (Hardcover), 978-981-4745-46-8 (eBook)
www.panstanford.com

Table 2.1 The Biopharmaceutics Classification System (BCS)

Biopharmaceutics class	Permeability	Solubility
I	High	High
II	High	Low
III	Low	High
IV	Low	Low

compound.[3] The Biopharmaceutics Classification System (BCS) is the scientific framework that allows the classification of drug substances based on their dissolution, aqueous solubility, and intestinal permeability.[1-4] A drug's bioavailability depends primarily on its solubility in the gastrointestinal (GI) tract and its permeability across the cell membranes upon oral administration.[5] This BCS system was proposed by the Food and Drug Administration (FDA) as a bioavailability/bioequivalence (BA/BE) regulatory guideline and assigns drugs into four groups illustrated in Table 2.1. According to the BCS framework, solubility is determined by obtaining the pH-solubility profile of the drug substance in question in an aqueous medium of pH range 1–8 at an established temperature of $37 \pm 1°C$.

A drug substance according to the solubility classification in the BCS is thus considered to be highly soluble when its highest dose strength proves to be soluble in 250 mL or less of an aqueous medium over the pH range of 1–8.[6-8] If not, the drug substance is considered as poorly soluble. The 250 mL volume estimate is from bioequivalence study protocols which prescribe drug product administration with a glass of water to fasting volunteers.[7] A drug substance is considered highly permeable when the extent of intestinal absorption is determined to be 90% or higher; if not, the drug substance is considered to be poorly permeable.[7] Permeability classification is thus based directly on the extent of intestinal absorption of a drug substance in humans or indirectly on the measurements of the rate of mass transfer across the human intestinal membrane.[7]

BCS class I compounds have high permeability and high solubility with examples including captopril, propranolol, diltiazem, metoprolol and antipyrine. The rate of drug dissolution limits the in vivo absorption of these drugs.[9] Class II compounds include naproxen, itraconazole, carbamazepine and piroxicam.[9-11] These drugs exhibit high permeability but fail to meet the criterion for high solubility across the physiological pH range of pH 1–8. Solubility, as such, limits the absorption flux. The restriction can be equilibrium-based, whereby either the composition or volume of the contents in the GI tract preclude complete dissolution of the drug. Alternatively, the volume and composition of the GI contents can, in theory, facilitate complete dissolution of the drug, but the dissolution rate is too slow in the appropriate absorption site in the intestine (kinetic-based). Class III compounds are those with high solubility, but low permeability. Drugs in this group include cimetidine, atenolol, and acyclovir, with permeability being the rate-determining step affecting absorption.[5,6,12-18] BCS class IV drugs are compounds exhibiting both low solubility and low permeability. In this group, the rate of the in vivo absorption of the drug depends on the relative interplay between the two and also whether the drug's low permeability is borderline or whether it is as a result of GI metabolism. Drugs in this group include cyclosporine and terfenadine.[12-18] Recent adaptations to the BCS include the Biopharmaceutical Drug Disposition Classification System (BDDCS) based on solubility and intestinal permeability rate, which is related to the extent of metabolism, to predict drug disposition and potential drug-drug interactions in the intestine and/or liver.[19] Both systems, although they differ in the criterion for permeability and have different purposes, are based on classifying drugs and new molecular entities into four categories using the same solubility criteria. The solubility of a drug is therefore a key consideration when systemic delivery is desired as low aqueous solubility can either delay or limit drug absorption.[5]

2.2 Drug Solubility Assay Development

Aqueous solubility and rate of dissolution are pivotal attributes in drug discovery and development as they affect both in vitro and in vivo assay results. Determination of these parameters will facilitate understanding the interplay among absorption,

Table 2.2 USP descriptive classification of drug solubility[24]

Definition	Parts of solvent required for one part of solute
Very soluble	<1
Freely soluble	1–10
Soluble	10–30
Sparingly soluble	30–100
Slightly soluble	100–1000
Very slightly soluble	1000–10,000
Practically insoluble	$>10,000$

distribution, metabolism, excretion, and toxicity (ADMET) parameters as insufficient solubility can interfere with pharmacokinetic and pharmacodynamic properties.[1] Along with other pharmacodynamic and pharmacokinetic parameters, one can utilize solubility to help prioritize early drug candidates, to not only accelerate the discovery and development of NCEs but also improve the attrition rate for drug candidates in late development.[4,20–22] The two major causes of poor aqueous solubility are high lipophilicity and strong intermolecular interactions, which render solubilization of the solid energetically unfavourable.[23] The United States Pharmacopeia (USP) classifies drug solubility using seven classes with definitions in Table 2.2.

In discovery and development phases, the determination of solubility, along with other physicochemical parameters, can allow screening out of unsuitable candidates and can also inform formulation optimization and salt selection. Solubility is considered a thermodynamic parameter being defined as the saturation concentration of a solute in solution under specific conditions, where an equilibrium between the solute and solvent is achieved thermodynamically. This equilibrium balances the energy of solvent and solute interacting with themselves against the energy of solvent and solute interacting with each other (Fig. 2.1).[3,20,25–30] The overall intent of measuring solubility in the drug development context is to determine an intrinsic property that influences the absorption of a potential compound. Solubility and permeability tend to be inversely related; hence close attention should be given to several physicochemical properties when increased oral absorption is desired.[31,32]

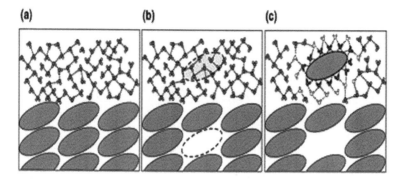

Figure 2.1 The intermolecular forces that determine thermodynamic solubility. (a) Solvent and solute are segregated; each interacts primarily with other molecules of the same type. (b) To move a solute molecule into solution, the interactions among solute molecules in the crystal (lattice energy) and among solvent molecules in the space required to accommodate the solute (cavitation energy) must be broken. The system entropy increases slightly because the ordered network of hydrogen bonds among solvent molecules has been disrupted. (c) Once the solute molecule is surrounded by solvent, new stabilizing interactions between the solute and solvent are formed (solvation energy), as indicated by the dark gray molecules. The system entropy increases owing to the mingling of solute and solvent (entropy of mixing), but also decreases locally owing to the new short-range order introduced by the presence of the solute, as indicated by the light gray molecules.[27]

The methods of determining aqueous solubility include

- thermodynamic or equilibrium (the concentration of a compound in saturated solution when excess of solid is present, and solution and solid are at equilibrium),
- semi-equilibrium (close to equilibrium), and
- kinetic methods (the concentration of a compound at the time when an induced precipitate first appears in the solution).

These terms are used commonly in pharmaceutical litera-ture.[25,33–36] In the general sense, assays that are equilibrium-based and start with the solid form of the compound are thermodynamic whereas assays that start with the compound predissolved in organic solvent generally have shorter incubation durations and are kinetic. The true thermodynamic solubility of a compound should be determined using the purified crystalline form of the compound because there can be very large differences in the solubility values between amorphous and crystalline forms.[32,37,38] Despite this, the solubilising and storage of compounds in solvents such as dimethyl sulfoxide (DMSO) and the acceptance of DMSO stocks in discovery for solubility testing is because the physical form of the compound at that stage of discovery is usually not the pure crystalline form.[4,32,39] Physical forms such as different polymorphs or solvates give rise to measurable differences in their physical properties due to the different lattice energies and entropies associated with them.[34] For example, salt formation will affect properties such as physical and chemical stability, melting point, electrostatic behaviour, crystal habit, refractive index, solubility, mechanical, dissolution rate, density, and biopharmaceutical properties, with the particular salt formed determining these properties.[37,40] However, thermodynamic or equilibrium solubility determinations tend not to be very feasible in the early stages of drug discovery owing to the fact that a relatively large sample is required, the sample preparation can be considered laborious, the compound's purity and physical form may have yet to be determined and there is a low throughput of the compound.[24,27,41] Despite this, it is becoming increasingly common to measure solubility as early as feasible in the discovery process due to the fact that the solid-state properties of compounds such as purity, degree of crystallinity, particle size, and polymorphism can be studied and characterized in detail.[1]

Typically more than one solubility determination methodology is applied to support the different stages of drug discovery and development. In the early phases of discovery, kinetic solubility or semi-equilibrium methods are often used facilitated by high-throughput (HT) assays for profiling large numbers of compounds, and medium-throughput (MT) and/or low-throughput (LT) equilibrium methods are often used for later stages of drug discovery and development. Although kinetic solubility can be used as an alternative for the thermodynamic method, it cannot, however, be used as a complete substitute since the method uses solvents, normally DMSO, meaning that any impact of the crystal lattice and the presence of polymorphic forms of the drug are lost. As such, kinetic methods tend to be most suitable for the drug discovery stage and the thermodynamic methods of determining solubility tend to be more appropriate at the drug development stage (Table 2.3).[1,24]

Table 2.3 Differences between the methods used for solubility measurement in drug discovery and drug development processes[1]

	Discovery	**Development**
Compounds tested		
Number	100–1000	10
Quantity available	A few mg	>g
Purity	Limited	Improved
Solid state	Amorphous or partially crystalline (not characterized)	Stable, crystalline material (characterized)
Distribution	Generally in DMSO stock solutions	Generally in solid form
Methods		
Type of solubility measured	Kinetic solubility (fully dependent on experimental conditions)	Thermodynamic solubility
Throughput	High	Low 25–50 compounds a week
Automation	Fully	Only partially automated
Format	96-well or 384-well microplates	Small scale (single tube)
Incubation time	Minutes	Hours or days
Detection	UV, turbidity	HPLC-UV, HPLC-MS
Media	Aqueous	>20 (aqueous, organic, biorelevant media, formulations, excipients …)
Data generated and intended purpose	Solubility in screening bioassay media to avoid misinterpretation Rank-order hits	Solubility and dissolution in biorelevant media Evaluation of formulations Characterization and optimization of solid state Selection of promising compounds Development of adequate strategies to overcome solubility problems

2.2.1 Media for Solubility Studies

2.2.1.1 Biorelevant media

Normally, to achieve desirable systemic exposure after oral dosing, the active drug substance or compound must be in solution in the aqueous environment of the intestine in order to cross the luminal wall; hence the importance of drug solubility and drug dissolution for orally administered drugs.[42] During the early screening process, drug candidates with low water solubility may be excluded from further development because failure in later stages is likely.[42] However, although drug candidates with solubility below an acceptable predetermined solubility may prompt a degree of concern, it is important to note that low aqueous solubility also may not always constitute a sufficient reason for the compound's elimination as it could still have relatively high intraluminal solubility.[3,42,43] The composition of GI fluids has a large impact on the solubility and dissolution of poorly soluble API in the GI tract, and hence a large influence on the drug absorption. The ability to predict solubility in the upper GI tract would thus be advantageous to drug discovery and development. Gastric and intestinal fluids under fasted and fed conditions vary with regards to pH, buffer capacity, osmolarity, surface tension, and lipid concentration of GI fluids.[44]

The two most widely implemented approaches for estimating drug solubility in the GI tract are determination of solubility in fluids aspirated from the human gastrointestinal tract, and determination of solubility in biorelevant media.[44,45] The solubilizing capacity of the GI environment may be much higher than aqueous buffer systems such as those used in pharmacopeial dissolution testing methods[46] and as such solubility measurements are also performed in biorelevant media to identify potential in vitro–in vivo *correlations* (IVIVCs). Bile salt composition is variable throughout the duodenum and jejenum (Fig. 2.2).[44] It is important that simulating media for solubility screening in early drug development reflect the solubilizing capacity of human intestinal fluid rather than having an identical composition to ensure functional relevance prevails over compositional biorelevance.[47]

This allows the evaluation of the impact of solubility on absorption. For example, Sunesen et al.[48] showed that the solubility

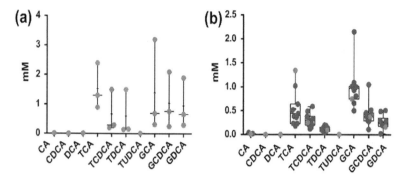

Figure 2.2 Bile salt composition in human (a) duodenum and (b) jejunum. CA: cholic acid; CDCA: chenodeoxycholic acid; DCA: deoxycholic acid; TCA: taurocholic acid; TCDCA: taurochenodeoxycholic acid; TDCA: taurodeoxycholic acid; TUDCA: tauroursodeoxycholic acid; GCA: glycocholic acid; GCDCA: glycochenodeoxycholic acid; GDCA: glycodeoxycholic acid. Box–whisker plots show minimum and maximum values, as well as 25, 50, and 75 percentile. The cross indicates the mean value. Each data point represents a group of participants ($n = 1$–10 gray, $n = 11$–20 dark gray) reported in one publication.[44] Reprinted from Bergström et al.,[44] with permission from Elsevier.

of danazol increased from 0.42 µg/mL in water to 1.61 and 2.04 µg/mL in gastric and intestinal fluids respectively, contributing in part to a higher bioavailability than predicted. Several authors have shown drug solubility to be enhanced in biorelevant media compared to aqueous buffers.[49–52] These authors together with Dressman et al.,[53] Naylor et al.,[54] Pederson et al.,[55] and Galia et al.[56] showed that bile salts (sodium taurocholate as a model) and lecithin can self-assemble and form micelles, as well as act as wetting agents, thereby increasing the solubility and dissolution rate of lipophilic compounds and are included in biorelevant media.

Jantratid et al.[57] updated the components or constituents of the biorelevant media taking into consideration fed and fasted states in man. Takács-Novák et al.[58] evaluated four poorly soluble drugs encompassing a range of lipophilicities and solubilities in biorelevant media and confirmed the primary role of ionization in governing the solubility in the biorelevant media. They also found the higher concentration of solubilizing agents in Fed State Simulated Intestinal Fluid (FeSSIF) to improve solubility. Using 17 model drugs, Clarysse et al.[42] evaluated the solubilizing capacity of Fasted State Simulated Intestinal Fluid (FaSSIF$_c$) and FeSSIF$_c$ (subscript indicates the use of crude taurocholate) and different concentrations of D-α-tocopheryl polyethylene glycol 1000 succinate (TPGS) in phosphate buffer and correlated it with the solubilizing capacity of human intestinal fluids (HIF) in the fasted and the early postprandial state. They found a good correlation between solubility in fasted HIF and FaSSIF and also between solubility in fed HIF and FeSSIF$_c$. Comparable values were also obtained for the 0.1% TPGS for the fasted state and 2% TPGS for the fed. They then concluded that FaSSIF and FeSSIF could be considered as biorelevant media for intestinal solubility estimations and the simpler TPGS-based system may be a valuable alternative with improved stability and lower cost. However, Ottaviani et al.[59] evaluated the solubility of 25 chemically diverse compounds in modified simulated intestinal fluid (FaSSIF-V2) and in aqueous phosphate and maleate buffers and found that the solubility of the ionized acids did not increase in FaSSIF-V2. They attributed this behaviour to electrostatic repulsions with the media components but found lipophilicity to play an important role mainly for charged bases with a $\log P > 4$ (or $\log D_{6.5} > 1.9$). They also found that when the aqueous solubility is mainly driven by lipophilicity, the FaSSIF-V2 components improved the solubility of basic compounds. This increase was less for compounds whose solubility is limited by crystal packing, leading them to conclude that ionization, lipophilicity, and crystal packing all play important but peculiar roles in controlling solubility in FaSSIF-V2 compared to that in aqueous buffer.

As the constituents of FaSSIF and FeSSIF are comparatively expensive and the preparation process is labour intensive, requiring fresh preparation daily,[60] their application in solubility studies has tended to be limited to later and advanced stages of the drug development process. However, an instant powder is available to facilitate the preparation of FaSSIF and FeSSIF, which may address some of these issues[42] and a high-throughput screening UV method for measuring biorelevant solubility showed excellent correlation with solubility data obtained using the shake-flask method without DMSO.[61] This may provide a reliable method for measuring intestinomimetic solubility during the early stages of drug discovery.

2.2.1.2 Organic solvents (DMSO)

Various methods have been developed for the measurement of solubility in the last two decades, including turbidity and UV plate scanner-based detection systems.[62] These are ranking assay methods for high-throughput screening where the buffered sample solutions are prepared by adding aliquots of 10 mM dimethyl sulfoxide (DMSO) stock solutions.[63] With regard to the turbidimetric methods, small aliquots of the stock solutions are added to pH 7 buffer solutions at fixed intervals until turbidity is detected. A plot of volume versus turbidity is extrapolated to zero turbidity after the addition of a few more aliquots to ascertain the onset of precipitation and anywhere from 50–300 compounds can be analyzed with this methodology.[64] The fast UV plate spectrophotometer method allows solubilities at a single or multiple pH values to be rapidly determined. In this method,

- a dry powder or a DMSO solution of a known quantity of sample is added to a known volume of a universal buffer solution of known pH (it is important that the amount of sample is sufficient to cause precipitation in the solution or an excess of solid),
- the saturated solution is allowed to reach equilibrium,
- the solid material or precipitate is removed by filtration,
- a suitable water-miscible cosolvent is added to the sample solution and the UV spectrum of the solution determined, and
- the compound's solubility is then calculated from the ratio of the reference spectrum and the sample spectrum taking the dilution factors into consideration.

If determining the compound's solubility at one pH, up to about 400 compounds can be analyzed in a day. Figure 2.3 shows the pH solubility profile of piroxicam determined using this method.[64]

As DMSO increases solubility, assays that use stock compounds in DMSO have a limited maximum solubility, owing to the specified %

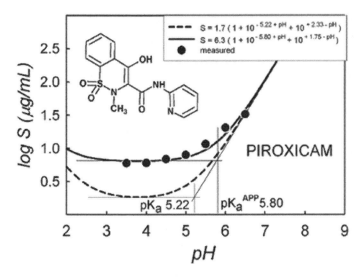

Figure 2.3 Solubility-pH profile of piroxicam, determined by the fast-UV method. The apparent curve (solid line) is shifted from the true aqueous curve (dashed line) due to the presumed formation of aggregates, as evidenced by the shift in the pK_a value of the drug. Reprinted from Avdeef and Testa,[64] with permission from Wiley.

DMSO in the assay and the DMSO stock concentration.[33] The DMSO stock solution can be removed by evaporation with subsequent buffer addition to the remaining material in an effort to minimize the contribution of DMSO to solubility measurement. The amount of DMSO used in solubility assays typically varies from 0 to 5% v/v and buffer pH is between 6.5 (intestine) and pH 7.4 (bioassay and blood). In addition to the use of DMSO, other issues cited with such assays include poor reproducibility for very sparingly water soluble compounds, differences between physical forms of candidates and an apparent lack of standardization, which limits comparisons between different sites.[64]

In drug discovery processes, kinetic solubility tends to be the most relevant parameter for rapidly evaluating whether the compound of interest, which is generally pre-dissolved in DMSO, stays in solution after dilution in specific screening media or not. It is important to note that incomplete solubility within the functional assay could lead to problems such as unreliable results, underestimated activity, reduced high-throughput screening (HTS) hit rates and an inaccurate structure–activity relationship (SAR) when not identified.[1,65] In the development phase of drug research, equilibrium solubility of drug-like molecules is measured by different methods, among them potentiometric methods such as Dissolution Template Titration (DTT) or CheqSol®.[35,36,66-69] The DTT procedure simulates the entire titration curve before the assaying begins using the pK_a and the estimated intrinsic solubility.[66,69] The basic method, however (against which new solubility methods are generally validated), remains the classical saturation shake-flask method.[25]

2.2.1.3 Detection methods

It is important to note that the different analytical tools used for compound detection, such light scattering/turbidity, UV plate readers, LC-UV, and LC–MS/LC-CLND (chemiluminescent nitrogen detection), can potentially impact assay results (Table 2.4). Other factors that can affect the ability to determine "true" thermodynamic solubility of compounds in solvents include the compound's purity; the reliability of compound's particle size measurement, shape factor, and particle growth, all of which can cause a change in the surface area. For example, Fini et al.[70] reported a linear relationship between the efficiency of dissolution and the shape factor of dissolving particles of a diclofenac salt crystallized from various solvents and mixtures of solvents suggesting importance of shape irregularity in affecting dissolution rate differences ($r^2 = 0.9077$); amount of residual solid, changes in solid properties of the compound (polymorph, amorphous, solvate formation); the crystallization solvent; solvent purity; mixing conditions which may affect the homogeneity of mixing and particle size reduction; the efficiency of temperature control within the host vessel; uncertainty regarding the time taken to establish equilibrium; degree of ionization; compound's stability in solution; pH of the solution at start and end of procedure; compound aggregation due to the formation of promiscuous aggregates; self-association of molecules or surface reduction by aggregates and filter pore or the effect on centrifugation on solid/liquid separation.[3,62,70-82]

Table 2.4 Advantages and limitations of common detection methods for solubility measurements[33]

Detection methods	Advantages	Limitations
Light scattering/ turbidity	Universal, fast, economical	Interference from certain colored compounds and impurities, sensitive to sedimentation and particle size, low sensitivity, measures precipitates rather than solution concentration
UV plate reader	High sample coverage, fast, economical, sufficient sensitivity for solubility measurement, good linearity over wide dynamic range	Requires UV chromophore, interference from impurities and matrix material
LC-UV	High sample coverage, less interference from impurities and matrix material, sufficient sensitivity for solubility measurement, good linearity over wide dynamic range	Requires UV chromophore, might need different HPLC methods for special compounds, not as fast and economical as UV plate readers
LC–MS	High sensitivity, high selectivity, low interference	Less universal, moderate sample coverage, low dynamic range for linearity, too sensitive for solubility measurement (need large dilution), high maintenance, costly
LC-CLND	No standard curve needed	Only for nitrogen containing compounds, interference from nitrogen containing solvents, significant signal loss for adjacent nitrogen

Kramer et al.[83] investigated kinetic solubility by nephelometry for over 700 drug-like compounds and built three classification criteria; insoluble (<20 μM), moderately soluble (20–200 μM) and soluble (>200 μM) using a forest algorithm for which the datasets' performance was evaluated to provide appropriate data for early phases of drug discovery. As such, the environment in which a solubility assay is run and the primary focus of the assay dictate how assays are currently set-up and performed in the drug discovery and development process (Fig. 2.4).[3] Although high-throughput kinetic solubility determination provides a cost-effective and fast solution, its application in projecting the impact of solubility in vivo for formulation is limited due to a poor correlation with conventional solubility data from equilibrated thermodynamic conditions.[20,84] Kinetic solubility, however, seems to becoming more prevalent due to reasons such as compounds being solubilized in DMSO for storage and distribution and allowing poorly soluble compounds to be available in the aqueous phase.[32] As both kinetic and thermodynamic solubility play crucial roles in drug discovery and development, it is important to remember that the effective solubility (the goal for high-throughput solubility assays) of compounds in the screening environment is kinetically driven, and the intrinsic solubility of the compound is driven by a thermodynamic process and with regards to the detection methods, they offer different advantages depending on the question being addressed.[32]

2.3 Drug Solubility Determination

2.3.1 Shake-Flask Method

The shake-flask (SF) method is a relatively simple procedure and is considered the method of reference for solubility measurement but it is time-consuming and requires a lot of manual processing. Due to potential compound degradation or polymorph formation, both of which can affect solubility, it is important that the compound's dissolution profile is investigated alongside solubility measurements. This allows the determination of the shortest time

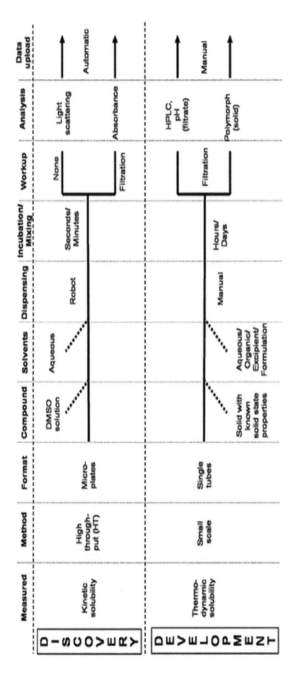

Figure 2.4 Solubility determination in drug discovery and development: key elements of traditional solubility assay workflows. Reprinted from Alsenz and Kansy,[3] with permission from Elsevier.

required for deducing the establishment of an equilibrium. It is also important to note that manipulations such as vortexing, sonicating, or the use of small glass microspheres prior to equilibration can be incorporated to offset problems of low wettability of poorly soluble compounds and also thus reduce equilibrium time. Although filtration is often used to separate phases for analysis, with the other main technique being centrifugation, the sorption of especially low soluble compounds on the filter can lead to errors in solubility values. Precautions such as pre-rinsing the filter with a saturated solution can reduce this error.

UV spectrophotometric analysis tends to be the most commonly used method in solubility determination; however, HPLC may often be a more suitable method, as it can detect impurities and degradation products if any are present.[24,85] As a result of lack of standardization of this method, there is a huge range of published literature showing variations not only in experimental conditions but also in separation techniques which can lead to differences between solubility data reported in literature for some compounds.[66,70,78,86–93]

This widely used thermodynamic or equilibrium solubility measurement method was proposed by Higuchi and Connors.[94] It involves the following steps:

- An excess of the compound of interest is added to a solubility-determining medium such that a saturated solution exists in equilibrium with some solid phase.
- The sample is then shaken at a controlled temperature for a specific time. Acidic or basic drugs should be dissolved using an unbuffered medium (usually water) as the solubility of compound of interest could change as a result of a change in pH arising from further addition of the compound. Thus the final pH of a saturated solution of the compound may be far from pH 7 due to self-buffering.[3]
- The sample is then filtered and the concentration of the compound in the filtrate determined after equilibrium (usually after 24 hours using HPLC). Data obtained can then be compared with literature to determine similarity or give an indication into the accuracy of the method used.

This method takes into consideration the effect of the crystal lattice and polymorphic forms of the compound being tested as these properties can greatly affect solubility. During solubility measurement, a compound may transform to a more stable polymorph. It is, therefore, especially important to isolate the solid at the end of the assay, to ensure no polymorphic transitions have occurred and the form of the compound is unaltered.[73,95,96] Pudipeddi and Serajuddin,[73] determined the solubilities of 55 compounds, with previously documented solubilities, each of which had multiple polymorphs. For the non-solvated crystal forms, there was only a two-fold difference between the most soluble and the least soluble polymorphs. The highest ratio of difference in solubilities between different forms of the active was 23, for premafloxacin. Solubility ratios of amorphous to crystalline forms were generally higher (about 10-fold). They proposed that "unstable solid forms with extraordinarily high free energies might have eluded detection due to rapid conversion to lower energy forms, thus limiting the observable range of measurable solubility differences."[73,97] As such thermodynamic solubility is often referred to as the "true" solubility of a compound.[1,63] Despite the benefits of this method of solubility measurement, it has several drawbacks:

- It is time-consuming and difficult to automate requiring an individual weighing step.
- It requires relatively high amounts of compounds (usually several mg).
- A relatively lengthy incubation time is needed for equilibrium attainment.
- The method cannot distinguish between soluble monomers and soluble aggregates of the drug molecules (which may range from dimers to micelles) unless more sophisticated experiments are performed.[23,98]
- Solubility measured at a fixed pH value may be highly dependent on the nature and concentration of the counter-ions present in the medium. This may be critical for poorly soluble compounds being highly ionized at the pH of the measurement.[23,98]

This method is therefore often used in the later development stages and advanced stages of lead compounds.

Baka et al.[25] evaluated the experimental conditions that affect equilibrium solubility by use of the classical saturation shake-flask method using hydrochlorothiazide as a model compound. They found modifications in temperature, sedimentation time, composition of aqueous buffer and the technique of separation of solid and liquid phases all strongly influence the equilibrium solubility results. Variations in the amount of excess solid and stirring time, however, were found to have less of an influence. A new shorter protocol was then developed by the same authors with the recommendations below to ensure a reduction in experimental error of solubility measurements to about 4%.

- The measurements must be carried out at controlled, standard temperature.
- Sörensen's phosphate buffer can be used between pH 3–7; Britton–Robinson buffer solution can be used between pH 2.5–11.5; HCl of appropriate concentration can be used below pH 2. (There is potential for salt formation with ionizable compounds which could affect the solubility hence the importance of isolating the solid for characterization as discussed in Section 2.2 and Section 2.3.2).
- To avoid difficulties in sampling, only a small excess (~5–10 mg/5 mL) of solid should be present.
- A minimum of 24 hours is necessary to reach the thermodynamic equilibrium; this time should consist of 6 hours of stirring plus 18 hours of sedimentation; but in case of very sparingly soluble compounds longer stirring time may be necessary for equilibrium, so in the most rigorous application of the shake-flask method, solubility would be measured after checking the required equilibration time from compound to compound.
- The recommended technique for phase separation is sedimentation, which assures the existence of a heterogeneous system until equilibrium has been achieved.

The equilibrium solubilities of five other drugs (Table 2.5) were measured using these recommendations and the results were in excellent agreement with the standard protocol used with the log S of a compound determined in less than one and a half days.[25]

Loftsson and Hreinsdottir[99] modified the classical shake-flask solubility method by shortening the equilibration time by application of a heating process to determine the solubility of 48 different drugs and pharmaceutical excipients in pure water at room temperature. Despite finding this modified shake-flask method generated reliable and reproducible data, the mean calculated solubilities were lower than experimental solubilities according to the Yalkowsky equation (see Section 2.3.2). This was due to the equation only being valid for non-electrolytes whereas many of the compounds tested were partly ionized in the aqueous solutions.

2.3.2 Miniaturized Shake-Flask Method

Despite application of Lipinski's rule of five as an early warning tool (see Section 2.3.4), the extensive use of organic solvents in the early stages of drug discovery can lead to a delay in determination of true solubility (see Section 2.2.1.2). With this in mind and also because of the large numbers of compounds screened and the very small amounts of compounds produced in the early development stage, Glomme et al.[85] developed a miniaturized shake-flask method for solubility determination that could be applied earlier in the discovery process. They determined the solubility of 21 compounds (solubility range 0.03–30 mg/mL) and compared them with those measured by a semi-automated potentiometric acid/base titrations and computational methods (Table 2.6). They showed their method's precision and throughput to be superior to the potentiometric method and that the miniaturized shake-flask method could be used for all compounds and a wide variety of media.[85]

2.3.2.1 Sample preparation and analysis

The theory and method is detailed in Glomme et al.,[85] but in brief, it is as follows:

Table 2.5 The solubility of compounds measured by the standard and the new shake-flask protocols[25]

Compound	MW	µg/mL	Solubility, S (M)	log S (M)	Solubility, S (µM)	log S (µM)	n
Standard protocol							
Hydrochlorothiazide	297.7	556 ± 13.2	0.001868	−2.73	1867	3.27	18
Furosemide	330.8	20.4 ± 2	0.000062	−4.21	61.7	1.79	8
Nitrofurantoin	238.2	109.5 ± 3	0.00046	−3.34	460	2.66	8
Piroxicam	331.4	5.95 ± 0.4	0.000018	−4.75	17.9	1.25	2
Quinine-HCl	360.4	201 ± 10	0.000558	−3.25	558	2.75	6
Trazodone	371.4	138 ± 10	0.000372	−3.43	370	2.57	6
New protocol							
Hydrochlorothiazide	297.7	571 ± 8.6	0.001918	−2.72	1918	3.28	12
Furosemide	330.8	18.7 ± 1.2	0.000057	−4.25	56.4	1.75	8
Nitrofurantoin	238.2	99 ± 4.1	0.000416	−3.38	416	2.62	8
Piroxicam	331.4	6.36 ± 0.04	0.000019	−4.72	19.2	1.28	3
Quinine-HCl	360.4	285 ± 30	0.000791	−3.10	717	2.86	5
Trazodone	371.4	176 ± 1.8	0.000474	−3.32	473	2.67	12

Note: Britton-Robinson (BR) buffer used for all samples except furosemide, which was measured in 0.01 M HCl solution.

Table 2.6 Solubility determined by the miniaturized shake-flask and pSol methods at several pH values for ionizable compounds[85]

		Method	
Compound	pH	Shake-flask solubility (μg/mL)	pSol solubility (μg/mL)
Dipyridamole	3.5	2199.4 ± 99.4	
	4.2	342.7	
	5	54.1 ± 2.2	39.9 ± 4.98
	6	10.5 ± 0.94	5.5 ± 0.66
	7	4.9 ± 0.13	2.1 ± 0.25
	7.8	6.0 ± 0.22	1.8 ± 0.22
Glyburide	2	0.07 ± 0.002	0.06[a]
	3	0.06 ± 0.01	0.06[a]
	5	0.10 ± 0.06	0.01[a]
	6	0.62 ± 0.15	0.56[a]
	7	5.62 ± 0.72	5.13[a]
	9	51.2 ± 0.29	51.2[a]
	9	98.6 ± 0.83	
	11.8	531.6 ± 9.60	
Mefenamic acid	2	0.06 ± 0.002	0.06 ± 0.02
	3	0.07 ± 0.01	0.06 ± 0.02
	5	0.65 ± 0.03	0.19 ± 0.08
	6	7.18 ± 0.15	1.41 ± 0.58
	7	67.2 ± 3.27	13.7 ± 5.62
	8	357.0 ± 2.79	
	9	486.2 ± 3.13	
Phenytoin	5		20.62 ± 2.67
	6		20.70 ± 2.69
	7	31.00 ± 2.03	21.49 ± 2.81
	8		29.52 ± 3.98
Levothyroxine	5	0.26 ± 0.1	0.22[a]
	6	0.27 ± 0.1	0.24[a]
	7	0.49 ± 0.1	0.45[a]

[a]Measured with cosolvent; pSol method is discussed in Section 2.3.3.

- It is recommended that the drug's approximate solubility is estimated using an in silico method outlined in Section 2.4 to ensure a minimal amount of drug is used.
- An amount of drug, corresponding to twice the calculated solubility, is weighed and filled into the Whatman UniPrep chamber (Fig. 2.5). 2 mL of appropriate medium is added

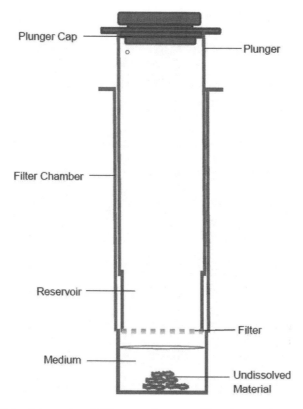

Figure 2.5 Schematic of Whatman Processor™. Adapted from Glomme et al.[85]

and chamber closed with a plunger. The sample is then shaken at 450 rpm and 37°C for 24 hours.

• With the exception of neutral compounds, after several hours, the pH and presence of precipitation should be checked. Additional sample can be added if dissolution is complete.

It is important to note that with regards to salts of strong acids and bases, an adjustment with 0.1 M NaOH/0.1 M HCl may be needed to maintain a constant pH value due to the small volume of medium used and a low buffering capacity.

- The plunger is inserted into the sample-containing chamber after shaking is completed and depressed, thereby forcing the filtrate into the reservoir of the plunger.
- The concentration of compound in the filtrate is determined after filtration using HPLC with UV detection, against a standard solution.

Bergstrom et al.[90] investigated the extent to which the Henderson–Hasselbalch equation could be used to predict the pH-dependent aqueous solubility of 25 weakly basic cationic drugs using a small-scale shake-flask method and found that the equations do not accurately predict the pH dependence at 25°C. This was attributed to the formation of aggregates/precipitates with the 0.15 M phosphate buffer used.[97] The authors deduced from solid state characterization of the aggregates/precipitates using differential scanning calorimetry (DSC) that the ionized samples had a higher melting point compared to that of the free base. It was concluded that the aggregates were as a result of the formation of phosphate salts. Procedures similar to this have been adapted to fast LC/MS detection and throughputs of about 50–200 compounds per day have been reported.[4, 100] Zhou et al.[20] developed a high-throughput equilibrium solubility assay method using a miniaturized shake-flask method for the early drug discovery process and found it to offer a fast, reliable, and a cost-effective screening tool for solubility assessment.

2.3.3 Potentiometric Titration Methods

Although equilibrium solubility measured by the classical shake-flask method is accurate if performed to a high standard, it is also slow to carry out. With advances in high-throughput screening for discovery of NCEs, kinetic solubility determinations may provide a faster alternative. Kinetic solubility measurements are based on the precipitation of a compound pre-dissolved in a co-solvent or aqueous medium by pH adjustment for ionizable compounds after dilution in a given medium (Fig. 2.6).[1,3,101] Determinations are based on the shift in pH values caused by loss of compound due to precipitation. It is important to note that kinetic solubility

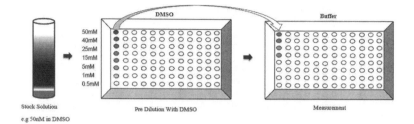

Figure 2.6 Principle of kinetic solubility determination exemplified by the Roche in-house Parallel Incremental Solubility Assay (PISA) adapted from Alsenz and Kansy.[3]

measurement is not a direct substitute for evaluation of the thermodynamic solubility, for reasons such as solubility is not measured at equilibrium, the appearance of a precipitate is strongly time-dependent and the disruption of the crystal lattice is not taken into consideration in this process.

The potentiometric titration method generating a plot of pH against the titrant volume added (Fig. 2.7a) was introduced by Avdeef[35, 36, 66] and is the reference method for kinetic measurements. The method of potentiometric measurements is detailed in[9] but in brief;

- 2–20 mL of water or mixed solvent comprising water and an organic water-miscible co-solvent such as DMSO, acetonitrile, 1,4-dioxane or methanol is used to dissolve about 50–500 μM of the sample to be assayed.
- Known volumes of strong standardized acids such as HCl or base such as KOH or NaOH are added to the ionizable compound solution of interest under vigorous stirring.
- Using a precision combination glass electrode, pH is continuously measured between the intervals of pH values 1.5–12.5.

To improve the precision of the measurement and to better mimic physiological states, an inert water-soluble salt (such as 0.15 M of KCl or NaCl) can be added to the reaction vessel which is thermostated at 25°C whilst the solution surface is bathed with a heavy inert gas (normally argon).

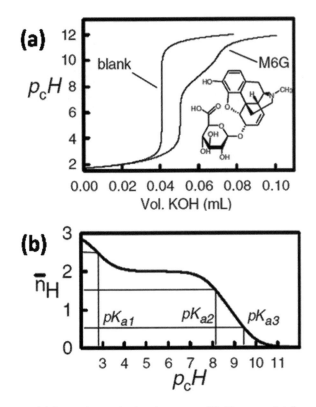

Figure 2.7 (a) Potentiometric titration curve (b) Bjerrum plot for a three-pK_a molecule.[9] M6G is morphine-6-glucuronide.

It is important to note that the shape of the curve produced can give an indication of the amount of substance present and its acid-base characteristic properties.[9] The solubility determinations are based on the difference between the aqueous drug or compound's pK_a measured in the absence of a solid phase and the apparent pK_a determined in the presence of an excess of solid compound. The shift produced or observed is thus proportional to the loss of compound, and thus to the solubility.[1] The advantages of this method include its ability to provide tangible solubility screening data in early stages, it is very economical (approximately 100 mg of a poorly soluble compound is needed) and is able to create a pH/solubility profile with one single determination, but is limited to ionizable compounds. It is used for calibrating high-throughput solubility methods and computational methods due to its ability to provide reliable data without the use of co-solvent, providing a complete solubility pH profile with a limited number of experiments and providing a better insight into potential solubility behaviour through the GI tract.[21,66,85,87,102] However, the simplified protocol for the determination of kinetic solubility is prone to inaccuracies and variation, e.g., insufficient incubation time, improper handling of the phase separation process, presence of DMSO, and indirect readout of solubility data can result in false positives or negatives due to discrepancies in solubility values sometimes up to 1–3 times in order of magnitude.

Figure 2.7a suggests that the compound, M6G, has a pK_a of ∼8.8; however, the wrong conclusions can be deduced from a simplistic reading or interpretation of a titration curve and as such, Bjerrum plots which can be obtained from the transformation of titration curves are necessary.[103–108]

Knowledge of the amount of strong acid and base and how many dissociable protons the compound of interest brings means that, irrespective of the on-going equilibrium reaction, one would know the total hydrogen ion concentration in solution. The concentration of the bound hydrogen ions is the difference between the total and the free concentrations of the hydrogen ion concentration. Dividing this concentration by the concentration of the sample, the average number of bound hydrogen atoms per molecule of substance, $\bar{n}_H$ can be calculated. The Bjerrum curve is thus a plot of $\bar{n}_H$ against the pH scale based on hydrogen ion concentration (p_cH). Bjerrum plots are an important graphical tool in the initial stages of titration data analysis[9] and can be obtained by:

- Subtracting the "blank" (titration curve with no sample) from a titration curve with sample present at fixed pH values (Fig. 2.7a).
- The difference is plotted and then rotated to show emphasis on $\bar{n}_H$ and pH being the dependent and independent variables, respectively (Fig. 2.7b).

Figure 2.7b shows the three pK_a values of M6G as p_cH values at half integral $\bar{n}_H$ positions, thus demonstrating the capability

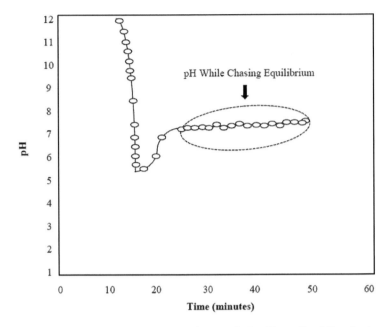

Figure 2.8 pH versus time curve for warfarin. The pH while chasing equilibrium corresponds with the pH at the CheqPoint. Adapted from Box et al.[26]

of the technique for studying compounds with overlapping pK_a values. Two commercial systems for potentiometric solubility measurement, pSol® and CheqSol®, are currently available.

Box et al.[26] developed a novel potentiometric procedure for rapid measurement of equilibrium aqueous solubility values of organic acids, bases, and ampholytes that form supersaturated solutions. In this procedure, the equilibrium solubility is actively sought by changing the concentration of the neutral form by adding HCl or KOH titrants and monitoring the rate of change of pH due to precipitation or dissolution in a process called Chasing Equilibrium (Fig. 2.8).[68] The results were reported in terms of intrinsic solubility, which for the acids and bases represented the concentration of their unionized species in a saturated solution at the pH where the compound was fully unionized. The Chasing Equilibrium method developed was used in the determination of the kinetic and equilibrium solubilities of the same 16 compounds.[26] Out of the 16 small organic compounds ranging from high to moderate to poorly soluble, 10 of the compounds showed an excellent correlation between the aqueous equilibrium solubility measured by the shake-flask protocol and the Chasing Equilibrium solubility method (Table 2.7). The kinetic solubilities, however, were higher than the equilibrium solubilities for the compounds that "chased equilibrium" but correlated well with the "non-chasers." The authors also found this new method to provide equilibrium solubility values in about an hour per sample.[26]

This chasing equilibrium technique, which is based on the same principle as the traditional potentiometric titration method, proved to be faster since the intrinsic solubility of the compound is determined rather than the entire pH/solubility profile. The Henderson–Hasselbach relationships can then be used in determining the approximate pH/solubility profile.[1]

2.3.4 High-throughput Solubility Determinations

Experimental and computation approaches for the estimation of permeability and solubility followed the establishment of Lipinski's so-called Rules of Five, according to which a compound is "drug-like" if its molecular weight is <500 Da, its octanol/water partition coefficient <5 (log scale), and it possesses <10 hydrogen bond acceptors and <5 hydrogen bond donors after examining the compounds on the market.[63] This provided an impetus for the development of the high-throughput approach for solubility determinations and allowed the estimation of aqueous solubility in the early stages of drug discovery to become a critical parameter in lead selection and optimization. The collection of such data via high-performance liquid chromatography (HPLC) or liquid chromatography mass spectrometry (LC/MS) was commonly carried out in manual format thus being time-consuming, labour-intensive, and costly. Its use is often restricted to late discovery or early development phases to test a few, selected compounds. As there are large number of compounds and small samples of compounds available at the early stages of drug discovery, small-scale, compound-sparing and fully automated methods of determining drug solubility are often desirable. A first estimation of solubility, to help define the absorption, distribution, metabolism, excretion, and toxicity (ADMET) space of a compound

Table 2.7 Summary of results from SF and Chasing equilibrium study[26]

Compound	Shake-flask measurement pH [unionized percent]	Spectroscopic data: $A^{1\%}_{1cm}$ @ λ_{max} (nm)	Intrinsic solubility (μg/mL)	Intrinsic solubility (μM)	Intrinsic solubility (log S) (mol/L)	Potentiometric range of sample weights (mg) (6 assays)	Time taken (min)	Kinetic solubility (μg/mL)	Intrinsic solubility (μg/mL)	Intrinsic solubility (μM)	Intrinsic solubility (log S) (mol/L)
Chasers											
Amodiaquin (ampholyte)	10	†	†	†	†	9.0–12.5	40±3	9±4	0.4±0.1	1.2±0.1	−5.94±0.04
Flumequine (acid)	2.50 [99.99]	403 @ 311	20.7±0.8	79.2±3	−4.10±0.02	9.3–12.4	48±2	121±10	34.2±2.1	130.8±8.2	−3.88±0.03
Furosemide (diacid)	2.00 [97.55]	1225 @ 234	20.4±2	61.7±6	−4.21±0.04	9.7–11.2	46±1	96±17	19.7±0.8	59.5±2.4	−4.23±0.02
Maprotiline (base)	11.50 [93.66]	45.3 @ 271	8.05±3ᶜ	29±9	−4.54±0.12	9.8–10.8	74±25	26.8±4.7	5.8±0.3	20.9±1.1	−4.68±0.02
Miconazole (base)e	11.5	†	†	†	−5.85ⁱ	6.2–26.4	55±8	~ 2.3ᶠ	1.0ᵈ±0.2	2.3ᵈ±0.5	−5.63±0.09
Niflumic acid (ampholyte)	3.35 [86.02]	408 @ 254	29.5±3	104.5±9	−3.98±0.04	10.3–11.5	45±2	59±12	9.5±0.5	33.8±1.8	−4.47±0.03
Nitrofurantoin (acid)	2.50 [99.99]	510 @ 265	109.5±3	460±10	−3.34±0.01	9.7–16.8	70±7	441±81	112±7	472±28	−3.33±0.03
Phthalic acid (diacid)h	0.50 [99.37]	65 @ 275	5950±200	35800±1000	−1.45±0.02	506–549	78±14	10104±5223	5330±47	32080±287	−1.49±0.01
Piroxicam (ampholyte)	3.73 [96.03]	640 @ 361	5.95±0.4	17.9±1	−4.75±0.03	9.8–13.4	54±2	234±88	5.9±0.9	17.9±2.8	−4.75±0.06
Warfarin (acid)	2.50 [99.63]	430 @ 282ᵇ	5.25±0.2	17.0±0.7	−4.77±0.02	10.1–11.3	61±9	119±2	5.3±0.2	17.2±0.7	−4.77±0.02
Non-chasers											
Chlorpromazine (base)	11.50 [99.45]	1045 @ 254	2.41±0.3	7.56±0.9	−5.12±0.05	9.7–10.2	84±15	2.7±0.1	2.7±0.1	8.5±0.2	−5.07±0.01
Imipramine (base)	11.50 [98.92]	261 @ 251ᵃ	21.7±3ᶜ	77.6±10	−4.11±0.06	9.5–12.9	77±3	17.3±0.3	17.2±0.4	61.3±1.5	−4.21±0.01
Nortriptyline (base)	11.50 [95.12]	384 @ 238	49.3±2	187±6	−3.73±0.01	9.9–10.9	95±11	27.3±0.9	27.0±1.4	102.6±5.3	−3.99±0.02

Quinine (dibase)	11.50 [99.89]	122 @ 329	201 ± 10	558 ± 40	−3.25 ± 0.03	24.5-34.5	154 ± 8	391 ± 5	363 ± 11	1118 ± 32	−2.95 ± 0.01
Verapamil (base)	11.50 [99.83]	260 @ 228^b	48.5 ± 2	106 ± 5	−3.97 ± 0.02	9.2-10.8	51 ± 6	47.8 ± 1.7	48.5 ± 2.0	107 ± 4	−3.97 ± 0.02
Trazodone (base) Change in state	11.50 [99.99]	278 @ 246	138 ± 10	370 ± 30	−3.43 ± 0.03	9.6-11.7	147 ± 15	435 ± 25	135^g ± 8	362^g ± 22	−3.44^g ± 0.02

Results presented in three units (μg/mL, μM, and mol/L (as log S) to aid comparison with other published values. ($n = 6$, Errors represent ± one standard deviation; All measurements made at 25 ± 0.2° C; All potentiometric measurements made at ionic strength 0.156 ± 0.003 M except miconazole and phthalic acid).

(a) Measured at pH 5.
(b) Measured in the presence of methanol.
(c) Opalescence present.
(d) Result extrapolated from measurements in 8–45% methanol, $p_s K_a s$ were measured in similar conditions.
(e) Reduced ionic strength 0.066 ± 0.03 M to prevent precipitation of miconazole hydrochloride salt.
(f) Estimated kinetic value based on average supersaturation level in 8–45% methanol.
(g) Final settled stable value.
(h) Ionic strength 0.346 ± 0.012 M is a consequence of high sample weight and high HCl concentrations at the low pH of experiments.
(i) Published value, measured by Dissolution Template Titration (DTT)[109]
†Could not be measured.

series or to anticipate potential issues, is often conducted using the a high-throughput solubility method. Solubility data based on DMSO stock solutions can help with interpretation of activity test results in screening and subsequently improve structure activity relations. The solubility data also contribute to medicinal chemistry where already in the early stage of lead identification, the problem of poor oral absorption could be encountered.[63,110–112]

As such there is a drive towards finding cost-effective, high-throughput equilibrium solubility assaying methods.[1,33] DMSO is generally the solvent of choice for kinetic solubility testing, acting as a cosolvent with water due to its amphipathic structure and as such can induce higher aqueous solubility for compounds.[113] Kinetic solubility measurements evaluated using HTS methods determine whether compounds dissolved in DMSO remain soluble after dilution in aqueous screening media. Turbidimetric methods are often used in the detection of precipitate formation used as an indicator of the solubility limit.[63] Such methods require addition of small aliquots of stock solutions of DMSO every minute and measuring light scattering by the use of a nephelometric turbidity detector in the 620–820 nm range to determine the presence or formation of precipitates. Authors such as Bevan and Lloyd,[102] Pan et al.,[62] Dehring et al.,[114] and Fligge and Schuler[115] successfully managed to scale down the methodology to use a 96-well plate format keeping the DMSO concentration in the buffer constant. These approaches, however, are not free of drawbacks. Compounds with very low aqueous solubilities (<20 μM) are not always ranked due to the relatively low sensitivity of the detection method and a lack of sensitivity to impurities.[115] LC/MS detection systems can be employed as a means to address this resulting in throughput of about 200 compounds a day.[62,64,100]

Because incomplete solubility can lead to unreliable results, adaptations to this method have been reported requiring small aliquots of DMSO stock solutions being diluted with aqueous buffer in a filter microplate until a final DMSO percentage varying from 1 to 5% (v/v) is obtained. Experimental conditions have been adapted to require DMSO concentrations as low as 0.33%.[116] Other authors have reported DMSO values ranging from 0.5 to 1%.[61,62,114,115,117,118] This method has been proven to be more

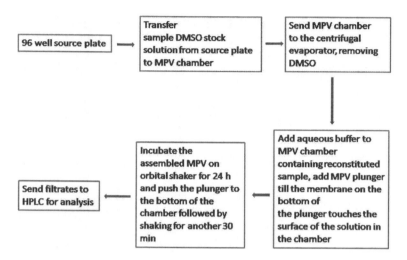

Figure 2.9 Schematic diagram of the HT equilibrium solubility assay adapted from Zhou et al.[20] *Abbreviations*: MPV is mini-prep vial.

sensitive and less subject to errors caused by impurities compared to turbidimetric methods. Precipitates that occur when a compound is poorly soluble are removed by either filtration or centrifugation and solubilities are determined by the measurement of the solubilized fractions of the compound in questions concentration using either UV detection, HPLC, or UHPLC-MS.[62,115,119,120]

Zhou et al.[20] introduced a high-throughput equilibrium solubility (HT-Eq sol) assay method by adapting the miniaturized shake-flask approach and using inline HPLC for analysis (Fig. 2.9). The assay was validated and optimized using a test set of 85 marketed drugs and Novartis internal compounds and there was excellent correlation between the new assay method and the conventional shake-flask thermodynamic solubility data generated in-house and the equilibrium solubility results reported in literature with an R^2 value in the rage 0.95–0.96. They concluded that their method was a fast, reliable, and cost-effective screening tool for solubility assessment in early drug discovery, allowing for prioritization of drug candidates using aqueous solubility in conjunction with other profiling information and efficacy data.

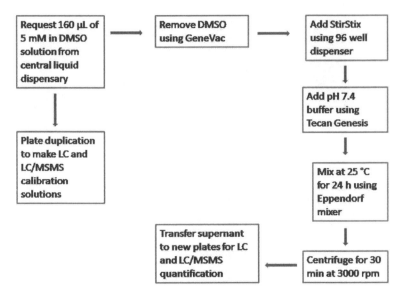

Figure 2.10 Flow chart of the high-throughput Dried-DMSO solubility measurement adapted from Alelyunas et al.[121] It is important to note that the use of co-solvents and supersaturated solutions mean that although HTS methods are very useful for evaluating solubility in screening bioassay media, solubility values generated are generally overestimated in comparison with thermodynamic solubility and also that solubility data obtained from different kinetic methods involving different experimental conditions are not intended to be comparable.[61,68,99,115,123]

Alelyunas et al.,[121] developed a rapid throughput Dried-DMSO equilibrium method as the primary method for solubility profiling of discovery drug candidates. This method is summarized in a flow chart in Fig. 2.10. By removing DMSO and generating solid material, the method minimizes concern of potential DMSO effects on measured solubility.[20,122] Solubility was determined for 31 commercially available compounds with differing ionization states using both Dried-DMSO and solid methods. Data obtained showed good agreement for greater than 80% of compounds studied with the ratio of Dried-DMSO over solid within three times experimental reproducibility.

The large differences observed for a few of the compounds, with the Dried-DMSO value typically higher than the solid value, suggested that perhaps stirring times needed to be extended beyond 24 hours to obtain equilibrium solubility, especially when starting with dry powder. The Dried-DMSO solubility data were consistent from batch-to-batch and were generally independent of salt forms. The authors also showed Dried-DMSO data could be complementary to solid values by providing solubility from a solid-state that is different from initial dry powder.[121]

The pursuit of overcoming such drawbacks with the discrepancies between thermodynamic and kinetic solubility values meant researchers have had to devise solutions to measuring thermodynamic solubility during the drug discovery process. Several adaptations have been made to the HTS solubility method including using other solvents and polarized light microscopy for analysing the solid form of the precipitants. Such changes have improved the quality of data in the early drug discovery phase and given solubility values in acceptable agreement with the thermodynamic solubilities.[3, 61, 74, 124–128] Fligge et al.[110] integrated a commercial nephelometer into an automated liquid-handling system using 384-well plates for rapid solubility classification and linked the system to an LC/MS machine for hit validation and analysis routinely performed for the confirmation of a compound's integrity. This meant that both solubility measurements and hit validation could be conducted within the same sample plates, negating further sample processing and additional material consumption. Seadeek et al.[124] designed and validated a novel high-throughput system for thermodynamic solubility determination that required only 5 mg of sample and coupled solubility with final pH and crystallinity. A favourable comparison was found after validation and a comparison with a subset of 10 compounds from the Pfizer discovery compounds run on the automated and more traditional shake-flask method employing HPLC. Sugano et al.,[74] investigated their experimental DMSO stock solutions when the DMSO content was 1% and found that when the incubation time was 20 hours and the precipitant was crystalline, the DMSO stock solubility was similar to the solubility using powder material. When the incubation time was 10 minutes and/or the precipitant not crystalline, the DMSO stock solubility was higher than the powder material's solubility. The results highlight

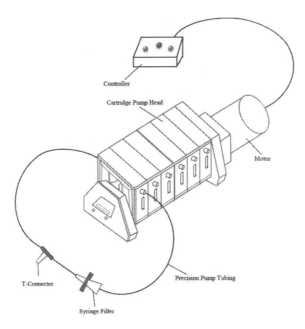

Figure 2.11   Schematic representation of the miniature device adapted
from Chen and Venkatesh.[129]

the importance of the solid form of the precipitant solubility data
interpretation.

Chen and Venkatesh[129] developed a miniaturized device
(Fig. 2.11) for measuring aqueous and non-aqueous solubility
during the drug discovery process and found the device to provide
a unique testing method for equilibrium solubility in a microscale
setting, requiring as little as ~1 mg of a testing compound to produce
the entire pH-solubility profile. As with Glomme et al.,[85] there is a
relatively low throughput for such methodologies due to problems
with automation of accurate distribution of solid compounds into
vials or wells of the microplate.[3, 129, 130] Alsenz et al.[131] developed
a medium-throughput, compound-saving, thermodynamic solubility
assay also known as 'partially automated solubility testing' (PASS)
for early drug development. Results showed the PASS assay
correlated well with reported solubility values ($r^2 = 0.882$). Such
methods are useful where maximal solubilities in many commonly
used solvents have to be determined. Dehring et al.[114] developed

a fully automated, laser-based nephelometry system for high-throughput kinetic aqueous solubility measurement that integrated a robotic liquid/plate handler with nephelometry detection. This system could determine the kinetic solubility of up to 1800 compounds within 6 days, and the solubility values determined are comparable to the literature values determined by other methods.

2.4 In silico Predictions

Poor solubility has been identified as the cause of numerous drug development failures and, as such, several strategies to either improve a compound's "druggability" or limited solubility using solubilization techniques have been developed.[23,63,132,133] However, due to the high lipophilicity and strong intermolecular interactions of poorly aqueous soluble solid compounds, it renders solubilising energetically costly.[23] As such strategies including the application of in silico methods during structural design to predict solubility risk, screening solubility to identify potential issues early, modifying structure to improve solubility and developing formulations to increase solubility and dissolution rate have been developed.[33] However, data scattering due to assay limitations, the imprecise definition of the term "solubility" (analogous to log P and log D, there is the need to distinguish intrinsic solubility, S_0, from solubility measured at a given pH value in a defined medium as intrinsic solubility refers to the solubility of the unionized species) and all the issues affecting experimental determinations described in previous sections, largely impact on the ability to create predictive models for solubility.[23] There are, however, several models for predicting solubility along with commercial software products and these predictive in silico solubility models have several benefits.[33,134–141]

These benefits include:

- usefulness in medicinal chemistry optimization
- structure–property relationships
- projecting the solubility impact on preclinical/clinical PK
- foreseeing the risk of development candidates.

Good and accurate prediction of solubility can be obtained if the solubility of a compound is mostly governed by lipophilicity. However, if the solubility of a compound is mostly controlled by crystal packing, it is much more difficult to obtain an accurate prediction. As such, predicting crystal packing is the major challenge in accurately predicting solubility.[139]

- Approaches for predicting solubility in silico include fragment-based models which predict solubility as a sum of substructure contributions taking the contributions of atoms, bonds or larger substructures in consideration.[142-144]

Hansch developed the first correlation between log P and solubility equation[145] and it can be applied to many organic lipids. This was one of the early solubility calculation approaches.[145]

$$\log S = A\log K_{ow} + B \tag{2.1}$$

where $\log K_{ow} = $ octanol/water partition coefficient
A and $B = $ constants (depending on lipid under consideration)

Other models based on log P are founded on Hansch's equation. Equation 2.2, the Yalkowsky equation, was derived from equation 2.1 and combines two fundamental parameters important to solubility, i.e., melting point and log P.[146]

$$\log S = 0.8 - \log P - 0.01(\text{Mp} - 25) \tag{2.2}$$

where Mp = melting point (25°C for liquids)

A general solubility equation (GSE) (2.3) was proposed by Yalkowsky and co-workers in which the correlation with log P was improved further by the addition of an experimental melting point (MP).[147] Models based on solvation properties take the solute-solvent interactions into consideration such as the equation derived by Abraham and Le (2.4).[148]

$$\log S_0 = 0.5 - 0.01(\text{MP} - 25) - \log P \tag{2.3}$$

$$\log S_0 = 0.52 - 1.00R_2 + 0.77\pi^H + 2.17\Sigma\alpha^H - 4.24\Sigma\beta^H$$
$$-3.36\Sigma\alpha^H\Sigma\beta^H - 3.99V_x \tag{2.4}$$

where

$\log S_0$ is the intrinsic solubility,

R_2 is excess molar refraction,

π^H is dipolarity/polarizability,

α^H and β^H are hydrogen-bond acidity and basicity, respectively, and

V_x is the McGowan's molecular volume (which also characterizes the hydrophobicity of the solute).

The additional hydrogen-bonding cross-correlation terms bring about improvements in the correlation with solubility. The ABSOLV module of the ADME Boxes software provides an alternative by calculating the parameters for the Abraham and Lee equation for complex multifunctional drug-like structures from fragment contributions.[23] Hybrid models for in silico predictions include polar surface area (PSA), which characterizes molecule polarity and hydrogen bonding features, and quantum chemical methods (Conductor like Screening Model for Realistic Solvents, COSMO-RS).[134,149–151] Statistical and data-mining techniques including the classical linear regression, Partial Least Squares (PLS), neural networks, support vector machines, Monte Carlo simulations, genetic algorithms and cellular automata have been applied in the field of solubility prediction.[33,137,143,152–155] Hill and Young[156] analyzed the relationship between calculated and measured hydrophobicity and about 100,000 measured kinetic solubility values and proposed a "Solubility Forecast Index" (SFI). Although the proposed SFI was simple (2.5), they found it to be an effective guide to predicting solubility. They concluded that keeping SFI < 5 provides a powerful guide for compound design in drug discovery, which gives a high probability of securing favourable physical properties.

$$\text{SFI} = c\log D_{\text{pH }7.4} + \#\text{Ar} \qquad (2.5)$$

where #Ar is the number of aromatic rings and c Log D is calculated Log D.

Salahinejad et al.[157] calculated values for lattice energy and sublimation enthalpy of organic molecules as descriptors to improve the accuracy of the aqueous solubility models. Using a large data set of 4558 compounds and multiple statistical analysis, they found the Bayesian regularized artificial neural network with a

Laplacian prior (BRANNLP) to have the best statistical results with squared correlation coefficients of 0.90. However, incorporation of the descriptors that captured crystal lattice interactions showed no significant improvement in the quality of the aqueous solubility models.

2.5 Other Physicochemical Determinations

2.5.1 pK_a

The pK_a or ionization constant affects drug properties such as solubility, permeability, Log D and oral absorption through the modulation of the distribution of neutral and charged species. Generally, acidic compounds tend to be more soluble and less permeable at high pH and basic compounds tend to be more soluble and less permeable at low pH. Information on pK_a facilitates salt selection processes. As more than 60% of marketed drugs are ionizable, it is important to note that pK_a impacts biological activity and metabolism through electrostatic interactions.[158] Potentiometric and spectrophotometric methods are the most commonly used or traditional methods for measuring pK_a values with high-throughput methods for pK_a measurement, including spectral gradient analysis and capillary electrophoresis (Table 2.8).[159,160] Useful software for pK_a prediction is the ACD pK_a Laboratory software.[22,161]

With regards to the potentiometric titration method, the pK_a is determined from the difference between the potentiometric titration curve of the tested compound derived from the pH of a vigorously stirred solution and a blank aqueous titration.[1] The pH is continuously measured with a glass electrode as precisely known volumes of a standardized strong acid or base are added.[67] Co-solvent can be added where there is insufficient solubility in aqueous media with titrations at different co-solvent concentrations allowing an extrapolation to 0% co-solvent.[105,106,162,163] The Sirius GLpK_a allows this to be done in an automated set-up and a DPAS option allows for spectrophotometric titrations.[4]

The determination of spectrophotometric pK_a is based on the change in the absorption of a drug's chromophore near ionizable

Table 2.8 Experimental techniques for pK_a determination[1]

Method	Potentiometry	Spectrophotometry	CZE-UV
Measurement	Ionization profile	Ionization profile	Single points
Quantity	2–10 mg	2–10 mg	<<<1 mg
High purity	Necessary	Necessary	Not necessary
Low/medium-throughput instrumentation	Automatic titrator: GLpKa (Sirius Analytical Instruments, Forest Row, UK)	DAD-UV spectrophotometer coupled to an automatic titrator (various vendors)	Capillary electrophoresis unit (various vendors)
High-throughput instrumentation with 96-well plate technology	—	Spectral gradient analysis, SGA (Sirius Analytical Instruments, Forest Row, UK)	Multiplexed capillary electrophoresis: cePRO 9600 (Advanced Analytical Technologies, Ames, IA, USA) or MCE 2000 (Pfizer Laboratory, Groton, CT, USA)
Miscellaneous	—	Only for compounds with a measurable chromophore that is dependent on the pH	—

moieties. A plot of the absorbance measured at an appropriate wavelength as a function of pH is how the ionization constants are determined and this technique has been used successfully in determining the pK_a values for sartans and lansoprazole.[4,164,165] Spectrophotometric pK_a determinations are usually more sensitive than the potentiometric methods.[1] Several other spectrophotometric methods have been developed, including a multi-wavelength spectrophotometric titration, which requires a diode array detector coupled to an automated pH titrator with a more complex data treatment analysis;[166–169] the microscale spectrophotometric method;[170] and the spectral gradient analysis (SGA) method, which increases throughput (a commercial SGA instrument is available from Sirius Analytical Instruments). Despite being capable of pK_a determination for a compound in 4 minutes, this technique requires a measurable UV chromophore that is dependent on the pH and has difficulty dealing with impure and unstable compounds.[4,67,171]

The numerous advantages of the capillary zone electrophoresis (CZE) method means it has emerged as a popular method for pK_a determination.[1] These advantages include the small sample and solvent consumption, the capacity for coupling with different detection systems such as mass spectrometry (MS),[4,64,172–179] the pK_a values of non-UV-absorbing compounds can be determined by indirect UV detection,[180,181] amperometric detection,[182] or contactless conductivity detection (CCD).[183–185] Also the sensitivity of the UV detector on the CZE means pK_a determination is possible for the majority of pharmaceutical compounds. The UV detector also enables a simultaneous spectrophotometric determination of pK_a values.[159,181,186–189] To aid throughput, methods such as pressure-assisted capillary electrophoresis (PACE), PACE coupled with MS and a vacuum-assisted multiplexed 96-channel capillary electrophoresis with UV detection have been used for pK_a determination with some of the methods screening up to 24 compounds per hour.[118,159,172,178,184,190–197]

2.5.2 Lipophilicity

Lipophilicity is often expressed as a partition or distribution coefficient (Log P or Log D) between octanol and aqueous phases and usually reported as the logarithm of the partition coefficient. It impacts significantly on ADMET properties and is widely used in drug discovery for quantitative SAR (QSAR) and quantitative structure-property relationship (QSPR) and is valid for single electrical species.[198–200] The classical shake-flask technique still remains the reference method for lipophilicity measurement and lipophilicity can be increased by increasing molecular size and decreasing hydrogen-bonding capacity.[201] Several high-throughput methods developed to speed up, automate and miniaturize the shake-flask method have been developed and are available for Log D determination.[202–210] Capillary electrophoresis is also emerging as an interesting method for log P_{oct} determination.[172,175,176,179,210–216] Also, different capillary electrophoresis modes, including pseudo-stationary phases namely MEKC, MEEKC, and VEKC/LEKC (micellar, microemulsion, and vesicle/liposome electrokinetic chromatography, respectively) have been reported and their pseudo-stationary phases are considered to mimic biological membranes better than 1-octanol.[174,192,217–237]

Other authors have attempted to develop methods for the simultaneous measurement of pK_a and Log P[238–246] with the obtained log k_w values correlating well with log P_{oct} values. Agreement with literature pK_a values was poor and improvements in the methodology generated lower log P_{oct} values as compared to the previous method thereby generating a moderate agreement with literature values. In silico methods are also quite effective in predicting Log D, especially PrologD^{TM} from CompuDrug Chemistry Ltd based on calculations performed using the PrologD^{TM} version 1.3 and comparison with data from literature for clonidine derivatives, clonidine blockers, and amphoprotic fluoroquinolone-forming zwitterions.[247]

2.6 Summary

Aqueous solubility is a significant factor influencing several aspects of the pharmacokinetic profile of a drug and many publications report different methodologies for the development of reliable measurements and computational models for the prediction of solubility. The solubility in the body can often be different from that determined in common in vitro buffer systems used for quality control processes. The quality of computational models is affected by the accuracy of the employed experimental solubility data. High-throughput discovery processes have driven the development of high-throughput screening processes for measurement of physicochemical determinations including lipohilicity, pK_a, and solubility, but the accuracy of these is often compromised by the requirement for speed and the inclusion of non-biorelevant media in the determination. Current practice is to use high-throughput screening processes and prediction software at earlier stages but to incorporate an accurate determination of solubility as early as feasible within the development process.

Acknowledgements

We would like to acknowledge the contribution of Mohamed Hajamohaideen for production of the images.

References

1. Henchoz, Y., Bard, B., Guillarme, D., Carrupt, P. A., Veuthey, J. L., & Martel, S. (2009). Analytical tools for the physicochemical profiling of drug candidates to predict absorption/distribution. *Anal. Bioanal. Chem.*, 394(3), 707–729.
2. Kennedy, T. (1997). Managing the drug discovery/development interface. *Drug Discov. Today*, 2(10), 436–444.
3. Alsenz, J., & Kansy, M. (2007). High throughput solubility measurement in drug discovery and development. *Adv. Drug Deliv. Rev.*, 59(7), 546–567.
4. Kerns, E. H. (2001). High throughput physicochemical profiling for drug discovery. *J. Pharm. Sci.*, 90, 1838–1858.
5. Amidon, G. L. (1995). The rationale for a biopharmaceutics drug classification, in Biopharmaceutics Drug Classification and International Drug Regulation, Capsugel Library, 179–194.
6. Amidon, G. L., Lennernäs, H., Shah, V. P., & Crison, J. R. (1995). A theoretical basis for a biopharmaceutic drug classification: the correlation of *in vitro* drug product dissolution and *in vivo* bioavailability. *Pharm. Res.*, 12(3), 413–420.
7. Yu, L. X., Amidon, G. L., Polli, J. E., Zhao, H., Mehta, M. U., Conner, D. P., Shah, V. P., & Hussain, A. S. (2002). Biopharmaceutics classification system: the scientific basis for biowaiver extensions. *Pharm. Res.*, 19(7), 921–925.

8. Polli, J. E., Yu, L. X., Cook, J. A., Amidon, G. L., Borchardt, R. T., Burnside, B. A., Burton, P. S., & Zhang, G. (2004). Summary workshop report: Biopharmaceutics classification system—implementation challenges and extension opportunities. *J. Pharm. Sci.*, 93(6), 1375–1381.

9. Avdeef, A. (2012). *Absorption and Drug Development: Solubility, Permeability, and Charge State*, 2nd ed. John Wiley & Sons Inc. New York, NY.

10. Al-Hamidi, H., Edwards, A. A., Mohammad, M. A., & Nokhodchi, A. (2010). To enhance dissolution rate of poorly water-soluble drugs: glucosamine hydrochloride as a potential carrier in solid dispersion formulations. *Colloids Surf. B*, 76(1), 170–178.

11. Al-Hamidi, H., Edwards, A. A., Douroumis, D., Asare-Addo, K., Nayebi, A. M., Reyhani-Rad, S., Mahmoudi, J., & Nokhodchi, A. (2012). Effect of glucosamine HCl on dissolution and solid state behaviours of piroxicam upon milling. *Colloids Surf. B*, 103, 189–199.

12. Grundy, J. S., Anderson, K. E., Rogers, J. A., & Foster, R. T. (1997). Studies on dissolution testing of the nifedipine gastrointestinal therapeutic system. I. Description of a two-phase in vitro dissolution test. *J. Control. Release*, 48(1), 1–8.

13. Amidon, G. L., & Walgreen, C. R. (1998). Rationale and implementation of a biopharmaceutics classification system (BCS) for new drug regulation, in Biopharmaceutics Drug Classification and International Drug Regulation, Capsugel Library, 13–27.

14. Hussain, A. S. (1998). Methods for permeability determination: A regulatory perspective. AAPS Workshop on Permeability Definitions and Regulatory Standards for Bioequivalence, Arlington, VA, Aug. 17–19, 1998.

15. Blume, H. H., & Schug, B. S. (1999). The biopharmaceutics classification system (BCS): class III drugs—better candidates for BA/BE waiver? *Eur. J. Pharm. Sci.*, 9(2), 117–121.

16. Lentz, K. A., Hayashi, J., Lucisano, L. J., & Polli, J. E. (2000). Development of a more rapid, reduced serum culture system for Caco-2 monolayers and application to the biopharmaceutics classification system. *Int. J. Pharm.*, 200(1), 41–51.

17. Chen, M. L., Shah, V., Patnaik, R., Adams, W., Hussain, A., Conner, D., Mehta, M., Malinowski, H., Lazor, J., Huang, S. M., Hare, D., Lesko, L., Sporn, D., & Williams, R. (2001). Bioavailability and bioequivalence: an FDA regulatory overview. *Pharm. Res.*, 18(12), 1645–1650.

18. Rege, B. D., Yu, L. X., Hussain, A. S., & Polli, J. E. (2001). Effect of common excipients on Caco-2 transport of low-permeability drugs. *J. Pharm. Sci.*, 90(11), 1776–1786.

19. Chen, M-L., Amidon, G. L., Benet, L. Z., Lennernas, H., & Yu, L. X. (2011). The BCS, BDDCS and regulatory guidance. *Pharm. Res.*, 28, 1774–1778

20. Zhou, L., Yang, L., Tilton, & S., Wang, J. (2007). Development of a high throughput equilibrium solubility assay using miniaturized shake-flask method in early drug discovery. *J. Pharm. Sci.*, 96, 3052–3071.

21. Wang, J., & Urban, L. (2004). The impact of early ADME profiling on drug discovery and development strategy. *Drug Discov. World*, 5, 73–86.

22. Di, L., & Kerns, E. H. (2003). Profiling drug-like properties in discovery research. *Curr. Opin. Chem. Biol.*, 7, 402–408.

23. Faller, B., & Ertl, P. (2007). Computational approaches to determine drug solubility. *Adv. Drug Deliv. Rev.*, 59(7), 533–545.

24. Jouyban, A. (2010). Handbook of solubility data for pharmaceuticals. CRC Press.

25. Baka, E., Comer, J. E., & Takács-Novák, K. (2008). Study of equilibrium solubility measurement by saturation shake-flask method using hydrochlorothiazide as model compound. *J. Pharm. Biomed. Anal.*, 46(2), 335–341.

26. Box, K. J., Völgyi, G., Baka, E., Stuart, M., Takács-Novák, K., & Comer, J. E. (2006). Equilibrium versus kinetic measurements of aqueous solubility, and the ability of compounds to supersaturate in solution—a validation study. *J. Pharm. Sci.*, 95(6), 1298–1307.

27. Bhattachar, S. N., Deschenes, L. A., & Wesley, J. A. (2006). Solubility: it's not just for physical chemists. *Drug Discov. Today*, 11(21), 1012–1018.

28. Jain, N., & Yalkowsky, S. H. (2001). Estimation of the aqueous solubility I: Application to organic nonelectrolytes. *J. Pharm. Sci.*, 90(2), 234–252.

29. Yalkowsky, S. H., & Valvani, S. C. (1980). Solubility and partitioning I: solubility of nonelectrolytes in water. *J. Pharm. Sci.*, 69(8), 912–922.

30. Martin, A. N., Swarbrick, J., & Cammarata, A. (1993). Physical Pharmacy (pp. 423–452). Philadelphia: Lea & Febiger.

31. Hendriksen, B. A., Felix, M. V. S., & Bolger, M. B. (2003). The composite solubility versus pH profile and its role in intestinal absorption prediction. *AAPS Pharm. Sci.*, 5(1), 35–49.

32. Goodwin, J. J. (2006). Rationale and benefit of using high throughput solubility screens in drug discovery. *Drug Discov. Today*, 3(1), 67–71.

33. Di, L., Fish, P. V., & Mano, T. (2012). Bridging solubility between drug discovery and development. *Drug Discov. Today*, 17(9), 486–495.

34. Grant, D. J. W., & Higuchi, T. (1990). *Solubility Behavior of Organic Compounds* (p. liii). Techniques of Chemistry 51, Wiley InterScience, New York, NY.

35. Avdeef, A. (1998). pH-metric Solubility. 1. Solubility-pH profiles from Bjerrum plots. Gibbs buffer and pK_a in the solid state. *Pharm. Pharmacol. Commun.*, 4(3), 165–178.

36. Avdeef, A., Berger, C. M., & Brownell, C. (2000). pH-metric solubility. 2: correlation between the acid-base titration and the saturation shake-flask solubility-pH methods. *Pharm. Res.*, 17(1), 85–89.

37. Huang, L. F., & Tong, W. Q. T. (2004). Impact of solid state properties on developability assessment of drug candidates. *Adv. Drug Deliv. Rev.*, 56(3), 321–334.

38. Hancock, B. C., & Parks, M. (2000). What is the true solubility advantage for amorphous pharmaceuticals? *Pharm. Res.*, 17(4), 397–404.

39. Lipinski, C. A., Lombardo, F., Dominy, B. W., & Feeney, P. J. (2012). Experimental and computational approaches to estimate solubility and permeability in drug discovery and development settings. *Adv. Drug Deliv. Rev.*, 64, 4–17.

40. Šupuk, E., Ghori, M. U., Asare-Addo, K., Laity, P. R., Panchmatia, P. M., & Conway, B. R. (2013). The influence of salt formation on electrostatic and compression properties of flurbiprofen salts. *Int. J. Pharm.*, 458(1), 118–127.

41. Tietgen, H., & Walden, M. (2013). Physicochemical properties. In *Drug Discovery and Evaluation: Safety and Pharmacokinetic Assays* (pp. 1125–1138). Springer Berlin Heidelberg.

42. Clarysse, S., Brouwers, J., Tack, J., Annaert, P., & Augustijns, P. (2011). Intestinal drug solubility estimation based on simulated intestinal fluids: Comparison with solubility in human intestinal fluids. *Eur. J. Pharm. Sci.*, 43(4), 260–269.

43. Lipinski, C. A. (2004). Solubility in water and DMSO: issues and potential solutions. In *Pharmaceutical Profiling in Drug Discovery for Lead Selection*, (pp. 93–125) AAPS Press, Virginia.

44. Bergström, C. A., Holm, R., Jørgensen, S. A., Andersson, S. B., Artursson, P., Beato, S., Borde, A., Box, K., Brewster, M., Dressman, J., Fengi, K-I., Halbert, G., Kostewicz, E., McAllister, M., Muenster, U., Thinnes, J., Taylor, R., & Mullertz, A. (2014). Early pharmaceutical profiling to predict oral drug absorption: current status and unmet needs. *Eur. J. Pharm. Sci.*, 57, 173–199.

45. Dressman, J. B., Vertzoni, M., Goumas, K., & Reppas, C. (2007). Estimating drug solubility in the gastrointestinal tract. *Adv. Drug Deliv. Rev.*, 59(7), 591–602.

46. Clarysse, S., Tack, J., Lammert, F., Duchateau, G., Reppas, C., & Augustijns, P. (2009). Postprandial evolution in composition and characteristics of human duodenal fluids in different nutritional states. *J. Pharm. Sci.*, 98(3), 1177–1192.

47. Augustijns, P., Wuyts, B., Hens, B., Annaert, P., Butler, J., & Brouwers, J. (2014). A review of drug solubility in human intestinal fluids: implications for the prediction of oral absorption. *Eur. J. Pharm. Sci.*, 57, 322–332.

48. Sunesen, V. H., Vedelsdal, R., Kristensen, H. G., Christrup, L., & Müllertz, A. (2005). Effect of liquid volume and food intake on the absolute bioavailability of danazol, a poorly soluble drug. *Eur. J. Pharm. Sci.*, 24(4), 297–303.

49. Mithani, S. D., Bakatselou, V., TenHoor, C. N., & Dressman, J. B. (1996). Estimation of the increase in solubility of drugs as a function of bile salt concentration. *Pharm. Res.*, 13(1), 163–167.

50. Bakatselou, V., Oppenheim, R. C., & Dressman, J. B. (1991). Solubilization and wetting effects of bile salts on the dissolution of steroids. *Pharm. Res.*, 8(12), 1461–1469.

51. Wiedmann, T. S., Liang, W., & Kamel, L. (2002). Solubilization of drugs by physiological mixtures of bile salts. *Pharm. Res.*, 19(8), 1203–1208.

52. Glomme, A., März, J., & Dressman, J. B. (2007). Predicting the intestinal solubility of poorly soluble drugs. In *Pharmacokinetic Profiling in Drug Research: Biological, Physicochemical, and Computational Strategies*, 259–280.

53. Dressman, J. B., Berardi, R. R., Dermentzoglou, L. C., Russell, T. L., Schmaltz, S. P., Barnett, J. L., & Jarvenpaa, K. M. (1990). Upper gastrointestinal (GI) pH in young, healthy men and women. *Pharm. Res.*, 7(7), 756–761.

54. Naylor, L. J., Bakatselou, V., & Dressman, J. B. (1993). Comparison of the mechanism of dissolution of hydrocortisone in simple and mixed micelle systems. *Pharm. Res.*, 10(6), 865–870.

55. Pedersen, B. L., Müllertz, A., Brøndsted, H., & Kristensen, H. G. (2000). A comparison of the solubility of danazol in human and simulated gastrointestinal fluids. *Pharm. Res.*, 17(7), 891–894.

56. Galia, E., Nicolaides, E., Hörter, D., Löbenberg, R., Reppas, C., & Dressman, J. B. (1998). Evaluation of various dissolution media for

predicting in vivo performance of class I and II drugs. *Pharm. Res.*, 15(5), 698–705.

57. Jantratid, E., Janssen, N., Reppas, C., & Dressman, J. B. (2008). Dissolution media simulating conditions in the proximal human gastrointestinal tract: an update. *Pharm. Res.*, 25(7), 1663–1676.

58. Takács-Novák, K., Szöke, V., Völgyi, G., Horváth, P., Ambrus, R., & Szabó-Révész, P. (2013). Biorelevant solubility of poorly soluble drugs: rivaroxaban, furosemide, papaverine and niflumic acid. *J. Pharm. Biomed. Anal.*, 83, 279– 285.

59. Ottaviani, G., Gosling, D. J., Patissier, C., Rodde, S., Zhou, L., & Faller, B. (2010). What is modulating solubility in simulated intestinal fluids? *Eur. J. Pharm. Sci.*, 41(3), 452–457.

60. Söderlind, E., Karlsson, E., Carlsson, A., Kong, R., Lenz, A., Lindborg, S., & Sheng, J. J. (2010). Simulating fasted human intestinal fluids: understanding the roles of lecithin and bile acids. *Mol. Pharm.*, 7(5), 1498–1507.

61. Bard, B., Martel, S., & Carrupt, P. A. (2008). High throughput UV method for the estimation of thermodynamic solubility and the determination of the solubility in biorelevant media. *Eur. J. Pharm. Sci.*, 33(3), 230– 240.

62. Pan, L., Ho, Q., Tsutsui, K., & Takahashi, L. (2001). Comparison of chromatographic and spectroscopic methods used to rank compounds for aqueous solubility. *J. Pharm. Sci.*, 90(4), 521–529.

63. Lipinski, C. A., Lombardo, F., Dominy, B. W., & Feeney, P. J. (1997). Experimental and computational approaches to estimate solubility and permeability in drug discovery and development settings. *Adv. Drug Deliv. Rev.*, 23(1), 3–25.

64. Avdeef, A., & Testa, B. (2002). Physicochemical profiling in drug research: a brief survey of the state-of-the-art of experimental techniques. *Cell Mol. Life Sci.*, 59(10), 1681–1689.

65. Di, L., & Kerns, E. H. (2006). Biological assay challenges from compound solubility: strategies for bioassay optimization. *Drug Discov. Today*, 11(9), 446–451.

66. Avdeef, A., & Berger, C. M. (2001). pH-metric solubility.: 3. Dissolution titration template method for solubility determination. *Eur. J. Pharm. Sci.*, (4), 281–291.

67. Avdeef, A. (2003). *Absorption and Drug Development: Solubility, Permeability and Charge State*, Wiley-Interscience, New York, NY.

68. Stuart, M., & Box, K. (2005). Chasing equilibrium: measuring the intrinsic solubility of weak acids and bases. *Anal. Chem.*, 77, 983–990.

69. Faller, B., & Wohnsland, F. (2001). Physicochemical parameters as tools in drug discovery and lead optimization. In *Pharmacokinetic Optimization in Drug Research: Biological, Physicochemical, and Computational Strategies* (eds Testa, B., van de Waterbeemd, H., Folkers, G., & Guy, R.), Verlag Helvetica Chimica Acta, Zürich, 257–274.

70. Fini, A., Fazio, G., Fernandez-Hervas, M. J., Holgado, M. A., & Rabasco, A. M. (1995). Influence of crystallization solvent and dissolution behavior for a diclofenac salt. *Int. J. Pharm.*, 121(1), 19–26.

71. Dokoumetzidis, A., & Macheras, P. (2006). A century of dissolution research: from Noyes and Whitney to the biopharmaceutics classification system. *Int. J. Pharm.*, 321(1–2), pp. 1–11.

72. Semin, D. J., Malone, T. J., Paley, M. T., & Woods, P. W. (2005). A novel approach to determine water content in DMSO for a compound collection repository. *J. Biomol. Screen.*, 10(6), 568–572.

73. Pudipeddi M., & Serajuddin, A. (2005). Trends in solubility of polymorphs. *J. Pharm. Sci.*, 94(5), 929–939.

74. Sugano, K., Kato, T., Suzuki, K., Keiko, K., Sujaku, T., & Mano, T. (2006). High throughput solubility measurement with automated polarized light microscopy analysis. *J. Pharm. Sci.*, 95(10), 2115–2122.

75. Lindenberg, M., Wiegand, C., & Dressman, J. B. (2005). Comparison of the adsorption of several drugs to typical filter materials. *Dissolut. Technol.*, 12(1), 22–25.

76. Thoma, K., & Ziegler, I. (1998). Development of an automated flow-through dissolution system for poorly soluble drugs with poor chemical stability in dissolution media. *Pharmazie*, 53(11), 784–790.

77. Merisko-Liversidge, E., Liversidge, G. G., & Cooper, E. R. (2003). Abstracts of Papers, 226th ACS National Meeting, New York, NY, United States, September 7–11, 2003, p. PMSE–444.

78. Fini, A., Fazio, G., & Feroci, G. (1995). Solubility and solubilization properties of non-steroidal anti-inflammatory drugs. *Int. J. Pharm.*, 126(1), 95–102.

79. McGovern, S. L., Caselli, E., Grigorieff, N., & Shoichet, B. K. (2002). A common mechanism underlying promiscuous inhibitors from virtual and high-throughput screening. *J. Med. Chem.*, 45(8), 1712–1722.

80. Taboada, P., Gutierrez-Pichel, M., & Mosquera, V. (2004). Effects of self-aggregation on the dehydration of an amphiphilic antidepressant drug in different aqueous media. *J. Chem. Phys.*, 298, 65–74.

81. Wang, Z., Burrell, L., & Lambert, W. (2002). Solubility of E2050 at various pH: a case in which apparent solubility is affected by the amount of excess solid. *J. Pharm. Sci.*, 91(6), 1445–1455.

82. Land, L., Li, P., & Bummer, P. (2005). The influence of water content of triglyceride oils on the solubility of steroids. *Pharm. Res.*, 22(5), 784–788.

83. Kramer, C., Heinisch, T., Fligge, T., Beck, B., & Clark, T. (2009). A consistent dataset of kinetic solubilities for early-phase drug discovery. *ChemMedChem*, 4(9), 1529–1536.

84. Wang, J., & Bell, L. (2012). Technical challenges and recent advances of implementing comprehensive ADMET tools in drug discovery. In *ADME-Enabling Technologies in Drug Design and Development* 131.

85. Glomme, A., März, J., & Dressman, J. B. (2005). Comparison of a miniaturized shake-flask solubility method with automated potentiometric acid/base titrations and calculated solubilities. *J. Pharm. Sci.*, 94(1), 1–16.

86. Yalkowsky, S. H., & He, Y. (2003). *Handbook of Aqueous Solubility Results*, CRC Press, Boca Raton.

87. Avdeef, A. (2001). High-throughput measurements of solubility profiles. In *Pharmacokinetic Optimization in Drug Research: Biological, Physicochemical, and Computational Strategies* (eds Testa, B., van de Waterbeemd, H., Guy, F. R.), Verlag Helvetica Chimica Acta, Zürich, pp. 304–325.

88. Pinsuwan, S., Myrdal, P. B., Lee, Y. C., & Yalkowsky, S. H. (1997). AQUAFAC 5: aqueous functional group activity coefficients; application to alcohols and acids. *Chemosphere*, 35(11), 2503–2513.

89. Pinsuwan, S., Li, A., & Yalkowsky, S. H. (1995). Correlation of octanol/water solubility ratios and partition coefficients. *J. Chem. Eng. Data*, 40(3), 623–626.

90. Bergström, C. A., Luthman, K., & Artursson, P. (2004). Accuracy of calculated pH-dependent aqueous drug solubility. *Eur. J. Pharm. Sci.*, 22(5), 387–398.

91. Powell, M. F. (1986). *Analytical Profiles of Drug Substances* (ed Florey, K.), 15, Academic Press, San Diego, 1986, pp. 761–779.

92. Kawakami, K., Miyoshi, K., & Ida, Y. (2005). Impact of the amount of excess solids on apparent solubility. *Pharm. Res.*, 22(9), 1537–1543.

93. Madan, D. K., & Cadwallader, D. E. (1973). Solubility of cholesterol and hormone drugs in water. *J. Pharm. Sci.*, 62(9), 1567–1569.

94. Higuchi, T., & Connors, K. A. (1965). Phase-solubility techniques. *Adv. Anal. Chem. Instrum.*, 4(2), 117–212.

95. Hancock, B. C., & Zografi, G. (1997). Characteristics and significance of the amorphous state in pharmaceutical systems. *J. Pharm. Sci.*, 86(1), 1–12.

96. Brittain, H. G. (1997). Spectral methods for the characterization of polymorphs and solvates. *J. Pharm. Sci.*, 86(4), 405–412.

97. Avdeef, A. (2007). Solubility of sparingly-soluble ionizable drugs. *Adv. Drug Deliv. Rev.*, 59(7), 568–590.

98. Avdeef, A., Voloboy, D., & Foreman, A. (2007). Dissolution and solubility, in *Comprehensive Medicinal Chemistry II*, 5, 399–423.

99. Loftsson, T., & Hreinsdóttir, D. (2006). Determination of aqueous solubility by heating and equilibration: a technical note. *AAPS PharmSciTech.*, 7(1), E29–E32.

100. Hayward, M. J., Hargiss, L. O., Munson, J. L., Mandiyan, S. P., & Wennogle, L. P. (2000). Validation of solubility measurements using ultra-filtration liquid chromatography mass spectrometry (UF-LC/MS). In American Society for Mass Spectrometry: 48th Annual Conference on Mass Spectrometry and Allied Topics, Long Beach, CA, TOB AM10 (Vol. 55).

101. Alsenz, J. (2012). The impact of solubility and dissolution assessment on formulation strategy and implications for oral drug disposition. In *Encyclopedia of Drug Metabolism and Interactions*. John Wiley and Sons, New York, 493–562.

102. Bevan, C. D., & Lloyd, R. S. (2000). A high-throughput screening method for the determination of aqueous drug solubility using laser nephelometry in microtiter plates. *Anal. Chem.*, 72(8), 1781–1787.

103. Bjerrum, J. (1941). *Metal-Ammine Formation in Aqueous Solution*, Haase, Copenhagen.

104. Avdeef, A., Kearney, D. L., Brown, J. A., & Chemotti, A. R. Jr. (1982). Bjerrum plots for the determination of systematic concentration errors in titration data. *Anal. Chem.* 54, 2322–2326.

105. Avdeef, A., Comer, J. E., & Thomson, S. J. (1993). pH-Metric log P. 3. Glass electrode calibration in methanol-water, applied to pK_a determination of water-insoluble substances. *Anal. Chem.*, 65(1), 42–49.

106. Avdeef, A. (1993). *Applications and Theory Guide to pH-Metric pKa and log P Measurement*, Sirius Analytical Instruments Ltd. Forest Row, UK.

107. Avdeef, A. (1993). pH-Metric logP. 2. Refinement of partition coefficients and ionization constants of multiprotic substances. *J. Pharm. Sci.*, 82, 183–190.

108. Avdeef, A., Bucher, J. J. (1978). Accurate measurements of the concentration of hydrogen ions with a glass electrode: calibrations using the Prideaux and other universal buffer solutions and a computer-controlled automatic titrator, *Anal. Chem.*, 50, 2137–2142.

109. Sasaki K., Suzuki H., & Nakagawa H. (1993). Physicochemical characterization of trazodone hydrochloride tetrahydrate. *Chem. Pharm. Bull.*, 41, 325–328.

110. Fligge, T. A., & Schuler, A. (2006). Integration of a rapid automated solubility classification into early validation of hits obtained by high throughput screening. *J. Pharm. Biomed.*, 42(4), 449–454.

111. Kerns, E. H., Di, L., Petusky, S., Kleintop, T., Huryn, D., McConnell, O., & Carter, G. (2003). Pharmaceutical profiling method for lipophilicity and integrity using liquid chromatography–mass spectrometry. *J. Chromatogr. B*, 791(1), 381–388.

112. Kerns, E. H., & Di, L. (2002). Multivariate pharmaceutical profiling for drug discovery. *Curr. Top. Med. Chem.*, 2(1), 87–98.

113. Santos, N. C., Figueira-Coelho, J., Martins-Silva, J., & Saldanha, C. (2003). Multidisciplinary utilization of dimethyl sulfoxide: pharmacological, cellular, and molecular aspects. *Biochem. Pharmacol.*, 65(7), 1035–1041.

114. Dehring, K. A., Workman, H. L., Miller, K. D., Mandagere, A., & Poole, S. K. (2004). Automated robotic liquid handling/laser-based nephelometry system for high throughput measurement of kinetic aqueous solubility. *J. Pharm. Biomed.*, 36(3), 447–456.

115. Shen, H., & Zhu, C. (2002). Evaluation of a method for high throughput solubility determination using a multi-wavelength UV plate reader. *Comb. Chem. High Throughput Screen*, 5(7), 575–581.

116. Martel, S., Castella, M. E., Bajot, F., Ottaviani, G., Bard, B., Henchoz, Y., Valloton, B. G., Reist, M. & Carrupt, P. A. (2005). Experimental and virtual physicochemical and pharmacokinetic profiling of new chemical entities. *CHIMIA*, 59(6), 308–314.

117. Taub, M. E., Kristensen, L., & Frokjaer, S. (2002). Optimized conditions for MDCK permeability and turbidimetric solubility studies using compounds representative of BCS classes I–IV. *Eur. J. Pharm. Sci.*, 15(4), 331–340.

118. Kibbey, C. E., Poole, S. K., Robinson, B., Jackson, J. D., & Durham, D. (2001). An integrated process for measuring the physicochemical properties of drug candidates in a preclinical discovery environment. *J. Pharm. Sci.*, 90(8), 1164–1175.

119. Yamashita, T., Dohta, Y., Nakamura, T., & Fukami, T. (2008). High-speed solubility screening assay using ultra-performance liquid chromatography/mass spectrometry in drug discovery. *J. Chromatogr. A*, 1182(1), 72–76.

120. Millipore Corp. (2003). MultiScreen Solubility Filter Plate. Millipore Corporation, Billerica.

121. Alelyunas, Y. W., Liu, R., Pelosi-Kilby, L., & Shen, C. (2009). Application of a dried-DMSO rapid throughput 24-h equilibrium solubility in advancing discovery candidates. *Eur. J. Pharm. Sci.,* 37(2), 172–182.

122. Lipinski, C., 2008. Drug solubility in water and dimethylsulfoxide. In *Methods and Principles in Medicinal Chemistry,* vol. 37, pp. 257–282, Wiley-VCH Weinheim, Germany.

123. Wan, H., & Holmen, A. G. (2009). High throughput screening of physicochemical properties and *in vitro* ADME profiling in drug discovery. *Comb. Chem. High Throughput Screen,* 12(3), 315–329.

124. Seadeek, C., Ando, H., Bhattachar, S. N., Heimbach, T., Sonnenberg, J. L., & Blackburn, A. C. (2007). Automated approach to couple solubility with final pH and crystallinity for pharmaceutical discovery compounds. *J. Pharm. Biomed.,* 43(5), 1660–1666.

125. Hewitt, M., Madden, J. C., Rowe, P. H., & Cronin, M. T. D. (2007). Structure-based modelling in reproductive toxicology:(Q) SARs for the placental barrier. *SAR QSAR Environ. Res.,* 18(1-2), 57–76.

126. Roy, D., Ducher, F., Laumain, A., & Legendre, J. Y. (2001). Determination of the aqueous solubility of drugs using a convenient 96-well plate-based assay. *Drug Dev. Ind. Pharm.,* 27(1), 107–109.

127. Tan, H., Semin, D., Wacker, M., & Cheetham, J. (2005). An automated screening assay for determination of aqueous equilibrium solubility enabling SPR study during drug lead optimization. *J. Lab. Autom.,* 10(6), 364–373.

128. Dai, W. G., Pollock-Dove, C., Dong, L. C., & Li, S. (2008). Advanced screening assays to rapidly identify solubility-enhancing formulations: high-throughput, miniaturization and automation. *Adv. Drug Deliv. Rev.,* 60(6), 657–672.

129. Chen, X. Q., & Venkatesh, S. (2004). Miniature device for aqueous and non-aqueous solubility measurements during drug discovery. *Pharm. Res.,* 21(10), 1758–1761.

130. Cherng, J. P. J., Gonzalez-Zugasti, J., Kane, N. R., Cima, M. J., & Lemmo, A. V. (2004). Integration of an opto-mechanical mass sensor with a powder-dispensing device for microgram sensitivity. *J. Lab. Autom.,* 9(4), 228–237.

131. Alsenz, J., Meister, E., & Haenel, E. (2007). Development of a partially automated solubility screening (PASS) assay for early drug development. *J. Pharm. Sci.,* 96(7), 1748–1762.

132. Di, L., Kerns, E. H., & Carter, G. T. (2009). Drug-like property concepts in pharmaceutical design. *Curr. Pharm. Design,* 15(19), 2184–2194.

133. Venkatesh, S., & Lipper, R. A. (2000). Role of the development scientist in compound lead selection and optimization. *J. Pharm. Sci.,* 89(2), 145–154.

134. Ertl, P., Rohde, B., & Selzer, P. (2000). Fast calculation of molecular polar surface area as a sum of fragment-based contributions and its application to the prediction of drug transport properties. *J. Med. Chem.,* 43(20), 3714–3717.

135. Bergström, C. A. (2005). Computational models to predict aqueous drug solubility, permeability and intestinal absorption. *Expert Opin. Drug Metab. Toxicol.,* 1(4), 613–627.

136. Delaney, J. S. (2005). Predicting aqueous solubility from structure. *Drug Discov. Today,* 10(4), 289–295.

137. Taskinen, J., & Norinder, U. (2007). *In silico* prediction of solubility. ADME/Tox Approaches, 5, 627–648.

138. Balakin, K. V., Savchuk, N. P., & Tetko, I. V. (2006). *In silico* approaches to prediction of aqueous and DMSO solubility of drug-like compounds: trends, problems and solutions. *Curr. Med. Chem.,* 13(2), 223–241.

139. Johnson, S. R., Chen, X. Q., Murphy, D., & Gudmundsson, O. (2007). A computational model for the prediction of aqueous solubility that includes crystal packing, intrinsic solubility, and ionization effects. *Mol. Pharm.,* 4(4), 513–523.

140. Jorgensen, W. L., & Duffy, E. M. (2002). Prediction of drug solubility from structure. *Adv. Drug Deliv. Rev.,* 54(3), 355–366.

141. Kerns, E., & Di, L. (2008). Drug-like properties: concepts, structure design and methods: from ADME to toxicity optimization. Academic Press.

142. Huuskonen, J. (2000). Estimation of aqueous solubility for a diverse set of organic compounds based on molecular topology. *J. Chem. Inf. Comp. Sci.,* 40(3), 773–777.

143. Tetko, I. V., Tanchuk, V. Y., Kasheva, T. N., & Villa, A. E. (2001). Estimation of aqueous solubility of chemical compounds using E-state indices. *J. Chem. Inf. Comp. Sci.,* 41(6), 1488–1493.

144. Klopman, G., Wang, S., & Balthasar, D. M. (1992). Estimation of aqueous solubility of organic molecules by the group contribution approach. Application to the study of biodegradation. *J. Chem. Inf. Comp. Sci.,* 32(5), 474–482.

145. Hansch, C., Quinlan J. E., & Lawrence, G. L. (1968). Linear free energy relationships between partition coefficients and the aqueous solubility of organic liquids. *J. Org. Chem.,* 33, 347–350.

146. Yalkowsky, S. H. (1999). Solubility and partial miscibility. In *Solubility and Solubilization in Aqueous Media* (ed Yalkowsky, S. H.). Washington, DC: American Chemical Society, pp. 49–80.

147. Ran, Y., Jain, N., & Yalkowsky, S. H. (2001). Prediction of aqueous solubility of organic compounds by the general solubility equation (GSE). *J. Chem. Inf. Comp. Sci.,* 41(5), 1208–1217.

148. Abraham, M. H., & Le, J. (1999). The correlation and prediction of the solubility of compounds in water using an amended solvation energy relationship. *J. Pharm. Sci.,* 88(9), 868–880.

149. Klamt, A., Eckert, F., Hornig, M., Beck, M. E., & Bürger, T. (2002). Prediction of aqueous solubility of drugs and pesticides with COSMO-RS. *J. Comp. Chem.,* 23(2), 275–281.

150. Hornig, M., & Klamt, A. (2005). COSMOfrag: a novel tool for high-throughput ADME property prediction and similarity screening based on quantum chemistry. *J. Chem. Inf. Model.,* 45(5), 1169–1177.

151. Clark, T. (2000). Quantum cheminformatics: an oxymoron. In Proceedings of the Beilstein Institute Workshop: Chemical Data Analysis in the Large, Bozen, Italy, May 22r26 (pp. 88r-99).

152. Kier, L. B., & Cheng, C. K. (1994). A cellular automata model of an aqueous solution. *J. Chem. Inf. Comp. Sci.,* 34(6), 1334–1337.

153. Wegner, J. K., & Zell, A. (2003). Prediction of aqueous solubility and partition coefficient optimized by a genetic algorithm based descriptor selection method. *J. Chem. Inf. Comp. Sci.,* 43(3), 1077–1084.

154. Jorgensen, W. L., & Duffy, E. M. (2000). Prediction of drug solubility from Monte Carlo simulations. *Bioorg. Med. Chem. Lett.,* 10(11), 1155–1158.

155. Lind, P., & Maltseva, T. (2003). Support vector machines for the estimation of aqueous solubility. *J. Chem. Inf. Comp. Sci.*, 43(6), 1855–1859.

156. Hill, A. P., & Young, R. J. (2010). Getting physical in drug discovery: a contemporary perspective on solubility and hydrophobicity. *Drug Discov. Today*, 15(15), 648–655.

157. Salahinejad, M., Le, T., & Winkler, D. A. (2013). Aqueous solubility prediction: do crystal lattice interactions help? *Mol. Pharm.*, 10 (7) 2757–2766.

158. Lewis, D. F. (2000). On the recognition of mammalian microsomal cytochrome P450 substrates and their characteristics: towards the prediction of human p450 substrate specificity and metabolism. *Biochem. Pharmacol.*, 60(3), 293–306.

159. Ishihama, Y., Nakamura, M., Miwa, T., Kajima, T., & Asakawa, N. (2002). A rapid method for pKa determination of drugs using pressure-assisted capillary electrophoresis with photodiode array detection in drug discovery. *J. Pharm. Sci.*, 91(4), 933–942.

160. Box, K., Bevan, C., Comer, J., Hill, A., Allen, R., & Reynolds, D. (2003). High-throughput measurement of pKa values in a mixed-buffer linear pH gradient system. *Anal. Chem.*, 75(4), 883–892.

161. Jouravleva D: ACD/PhysChem: product overview. AAPS - ACD Users' Meeting 2002, Nov. Toronto http://www.acdlabs.com/publish/.

162. Avdeef, A., Box, K. J., Comer, J. E. A., Gilges, M., Hadley, M., Hibbert, C., Patterson, W., & Tam, K. Y. (1999). PH-metric logP11. pKa determination of water-insoluble drugs in organic solvent–water mixtures. *J. Pharm. Biomed.*, 20(4), 631–641.

163. Takács-Novák, K., Box, K. J., & Avdeef, A. (1997). Potentiometric pKa determination of water-insoluble compounds: validation study in methanol/water mixtures. *Int. J. Pharm.*, 151(2), 235–248.

164. Kristl, A. (2009). Acido-basic properties of proton pump inhibitors in aqueous solutions. *Drug Dev. Ind. Pharm.*, 35(1), 114–117.

165. Tosco, P., Rolando, B., Fruttero, R., Henchoz, Y., Martel, S., Carrupt, P. A., & Gasco, A. (2008). Physicochemical profiling of sartans: a detailed study of ionization constants and distribution coefficients. *Helv. Chim. Acta*, 91(3), 468–482.

166. Allen, R. I., Box, K. J., Comer, J. E. A., Peake, C., & Tam, K. Y. (1998). Multiwavelength spectrophotometric determination of acid dissociation constants of ionizable drugs. *J. Pharm. Biomed.*, 17(4), 699–712.

167. Mitchell, R. C., Salter, C. J., & Tam, K. Y. (1999). Multiwavelength spectrophotometric determination of acid dissociation constants: Part III. Resolution of multi-protic ionization systems. *J. Pharm. Biomed.,* 20(1), 289–295.

168. Takács-Novák, K., & Tam, K. Y. (2000). Multiwavelength spectrophotometric determination of acid dissociation constants: Part V: microconstants and tautomeric ratios of diprotic amphoteric drugs. *J. Pharm. Biomed.,* 21(6), 1171–1182.

169. Tarn, K. Y., & Takács-Novák, K. (1999). Multiwavelength spectrophotometric determination of acid dissociation constants: Part II. First derivative vs. target factor analysis. *Pharm. Res.,* 16(3), 374–381.

170. Morgan, M. E., Liu, K., & Anderson, B. D. (1998). Microscale titrimetric and spectrophotometric methods for determination of ionization constants and partition coefficients of new drug candidates. *J. Pharm. Sci.,* 87(2), 238–245.

171. Mannhold, R. (2008). *Molecular Drug Properties: Measurement and Prediction.* Wiley-VCH, Weinheim.

172. Pang, H. M., Kenseth, J., & Coldiron, S. (2004). High-throughput multiplexed capillary electrophoresis in drug discovery. *Drug Discov. Today,* 9(24), 1072–1080.

173. Poole, S. K., Patel, S., Dehring, K., Workman, H., & Poole, C. F. (2004). Determination of acid dissociation constants by capillary electrophoresis. *J. Chromatogr. A,* 1037(1), 445–454.

174. Jia, Z. (2005). Physicochemical profiling by capillary electrophoresis. *Curr. Pharm. Anal.,* 1(1), 41–56.

175. Wan, H., Thompson, R. A. (2005). Drug Discov Today TechnolJia, Z. (2005). Physicochemical profiling by capillary electrophoresis. *Curr. Pharm. Anal.,* 1(1), 41–56.

176. Wan, H., & Ulander, J. (2006). High-throughput pKa screening and prediction amenable for ADME profiling. *Expert Opin. Drug Metab. Toxicol.,* 2, 139–155.

177. Babić, S., Horvat, A. J., Mutavdžić Pavlović, D., & Kaštelan-Macan, M. (2007). Determination of pKa values of active pharmaceutical ingredients. *Trac-Trend Anal. Chem.,* 26(11), 1043–1061.

178. Wan, H., Holmén, A. G., Wang, Y., Lindberg, W., Englund, M., Någård, M. B., & Thompson, R. A. (2003). High-throughput screening of pKa values of pharmaceuticals by pressure-assisted capillary electrophoresis and mass spectrometry. *Rapid Commun. Mass Sp.,* 17(23), 2639–2648.

179. Krishna, M. V., Srinath, M., & Sankar, D. G. (2008). Principles and applications of capillary electrophoresis in new drug discovery. *Curr. Trends Biotechnol. Pharm.*, 2(1), 142–155.

180. Lalwani, S., Tutu, E., & Vigh, G. (2005). Isoelectric buffers, part 3: Determination of pKa and pI values of diamino sulfate carrier ampholytes by indirect UV-detection capillary electrophoresis. *Electrophoresis*, 26(13), 2503–2510.

181. Mercier, J. P., Morin, P., Dreux, M., & Tambute, A. (1998). Determination of weak (2.0–2.5) dissociation constants of non-UV absorbing solutes by capillary electrophoresis. *Chromatographia*, 48(7-8), 529–534.

182. Hu, Q., Hu, G., Zhou, T., & Fang, Y. (2003). Determination of dissociation constants of anthrocycline by capillary zone electrophoresis with amperometric detection. *J. Pharm. Biomed.*, 31(4), 679–684.

183. Zusková, I., Novotná, A., Včeláková, K., & Gaš, B. (2006). Determination of limiting mobilities and dissociation constants of 21 amino acids by capillary zone electrophoresis at very low pH. *J. Chromatogr. B*, 841(1), 129–134.

184. Včeláková, K., Zuskova, I., Kenndler, E., & Gaš, B. (2004). Determination of cationic mobilities and pKa values of 22 amino acids by capillary zone electrophoresis. *Electrophoresis*, 25(2), 309–317.

185. Henchoz, Y., Schappler, J., Geiser, L., Prat, J., Carrupt, P. A., & Veuthey, J. L. (2007). Rapid determination of pKa values of 20 amino acids by CZE with UV and capacitively coupled contactless conductivity detections. *Anal. Bioanal. Chem.*, 389(6), 1869–1878.

186. Cleveland Jr, J. A., Martin, C. L., & Gluck, S. J. (1994). Spectrophotometric determination of ionization constants by capillary zone electrophoresis. *J. Chromatogr. A*, 679(1), 167–171.

187. Wang, D., Yang, G., & Song, X. (2001). Determination of pKa values of anthraquinone compounds by capillary electrophoresis. *Electrophoresis*, 22(3), 464–469.

188. Wu, X., Gong, S., Bo, T., Liao, Y., & Liu, H. (2004). Determination of dissociation constants of pharmacologically active xanthones by capillary zone electrophoresis with diode array detection. *J. Chromatogr. A*, 1061(2), 217–223.

189. Perez-Urquiza, M., & Beltrán, J. L. (2001). Determination of the dissociation constants of sulfonated azo dyes by capillary zone electrophoresis and spectrophotometry methods. *J. Chromatogr. A*, 917(1), 331–336.

190. Jankowsky, R., Friebe, M., Noll, B., & Johannsen, B. (1999). Determination of dissociation constants of 99mTechnetium radiopharmaceuticals by capillary electrophoresis. *J. Chromatogr. A*, 833(1), 83–96.

191. Szakács, Z., & Noszál, B. (2006). Determination of dissociation constants of folic acid, methotrexate, and other photolabile pteridines by pressure-assisted capillary electrophoresis. *Electrophoresis*, 27(17), 3399–3409.

192. Marsh, A., & Altria, K. (2006). Use of multiplexed CE for pharmaceutical analysis. *Chromatographia*, 64(5-6), 327–333.

193. Gong, X., Figus, M., Plewa, J., Levorse, D. A., Zhou, L., & Welch, C. J. (2008). Evaluation of multiplexed CE with UV detection for rapid pKa estimation of active pharmaceutical ingredients. *Chromatographia*, 68(3-4), 219–225.

194. Miller, J. M., Blackburn, A. C., Shi, Y., Melzak, A. J., & Ando, H. Y. (2002). Semi-empirical relationships between effective mobility, charge, and molecular weight of pharmaceuticals by pressure-assisted capillary electrophoresis: applications in drug discovery. *Electrophoresis*, 23(17), 2833–2841.

195. Jia, Z., Ramstad, T., & Zhong, M. (2001). Medium-throughput pKa screening of pharmaceuticals by pressure-assisted capillary electrophoresis. *Electrophoresis*, 22(6), 1112–1118.

196. Zhou, C., Jin, Y., Kenseth, J. R., Stella, M., Wehmeyer, K. R., & Heineman, W. R. (2005). Rapid pKa estimation using vacuum-assisted multiplexed capillary electrophoresis (VAMCE) with ultraviolet detection. *J. Pharm. Sci.*, 94(3), 576–589.

197. Wan, H., Holmén, A., Någård, M., & Lindberg, W. (2002). Rapid screening of pKa values of pharmaceuticals by pressure-assisted capillary electrophoresis combined with short-end injection. *J. Chromatogr. A*, 979(1), 369–377.

198. van de Waterbeemd, H., Smith, D. A., & Jones, B. C. (2001). Lipophilicity in PK design: methyl, ethyl, futile. *J. Comput. Aided Mol. Des.*, 15(3), 273–286.

199. Caron, G., Reymond, F., Carrupt, P. A., Girault, H. H., & Testa, B. (1999). Combined molecular lipophilicity descriptors and their role in understanding intramolecular effects. *Pharm. Sci. Technol. Today*, 2(8), 327–335.

200. Comer, J. E. (2003). Comer, J. E. A. (2003). High-throughput measurement of log D and pKa. In *Drug Bioavailability: Estimation of Solubility, Permeability, Absorption and Bioavailability* (eds van de Waterbeemd,

H., Lennernäs, H., & Artursson, P.), Wiley-VCH Verlag GmbH & Co. KGaA, Weinheim, 21–45.

201. Gocan, S., Cimpan, G., & Comer, J. (2006). Lipophilicity measurements by liquid chromatography. In *Advances in Chromatography*, Taylor & Francis, Boca Raton, NY, 44, 79–176.

202. Hitzel, L., Watt, A. P., & Locker, K. L. (2000). An increased throughput method for the determination of partition coefficients. *Pharm. Res.,* 17(11), 1389–1395.

203. Gulyaeva, N., Zaslavsky, A., Lechner, P., Chait, A., & Zaslavsky, B. (2003). pH dependence of the relative hydrophobicity and lipophilicity of amino acids and peptides measured by aqueous two-phase and octanol–buffer partitioning. *J. Pept. Res.,* 61(2), 71–79.

204. Gulyaeva, N., Zaslavsky, A., Lechner, P., Chlenov, M., McConnell, O., Chait, A., Kipnis, V., & Zaslavsky, B. (2003). Relative hydrophobicity and lipophilicity of drugs measured by aqueous two-phase partitioning, octanol-buffer partitioning and HPLC. A simple model for predicting blood–brain distribution. *Eur. J. Med. Chem.,* 38(4), 391–396.

205. Gulyaeva, N., Zaslavsky, A., Lechner, P., Chlenov, M., Chait, A., & Zaslavsky, B. (2002). Relative hydrophobicity and lipophilicity of β-blockers and related compounds as measured by aqueous two-phase partitioning, octanol–buffer partitioning, and HPLC. *Eur. J. Pharm. Sci.,* 17(1), 81–93.

206. Valko, K., Bevan, C., Reynolds, D. P., & Abraham, M. H. (2001). Rapid method for the estimation of octanol/water partition coefficient (log P(oct)) from gradient RP-HPLC retention and a hydrogen bond acidity term (zetaalpha(2)(H)). *Curr. Med. Chem.,* 8(9), 1137–1146.

207. Lombardo, F., Shalaeva, M. Y., Tupper, K. A., & Gao, F. (2001). ElogDoct: a tool for lipophilicity determination in drug discovery. 2. Basic and neutral compounds. *J. Med. Chem.,* 44(15), 2490–2497.

208. Poole, S. K., Durham, D., & Kibbey, C. (2000). Rapid method for estimating the octanol–water partition coefficient (logPow) by microemulsion electrokinetic chromatography. *J. Chromatogr. B,* 745(1), 117–126.

209. Wilson, D. M., Wang, X., Walsh, E., & Rourick, R. A. (2001). High throughput log D determination using liquid chromatography-mass spectrometry. *Comb. Chem. High Throughput Screen,* 4(6), 511–519.

210. Martel, S., Guillarme, D., Henchoz, Y., Galland, A., Veuthey, J. L., Rudaz, S., Carrupt, P. A. (2008). Drug properties: measurement and computation. Wiley-VCH, Weinheim.

211. Poole, S. K., & Poole, C. F. (2003). Separation methods for estimating octanol–water partition coefficients. *J. Chromatogr. B*, 797(1), 3–19.

212. Huie, C. W. (2006). Recent applications of microemulsion electrokinetic chromatography. *Electrophoresis*, 27(1), 60–75.

213. Kaliszan, R. (2007). QSRR: quantitative structure-(chromatographic) retention relationships. *Chem. Rev.*, 107(7), 3212–3246.

214. Nasal, A., & Kaliszan, R. (2006). Progress in the use of HPLC for evaluation of lipophilicity. *Curr. Comput. Aided Drug Des.*, 2(4), 327–340.

215. Kaliszan, R., Nasal, A., & Markuszewski, M. J. (2003). New approaches to chromatographic determination of lipophilicity of xenobiotics. *Anal. Bioanal. Chem.*, 377(5), 803–811.

216. Nasal, A., Siluk, D., & Kaliszan, R. (2003). Chromatographic retention parameters in medicinal chemistry and molecular pharmacology. *Curr. Med. Chem.*, 10(5), 381–426.

217. Takeda, S., Wakida, S., Yamane, M., Kawahara, A., & Higashi, K. (1993). Migration behavior of phthalate esters in micellar electrokinetic chromatography with or without added methanol. *Anal. Chem.*, 65(18), 2489–2492.

218. Rosés, M., Ràfols, C., Bosch, E., Martínez, A. M., & Abraham, M. H. (1999). Solute–solvent interactions in micellar electrokinetic chromatography: Characterization of sodium dodecyl sulfate–Brij 35 micellar systems for quantitative structure–activity relationship modelling. *J. Chromatogr. A*, 845(1), 217–226.

219. Hanna, M., de Biasi, V., Bond, B., Salter, C., Hutt, A. J., & Camilleri, P. (1998). Estimation of the partitioning characteristics of drugs: a comparison of a large and diverse drug series utilizing chromatographic and electrophoretic methodology. *Anal. Chem.*, 70(10), 2092–2099.

220. Ferguson, P. D., Goodall, D. M., & Loran, J. S. (1998). Systematic approach to links between separations in MEKC and reversed-phase HPLC. *Anal. Chem.*, 70(19), 4054–4062.

221. Morin, P., Archambault, J. C., Andre, P., Dreux, M., & Gaydou, E. (1997). Separation of hydroxylated and methoxylated flavonoids by micellar electrokinetic capillary chromatography: determination of analyte partition coefficients between aqueous and sodium dodecyl sulfate micellar phases. *J. Chromatogr. A*, 791(1), 289–297.

222. Muijselaar, P. G., Claessens, H. A., & Cramers, C. A. (1994). Application of the retention index concept in micellar electrokinetic capillary chromatography. *Anal. Chem.*, 66(5), 635–644.

223. Müller, L., Bednář, P., Barták, P., Lemr, K., & Ševčík, J. (2005). Estimation of partition coefficients by MEKC Part 2: Anthocyanins. *J. Sep. Sci.,* 28(12), 1285–1290.

224. Annoura, H., Nakanishi, K., Uesugi, M., Fukunaga, A., Miyajima, A., Tamura-Horikawa, Y., & Tamura, S. (1999). A novel class of Na+ and Ca2+ channel dual blockers with highly potent anti-ischemic effects. *Bioorg. Med. Chem. Lett.,* 9(20), 2999–3002.

225. Detroyer, A., Vander Heyden, Y., Cambre, I., & Massart, D. L. (2003). Chemometric comparison of recent chromatographic and electrophoretic methods in a quantitative structure–retention and retention–activity relationship context. *J. Chromatogr. A,* 986(2), 227–238.

226. Taillardat-Bertschinger, A., Carrupt, P. A., & Testa, B. (2002). The relative partitioning of neutral and ionised compounds in sodium dodecyl sulfate micelles measured by micellar electrokinetic capillary chromatography. *Eur. J. Pharm. Sci.,* 15(2), 225–234.

227. Mrestani, Y., Marestani, Z., & Neubert, R. H. (2001b). Characterization of micellar solubilization of antibiotics using micellar electrokinetic chromatography. *J. Pharm. Biomed.,* 26(5), 883–889.

228. Mrestani, Y., Marestani, Z., & Neubert, R. H. (2001a). The effect of a functional group in penicillin derivatives on the interaction with bile salt micelles studied by micellar electrokinetic chromatography. *Electrophoresis,* 22(16), 3573–3577.

229. Trone, M. D., Leonard, M. S., & Khaledi, M. G. (2000). Congeneric behavior in estimations of octanol-water partition coefficients by micellar electrokinetic chromatography. *Anal. Chem.,* 72(6), 1228–1235.

230. Mrestani, Y., Janich, M., Rüttinger, H. H., & Neubert, R. H. (2000). Characterization of partition and thermodynamic properties of cephalosporins using micellar electrokinetic chromatography in glycodeoxycholic acid solution. *J. Chromatogr. A,* 873(2), 237–246.

231. García-Ruiz, C., García, M., & Marina, M. L. (2000). Separation of a group of N-phenylpyrazole derivatives by micellar electrokinetic chromatography: application to the determination of solute-micelle association constants and estimation of the hydrophobicity. *Electrophoresis,* 21(12), 2424–2431.

232. Maeder, C., Beaudoin, G. M., Hsu, E. K., Escobar, V. A., Chambers, S. M., Kurtin, W. E., & Bushey, M. M. (2000). Measurement of bilirubin partition coefficients in bile salt micelle/aqueous buffer solutions by

micellar electrokinetic chromatography. *Electrophoresis*, 21(4), 706–714.

233. Yang, S., Bumgarner, J. G., Kruk, L. F., & Khaledi, M. G. (1996). Quantitative structure-activity relationships studies with micellar electrokinetic chromatography influence of surfactant type and mixed micelles on estimation of hydrophobicity and bioavailability. *J. Chromatogr. A*, 721(2), 323–335.

234. Garcia, M. A., Diez-Masa, J. C., & Marina, M. L. (1996). Correlation between the logarithm of capacity factors for aromatic compounds in micellar electrokinetic chromatography and their octanol-water partition coefficients. *J. Chromatogr. A*, 742(1), 251–256.

235. Herbert, B. J., & Dorsey, J. G. (1995). n-Octanol-water partition coefficient estimation by micellar electrokinetic capillary chromatography. *Anal. Chem.*, 67(4), 744–749.

236. Wehmeyer, K. R., Jian, T., Yingkun, J., King, S., Stella, M., Stanton, D. T., Strasburg, R., Kenseth, J., & Wong, K. S. (2003). The application of multiplexed microemulsion electrokinetic chromatography for the rapid determination of log Pow values for neutral and basic compounds. *LC-GC N. Am.*, 21(11), 1078–1088.

237. Wong, K. S., Kenseth, J., & Strasburg, R. (2004). Validation and long-term assessment of an approach for the high throughput determination of lipophilicity (log POW) values using multiplexed, absorbance-based capillary electrophoresis. *J. Pharm. Sci.*, 93(4), 916–931.

238. Wiczling, P., Kawczak, P., Nasal, A., & Kaliszan, R. (2006). Simultaneous determination of pKa and lipophilicity by gradient RP HPLC. *Anal. Chem.*, 78(1), 239–249.

239. Wiczling, P., Markuszewski, M. J., Kaliszan, M., & Kaliszan, R. (2005). pH/organic solvent double-gradient reversed-phase HPLC. *Anal. Chem.*, 77(2), 449–458.

240. Kaliszan, R., & Wiczling, P. (2005). Theoretical opportunities and actual limitations of pH gradient HPLC. *Anal. Bioanal. Chem.*, 382(3), 718–727.

241. Wiczling, P., Markuszewski, M. J., & Kaliszan, R. (2004). Determination of pKa by pH gradient reversed-phase HPLC. *Anal. Chem.*, 76(11), 3069–3077.

242. Kaliszan, R., Wiczling, P., & Markuszewski, M. J. (2004b). pH gradient high-performance liquid chromatography: theory and applications. *J. Chromatogr. A*, 1060(1), 165–175.

243. Kaliszan, R., Haber, P., Bączek, T., Siluk, D., & Valko, K. (2002). Lipophilicity and pKa estimates from gradient high-performance liquid chromatography. *J. Chromatogr. A*, 965(1), 117–127.

244. Kaliszan, R., Haber, P., Baczek, T., & Siluk, D. (2001). Gradient HPLC in the determination of drug lipophilicity and acidity. *Pure Appl. Chem.*, 73(9), 1465–1475.

245. Wiczling, P., Markuszewski, M. J., Kaliszan, M., & Kaliszan, R. (2005). pH/organic solvent double-gradient reversed-phase HPLC. *Anal. Chem.*, 77(2), 449–458.

246. Kaliszan, R., Wiczling, P., & Markuszewski, M. J. (2004a). pH gradient reversed-phase HPLC. *Anal. Chem.*, 76(3), 749–760.

247. Tsantili-Kakoulidou, A., Panderi, I., Piperaki, S., Csizmadia, F., & Darvas, F. (1999). Prediction of distribution coefficients from structure. Comparison of calculated and experimental data for various drugs. *Eur. J. Drug Metab. Pharmacokinet.*, 24(3), 205–212.

Chapter 3

Use of Surfactants in Dissolution Testing

Amit Gupta

Zydus-Cadila Healthcare Ltd., Pharmaceutical Technology Centre,
Sarkhej-Bavla N.H. No. 8A, Moraiya, Taluka Sanand,
Ahmedabad 382 210, India
amit.gupta@zydusmail.com, amitopgupta@gmail.com

3.1 Introduction

Dissolution testing began at a time when pharmaceuticals were both soluble and permeable. The initial requirement for these immediate release formulations was testing typically in media consisting of 0.1 N HCl, 50 mM acetate buffers at pH 4.5, or 50 mM phosphate buffers at pH 6.8 alone or in various combinations. The development and use of these media conditions is well documented.[1,2]

Today, oral dosage forms in development have varying levels of solubility in aqueous media. In order to facilitate the dissolution testing of drugs that have lower aqueous solubility, surfactants at low concentrations are allowed by regulatory agencies to enhance solubility.[3] Organic solvents, though sometimes used, are not recommended as dissolution media with enhanced solubility since they typically are not representative of conditions in the gastrointestinal tract (GIT). However, bile acids and natural surfactants do exist in the GIT.

Poorly Soluble Drugs: Dissolution and Drug Release
Edited by Gregory K. Webster, J. Derek Jackson, and Robert G. Bell
Copyright © 2017 Pan Stanford Publishing Pte. Ltd.
ISBN 978-981-4745-45-1 (Hardcover), 978-981-4745-46-8 (eBook)
www.panstanford.com

For the pharmaceutical industry, it is now quite commonplace to find laboratory methods that incorporate surfactants into dissolution media. This addition is done primarily to enhance drug solubility in the test media to achieve sink conditions. As a significant number of pharmaceuticals in development are poorly soluble (collectively BCS class II and class IV) it is important to note that the addition of surfactants to dissolution media is not the only choice of the method development chemist to increase solubility, and conversely the presence of surfactant in the dissolution media does not necessarily indicate that drug solubility is a challenge. This understanding will be highlighted in case studies later in the chapter.

3.2 Surfactants

"Surfactant" is the common term applied to a group of chemicals referred to as "surface active agents." Surfactants are organic compounds with amphiphilic attributes (affinity toward both hydrophilic and hydrophobic environments) due to hydrophobic groups (tails) and hydrophilic groups (heads) present in the surfactant molecule. Therefore, a surfactant molecule contains both a water-insoluble (oil-soluble) and a water-soluble component. Surfactant molecules will migrate to the water surface, where the insoluble hydrophobic group may extend out of the bulk water phase, either into the air or, if water is mixed with an oil, into the oil phase, while the water-soluble head group remains in the water phase. This alignment and aggregation of surfactant molecules at the surface acts to alter the surface properties of water at the water/air or water/oil interface (Fig. 3.1).

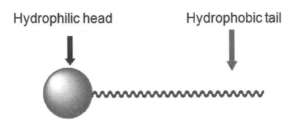

Hydrophilic head   Hydrophobic tail

Figure 3.1   Surfactant monomer.

Excellent reviews for the chemistry of surfactants can be found in the literature.[4]

3.2.1 Use of Surfactants in Dissolution Media Development

When surfactant is added to dissolution media, the hydrophilic end will associate with the aqueous media, with the hydrophobic tail encountering repulsive forces, effectively seeking alternative phases with which to associate. The "push and pull" between phases decreases the intermolecular forces within the aqueous phase, thereby reducing both surface and interfacial tension. Indeed, it is the reduction in interfacial tension that is the key initial driver for surfactant-based solubility enhancements.

Consider a case where a drug does not dissolve in water or dissolution media due to high hydrophobicity. A surfactant is added and dissolved into the media where it exists as an extended/linear monomer or self-associated spherical form, which is distributed within the media. Further increases in concentration of the surfactant will eventually result in micelles, self-association of multiple surfactant molecules creating a new colloidal phase of a hydrophobic core of surfactant tail. The concentration at which this phase change occurs is called the critical micelle concentration (CMC). In the presence of pure aqueous phase, the interaction of solvent and the surface of any hydrophobic is not energetically favored, resulting in poor wetting and low solubility. Interaction between a hydrophobic solid (insoluble drug) with the hydrophobic tails of dissolved surfactant decreases the energy required to wet and dissolve the solid, thereby increasing drug solubility. Further enhancement of solubility can be achieved through subsequent partitioning of dissolved species into the hydrophobic core of surfactant micelles. Selecting the optimal surfactant concentration in method development must take into consideration any impact the presence or absence of micelles might have on the fundamental mechanisms of in vitro release.

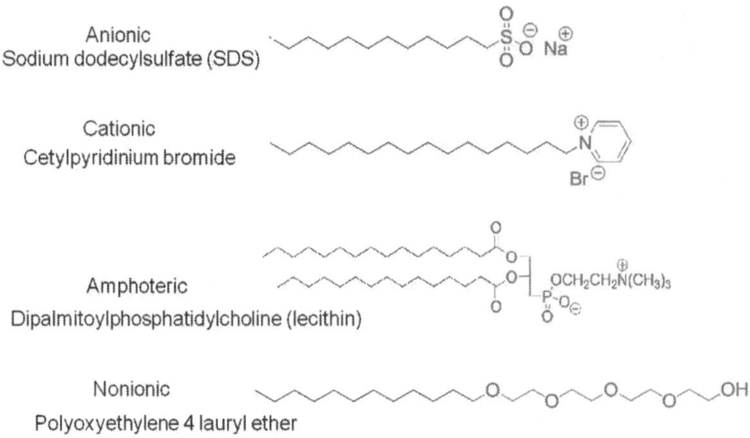

Figure 3.2 Types of surfactants.

3.2.2 Types of Surfactants in Dissolution Development

In dissolution method development, surfactants can be classified by their ionic charge into four major classes for screening purposes (Fig. 3.2):

- Anionic: e.g., sodium lauryl/sodium dodecyl sulfate (SLS/SDS)
- Cationic: e.g., cetyl trimethylammonium bromide (CTAB)
- Nonionic: e.g., polysorbate 20 & 80, polaxomer
- Amphoteric/zwitterionic: e.g., lecithin, cocamidopropyl betaine

In addition, for in vitro evaluation of performance in the GIT, more complicated "biorelevant" surfactant media systems may be considered. These formulations mimic the fasted (FaSSIF) and fed state (FeSSIF) environment in the human GIT.[5] FaSSIF and FeSSIF media formulations are commercially available.

3.2.3 Surfactant Concentration in Dissolution Media

As understood above, the solubility enhancement of surfactant based media is concentration dependent and while higher concentrations of surfactant will dissolve more drug,[6] the surfactant concentration must be optimized in order to balance solubility and sink conditions with the discriminating power of the method to detect manufacturing or stability changes. Typically the goal in setting surfactant concentration is to use the least amount of surfactant possible in the dissolution media to achieve the desired sink condition and method robustness while achieving and maintaining discrimination with respect to critical quality attributes of the drug product.

Solubility and sink condition can be evaluated during early development, but the discriminating character of the method tends to be revealed in the later stages of development, such as during verification of method reliability to detect deliberate changes in the formulation/process.

Additionally, two factors should be considered for surfactant-based dissolution media: (i) Surfactants media systems should be buffered to ensure method transferability. Various sources of surfactants sometimes lead to variable pH upon preparation. This is particularly true for SDS media, as this surfactant is typically derived from a ethoxylation neutralization process. (ii) The pH of the buffers used in surfactant media needs to be adjusted prior to the addition of the surfactant. As surfactants change the surface environment of the electrode, the resulting solution should be considered an apparent pH.

3.2.4 Other Non-surfactant Media Adjuvants

Apart from the selection of a suitable pH for the dissolution media and surfactant selection, secondary agent addition may be used in conjunction with surfactants such as polymers (e.g., HPC) and enzymes.

Secondary dissolution media components not only can aid in solubilization of actives, but are often used to address followings:

- adsorption of actives to materials and apparatus surfaces,
- possible food effect with the drug product, and
- cross-linking in gelatin capsules.

A limited representation of surfactant media methods found on the FDA Internet portal[7] are listed in Table 3.1. Solubility

Table 3.1 Representative dissolution methods containing surfactants recommended by FDA

Formulation	Dissolution media	Solubility attributes (in water)
Aprepitant cap	2.2% SDS in distilled water	insoluble
Bicalutamide tablet	1% SLS in water 1000	very slightly soluble
Cilostazol tablet	0.3% SLS in water 900	practically insoluble
Efavirenz tablet	2% SLS in water 1000	practically insoluble
Ergocalciferol capsule	0.5 N NaOH with 10% Triton-X-100	insoluble
Fenofibrate capsule	Phosphate buffer w/ 2% Tween 80 and 0.1% pancreatin, pH 6.8	insoluble
Isradipine capsule	0.1% Lauryl Dimethylamine Oxide (LDAO) in water	insoluble
Paricalcitol capsule	4 mg/mL (0.4%) Lauryldimethylamine N-oxide (LDAO)	insoluble
Roflumilast tablet	pH 6.8 phosphate buffer with 1% SLS	insoluble

data in water has been summarized from FDA approved "Label Information."[8, 9]

3.3 Impact of Surfactants on Dissolved Gases

As previously mentioned, the presence of surfactants in the dissolution media changes the surface and interfacial tension of the media. This results in changes in the solubility of dissolved oxygen in media.

Fliszar et al.[10] evaluated the effect of dissolved oxygen in surfactant-containing dissolution media. Using, aqueous (without surfactant) media and dissolution media containing 0.5% SLS, 2.0% SLS, and 0.5% Tween 80, the effect of oxygen upon the dissolution of several standard formulations was studied. In this study, the oxygen content of the surfactant-containing media was found to be 7.5–8.5 mg/mL due to the reduction in surface tension. However, the aqueous media without surfactant was 5.5 mg/mL lower. Irrespective to the deaeration method used (stirring under vacuum,

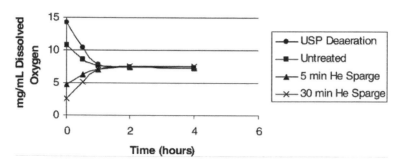

Figure 3.3 Dissolved oxygen equilibrium in 0.5% Tween media. Reprinted with permission from Fliszar et al.[10]

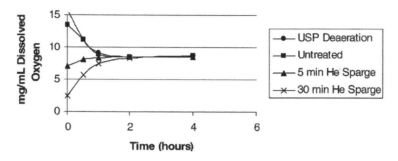

Figure 3.4 Dissolved oxygen equilibrium in 0.5% SDS media. Reprinted with permission from Fliszar et al.[10]

heating, sonication, helium sparging, and membrane filtration), all the media preparations regained or reaerated once deaeration was done. The initial oxygen content and the duration of aeration to reach equilibrium depended upon the method used for deaeration (Figs. 3.3–3.5). This increase in oxygen content was evaluated for its impact upon dissolution. The study confirmed that no resulting values were found (within error) to be slower at initial time points for surfactant-containing media (Figs. 3.6 and 3.7). In addition, a compound known to be susceptible for dissolved oxygen (prednisone) showed significant change in its dissolution profile

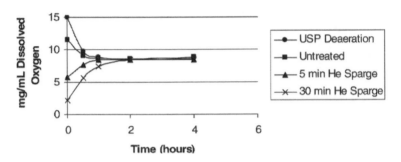

Figure 3.5 Dissolved oxygen equilibrium in 2.0% SDS media. Reprinted with permission from Fliszar et al.[10]

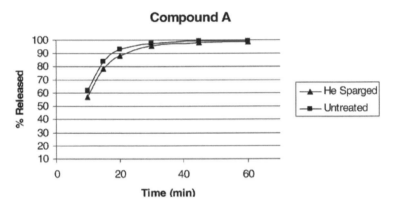

Figure 3.6 Release profile of compound A in helium sparged and untreated 1.0% SDS media. Reprinted with permission from Fliszar et al.[10]

in response to the aeration and deaeration (or oxygen content in other words), as illustrated in Fig. 3.8. It can be concluded from this work that surfactant-containing media regain their equilibrium oxygen content rapidly and variation is of minimal error. This study confirmed that it is important that dissolved gases in the media reach equilibrium before the experiment begins.

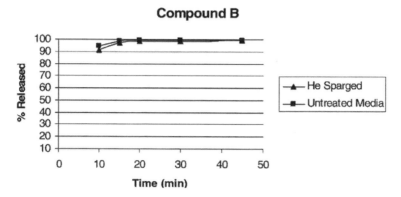

Figure 3.7 Release profile of compound B in helium sparged and untreated 0.5% SDS media. Reprinted with permission from Fliszar et al.[10]

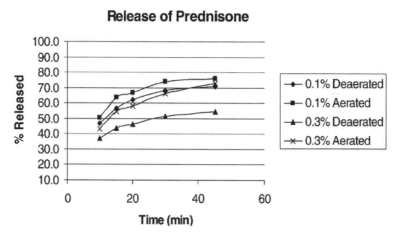

Figure 3.8 Release profile of Prednisone in 0.3% and 0.1% SDS. Reprinted with permission from Fliszar et al.[10]

3.4 Case Studies

To understand the application of surfactants in dissolution media, the following case studies are presented.

3.4.1 Roflumilast

Roflumilast

Molecular formulation	$C_{17}H_{14}Cl_2F_2N_2O_3$
Molecular weight	403.22
pK_a	8.7
Dosage	500 mcg
Solubility	Practically insoluble in water and hexane, sparingly soluble in ethanol and freely soluble in acetone

Roflumilast is a weak acid with a pK_a of 8.74 and pH-dependent solubility. Its solubility in water is 0.52–0.56 μg/mL at 22°C.[11] The solubility of roflumilast increases from approximately 0.8 μg/mL at neutral pH to approximately 35.8 μg /mL at pH 10.[11]

Solubility of the drug substances is pH independent across the physiologica pH range (pH 1.2–7.5) but does not achieve sink conditions, so a purely aqueous media is not suitably robust for use as the dissolution media. The FDA-recommended dissolution method (Table 3.2) utilizes a media consisting of 1.0% of SLS in pH 6.8 phosphate buffer. This media should be greater than 3×

Table 3.2 FDA-recommended dissolution method for roflumilast[7]

Drug name	Dosage form	USP apparatus	Medium
Roflumilast	Tablet	II (Paddle) at 50 rpm	1000 mL, 1.0% SDS (sodium dodecyl sulfate) in phosphate buffer, pH 6.8

of solubility (sink), and hence, robust and complete dissolution of roflumilast at a strength of 500 mcg. On the other hand, the drug product formulation of roflumilast utilizes enabling technology and excipients to enhance dissolution and bioavailability required.[12]

3.4.2 Fingolimod

Fingolimod	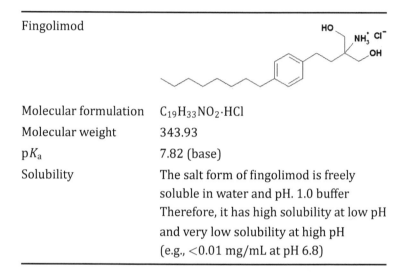

Molecular formulation	$C_{19}H_{33}NO_2 \cdot HCl$
Molecular weight	343.93
pK_a	7.82 (base)
Solubility	The salt form of fingolimod is freely soluble in water and pH. 1.0 buffer Therefore, it has high solubility at low pH and very low solubility at high pH (e.g., <0.01 mg/mL at pH 6.8)

Fingolimod hydrochloride is very poorly soluble at neutral pH, with higher solubility in the acidic pH range of more than 100 mg/mL. The use of 0.1 N hydrochloric acid media provides >3× (sink condition) without the need for surfactant to achieve sink condition. However, fingolimod has a tendency to adsorb onto various surfaces, including dissolution vessels, leading to an incomplete recovery during dissolution method validation.[9,22] To attain the complete recovery, the use of 0.2% SLS (w/w) has been used and accepted by FDA. This case represents a case where surfactant is not required for solubility enhancement but instead is used to facilitate recovery in the experimental conditions.

3.4.3 Lamotrigine

Lamotrigine

Molecular formulation	$C_9H_7N_5Cl_2$
Molecular weight	256.09
pK_a	5.7
Dosage	25–200 mg
Solubility	Very slightly soluble in water (0.17 mg/mL at 25°C) and slightly soluble in 0.1 M HC1 (4.1 mg/mL at 25°C)

Lamotrigine[15] exhibits poor solubility in water with some pH dependency (Fig. 3.9). The solubility of lamotrigine is higher in acidic media as lamotrigine becomes protonated and shows high solubility.[10] The highest dose of the lamotrigine is 200 mg, dissolved into 900 ml yields a concentration of 0.222 mcg/mL. In order to achieve theoretical sink condition in dissolution (3×), the solubility required would be at least 0.222 mcg/mL × 3, or 0.666 mcg/mL. Any media having solubility of more than 0.666 mcg/mL would suffice the need of in vitro sink condition in 900 mL for lamotrigine.

This pH dependency can be used to develop dissolution method. This was done in the development of the lamotrigine tablet (regular and chewable) where 0.1 N HCl was selected as dissolution media (Table 3.4).

Yet, during development of the dissolution method for lamotrigine extended release tablets, a realistic method mimicking the in vivo dynamic was explored. The initial acid phase for 2 hours (stomach condition) was followed by alkaline pH 6.8 (intestinal condition). Having low solubility at neutral and alkaline pH, the addition of SLS

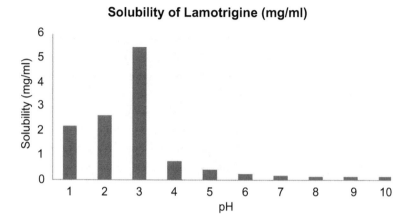

Figure 3.9   Solubility of lamotrigine across pH.

Table 3.3   FDA-recommended dissolution method for fingolimod

Drug name	Dosage form	USP apparatus	Medium
Fingolimod	Capsule	I (Basket) at 100 rpm	500 mL, 0.1 N HCl with 0.2% SDS (sodium dodecyl sulfate)

Table 3.4   FDA-recommended dissolution methods for lamotrigine

Drug name	Dosage form	USP apparatus	Medium
Lamotrigine	Tablet (extended release)	II (Paddle) at 50 rpm	700 mL acid stage: 0.01 M HCl; buffer stage: 900 mL phosphate buffer, pH 6.8 + 0.5% SLS
Lamotrigine	Tablet (regular)	II (Paddle) at 50 rpm	900 mL, 0.1 N HCl
Lamotrigine	Tablet (chewable dispersible)	II (Paddle) at 50 rpm	900 mL, 0.1 N HCl

was required to maintain sink conditions. This latter condition was used to mimic bioavailability.

3.4.4 Tacrolimus

Tacrolimus

Molecular formulation	$C_{44}H_{69}NO_{12} \cdot H_2O$
Molecular weight	822
Dosage	0.5, 1, and 5 mg base
Solubility	Practically insoluble in water (about 0.001%)[12] and in hexane, freely soluble in ethanol, and very soluble in methanol. The partition coefficient in *n*-octanol/water system is greater than 1000.

The absolute bioavailability of tacrolimus from Prograf immediate release capsules is 20–25%.[16] In order to discriminate better between formulations and during stability, complete drug release is required. However, tacrolimus is insoluble in water and has solubility of approximately 0.001%. Despite its poor solubility, the low maximum dose (5 mg) can be dissolved in standard dissolution volumes, though the dissolution rate is slow.

More problematic is that tacrolimus tends to adsorb to surfaces.[17] This adsorption is problematic due to the low-dose

Table 3.5   FDA-recommended dissolution method for tacrolimus

Drug name	Dosage form	USP apparatus	Medium
Tacrolimus	Capsule	II (Paddle) at 50 rpm	900 mL, hydroxypropyl cellulose solution (1 in 20,000). Adjust to pH 4.5 by phosphoric acid

masses; a small mass loss to surfaces represents a large percentage loss with respect to the total drug amount. This leads to the tacrolimus becoming immobilized at the surface of vessel and lowers the concentration of the drug available to be assayed.[18]

Hence, a secondary agent is needed to address the slow dissolution rate in media and surface adsorption. As seen in Table 3.5, hydroxypropyl cellulose has been found to address both issues. It acts as the surfactant/solubilizer.

3.4.5 Amiodarone

Strengths: 200 mg

Amiodarone

Molecular formulation	$C_{25}H_{29}I_2NO_3 \cdot HCl$
Molecular weight	691
pK_a	6.56 for base[25]
Dosage	200 mg
Solubility	Soluble in water (700 mg/L at 25°C). The solubility of amiodarone hydrochloride in water is reportedly highly temperature dependent. The solubility ranges from 0.3 to 0.5 mg/mL at 20°C. to about 7 mg/mL at 50°C. At about 60°C, the solubility increases to greater than 100 mg/mL.[25]

Table 3.6   FDA-recommended dissolution method for amiodarone

Drug name	Dosage form	USP apparatus	Medium
Amiodarone HCl (Test 1)	Tablet	II (Paddle) at 100 rom	1000 mL, 1% SLS in water
Amiodarone HCl (Test 2)	Tablet	I (Basket) at 50 rpm	900 mL, acetate buffer, pH 4.0, with 1% Tween 80

Amiodarone[19] has poor intrinsic water solubility resulting in slow dissolution of API. As shown in Table 3.6, use of sufficient quantity of surfactant has been recommended. Additionally, amiodarone has the tendency to adsorb onto surfaces[20] and may lead to incomplete recovery. The use of a surfactant eliminates this issue.

3.4.6 Enzyme Addition

Capsules are amongst the most popular solid oral dosage form owing to its manufacturing and administration ease by the patient. Gelatin capsules are the most widely used but suffer from its inherent property of cross-linking in the presence of aldehydes and high humidity upon storage. Gelatin, once cross-linked, forms a pellicle barrier, which does not allow the gelatin shell to rupture, thereby limiting release of the content fill inside during in vitro analysis.

The true in vivo disintegration impact of gelatin cross-linking has been evaluated by Brown et al.[21] Using scintigraphic investigation technique, the actual in vivo disintegration time for stressed and unstressed acetaminophen capsules was observed. Moderately stressed capsules disintegrated in 10 minutes, compared to 8 minutes for the unstressed capsule. This study confirmed that there is not a significant difference between the in vivo disintegration properties of moderately stressed and unstressed capsules.

Meyer et al.[22] studied both hard and soft acetaminophen capsules stressed exposed to formaldehyde to initiate cross-linking of the gelatin capsule. Cross-linking was confirmed by in vitro dissolution studies where capsules failed in USP dissolution test conditions. To identify the real impact on an in vivo study, a three-way crossover bioequivalence study in 24 subjects between unstressed vs. stressed capsules was used. These findings have led to the acceptance of a two-tier dissolution test using enzymes.[23]

Based on satisfactory dissolution results obtained for products by adding proteolytic enzymes to the dissolution medium, two-tier dissolution testing was included in USP 25.[24]

The current text for the two-tier dissolution testing in USP 35 is:

> For hard or soft gelatin capsules and gelatin-coated tablets that do not conform to the Dissolution specification, repeat the test as follows. Where water of a medium with a pH of less than 6.8 is specified as the Medium in the individual monograph, the same Medium specified may be used with the addition of purified pepsin that result in an activity of 750,000 Units or less per 1000 mL. For media with a pH of 6.8 or greater, pancreatin can be added to produce not more than 1750 USP Units of protease activity per 1000 mL.

3.5 Regulatory Concerns

For water-insoluble or sparingly water-soluble drug products, use of a surfactant such as sodium lauryl sulfate is recommended.[18] The need for and the amount of the surfactant should be justified. However, general recommendations indicate levels not to exceed 2.0%.[19] Use of a hydroalcoholic medium is discouraged by FDA.[18]

USP states that dissolution method should reflect "relevant" changes in the drug product caused by stress (light, temperature, humidity, etc.). In other words, it should be stability-indicating. On the other hand, the same chapter suggests reducing variability in dissolution whenever possible. The balance between these two statements is often overlooked. So far, FDA or USP has not bound any maximum quantity of surfactant to be used, but any quantity used is required to be justified with the above criteria.

References

1. Banakar, U. V. *Pharmaceutical Dissolution Testing* (Drugs and the Pharmaceutical Sciences). New York: Marcel Dekker, 1992.

2. Hanson, R., Gray, V. *Handbook for Dissolution Testing*, 3rd ed. Hockessin, DE: Dissolution Technologies, 2004.

3. Noory, C., Tran, N., Ouderkirk, L., Shah, V. Steps for development of a dissolution test for sparingly water-soluble drug products. *Dissolut. Technol.*, 2000, 7(1), 16–18.

4. Attwood, D. Surfactants, chapter 12 in *Modern Pharmaceutics*, 5th ed., A. T. Florence and J. Siepmann (eds.). Boca Raton, FL: CRC Press, 2009.

5. Bhagat, N. B., Yadav, A. B., Mail, S. S., Khutale, R. A., Hajare, A. A., Salunkhe, S. S., Nadaf, S. J. A review on development of biorelevant dissolution medium. *J. Drug Deliv. Ther.*, 2014, 4(2), 140–148.

6. Shah, V. P., Konecny, J. J., Everett, R. L., McCullough, B., Noorizadeh, A. C., Skelly, J. P. In vitro dissolution profile of water-insoluble drug dosage forms in the presence of surfactant. *Pharm. Res.*, 1989, 6(7), 612–618.

7. Dissolution method: List of all Drugs in the Database. Available at <http://www.accessdata.fda.gov/scripts/cder/dissolution/dsp_Search Results_Dissolutions.cfm?PrintAll=1>[05 October 2014].

8. Drugs@FDA, FDA-approved drug product. Available at <http://www.accessdata.fda.gov/scripts/cder/drugsatfda/ >[05 October 2014].

9. United State Pharmacopeia, USP 29/NF 24 <http://www.pharmacopeia.cn/>[05 October 2014].

10. Fliszar, K. A., Forsyth, R. J., Zhong, L., Martin, G. P. Effects of dissolved gases in surfactant dissolution media. *Dissolut. Technol.*, 2005, 12(3), 6–10.

11. WO/2006/032676. Pharmaceutical compositions comprising roflumilast or the N-oxide of roflumilast. Altana Pharma.

12. Dietrich, R., Eistetter, K., Ney, H. Oral dosage form containing a PDE 4 inhibitor as an active ingredient and polyvinylpyrrolidone as excipient. US Patent, US 8431154 B2, 2013.

13. Clinical Pharmacology and Biopharmaceutics Review(S), Application Number 22-527, 63–64 <http://www.accessdata.fda.gov/drugsatfda_docs/nda/2010/022527orig1s000clinpharmr.pdf>[05 October 2014].

14. Taormina, D., Abdullah, H. Y., Venkataramanan, R. Stability and sorption of FK 506 in 5% dextrose injection and 0.9% sodium chloride injection

in glass, polyvinyl chloride, and polyolefin containers. *Am. J. Hosp. Pharm.*, 1992, 49, 119–123.

15. Brittain, H. G. *Profiles of Drug Substances, Excipients and Related Methodology*, vol. 37. Salt Lake City, UT: Academic Press, 2012.

16. Australian Public Assessment Report for Tacrolimus, Prograf-XL, PM-2008-03783-3-2, Australian government—Department of health and ageing—therapeutic goods administration <https://www.tga.gov.au/pdf/auspar/auspar-prograf-xl.pdf>[05 October 2014].

17. Jacobson, P. A., Johnson, C. E., West, N. J., Foster, J. A. Stability of tacrolimus in an extemporaneously compounded oral liquid. *Am. J. Health Syst. Pharm.*, 1997, 54, 178–180.

18. Trissel, L. A. *Handbook on Injectable Drugs*, 9th ed. Bethesda: American Society of Pharmacists, 1996.

19. Weir, S. J., Myers, V. A., Bengtson, K. D., Ueda, C. T. Sorption of amiodarone to polyvinyl chloride infusion bags and administration sets. *Am. J. Hosp. Pharm.*, 1985, 42(12), 2679–2683.

20. Formulations containing amiodarone and sulfoalkyl ether cyclodextrin. US Patent, US 6869939 B2.

21. Brown, J., Madit, N., Cole, E. T., Wilding, I. R., Cadé, D. The effect of cross-linking on the in vivo disintegration of hard gelatin capsules. *Pharm. Res.*, 1986, 15(7), 1026–1030.

22. Meyer, M. C., Straughn, A. B., Mhatre, R. M., Hussain, A., Shah, V. P., Bottom, C. B., Cole, E. T., Lesko, L. L., Mallinowski, H., Williams, R. L. The effect of gelatin cross-linking on the bioequivalence of hard and soft gelatin acetaminophen capsules. *Pharm. Res.*, 2000, 17(8), 962–966.

23. US Pharmacopeial Convention, 2012, USP 35 – NF 30.

24. Singh, S., Rao, K. V. R., Venugopal, K., Manikandan, R. Alteration in dissolution characteristics of gelatin containing formulations. *Pharm. Technol.*, 2002, 26, 36–58.

Chapter 4

Intrinsic Dissolution Evaluation of Poorly Soluble Drugs

Michele Georges Issa and Humberto Gomes Ferraz

Department of Pharmacy, University of São Paulo, Rua do Lago,
250 Prédio Semi-Industrial Térreo, Sao Paulo, SP 05508-080, Brazil
sferraz@usp.br

4.1 Introduction

Evaluating the dissolution of a candidate molecule for an active pharmaceutical ingredient (API) before it is developed into a medicinal product has proven to be fundamental in the development of solid dosage forms for oral use that contain poorly soluble drugs (classes II and IV of the Biopharmaceutics Classification System [BCS]), since in these cases, dissolution is considered to be a limiting step in absorption.[1,2]

In the current scenario, the majority of molecules that are now available or under development present this characteristic and tests that enable to predict the release of the active substance can provide valuable data, facilitating the work of formulators, directly resulting

Poorly Soluble Drugs: Dissolution and Drug Release
Edited by Gregory K. Webster, J. Derek Jackson, and Robert G. Bell
Copyright © 2017 Pan Stanford Publishing Pte. Ltd.
ISBN 978-981-4745-45-1 (Hardcover), 978-981-4745-46-8 (eBook)
www.panstanford.com

in economy of resources and reduction in the time required to launching new medicinal products.[3-5]

Among the tests available, intrinsic dissolution enables the dissolution analysis of a pure drug, in a similar fashion to a conventional test. Accordingly, an apparatus that maintains a constant surface area of the drug exposed to the medium must be used, since, contrary to solubility, intrinsic dissolution is not related to equilibrium, but rather the speed with which the drug is released from the matrix to the dissolution medium.[6,7]

Intrinsic dissolution has been used for many years to characterize APIs, especially in preformulation studies, and it is especially useful because of the small quantity of sample that is necessary to conduct the tests, as well as the fact that it may be applied both in the stages prior to development (more precisely, in the prospection of new chemical entities), the selection of raw material at the preformulation stage and in the routine analysis of raw material in a quality assurance laboratory.[4,8] Table 4.1 lists some examples of intrinsic dissolution applications.

Furthermore, the intrinsic dissolution rate (IDR) may be a very useful parameter in quality by design (QbD) methodology, making a

Table 4.1 Use of the intrinsic dissolution test in different stages of pharmaceutical development and quality assurance[7,9-11]

Stage	Application
Obtaining a new drug candidate	• Salt selection • Cocrystal evaluation • Polymorph selection • Solubility evaluation
Preformulation	• Supplier selection • Polymorph and crystal habit evaluation • Crystal form stability during dissolution • Solubility evaluation – BCS classification • Drug dissolution evaluation in different media (pH variation/surfactants/biorelevant media)
Quality assurance	• Evaluation of raw material – equivalence between batches

significant contribution to knowledge of the critical variables of the raw material that, together with process variables, can outline the product design space.[12, 13]

With regards to apparatuses, the rotating disk method, also known as the Wood apparatus, and the fixed disk system are described in Pharmacopeias, although the possibility of using other configurations is also mentioned, since they are capable of providing a constant area of exposure of the drug to the dissolution medium.[6, 14, 15] Recently, new systems that use smaller sample amounts and reduced media volumes have been proposed.[8, 16, 17]

Accordingly, the purpose of this chapter is to analyze the aspects involved in the intrinsic dissolution assay, emphasizing its use in the evaluation of poorly soluble drugs. In the following text, the most commonly used apparatuses and the calculation of the intrinsic dissolution rate (IDR) will be presented, as well as some examples of applications and a brief explanation of the variables of the method and the use of experimental designs.

4.2 Apparatuses Used in Evaluating the Intrinsic Dissolution Rate

4.2.1 Rotating Disk System

This widely used system is easily coupled to a dissolution device, and it was the first apparatus to be cited in the U.S. Pharmacopeia. The term *rotating* comes from the stirring that is executed by the rod that secures the matrix with the exposed drug (Fig. 4.1). With an orifice of 0.8 cm in diameter, it provides for an exposure surface of 0.5 cm^2, and a hydraulic press is necessary to form the compact.[6, 11]

4.2.2 Fixed Disk System

Also described in the U.S. Pharmacopeia, this is very similar to the rotating disk system; however, the matrix that contains the drug is placed in the bottom of a flat-bottomed dissolution vessel and a paddle is used for stirring (Fig. 4.2).[6, 11]

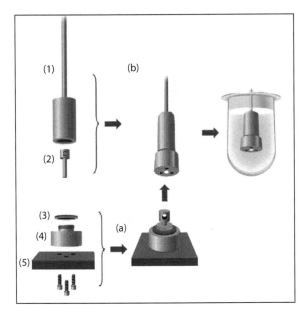

Figure 4.1 Rotating disk system: (1) rod, (2) steel punch, (3) sealing ring, (4) die, (5) base. (a) The configuration that should be taken to the press is shown; (b) Arrangement of the apparatus in the dissolution device.

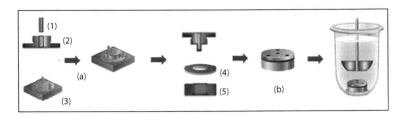

Figure 4.2 Fixed disk system: (1) steel punch, (2) die, (3) base, (4) sealing ring, (5) matrix support. (a) The configuration that should be taken to the press is shown. (b) Arrangement of the apparatus in the dissolution device.

For both systems, it is possible to work with sample quantities of between 100 and 700 mg, depending on drug solubility, since more soluble raw materials must be used in greater quantities in order to maintain the surface area of the compacted drug. With regards to results, a satisfactory correlation is obtained when the two systems are compared.[8, 11]

4.2.3 Flow Cell: Dissolution by UV Imaging

The characterization of a drug candidate molecule is essential in the use of methodologies such as High-throughput Screening and combinatorial chemistry;[1] however, the range of information obtained depends on the quantity of raw material available, which limits the number and types of tests that can be performed.

Accordingly, alternative methods for evaluating intrinsic dissolution have been conceived, including dissolution by UV imaging. Besides a necessity for smaller quantities of the API and reduced volumes of the dissolution medium, it is possible to visualize, in real time, what happens at the surface of the compacted drug, thus enabling monitoring of crystalline phase transitions that may occur during the assay.[1,18,19]

The image is obtained by the absorption map generated (Fig. 4.3) from the reading of a solution that passes over the surface of the compacted drug through a flow cell, reaching the UV detector at a certain wavelength. In this system, a smaller size matrix of around 2 mm is used, in which the dissolution medium passes under a controlled flow, providing favorable hydrodynamics, with no need to be concerned about the presence of bubbles, which are a significant factor in disk methods, nor with the sensitivity of the quantification system, which enables a considerable reduction in the analysis time.[1,20,21]

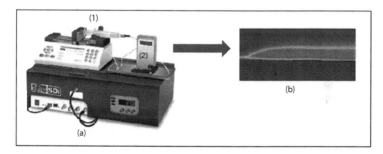

Figure 4.3 Equipment for dissolution by UV imaging (a) in (1) injection system of the dissolution medium, and in (2) compartment containing the die of the compacted drug and the detection system. (b) An example of an absorption map: the light gray coloring indicates the presence of dissolved drug.

In this system, the compacted drug is produced manually, using a torque wrench and it has an area of approximately 0.03 cm^2, with the amount of drug used being less than 10 mg.[20]

4.3 Calculating the Intrinsic Dissolution Rate (IDR)

Obtaining the IDR is based on the use of an adaptation of the Noyes & Whitney equation, in which it is possible to relate the dissolution rate (DR) with the surface area of the drug exposed to the dissolution medium.[6,7,22] In the following equations, V is the volume of the dissolution medium, dC is the concentration of dissolved drug, dt is the time, A is the surface area of the compacted drug, dm is the amount of accumulated dissolved drug and dm/dt is the dissolution rate:

$$\text{IDR} = \frac{V\,dC}{dt} \cdot \frac{1}{A} \Rightarrow \text{IDR} = \frac{dm}{dt} \cdot \frac{1}{A} \qquad (4.1)$$

Accordingly, the results of the accumulated dissolved quantity as a function of time must be plotted for calculation and with the linear regression of the points, the dissolution rate is obtained from the slope of the straight line (Fig. 4.4). When it is divided by the surface area of the compacted drug, this provides the IDR, generally expressed in mg·min^{-1}·cm^{-2} or mg·s^{-1}·cm^{-2}.[6,23]

In cases where there is a curvature, only the first points should be considered when calculating the IDR. A positive curvature (Fig. 4.5) may be related to an experimental problem, such as uneven wear of the compact, thus altering the surface area of the drug exposed to dissolution medium.[6]

On the other hand, a negative curvature (Fig. 4.6) may be the result of reduced drug solubility, either due to alteration in the crystal form at the surface of the compacted or when saturation occurs in the diffusion layer. Lehto et al.,[9] demonstrated the transition from the crystal phase of theophylline during an intrinsic dissolution assay, in which it was possible to ascertain the alteration in the slope of the straight line in the graph, from the moment in which drug hydration was observed.

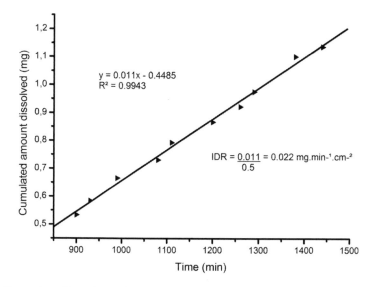

Figure 4.4 Hypothetical example of an intrinsic dissolution curve and the respective IDR calculation, considering a surface area of 0.5 cm^2.

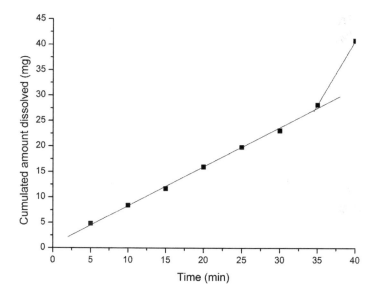

Figure 4.5 An example of a positive curvature due to altered surface area. Analysis of enrofloxacin conducted on 900 mL of HCl 0.01 M at 200 rpm in a rotating disk apparatus (data obtained in our laboratory).

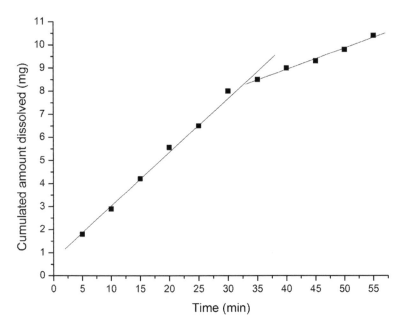

Figure 4.6 Hypothetical data for simulation of negative curvature during the intrinsic dissolution test.

4.4 Evaluation of Intrinsic Dissolution of Poorly Soluble Drugs

The understanding of how a poorly soluble drug behaves in dissolution testing can help in targeting formulation strategies and avoiding the risk of failure in the final stage of development.[3,4,24] Phenomena such as the alteration of crystal forms, hydration and amorphization may be monitored during the test and lead to initially unforeseen results with the original material, which would facilitate the comprehension of mechanisms that dictate the dissolution of the drug.[4,6,7,11]

Furthermore, although its concept is involved in the evaluation of a pure API, several studies describe the determination of the IDR of the drug in the presence of other components as a tool for evaluating improvements in solubility, such as cocrystals,[25-31] complexation with cyclodextrins[32] and solid dispersions.[33-35] The analysis of amorphous forms,[3,36,37] salts[4,24,38] and alternative crystal forms[5,39–41] are also widely reported.

On the basis of the fact that intrinsic dissolution may have a greater correlation with the *in vivo* dissolution dynamic, instead of equilibrium solubility, studies have evaluated and suggested the use of the IDR value for classifying the solubility of drugs in the BCS.[7,42] Since the dosing parameter does not feature in the calculation, the difficulty is in establishing a limit value for differentiation between high and low solubility.

Through the evaluation of different drugs with a surface area of 0.5 cm^2 in the rotating disk apparatus, Yu et al.[7] suggested a value of 0.1 mg·min^{-1}·cm^2 as a limit value for BCS classification of solubility, in accordance with the drugs tested. In another study, conducted by Zakeri-Milani et al.[42] in 2009, using the same experimental conditions, with the exception of the apparatus, which had a similar configuration to the fixed disk system, the drugs were compacted in a wax mold, resulting in a surface area of approximately 0.28 cm^2, it was suggested that drugs with IDR values below 1 mg·min^{-1}·cm^2 were considered poorly solubility. The difference between the two values adopted as a limit for classification may be related to the configurations of the apparatuses used.[23,42]

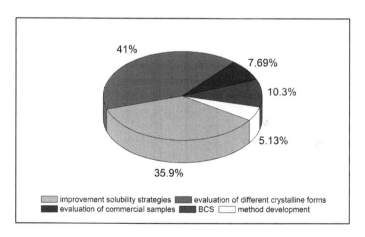

Figure 4.7 Applications of the determination of IDR for poorly soluble drugs. The construction of this graph was based on information originating from the 37 articles researched, pertaining to the 2001–2014 period.

Figure 4.7 was constructed based on articles that involve the intrinsic dissolution test and poorly soluble drugs. Accordingly, we observe that the test is very linked to the evaluation of the polymorphic forms and their respective phase transitions (41%), as well as to the solubility prediction (improvement solubility strategies, evaluation of commercial samples and BCS).

Among the APIs in which intrinsic dissolution was studied, carbamazepine (Table 4.2) may be highlighted, it is a class II drug under the Biopharmaceutics Classification System, which can be found in several crystal and solvate forms. In the following table, several examples of drugs are cited in which IDR has been determined, as well as their respective references.

4.5 Development of an Intrinsic Dissolution Test

Factors related to the drug and the test may influence the IDR outcome. The properties of the solid state, as well as the choice of dissolution medium, the quality of the compacted drug, which reflects on the surface area exposed to the medium and the hydrodynamics of the system, should be studied, so that the test can be used to comprehend the phenomena involved in the dissolution of the API.[6,23,48] When the purpose of the study is evaluating a poorly soluble drug that presents polymorphism, a well-designed and discriminative method can be of fundamental importance for the selection of the most suitable API, as the differences between batches or samples from different suppliers may be observed.[48]

4.5.1 Test Variables

4.5.1.1 Sample

Among the direct drug-related variables, particle form and size can be cited. Although one of the aspects explored in intrinsic dissolution is minimization of these effects due to compaction of the drug, studies suggest that the micronization and difference in crystal habit can lead to changes in the arrangement of the material at the surface that is exposed to the dissolution medium.[44,54,64,65]

Table 4.2 Examples of poorly soluble drugs in which IDR has been evaluated. Most of the examples cited pertain to the rotating disk and fixed disk systems

API	References
atorvastatin	[36]
carbamazepine	[7, 8, 10, 19, 34, 43–46]
celecoxib	[47]
deflazacort	[41]
dibenzyl sulfoxide	[26]
diflunisal	[25]
dipyridamol	[19]
efavirenz	[48]
felodipine	[35]
fenofibrate	[49]
furosemide	[7, 8, 19, 37, 42]
fusidic acid	[50]
glibenclamide	[8, 19]
griseofulvin	[7, 8, 19]
haloperidol	[4, 8, 19, 24]
hydrochlorothiazide	[7, 8, 19, 32, 42]
ibuprofen	[1, 11, 42, 51]
indometacin	[1, 5, 19, 52–54]
irbersartan	[39]
itraconazole	[29, 55]
ketoprofen	[7, 8, 19, 42, 56]
miconazole	[30]
naproxen	[3, 7, 8, 19, 21, 42, 57]
nevirapine	[58, 59]
phenazopyridine hydrochloride	[19]
piroxicam	[7, 8, 19, 40, 42]
puerarin	[60]
nimodipine	[61]
sodium diclofenac	[62]
rifampicin	[63]
tenoxicam	[28]
teophylline	[1, 9]
terbinafine hydrochloride	[64]

A single polymorph may present different crystal habits and each one can have a certain preferred orientation, leading to a texturized compact, which can facilitate dissolution. On the other hand, a reduction in particle size may generate a compact in which the particles are randomly arranged and thus result in a distinctive dissolution rate. It is worth noting that this preferred orientation phenomenon does not apply to amorphous drugs.[44, 65]

Based on their experiments, Modi et al.[47] suggest that the type of crystal habit may have an impact on the IDR of class II drugs. It has been observed that different formats of the polymorph III of celecoxib result in distinctive wettability rates, which reflect on the IDRs achieved.

4.5.1.2 Compaction pressure

In accordance with the U.S. Pharmacopeia,[6] the pressure applied on the formation of the compact must be adequate, in order to avoid any alteration in the crystal form or in the appearance of capillaries that may alter the surface area of the drug throughout the test. Pressures of around 2000 psi have proven to be satisfactory for most drugs.[6, 7] However, this parameter must be evaluated for each API.

Yu et al.[7] have reported that for piroxicam, pressures of above 1000 psi make the compact brittle, and the surface area was not maintained when the material was exposed to the dissolution medium. In the case of rifampicin, Agrawal et al.[63] observed that pressures lower than 500 psi create a brittle compact, while those above 2000 psi cause conversion from crystal form II to crystal form I.

In general, very close IDR values have been observed for different pressures,[7, 48] the great implication of this variable is related to the quality of the compacted drug and to the alteration of the polymorph, with regards to crystal forms. In the case of amorphous drugs, one study with indometacin demonstrated that the IDR can be influenced by compaction pressure.[54]

X-ray powder diffraction (XRPD), Fourier transform infrared spectroscopy (FTIR), differential scanning calorimetry (DSC) and Raman spectroscopy have all been used to monitor the influence of compaction pressure on polymorphic drugs.[46, 55, 63]

4.5.1.3 Dissolution medium

The same considerations adopted in a conventional dissolution test are applied in order to determine the IDR. The sink condition is fundamental for avoiding drug saturation in the diffusion layer close to the surface of the compact. In the case of poorly soluble drugs, surfactants may be required, and different concentrations of this agent must be evaluated.[6]

Another considerably important point is the de-aeration of the medium, considering the presence of bubbles may hamper or even avoid drug dissolution, thus generating a lower-than-expected IDR value.[6]

With regards to the volume, it is possible that, in some cases, since the non-saturated condition is respected, volumes of less than 900 mL may be used. This artifice may be useful for enabling more adequate readings, when a spectrophotometric method is used to quantify a poorly soluble drug.[7]

4.5.1.4 Stirring speed

Stirring speeds of between 60 and 500 rpm are suggested by the U.S. Pharmacopeia.[6] Because dissolution is dependent on stirring speed, one alternative to work with poorly soluble drugs is to use higher rotations, as long as loss of linearity of the results is not observed. This can be evaluated by the value of the coefficient of determination (R^2), which is an indication that the surface of the compacted drug has not undergone any alteration.[3, 66]

Another aspect to be considered is that, like the other selected parameters, it is important that the stirring speed does not interfere in the discrimination capacity of the method in the evaluation of different sample. The most common situation is one that requires rotation speeds of between 50 and 100 rpm.[48]

4.5.2 Design of Experiments (DOE)

Methodologies like the design of experiments and response surface methods have been widely employed in the evaluation of aspects related to the development of formulations; however, these approaches also may be explored in the development of dissolution methods, including intrinsic dissolution.[66]

As previously cited, the IDR is the result of variables inside and outside the drug. By challenging the method with a series of tests, according to an adequate experimental design, and obtaining results that are subject to statistical analysis, it is possible to evaluate the significant influence of these factors.[66,67]

In a study conducted by Issa et al.[66] and Giorgetti et al.,[67] fractional factorial designs were used to study the impact of variables such as particle size, dissolution medium, compaction pressure, and stirring speed in the intrinsic dissolution of metronidazol and amlodipine besylate.

Although highly soluble drugs have been studied in these cases, this methodology may be interesting for the study of factors involving the IDR of BCS class II and IV drugs, since the intrinsic dissolution test is much more applicable to this class of drugs.

4.6 Conclusion

The evaluation of the IDR is a very useful technique for the development of poorly soluble drug formulations. With an application in all the conception stages of a medicinal product, from the synthesis of a new chemical entity up to the quality assurance of the API, it is a tool that is capable, from small quantities of material, to supply the knowledge necessary for guiding the formulator to obtain a product with adequate dissolution and, consequently, bioavailability.

References

1. Hulse, W. L., Gray, J., and Forbes, R. T. (2012). A discriminatory intrinsic dissolution study using UV area imaging analysis to gain additional insights into the dissolution behaviour of active pharmaceutical ingredients. *Int. J. Pharm.*, **434**, 133–139.

2. Amidon, G. L., Lennernäs, H., Shah, V. P., and Crison, J. R. (1995). A theoretical basis for a biopharmaceutic drug classification: the

correlation of *in vitro* drug product dissolution and *in vivo* bioavailability. *Pharm. Res.,* **12**, 413–420.

3. Allesø, M., Chieng, N., Rehder, S., Rantanen, J., Rades, T., and Aaltonen, J. (2009). Enhanced dissolution rate and synchronized release of drugs in binary systems through formulation: amorphous naproxen-cimetidine mixtures prepared by mechanical activation. *J. Control. Release,* **136**, 45–53.

4. Ayres, C., Burke, W., Dickinson, P., Kirk, G., Pugh, R., Sharma-Singh, G., and Kittlety, R. (2007). Intrinsic dissolution rate determinations in early development and relevance to *in vivo* performance. *Am. Pharm. Rev.,* **10**, 74–78.

5. Sarnes, A., Østergaard, J., Jensen, S. S., Aaltonen, J., Rantanen, J., Hirvonen, J., and Peltonen, L. (2013). Dissolution study of nanocrystal powders of a poorly soluble drug by UV imaging and channel flow methods. *Eur. J. Pharm. Sci.,* **50**, 511–519.

6. United States Pharmacopeia (2014), 37th ed. (United States Pharmacopeial Convention, Rockville) 831–834.

7. Yu, L. X., Carlin, A. S., Amidon, G. L., and Hussain, A. S. (2004). Feasibility studies of utilizing disk intrinsic dissolution rate to classify drugs. *Int. J. Pharm.,* **270**, 221–227.

8. Avdeef, A., and Tsinman, O. (2008). Miniaturized rotating disk intrinsic dissolution rate measurement: effects of buffer capacity in comparisons to traditional Wood's apparatus. *Pharm. Res.,* **25**, 2613–2627.

9. Lehto, P., Aaltonen, J., Niemelä, P., Rantanen, J., Hirvonen, J., Tanninen, V. P., and Peltonen, L. (2008). Simultaneous measurement of liquid-phase transformation kinetic in rotating disc and channel flow cell dissolution devices. *Int. J. Pharm.,* **363**, 66–72.

10. Sehic, S., Betz, G., Hadzidedic, S., El-Arini, S. K., and Leuenberger, H. (2010). Investigation of intrinsic dissolution behavior of different carbamazepine samples. *Int. J. Pharm.,* **386**, 77–90.

11. Viegas, T. X., Curatella, R. U., Winkle, L. L., and Brinker, G. (2001). Measurement of intrinsic drug dissolution rates using two types of apparatus. *Pharm. Tech.,* **6**, 44–53.

12. Dickinson, P. A., Lee, Wang W., Stott, P. W., Townsend, A. I., Smart, J. P., and Ghahramani, P., Hammett, T., Billet, L., Behn, S., Gibb, R. C., and Abrahamsson, B. (2008). Clinical relevance of dissolution testing in quality by design. *AAPS J.,* **10**, 280–290.

13. Garala, K. C., Patel, J. M., Dhingani, A. P., and Dharamsi, A. T. (2013). Quality by design (QbD) approach for developing agglomerates

containing racecadotril and loperamide hydrochloride by crystallo-co-agglomeration. *Powder Technol.*, **247**, 128–146.

14. Brittish Pharmacopoeia (2010), **IV**, Appendix XIIB (Brittish Pharma-copoeia Comission), A317–A318.

15. European Pharmacopoeia (2008), 6th ed. (Council of Europe, 2008), 309–311.

16. Berger, C. M., Tsinman, O., Voloboy, D., Lipp, D., Stones, S., and Avdeef, A. (2007). Miniaturized intrinsic dissolution rate (Mini-IDR™) measurement of griseofulvin and carbamazepine. *Dissolut. Technol.*, **14**, 39–41.

17. Peltonem, L., Liljeroth, P., Heikkilä, T., Kontturi, K., and Hirvonen, J. (2003). Dissolution testing of acetylsalicylic acid by a channel flow method-correlation to USP basket and intrinsic dissolution methods. *Eur. J. Pharm. Sci.*, **19**, 395–401.

18. Laitinen, R., Lahtinen, J., Silfsten, P., Vartiainen, E., Jarho, P., and Ketolainen, J. (2010). An optical method for continuous monitoring of the dissolution rate of pharmaceutical powders. *J. Pharm. Biomed. Anal.*, **52**, 181–189.

19. Tsinman, K., Avdeef, A., Tsinman, O., and Voloboy, D. (2009). Powder dissolution method for estimating rotating disk intrinsic dissolution rates of low solubility drugs. *Pharm. Res.*, 26, 2093–2100.

20. Boetker, J. P., Savolainen, M., Koradia, V., Tian, F., Rades, T., Müllertz, A., Cornett, C., Rantanen, J., and Østergaard, J. (2011). Insights into early dissolution events of amlodipine using UV and Raman spectroscopy. *Mol. Pharm.*, 8, 1372–1380.

21. Niederquell, A., and Kuentz, M. (2014). Biorelevant dissolution of poorly soluble weak acids studied by UV imaging reveals ranges of fractal-like kinetics. *Int. J. Pharm.*, **463**, 38–49.

22. Carstensen, J. T. (2001). Advanced Pharmaceutical Solids, **110**, Chapter 12, "Dissolution from particles and surfaces", (Marcel Dekker, Inc., New York) 191–194.

23. Issa, M. G., and Ferraz, H. G. (2011). Intrinsic dissolution as a tool for evaluating drug solubility in accordance with the Biopharmaceutics Classification System. *Dissolut. Technol.*, **18**, 6–13.

24. Li, S., Wong, S., Sethia, S., Almoazen, H., Joshi, Y. M., and Serajuddin, A. T. M. (2005). Investigation of solubility and dissolution of a free base and two different salt forms as a function of pH. *Pharm. Res.*, **22**, 628–635.

25. Évora, A. O. L., Castro, R. A. E., Maria, Teresa M. R., Silva, M. R., Horst, J. H. ter, Canotilho, J., and Eusébio, M. E. S. (2014). A thermodynamic based

approach on the investigation of a diflunisal pharmaceutical co-crystal with improved intrinsic dissolution rate. *Int. J. Pharm.*, **466**, 68–75.

26. Grossjohann, C., Eccles, K. S., Maguire, A. R., Lawrence, S. E., Tajber, L., Corrigan, O. I., and Healy, A. M. (2012). Characterisation, solubility and intrinsic dissolution behavior of benzamide: dibenzyl sulfoxide cocrystal. *Int. J. Pharm.*, **422**, 24–32.

27. McNamara, D. P., Childs, S. L., Giordano, J., Iarriccio, A., Cassidy, J., Shet, M. S., Mannion, R., O'Donnell, E., and Park, A. (2006). Use of a glutaric acid cocrystal to improve oral bioavailability of a low solubility API. *Pharm. Res.*, **23**, 1888–1897.

28. Patel, J. R., Carlton, R. A., Needham, T. E., Chichester, C. O., and Vogt, F. G. (2012). Preparation, structural analysis, and properties of tenoxicam cocrystals. *Int. J. Pharm.*, **436**, 685–706.

29. Shevchenko, A., Bimbo, L. M., Miroshnyk, I., Haarala, J., Jelinkova, K., Syrjanen, K., Veen, B. van, Kiesvaara, J., Santos, H. A., and Yliruusi, J. (2012). A new cocrystal and salts of itraconazole: Comparison of solid-state properties, stability and dissolution behavior. *Int. J. Pharm.*, **436**, 403–409.

30. Tsutsumi, S., Iida, M., Tada, N., Kojima, T., Ikeda, Y., Moriwaki, T., Higashi, K., Moribe, K., and Yamamoto, K. (2011). Characterization and evaluation of miconazole salts and cocrystals for improved physicochemical properties. *Int. J. Pharm.*, **421**, 230–236.

31. Žegarac, M., Lekšić, E., Šket, P. Plavec, J., Bogdanović, M. D., Bučar, D.-K., Dumiće, M., and Meštrović, E. (2014). A sildenafil cocrystal based on acetylsalicylic acid exhibits an enhanced intrinsic dissolution rate. *Cryst. Eng. Comm.*, **16**, 32–35.

32. Pires, M. A. S., Santos, R. A. S., and Sinisterra, D. (2011). Pharmaceutical composition of hydrochlorothiazide: β-cyclodextrin: preparation by three different methods, physico-chemical characterization and *in vivo* diuretic activity evaluation. *Molecules*, **16**, 4482–4499.

33. Dong, Z., Chatterji, A., Sandhu, H., Choi, D. S., Chokshi, H., and Shah, N. (2008). Evaluation of solid state properties of solid dispersions prepared by hot-melt extrusion and solvent co-precipitation. *Int. J. Pharm.*, **355**, 141–149.

34. Sethia, S., and Squillante, E. (2004). Solid dispersion of carbamazepine in PVP K30 by conventional solvent evaporation and supercritical methods. *Int. J. Pharm.*, **272**, 1–10.

35. Tres, F., Treacher, K., Booth, J., Hughes, L. P., Wren, S. A. C., Aylott, J. W., and Burley, J. C. (2014). Real time Raman imaging to understand dissolution

performance of amorphous solid dispersions. *J. Control. Release*, **188**, 53–60.

36. Kim, M., Jin, S., Kim, J., Park, H. J., Song, H., Neubert, R. H. H., and Hwang, S. (2008). Preparation, characterization and *in vivo* evaluation of amorphous atorvastatin calcium nanoparticles using supercritical antisolvent (SAS) process. *Eur. J. Pharm. Biopharm.*, **69**, 454–465.

37. Nielsen, L. H., Gordon, S., Holm, R., Selen, A., Rades, T., and Müllertz, A. (2013). Preparation of an amorphous sodium furosemide salt improves solubility and dissolution rate and leads to a faster T_{max} after oral dosing rats. *Eur. J. Pharm. Biopharm.*, **85**, 942–951.

38. Parshad, H., Frydenvang, K., Lilijefors, T., and Larsen, C. S. (2003). Correlation of aqueous solubility of salts of benzylamine with experimentally and theoretically derived parameters. A multivariate data analysis approach. *Int. J. Pharm.*, **237**, 193–207.

39. Chawla, G., and Bansal, A. K. (2007). A comparative assessment of solubility advantage from glassy and crystalline forms of a water-insoluble drug. *Eur. J. Pharm. Sci.*, **32**, 45–57.

40. Lust, A., Laidmäe, I., Palo, M., Meos, A., Aaltonen, J., Veski, P., Heinämäki, J., and Kogermann, K. (2013). Solid-state dependent dissolution and oral bioavailability of piroxicam in rats. *Eur. J. Pharm. Sci.*, **48**, 47–54.

41. Paulino, A. S., Rauber, G., Campos, C. E. M., Maurício, M. H. P., Avillez, R. R., Capobianco, G., Cardoso, S. G., and Cuffini, S. L. (2013). Dissolution enhancement of deflazacort using hollow crystals prepared by antisolvent crystallization process. *Eur. J. Pharm. Sci.*, **49**, 294–301.

42. Zakeri-Milani, P., Barzegar-Jalali, M., Azimi, M., and Valizadeh, H. (2009). Biopharmaceutical classification of drugs using intrinsic dissolution rate (IDR) and rat intestinal permeability. *Eur. J. Pharm. Biopharm.*, **73**, 102–106.

43. Murphy, D., Rodríguez-Cintrón, F., Langevin, B., Kelly, R. C., and Rodríguez-Hornedo, N. (2002). Solution-mediated phase transformation of anhydrous to dehydrate carbamazepine and the effect of lattice disorder. *Int. J. Pharm.*, **246**, 121–134.

44. Tenho, M., Heinänen, P., Tanninen, V. P., Lehto, V. P. (2007). Does the preferred orientation of crystallites in tablets affect the intrinsic dissolution? *J. Pharm. Biomed. Anal.*, **43**, 1315–1323.

45. Tian, F., Sandler, N., Aaltonen, J., Lang, C., Saville, D. J., Gordon, K. C., Strachan, C. J., Rantanen, J., and Rades, T. (2007). Influence of polymorphic form, morphology, and excipient interactions on the dissolution of carbamazepine. *J. Pharm. Sci.*, **96**, 584–594.

46. Qiao, N., Wang, K., Schlindwein, Davies, A., and Li, M. (2013). *In situ* monitoring of carbamazepine-nicotinamide cocrystal intrinsic dissolution behavior. *Eur. J. Pharm. Sci.*, **83**, 415–426.

47. Modi, S. R., Dantuluri, A. K. R., Perumalla, S. R., Sun, C. C., and Bansal, A. K. (2014). Effect of crystal habit on intrinsic dissolution behavior of celecoxib due to differential wettability. *Cryst. Growth Des.*, **14**, 5283–5292.

48. Pinto, E. C., Cabral, L. M., and Sousa, V. P. (2014). Development of a discriminative intrinsic dissolution method for efavirenz. *Dissolut. Technol.*, **21**, May, 31–40.

49. Buch, P., Meyer, C., and Langguth, P. (2011). Improvement of the wettability and dissolution of fenofibrate compacts by plasma treatment. *Int. J. Pharm.*, **416**, 49–54.

50. Gilchrist, S. E., Letchford, K., and Burt, H. M. (2012). The solid-state characterization of fusidic acid. *Int. J. Pharm.*, **422**, 245–253.

51. Manish, M., Harshal, J., and Anant, P. (2005). Melt sonocrystallization of ibuprofen: Effect on crystal properties. *Eur. J. Pharm. Sci.*, **25**, 41–48.

52. Karmwar, P., Graeser, K., Gordon, K. C., Strachan, C. J., and Rades, T. (2012). Effect of different preparation methods on the dissolution behavior of amorphous indomethacin. *Eur. J. Pharm. Biopharm.*, **80**, 459–464.

53. Priemel, P. A., Grohganz, H., Gordon, K. C., Rades, T., and Strachan, C. J. (2012). The impact of surface-and nano-crystallisation on the detected amorphous content and dissolution bahaviour of amorphous indomethacin. *Eur. J. Pharm. Biopharm.*, **82**, 187–193.

54. Löbmann, K., Flouda, K., Qiu, D., Tsolakou, T., Wang, W., and Rades, T. (2014). The influence of pressure on the intrinsic dissolution rate of amorphous indomethacin. *Pharmaceutics*, **6**, 481–493.

55. Ghazal, H. S., Dyas, A. M., Ford, J. L., and Hutcheon, G. A. (2009). *In vitro* evaluation of the dissolution behavior of itraconazole in bio-relevant media. *Int. J. Pharm.*, **366**, 117–123.

56. Sheng, J. J., Kasim, N. A., Chandrasekharan, R., and Amidon, G. L. (2006). Solubilization and dissolution of insoluble weak acid, ketoprofen: Effects of pH combined with surfactant. *Eur. J. Pharm. Sci.*, **29**, 306–314.

57. Di Martino, P. Barthélémy, C., Palmieri, G. F., and Martelli, S. (2001). Physical characterization of naproxen sodium hydrate and anhydrate forms. *Eur. J. Pharm. Sci.*, **14**, 293–300.

58. Pereira, B. G., Fonte-Boa, F. D., Resende, J. A. L. C., Pinheiro, C. B., Fernandes, N. G., Yoshida, M. I., and Vianna-Soares, C. D. (2007).

Pseudopolymorphs and intrinsic dissolution of nevirapine. *Cryst. Growth Des.,* **7**, 2016–2023.

59. Oliveira, G. G. G., Ferraz, H. G., Severino, P., and Souto, E. B. (2013). Compatibility studies of nevirapine in physical mixtures with excipientes for oral HAART. *Mater. Sci. Eng. C,* **33**, 596–602.

60. Li, H., Dong, L., Liu, Y., Wang, G., Wang, G., and Qiao. (2014). Biopharmaceutics classification of puerarin and comparison of perfusion approaches in rats. *Int. J. Pharm.,* **466**, 133–138.

61. Riekes, M. K., Kuminek, G., Rauber, G. S., Cuffini, S. L., and Stulzer, H. K. (2014). Development and validation of an intrinsic dissolution method for nimodipine polymorphs. *Cent. Eur. J. Chem.,* **12**, 549–556.

62. Bartolomei, M., Bertocchi, P., Antoniella, E., and Rodomonte, A. (2006). Physico-chemical characterization and intrinsic dissolution studies of a new hydrate form of diclofenac sodium: comparison with anhydrous form. *J. Pharm. Biomed. Anal.,* **40**, 1105–1113.

63. Agrawal, S., Ashokraj, Y., Bharatam, P. V., Pillai, O., and Panchagnula, R. (2004). Solid-state characterization of rifampicin samples and its biopharmaceutic relevance. *Eur. J. Pharm. Sci.,* **22**, 127–144.

64. Kuminek, G., Rauber, G. S., Riekes, M. K., Campos, C. E. M., Montic, G. A. Bortoluzzi, A. J., Cuffini, S. L., and Cardoso, S. G. (2013). Single crystal structure, solid state characterization and dissolution rate of terbinafine hydrochloride. *J. Pharm. Biomed. Anal.,* **78–79**, 105–111.

65. Tenho, M., Aaltonen, J., Heinanen, P., Peltonen, L., and Lehto, V. (2007). Effect of texture on the intrinsic dissolution behavior of acetylsalicylic acid and tolbutamide compacts. *J. Appl. Crystallogr.,* **40**, 857–864.

66. Issa, M. G., Duque, M. D., Souza, F. M., and Ferraz, H. G. (2013). Evaluating the impact of different variables in the intrinsic dissolution of metronidazole. *Int. J. Pharm. Eng.,* **1**, 17–29.

67. Giorgetti, L., Issa, M. G., and Ferraz, H. G. (2014). The effect of dissolution medium, rotation speed and compaction pressure on the intrinsic dissolution rate of amlodipine besylate, using the rotating disk method. *Braz. J. Pharm. Sci.,* **50**, 513–520.

Chapter 5

Oral Delivery of Poorly Soluble Drugs

Dev Prasad,[a,*] Akash Jain,[b] and Sudhakar Garad[c]

[a] MCPHS University, 179 Longwood Avenue, Boston, MA 02115, USA
[b] Spero Therapeutics, Drug Product Development, 675 Massachusetts Ave., Cambridge, MA 02139, USA
[c] Chemical and Pharmaceutical Profiling, Novartis Institutes for BioMedical Research, 250 Massachusetts Ave., Cambridge, MA 02139, USA
dp1611@gmail.com, ajain@sperotherapeutics.com, sudhakar.garad@novartis.com

Advances in combinatorial chemistry and high-throughput screening in the past couple of decades has enabled discovery of new chemical entities (NCEs) for a variety of complex and diverse biological targets. As the majority of biological targets are highly lipophilic or hydrophobic in nature, there has been a significant increase in the number of NCEs with poor aqueous solubility. While oral delivery continues to be the most commonly used route of administration for NCEs, poor aqueous solubility can result in significant development challenges such as incomplete absorption, highly variable bioavailability, and highly variable pharmacokinetic profiles in preclinical species and humans. A number of conventional and enabling formulation technologies are now available to tackle poor solubility and the resulting poor biopharmaceutical performance of NCEs. However, a systematic evaluation and proactive selection

---

*Current address: Formulation Development, Fresenius Kabi USA, LLC, 8045 Lamon Ave., Skokie, Illinois 60077, USA

Poorly Soluble Drugs: Dissolution and Drug Release
Edited by Gregory K. Webster, J. Derek Jackson, and Robert G. Bell
Copyright © 2017 Pan Stanford Publishing Pte. Ltd.
ISBN 978-981-4745-45-1 (Hardcover), 978-981-4745-46-8 (eBook)
www.panstanford.com

of the optimal formulation technology is typically not available during discovery and early development stages due to limitations in both time and material. As a consequence, a number of promising NCEs are either terminated due to poor biopharmaceutics and lack of adequate exposure for preclinical safety assessment, or remain in the candidate selection phase for several years, resulting in significant delays for advancing new treatments into the clinic. The goal of this chapter is to describe a systematic approach for generating a cross-functional package of data comprised of physicochemical, biopharmaceutical, ADME, PK/PD, and delivery technology evaluations to enable oral delivery of poorly water-soluble NCEs.

5.1 Introduction

The oral route of administration continues to be the most commonly used delivery approach for NCEs due to convenience, ease of administration, large absorption site of the gastrointestinal (GI) tract and cost-effectiveness, especially for chronic indications (Zheng et al., 2012).

Oral delivery of NCEs is primarily a function of three physiological processes: absorption, pre-systemic metabolism and elimination. Each of these physiological processes is governed by the interplay of physicochemical properties (e.g., ionization constant (pK_a), lipophilicity ($\log P/D$), dissolution, solubility, and precipitation/recrystallization), biopharmaceutical properties (e.g., passive permeability, active transport, food effects, efflux and metabolism), and physiological properties (e.g., gastrointestinal pH, transit time, volume, blood flow). Of all the above mentioned properties, aqueous solubility of NCEs is one of the most important physicochemical properties for oral delivery, in particular oral absorption. When administered orally, a compound must dissolve in a predictable manner in the gastrointestinal fluids before it can permeate through the gastrointestinal membrane into systemic blood circulation. For compounds with inadequate dissolution or poor solubility, a number of formulations and enabling delivery technologies are available to enhance their dissolution and/or solubility and improve their oral absorption and/or oral bioavailability. It should be noted that the requirement of sufficient aqueous solubility and dissolution applies mainly to NCEs which require systemic absorption in order to exert their desired pharmacological effect when administered orally. Because the majority of NCEs being developed orally fall into this category, this chapter focuses on technologies and approaches that enable oral delivery for systemic exposure. NCEs and enabling technologies for localized oral delivery are not in the scope of this chapter. In addition, large molecule NCEs, such as peptides and proteins typically do not suffer from poor solubility or dissolution and are thus excluded from discussion herein.

5.2 Oral Delivery

Oral delivery of small molecule NCEs can be broken down into three primary processes:

(a) Oral absorption
(b) Pre-systemic metabolism
(c) Elimination

Oral bioavailability, also defined as the fraction of dose reaching systemic circulation (F), is a function of absorption as well as pre-systemic metabolism (Eq. 5.1) (Kwan, 1997).

$$F = F_a \times F_g \times F_h \tag{5.1}$$

where

F_a = fraction absorbed
F_g = fraction metabolized by gut enzymes
F_h = fraction metabolized by liver enzymes (first-pass effect)

Oral absorption is driven by multiple physicochemical and biopharmaceutical properties. The majority of formulation and enabling delivery technologies are aimed at improving physico-chemical properties and subsequently oral absorption. It should be noted that improving oral absorption (or F_a) does not necessarily translate into improving oral bioavailability (F) for NCEs that undergo significant pre-systemic metabolism (gut and/or liver). Therefore, it is critical for drug development teams to gain an early understanding of factors limiting oral bioavailability of any given NCE before engaging in a large formulation screening effort.

5.2.1 Physicochemical Properties

A thorough understanding of physicochemical properties of an NCE is a prerequisite for predicting the biopharmaceutical performance of its solid form as well as formulations in preclinical and clinical studies. A robust package of physicochemical characterization data using minimal amounts of time and material should be generated during early discovery stages, typically using in silico prediction tools and high-throughput screening technologies to minimize material requirements (Dearden, 2012).

5.2.1.1 Ionization

The ionization constant (pK_a) is a useful thermodynamic parameter to monitor the charge state of NCEs and their impact on oral absorption and oral bioavailability. On the basis of their pK_a, all NCEs can be classified into four major classes: acid, base, neutral and zwitter-ion. Each of these classes of molecules represents different challenges and opportunities in their biopharmaceutical performance upon oral delivery. Molecules with acidic pK_a (3.5–6.0), predominantly exist in their ionized and solubilized form at physiological intestinal pH (5.5–7.4). However, it has been reported and well documented that the ionized form of a molecule tends to have poor passive permeability (Prueksaritanont et al., 1998). As both the ionized and unionized form of a molecule exist in a dynamic/equilibrium process in vivo, the authors believe that a sufficient fraction absorbed (F_a) is achieved in vivo from acidic molecules. The amount absorbed is due to the passive diffusion of the unionized form from the site of absorption, thereby shifting the equilibrium between ionized and unionized moieties toward the generation of more unionized form of the drug molecule. The shift in equilibrium then leads to a higher F_a. Basic molecules demonstrate a much larger range of pK_a (2.0–9.0), depending on the type of functional groups present on the molecule. Strongly basic molecules ($pK_a > 6.0$) are considered ideal for improving solubility and dissolution rates and minimizing in vivo precipitation (and related complications) in the physiological pH range for both I.V. and oral delivery. Such molecules readily form salts with acidic counterions and remain solubilized at the site of absorption. On the other hand, weakly basic molecules ($pK_a < 6.0$) require stronger acidic counterions for salt formation and demonstrate improved solubility/dissolution rates only at pH 4.0 and below. At physiological pH for absorption (pH 5–7), the weakly basic molecules rapidly dissociate from their salt forms and convert to poorly soluble neutral forms resulting in poor and/or variable absorption and bioavailability. Zwitter-ionic molecules are the most complex in their physicochemical and biopharmaceutical behavior. Such molecules typically demonstrate high variability in oral absorption and bioavailability (Thomas et al., 2006). In addition, zwitter-ion pK_a values and solubility profiles are highly sensitive to the presence of additional functional groups in the vicinity of the ionizable groups. For instance, the presence of a strongly electron-withdrawing group next to the ionizable group can lead to a considerable shift in pK_a and result in poor solubility/dissolution. Given a preference, it is better to avoid zwitter-ionic molecules, especially for development, unless they demonstrate high intrinsic solubility and/or high potency (Hurst et al., 2007; Thomas et al., 2006). There are a variety of methods to determine pK_a including in silico and semi-empirical methods (Jelfs et al., 2007; Rupp et al., 2011). A range of experimental approaches with varying throughput, cycle time, sample requirement and cost are available for pK_a determination (Wan & Ulander, 2006; Wang & Faller, 2007; Wang et al., 2007). Some of the commonly used methods include potentiometric titration (Avdeef, 2003), capillary electrophoresis (CE) (Cleveland et al., 1993; Ishihama et al., 2002), Spectral Gradient analysis (Box et al., 2003) and use of the Sirius T3 (Sirius Analytical) instrument.

5.2.1.2 Aqueous solubility

NCE solubility is one of the most important physicochemical properties governing the drug absorption, distribution, metabolism and elimination (ADME) pharmacological and biopharmaceutical profile. Poor solubility and low dissolution rates are the most common rate-limiting factors for oral drug absorption, especially for NCEs with high permeability (Ku, 2008). A variety of solubility and dissolution measurements are performed throughout the discovery

Table 5.1 Solubility and dissolution experiments to guide oral formulation design

Experiment	Application(s)
Shake-flask solubility	pH solubility profile
Solubility in biorelevant media	Prediction of maximum absorbable dose, GastroPlus™ simulations, Build IVIVC, food effect prediction
Organic solvents	Salt and polymorph screening, crystallization process development
Solubility in co-solvents, surfactants, complexing agents and lipids	Develop solubilized formulations for preclinical and clinical studies; develop dissolution media for release testing
Intrinsic dissolution rate	Rank-order solid forms (e.g. salts, co-crystals, polymorphs)
Non-sink dissolution	Rank-order of prototype solid forms and formulations (e.g. solid dispersions, salts)
Dissolution (type II apparatus)	Release testing for solid dosage forms; biorelevant testing and food effect prediction for solid dosage forms

and development stages of NCEs (Avdeef, 2007; Galia et al., 1998; Glomme et al., 2005; Hariharan et al., 2003; Ingels et al., 2002; Kibbey et al., 2001; Lind et al., 2007; Vertzoni et al., 2005). A detailed list of solubility and dissolution experiments useful for oral formulation design that are typically performed during discovery and early development stages are shown in Table 5.1.

An important consideration for oral delivery of a poorly soluble NCE is its tendency to precipitate under physiological conditions. Depending on its solid form and pH-dependent solubility, a compound may precipitate out in stomach fluids (pH $\sim$2) or in small intestinal fluids (pH $\sim$5–7) during its transit through the GI tract. The precipitated compound may dissolve differently in the GI fluids than the original form of the drug dosed. Occasionally, for compounds with pH dependent solubility, certain forms/salts of a compound may dissolve in stomach fluids, and upon gastric emptying into the intestine may remain dissolved, thus yielding supersaturated solutions. In such cases, the compound may demonstrate much higher bioavailability. Thus, it is critical to evaluate dissolution characteristics of various solid forms before selecting a final solid form for further development. Dissolution studies should be conducted either with suspension formulations or with compounds filled in-capsule with or without formulation excipients. The physiological conditions can be simulated by using buffers (pH 1, 2, 4.5, and 6.8) and/or simulated fed/fasted gastric and intestinal media (Jantratid et al., 2008) at 37°C either as a single stage or as a pH-shift method (Mathias et al., 2013). The presence of formulation excipients can help simulate a solid dosage form and can reflect the effects of formulation excipients on the dissolution rate. For compounds with poor solubility, dissolution studies should be conducted in simulated gastric and intestinal media to determine the potential for a food effect.

5.2.1.3 Lipophilicity, permeability, and absorption

Lipophilicity, as expressed by the logarithm of the partition coefficient ($\log P$) or distribution coefficient ($\log D$) of NCEs between a lipophilic phase (e.g., octanol) and aqueous phase is an important physicochemical property in drug development. Compounds with very high lipochilicity ($\log P/D > 6$) or very low lipophilicity ($\log P/D < 1$) can present significant developability challenges, such as poor formulability, poor permeability, poor absorption and accumulation in tissues/organs. Therefore, a balance of lipophilicity in any compound is essential for its optimal oral delivery. Multiple techniques can be used to determine lipophilicity of a compound. While shake-flask is the conventional method for $\log P$ (or $\log D$) determination, the dual-phase potentiometric titration approach is also widely accepted, particularly during the late drug discovery phase (Avdeef, 2012). For NCEs lacking an ionizable group, the HPLC $\log P$ technique, also known as eLog$P$ can be applied (Lombardo et al., 2000). A variety of techniques that are suitable for early discovery include liposome chromatography, immobilized artificial membrane (IAM) chromatography, capillary electrophoresis (CE) (Avdeef, 2012), and artificial membrane preparations (Wohnsland & Faller, 2001).

Permeability plays a vital role in the absorption of orally delivered NCEs and is a complex phenomenon involving multiple mechanisms across the GI mucosa (Artursson & Tavelin, 2003).

Hence, it is important to identify the permeability behavior of NCEs early during lead optimization stages, especially when discovery and development teams are considering an oral route of administration. Unlike solubility, there are few formulation or delivery technologies that can overcome poor permeability of NCEs and provide significant improvement in their oral absorption or bioavailability (Maag, 2012; Maher & Brayden, 2012).

One of the simple parameters that correlate well with permeability is the molecular weight of an NCE. As per Lipinski's rule of 5, NCEs with molecular weights less than 500 Da, typically demonstrate medium to high passive permeability. On the other hand, NCEs with molecular weights greater than 500 Da usually tend to exhibit medium too poor to almost no passive permeability (Lipinski et al., 2001). Another in silico model used in pre-formulation screening is the one developed by Egan and coworkers (Egan et al., 2000). Their absorption model is based on polar surface area (PSA) and calculated $\log P$ (Clog$P$) and considers the physical processes involved in membrane permeability and the inter-relationships and redundancies between available descriptors. Based on their analysis, it was shown that the majority of highly permeable marketed drugs were populated in an area with PSAs $<$ 120 angstroms and Clog$P$s between -1 and 6.

In addition to the in silico methods, a number of in vitro models are available to predict permeability as well as to assess the contributions of active transporters in the permeation process (Balimane et al., 2006; Hämäläinen & Frostell-Karlsson, 2004). A few important assays commonly used in the last few decades for measuring permeability of NCEs with high speed, reliability and reproducibility include the parallel artificial membrane permeability assay (PAMPA), Caco-2 and MDRI-MDCK.

Given the advantages and limitations of each approach, the latest consensus appears to favor a strategy that combines all three approaches with in silico models to ensure high quality assessment of permeability in early discovery (Balimane et al., 2006; Faller & Ertl, 2007; Kerns et al., 2004). PAMPA should serve as a fast and high-throughput permeability ranking tool in particular for scaffolds using passive diffusion mechanisms. The Caco-2 model should be applied to challenging scaffolds involving active transport mechanisms or scaffolds with higher molecular weight (e.g., > 600 Da). Caco-2 mechanistic studies are valuable to identify the major transporters, such as Pgp, MRP2 and BCRP and to appraise the impact of shutting down active transporters via either inhibitory (Varma et al., 2003) or saturation mechanisms (Bourdet & Thakker, 2006). MDCK, expressed with a specific transporter, may be ideal to tackle the impact of an individual transporter subsequent to Caco-2 transporter assays (Varma et al., 2005).

5.2.1.4 Metabolism consideration

Metabolism, though not a physiochemical property, is an important intrinsic property that drives the clearance or elimination of NCEs. Metabolism is a predominant factor that differentiates oral absorption from oral bioavailability. For instance, a compound could have complete absorption ($\sim$100% F_a) through the GI tract upon oral dosing based on its high solubility and permeability. However, the oral bioavailability of the compound may be only $\sim$20–30% due to its high first-pass metabolism prior to reaching systemic circulation. This is an extremely important consideration when developing formulations for poorly soluble NCEs. The majority of solubility enhancing formulation technologies can potentially maximize the absorption (F_a%) of a given NCE, but not necessarily its oral bioavailability.

Two of the major sites of metabolism for orally administered compounds are the GI tract (sometimes referred to as pre-systemic metabolism) and the liver (also known as first-pass metabolism). Identifying compounds with the right balance of metabolic stability is critical for oral delivery of NCEs, as high metabolism (or rapid clearance) could result in poor bioavailability and poor efficacy. Alternatively, very low metabolism, along with entero-hepatic circulation, could lead to prolonged half-life and accumulation of NCEs in the body resulting in undesirable side-effects. Metabolism is a highly species-dependent phenomenon; metabolic rates can vary significantly among different species, due to the presence of unique metabolizing enzymes in each species, strain and gender (Martignoni et al., 2006). For example, CYP3A4 is the most important metabolizing enzyme in humans and current literature suggests that more than 50% of marketed drugs are metabolized by the CYP3A4 enzyme. However, CYP3A4 is not found in any pre-clinical species, including monkeys (Martignoni et al., 2006). The closest enzyme to human CYP3A4 is the mouse CYP3A11. Typically, rodents have a higher metabolic rate than dogs, monkeys, and humans (Kleiber, 1947). However, the rate and extent of metabolism and ranking of species can vary significantly from one chemical structure series to another. In particular, for compounds that exhibit significant species differences in their metabolism, it is very difficult to extrapolate the PK parameters in humans and obtain a reasonable estimate of human dose. In such instances, it is advisable to conduct micro dosing or exploratory studies in humans as early in development as possible. Screening of metabolic stability in multiple animal species early in drug discovery is very useful to guide structural modification and selection of compounds for in vivo studies. Metabolite identification is also very helpful for bioanalytical and pharmaceutical scientists to understand metabolically labile as well as chemically labile sites in an NCE.

5.2.2 Solid-State Properties

Understanding the form of a compound, and thereby the physical properties of that form in the solid state is critical to developing and formulating a drug. Solid-state characterization assays and technologies are well known to most scientists responsible for solid form selection in support of CMC development activities. Thermogravimetric analysis (TGA), differential scanning calorimetry (DSC) and powder and single-crystal X-ray diffraction are commonly used tools to characterize the solid forms of any given NCE using small amounts of material. Advanced calorimetric methods, such as modulated DSC (mDSC) provide higher sensitivity and can measure the heat of fusion of crystalline solids as well as glass transition events in amorphous solids. Hot stage and infrared microscopy are useful tools to study solid form conversions (e.g., polymorphs). Dynamic vapor sorption (DVS) is routinely used to measure hygroscopicity behavior of solid forms and provides useful information for solid form selection activities.

5.2.2.1 Morphology

Polymorphism studies require special attention during oral formulation development of poorly soluble NCEs because changes in physical form during development could result in significant impact on their dissolution rate, solubility and oral absorption and often, it is very difficult to reproduce the original form with optimal physicochemical and biopharmaceutical performance. Norvir® (Ritanovir) is a classic example of polymorphism affecting the bioavailability. Originally thought to have a single crystal form, this poorly absorbed molecule was formulated as a soft gel capsule containing an ethanol/water solution of the molecule. However, two years after market introduction, several batches failed dissolution specifications because a new crystal form precipitated out of solution with ∼50% lower intrinsic solubility. The product had to be withdrawn from market and reformulated in an oily vehicle (Bauer et al., 2001).

Amorphous materials have become more prevalent in the development pipeline as a means of overcoming the poor solubility of NCEs. Generation and stabilization of amorphous solid forms using polymeric matrices in solid dispersions provide an attractive formulation approach to improve the dissolution, solubility and oral bioavailability of these hydrophobic NCEs. Active pharmaceutical ingredients (APIs) can also be developed as amorphous forms if supported by a detailed understanding of the amorphous system and a robust scalable process for manufacturing. This approach often becomes inevitable, especially when NCEs are difficult to crystallize due to molecular complexity or the presence of trace impurities that act as crystallization inhibitors. Amorphous materials, with increased dissolution rate and aqueous solubility are chemically reactive and more hygroscopic than crystalline material (Byrn et al., 1999; Hancock & Parks, 2000; Hancock & Zografi, 1997). Amorphous materials exist in either the glassy state below their glass-transition temperature (T_g) or as a supercooled liquid above their T_g. Although the physical properties differ between each amorphous state, Arrhenius relationships are applicable below the T_g and allow for extrapolation to ambient storage/handling conditions. For such calculation, the T_g (in Kelvin) can be approximated to 2/3 (i.e., 0.67) of the melting point (in Kelvin) of crystalline material (Wolynes & Lubchenko, 2012). This provides a good estimate of the T_g and thereby feasibility and likelihood of successfully developing an amorphous API or formulation for a given NCE. Analysis of amorphous forms of drug candidates should include, in particular, the measurement of the T_g (most commonly using mDSC) and any changes in water solubility, hygroscopicity and solid-state stability relative to the crystalline form. Water solubility may be the most difficult parameter to measure for an amorphous material because of rapid crystallization to the more thermodynamically favored form (Hancock & Parks, 2000). In summary, the know-how and experience in development and characterization of amorphous APIs as well as solid dispersion formulations have progressed significantly in the last decade and more than a dozen products have been launched in the recent past using these technologies (Sanghvi et al., 2010).

5.2.2.2 Hygroscopicity

A detailed evaluation of the hygroscopicity of NCEs and various solid forms is essential for optimal physicochemical behavior (e.g., solubility, physical and chemical stability) as well as processability and manufacturing (e.g., control and reproducibility of desired polymorphic form). Hygroscopicity, simply defined as the study of moisture uptake as a function of percent relative humidity, can be measured in an automated manner with dynamic moisture sorption analyzers that quickly assess the material in a closed system at controlled temperature and ambient or controlled pressure. These instruments allow the measurement of the weight change kinetics and equilibration for small samples exposed to a stepwise change in humidity. Hygroscopicity evaluation should start with an independent determination of the initial moisture content (TGA, Karl Fischer, etc.). Analysis of XRPD patterns of solid forms before and after DVS measurement also provides very useful information pertaining to detecting solid form transitions, especially the commonly observed transitions of dehydration and/or hydrate formation.

5.2.2.3 Particle size, shape, and surface area

Particle size and distribution is an important solid state property that heavily impacts the dissolution behavior, flowability and processability of APIs. Low-magnification scanning electron microscopy (SEM) provides a simple record of particle size and shape. A number of automated particle-size measurement methods such as laser diffraction and dynamic light scattering (DLS) are now available, each with its inherent shape limitations. Usually, particle size measurement methods calculate particle size distributions by normalizing the shape to an equivalent spherical particle (Merkus, 2009). Based on the actual morphology of the particles (e.g., needle-shaped long crystals), such measurements can be limited in their accuracy. Nonetheless, these methods allow the counting of many particles in a short period of time and provide good quality control feedback on the reproducibility of a manufacturing process. Surface area analysis methods, such as Brunaur–Emmett–Teller (BET) also provide useful insight into changes in available surface area due to changes in chemical processing. The particle size recommendation for development is derived from the type of dosage form and the impact of particle size on dissolution and fraction absorbed/bioavailability. For compounds that have low bioavailability due to low permeability, reducing the particle size may have no effect on the percent of dose absorbed, but may be necessary from a manufacturing stand point in order to affect blend homogeneity and content uniformity of the formulated product. The most significant effect of particle size on absorption is typically observed for low dose–low solubility compounds where absorption may be dissolution limited. Particle size reduction and the corresponding increase in surface area should increase the rate of dissolution. A number of in silico methods (e.g., Noyes–Whitney equation, dissolution number using GastroPlus™) can be applied during discovery stages to identify the need for particle size reduction in order to improve dissolution and bioavailability. Such early guidance can be extremely useful to API and formulation development teams so that appropriate particle size reduction technologies can be incorporated in the manufacturing processes as early as IND-enabling or phase I stages. However, it should be noted that particle size recommendations may change as the clinical dose is refined based on the outcome of early clinical studies (phase I/IIa).

Overall, solid-state properties of any given NCE are very critical for understanding the NCE's biopharmaceutical performance and manufacturability in a suitable dosage form. Although sub-optimal solid state properties by themselves do not create a 'no-go' scenario for NCEs, it is important for research and development teams to realize that selecting a candidate with poor solid-state properties may require up-front investments of time and resources in selecting an optimal solid form for development and more often than not, result in higher risk and longer development timelines.

5.2.3 Biopharmaceutics and Developability Assessment

The interplay between physicochemical and solid state prop-erties with the biopharmaceutical performance of a compound must be considered when selecting the form and formulation for oral delivery of poorly soluble NCEs. With advancements in simulation/ modeling software, it is easy to simulate the in vivo absorption and exposure in humans as a function of dose based on physiochemical properties such as solubility, permeability, dissolution and PK data in animals (Dearden, 2012; Sliwoski et al., 2014). These are approaches, which could save time and valuable resources spent in actual experimentation. Moreover, these techniques could help identify potential issues early on. An example of such simulation software is GastroPlus^{TM}, which is useful in predicting the effect of particle size on absorption as a function of dose (Kuentz et al., 2006). The prediction could be valuable, especially for compounds with poor solubility and high permeability, where size reduction could lead to enhanced dissolution rate and oral bioavailability. Thus, such simulation tools could prove to be helpful for the formulator in evaluating the right candidate for particle size reduction techniques such as nano-/ micro-milling. Presently, several important simulation software programs, apart from the one already mentioned are at different stages of their evolution and validation (Kostewicz et al., 2014a; Kostewicz et al., 2014b). Use of simulation techniques can lead to higher productivity, better utilization of resources, faster screening and better prediction of possible issues in humans.

In addition to the more sophisticated modeling tools, simple calculations such as maximum absorbable dose (MAD) are very helpful tools for developability assessment to predict risk of poor oral absorption in humans and define formulation goals to overcome such issues. The MAD equation, as shown below (Eq. 5.2), takes into account the permeability, converted to an absorption rate constant (K_a, min^{-1}) (Hilgers et al., 2003), solubility (typically measured in biorelevant media, such as simulated intestinal fluid at 1 hour, 37°C), GI residence time (270 min in humans), and GI volume (250 mL in humans). Although the original MAD equation was developed to predict oral absorption limitations in humans, similar predictions can be made in different animal species based on their respective physiologies, especially GI residence time and GI volume. A list of such parameters for different animal species is shown in Table 5.2.

Table 5.2 Physiological parameters in multiple species to calculate maximum absorbable dose (MAD)

Parameters	Mouse	Rat	Dog	Monkey	Human
Intestinal Volume (mL)	1.5	11.23	480	230	250 (small intestine)
Transit time (min)	90	88	111	180	275
Average body weight (kg)	0.04	0.3	10	5	70

$$MAD = (K_a)\,(S)\,(SITT)\,(SIWV) \tag{5.2}$$

where

MAD = maximum absorbable dose (mg in human or mg/kg in animal species)

(Note: To calculate MAD for animal species, divide the MAD by animal body weight (kg) to obtain MAD in units of mg/kg)

K_a = absorption rate constant in min^{-1} (0.05 for medium to highly permeable compounds)
S = solubility in mg/mL (kinetic solubility is more relevant for absorption)

Table 5.3 Maximum absorbable dose (in mg) in preclinical species and human

Solubility (mg/mL)	Mouse	Rat	Dog	Monkey	Human
0.0001	0.016875	0.016471	0.02664	0.0414	0.34375
0.005	0.84375	0.823533	1.332	2.07	17.1875
0.01	1.6875	1.647067	2.664	4.14	34.375
0.015	2.53125	2.4706	3.996	6.21	51.5625
0.5	84.375	82.35333	133.2	207	1718.75
0.75	126.5625	123.53	199.8	310.5	2578.125
1	168.75	164.7067	266.4	414	3437.5

SIWV = small intestine volume (250 mL in human)
SITT = small intestine transit time (270 minutes in human)

The MAD as a function of compound solubility for preclinical species and humans is presented in Table 5.3.

5.3 Enabling Formulation Approaches for Poorly Soluble Drugs

Prior to the selection of a clinical formulation for a poorly soluble NCE, a number of enabling formulation approaches are usually explored for developing toxicology formulations. However, preclinical toxicological doses are very high and depending on clinical SAD/MAD doses, the formulation principles intended for clinical studies should be considered proactively. A simple approach based on pharmacokinetic evaluation of a compound at pharmacological dose with a solution and suspension formulation can serve as a guiding tool in selecting a clinical formulation principle (Table 5.4). On the basis of this evaluation, the enabling approaches described in the sections below can then be considered.

5.3.1 Conventional Approaches

5.3.1.1 pH adjustment and buffers

The Henderson–Hasselbalch equation describes the relationship between pH, pK_a, and relative concentrations of the ionized and

Table 5.4 Prediction of the clinical formulation principle from solution vs. suspension pre-clinical PK

Parameter	Scenario 1	Scenario 2	Scenario 3	Scenario 4
Oral Bioavailability from Solution	High (> 80%)	High (> 80%)	High (> 80%)	Low (< 30%)
Oral Bioavailability from Suspension	High (> 80%)	Low (< 30%)	Medium (40–60%)	Low (< 30%)
Conclusion	Easy to develop	Exposure limited by solubility	Exposure limited by dissolution rate	Permeability, efflux, metabolism
Clinical Formulation Principle (also consider MAD, food effects)	Powder-in-bottle, capsule or tablet	MEPC, solid solution	Salt, co-crystal, dispersions, micro or nano particulate systems	Low likelihood of improving exposure using formulation technologies
% NCE/BCS	5–15 (I/III)	60–75 (II)		< 5–10 (IV)

unionized forms of a compound in solution. The solubility of the ionized form is generally much greater than the solubility of the neutral form, and so the pH is modified in the direction of greater ionization in order to achieve solubilization. Chemical stability of the compound at the desired pH should also be verified. For compounds with high pH-dependent solubility in the pH range of the gut (pH 2–7), buffered systems may be used for short-term studies to assess and minimize the risk of precipitation within the GI tract. Dilution tests are typically performed using 1 part of formulation: 3 parts of physiological buffer for oral assessment and 1 part of formulation: 20 parts of physiological buffer for IV assessment or a serial dilution can be performed to access the precipitation behavior (Li & Zhao, 2007). For parenterally administered formulations with a pH outside of the physiological range of pH ∼6–8, it is important to make sure that the buffer capacity of the formulation is sufficiently low (less than 50 mM) so as not to cause venous irritation. For basic compounds passing from the region of pH_{max}

(pH of maximum solubility) to regions of higher pH during GI transit, the equilibrium solubility at the corresponding pH could induce precipitation. However, depending on the intrinsic properties of the compound, or the composition of the local environment, it is fairly common for compounds to remain in a metastable state, known as supersaturation for a significant duration of time (Brouwers et al., 2009). Supersaturation increases the thermodynamic activity in the GI tract leading to enhanced flux and subsequent exposure (Pole, 2008). However, supersaturation can also lead to precipitation of an acidic drug in the stomach, and a basic drug in the intestine, possibly leading to low and/or variable exposure. Another approach for obtaining a solution formulation via pH adjustment is the formation of an in situ salt, which is accomplished by adding molar equivalents of an acidic or basic counterion to the free form of an ionizable compound. This technique takes advantage of the different equilibrium constants (K_a) that arise when different ions are present in a saturated solution of an ionic compound (Tong & Whitesell, 1998). It is also useful when compounds prove difficult to formulate due to solid state issues encountered with the free form, such as poor suspendability or stickiness of the material in the formulation media. While the selection of an appropriate salt form is an important aspect of drug development, it is not usually practical to conduct a traditional salt screen in the early discovery hit to lead phase when only 5–20 mgs of API is available. This challenge is easily overcome by the formation of an in situ salt. Typically, the pK_a of the counterion selected should be approximately 2 pK_a units away from that of the free form. For a basic compound, the counterion should have a pK_a that is at least 2 units lower than the free form while for an acidic compound, the pK_a should be at least 2 units higher than the free form. Due to the small amounts of material required, a number of different counterions can be screened rapidly in order to select the most desirable for use.

5.3.1.2 Co-solvents and surfactants

Co-solvents and surfactants are very commonly used in formulation vehicles, especially for initial pharmacokinetic and pharmacology studies. The high amounts (>40–60%) of surfactants and co-solvents necessary to achieve solubility enhancements typically required for very high dose pharmacokinetic, pharmacology, and toxicity studies are generally poorly tolerated due to local GI effects and consequent effects on electrolytes and body weight over time. Therefore, their use here has been limited to short-term exploratory studies. Co-solvent or surfactant based formulations may be used for shorter term studies that generally do not exceed 3–5 days, but these excipients are not preferred. However, excipients are readily available and fairly simple to prepare. The use of excipients must always be based on a good understanding of the risk of drug precipitation upon oral as well as parenteral administration.

Co-solvents alter the polarity (dielectric constant) of aqueous systems to provide a more favorable solubilization environment for nonpolar/less hydrophobic molecules. Most co-solvents are characterized by hydrogen bond donor and acceptor groups that interact strongly with water and help ensure mutual miscibility in any proportion. Co-solvents also have small hydrocarbon regions that do not interact strongly with water which may help to interact with the hydrocarbon region of the molecule. As a result, co-solvency is a highly versatile and powerful means of solubilizing nonpolar solutes in aqueous media (Williams, 2000). Typical co-solvents include propylene glycol, N-methyl pyrrolidone (NMP), 2-pyrrolidone, dimethyl sulfoxide (DMSO), polyethylene glycol (PEG) 200, 300 and 400, and dimethylacetamide/dimethyl formamide (DMA/DMF). The applications and possible side effects of the common co-solvents are discussed in detail in the literature (Kawakami et al., 2006; Rubino, 2006; Spiegel & Noseworthy, 1963).

Surfactants are amphiphilic molecules containing both hydrophilic and hydrophobic regions. In aqueous solutions at concentrations above a critical micelle concentration (CMC), they form aggregates, such as micelles, where the hydrophilic region is oriented toward the bulk media and the hydrophobic region is oriented toward the core. Surfactants can be useful additives when compound hydrophobicity is the limiting factor in solvation in aqueous media, or when the molecule itself is amphiphilic. In certain cases, the combination of surfactants may be used to enhance the drug solubility as high concentrations of single surfactants can possibly cause adverse effects. It is common to investigate systems containing two or more structurally unrelated surfactants since their combination can lead to the formation of mixed micelles with drug solubilization properties that are greater than the simple additive effect of the two surfactants. Surfactants are also used as a component of lipid dosage forms and nanosuspension formulations as well as in combination with other techniques such as co-solvent and pH adjustment. Therefore, surfactants can play a critical role in liquid as well as solid dosage formulations to maintain supersaturation and avoid precipitation. Surfactant formulations are also prone to precipitation as they are diluted in situ after administration or diluted before administration (IV formulations). Therefore, understanding the potential for loss of solubilization capacity upon dilution is an important aspect of developing these formulations. The dilution can cause drug precipitation when the concentration of surfactant approaches the CMC and when the formulation contains a level of drug close to the maximum solubilization capacity of the formulation (Lawrence, 1994). Overall, surfactant formulations are much less susceptible to precipitation and are commonly added to both co-solvent and buffered formulations to reduce the risk of precipitation, regardless of an effect on intrinsic solubility. A few commonly used surfactants include polysorbates (e.g., Tween 80), polyoxyl castor oil (Cremophor EL), and sodium lauryl sulfate (SLS) (Kawakami et al., 2006). In general, the primary challenge with the use of surfactants in preclinical formulations is the large amount of excipient required for solubilization, which can lead to tolerability issues. Hence it is important to use allowable/tolerable concentrations depending on the highest dose required for the studies.

5.3.1.3 Cyclodextrins

Cyclodextrins are cyclic sugars and their use in drug solubilization has been reviewed extensively (Kurkov & Loftsson, 2013; Loftsson & Brewster, 2010; Stella & He, 2008). Three different types of cyclodextrin (alpha, beta, and gamma) are available; the most commonly used type is the beta class. Cyclodextrins possess a hydrophilic exterior and a hydrophobic core, and therefore the primary mechanism of solubilization is due to the ability of these agents to form noncovalent, reversible inclusion complexes with hydrophobic/lipophilic drugs. If the cyclodextrin–drug complex results from a 1:1 interaction, solubility increases linearly as a function of cyclodextrin concentration. The primary advantage of complexation is a reduced risk of drug precipitation upon dilution. Upon administration, dilution and competitive binding with plasma components are the major driving forces for dissociation of the complex (Kurkov et al., 2012; Okimoto et al., 1999). In most cases, dissociation is complete, providing rapid release of the drug. However in a few cases, where the drug–cyclodextrin binding constant (K) is reported to be very high ($> 1 \times 10^5$ M^{-1}), an effect on drug disposition has been observed (Charman et al., 2006). The importance of the binding constant has been detailed in an excellent review (Carrier et al., 2007) and the authors state that most poorly soluble drugs will have increased oral bioavailability when dosed as a cyclodextrin complex, provided the binding constant is low ($< 10^4$ M^{-1}). Cyclodextrins most commonly used in discovery and development are 2-hydroxypropyl-β-cyclodextrin (HPβCD) and sulfobutylether-β-cyclodextrin (SBEβCD). These cyclodextrins are highly water-soluble with solubility of > 500 mg/mL and have been extensively characterized with regard to safety and physicochemical properties. While both cyclodextrins are found in US marketed parenteral formulations, preclinical data suggests SBEβCD may be preferred for parenteral administration due to lower in vitro hemolysis compared with HPβCD (Shiotani et al., 1995). As previously mentioned, the mechanism by which cyclodextrin improves solubility of a drug is by complexation. In some cases, complexation also helps to improve the stability of the drug. Through the use of nuclear magnetic resonance (NMR), it has been shown that the hydrophobic portion of the NCE with the appropriate molecular size and polarity is encapsulated by the cyclodextrin molecule and the hydrophilic portion of the cyclodextrin remains available for hydration, thereby solubilizing the NCE (Jullian et al., 2010; Uekama et al., 1998). The complex formation is reversible in solution and thus precipitation of the NCE during dissolution is very rare. There are several techniques to prepare inclusion complexes, including solvent evaporation, co-precipitation, slurry complexation, extrusion, damp mixing and heating, dry mixing and paste complexation. The key factor in formulating drug-cyclodextrin complexation is the contact between the active and the cyclodextrin. The time required for the formation of the complex varies from several minutes to several hours. In the case of solvent evaporation, both the API as well as cyclodextrin should dissolve and form a clear solution. Removal of solvent by slow evaporation results in formation of the complex. Further drying removes traces of residual solvents. In this technique, both active and cyclodextrin should have adequate solubility in the selected solvent system, such as methanol or ethanol or a combination of the two (Jantarat et al., 2014). The resulting complex can be filled into a capsule or compressed into a tablet. A second method for complex formation is co-precipitation, where cyclodextrin is dissolved in water and the active is added to the solution with stirring. Solubility of the cyclodextrin/active increases as the reaction proceeds, and the resultant solid is filtered after cooling, washed with water and dried. Unfortunately, this method has limited scale-up potential due to the limited solubility of cyclodextrin; a large volume of water is required. The third technique is slurry complexation where 50–60% solid is added and stirred over a period of time. As the complex is formed it goes into solution, the aqueous phase gets saturated and the complex precipitates out. The precipitate is filtered and washed with solvent. It is very important to confirm complex formation during this process. Techniques such as thermal analysis (DSC) or spectroscopic analysis (IR) can be used for confirmation of the solid state. NMR, Circular Dichroism (CD) spectroscopy or simpler techniques, such as solubility studies and pH-potentiometric titration can be used to study inclusion complexes in solution (Loftsson & Brewster, 1996; Singh et al., 2010). The main advantage of these techniques over co-precipitation is that less water is required. Typical cyclodextrin concentrations in both preclinical and clinical formulations are approximately 2–20 % w/v. For example, the amount of cyclodextrin required for solubilization, given a target drug solubility of 10 mg/mL, a drug molecular weight of 500 Da, and a 1:1 complex formulation, is 5% w/v SBEβCD (average MW 2,163 Da) or 2.8% w/v HPβCD (average MW 1,400 Da). Cyclodextrin complexation is often used along with complementary approaches such as pH adjustment and low levels of water-soluble polymers such as PVP,

PEGs, and HPMC to improve the extent of solubilization (Challa et al., 2005). In particular, SBEβCD carries a negative charge at physiological pH due to the low pK_a of the sulfonic acid groups. As a result, through charge attraction, the cationic form of a drug may bind better to SBEβCD than the neutral form.

5.3.2 Salts and Co-crystals

As discussed in Section 5.2.2 the form of a molecule can have a significant impact on the physicochemical and physical properties of a material. Arrangement of molecules within a solid is responsible for the form the material. Alteration of the molecular arrangement or interactions results in polymorphs which can change the properties of the material. The form can also be altered by introducing interactions with other moieties resulting in salts, hydrates, solvates, and co-crystals. By changing the physical form of the drug, properties such as solubility can be enhanced.

Traditionally, for ionizable drugs (anionic, cationic and zwitter ionic), salt formation is the simplest and most cost-effective approach to addressing poor water solubility and thereby enhancing bioavailability. For non-ionizable or weakly ionic compounds, modification of physical properties through form change can be achieved through formation of a co-crystal.

Formation of a salt or co-crystal form can address numerous problems including poor solubility, stability issues, poor crystallinity, flow properties and/or compressibility of the free form of the drug candidate. Identification of a superior form early in the product development life cycle can greatly improve the efficiency of drug development. A summary of salt or co-crystal selection criteria is provided in Table 5.5. Additionally, identification of the preferred and/or superior form for drug development can strengthen intellectual property and freedom to operate. Both the salt and co-crystal approaches are discussed below.

5.3.2.1 Salts

The term pharmaceutical salt is used to refer to an ionizable drug that has been combined with a counterion to form a neutral

Table 5.5 Selection criteria for salt or co-crystal development

Salt/co-crystal criteria	Improvement as compared to free form	Comments
Crystallinity	Greater than 99.5% crystalline	
Kinetic solubility/ dissolution rate	Minimum 5–10-fold	
Equilibrium solubility at physiological pH (4.5–7.4)	Depending on pK_a of a molecule, minimum 2 fold	
Two step dissolution in biorelevant media	• Should maintain at least twofold higher solubility after pH change for minimum 30 min • It should crash out as an amorphous nanoparticles of a free form, which may re-dissolve further	
Fasted/fed solubility	Ideally no difference in fasted and fed solubility	Counterion may change pH, need to make sure difference is not due to change in media pH
Stability	Stable at accelerated conditions	
Hygroscopicity	Less than 2% up to 50% RH	
Polymorphism	If compound exists in multiple polymorph or hydrate forms, need to select the most stable, reproducible polymorph	
Melting point	Above 80°C	For lower MP enthalpy should be high > 110 J/g (e.g. Ibuprofen)
Crystal Habit/size	Needle, prism, spherical	Can be controlled during crystallization
Compressibility	Good compressibility	
PK data (rodent/non-rodent)	Minimum 2-fold	
PK dose linearity (non-rodent)	Low dose: $r > 0.8$ High dose: $r > 0.6$	

complex through an ionic bond. Pharmaceutical salts provide a simple approach to overcoming undesirable physicochemical or biopharmaceutical properties of a drug candidate. A salt form is useful when the free acid/base has undesirable properties such as low solubility in water, poor crystallinity and form control, low melting point, high hygroscopicity, low chemical stability, etc. Use of salt forms can provide an effective method for increasing solubility and dissolution rates of drugs with ionizable functional groups. Salts of acidic and basic drugs have, in general, higher solubility than their corresponding acid or base forms. Salts are formed via proton transfer from an acid to a base by pairing a basic or acidic drug molecule with a counterion. In a solution, a stable ionic bond can be formed when the difference of pK_a between an acid and a base is greater than 3 (Sarma et al., 2009). Salts are formed when a compound that is ionized in solution forms a strong ionic interaction with an oppositely charged counterion, leading to crystallization of the salt. The ionic bond formed between the drug and counterion leads to different solid-state and physicochemical properties and upon dissociation in polar aqueous solvents, leads to stronger interactions between the charged API and solvent. The solubility and dissolution rates of a salt are governed by the solubility products of the API and the counterion.

The most effective salt is determined through a screening process in tangent with polymorph screening for pharmaceutical development. Potential counterions are chosen based on pK_a differences and toxicological/safety considerations. Currently, more than 50% of the oral and parenteral products on the market are manufactured as a salt form of the active ingredient (Kumar et al., 2008). Around 69 cations and 21 anions are generally recognized as safe (GRAS): the most common cations being sodium, calcium, potassium and magnesium and the most common anions being hydrochloride, sulphate, hydrobromide, tartrate, mesylate (methansulfonate), maleate, and citrate. Generally, salt screening is done in a two-step process: first a large number of counterions are screened on microscopic scale followed by characterization to identify promising candidates. Desirable properties of the salts include crystallinity, high water solubility, low hygroscopicity, good chemical stability, and high melting point. After the selection, larger

Table 5.6 Compound ABC-123456-A (weak base) solubility in 1 M acidic counterion solutions in water

Counterions	Approx. solubility (mg/mL)
Hydrochloric acid	0.21
Sulfuric acid	0.38
Methane sulfonic acid	3.43
Ethane sulfonic acid	1.5
Phosphoric acid	0.12
Toluene sulfonic acid	1.6
Benzene sulfonic acid	1.8
Maleic acid	0.23
Fumaric acid	0.16
Oxalic acid	0.06

batches are prepared for more detailed characterization. Screening of in situ salts often is readily achieved by evaluating the solubility of a given compound in a solution of acidic or basic counterions at 0.1 M or 1.0 M stock solution. Examples of solubility enhancement for a proprietary compound are shown in Table 5.6 and Table 5.7 using acidic and basic counterions respectively. In situ assessment allows for rapid selection of counterions for use in preclinical studies and also provides insight into which counterions to pursue for further scale-up and development of a solid form.

Table 5.7 Compound ABC-123456-B (weak acid) solubility in 1 M basic counterion solutions in water

Counterions	Approx. solubility (mg/mL)
Sodium hydroxide	300
Potassium hydroxide	380
Calcium hydroxide	5.0
Magnesium hydroxide	3.8
Arginine	160
Lysine	180
Glycine	220
Diethylamine	48
Tert-butylamine	62
Ethylenediamine	42
Benzathine	35

5.3.2.2 Co-crystals

The discussion on how to properly define co-crystals, as this is a newer approach in pharmaceutical development, is still ongoing. However, there is consensus that the definition of a co-crystal is crystalline material consisting of at least two different solid components at room temperature (Childs et al., 2007; Schultheiss & Newman, 2009). In recently issued FDA guidelines on pharmaceutical co-crystals, the FDA defines co-crystals as "solids that are crystalline materials composed of two or more molecules in the same crystal lattice." Furthermore, the regulatory agency considers a co-crystal as a special form of a given molecule and not as a new API (FDA, April 2013). Structurally, co-crystals consist of an API and a stoichiometric amount of a pharmaceutically acceptable co-crystal former (known as co-former). There can be various types of interactions between an API and co-formers including hydrogen bonding, pi-stacking, and van der Waals forces. Co-crystals provide an opportunity to address physical property issues such as solubility and stability without changing the composition of an API. Co-crystals can also provide a more stable form of the API compared to other techniques, such as amorphous formation. The difference between salts and co-crystals is subtle; in salts, the carboxyl proton is transferred to the hydrogen atom of the base while in co-crystals the carboxyl of the acid remains protonated (i.e., for salts, the components in the crystal lattice are in an ionized state while in the case of a co-crystal, the components are in a neutral state and interact via non-ionic interactions) (Elder et al., 2013).

Evaluation of API properties such as the number of hydrogen bond donors and acceptors, propensity for salt formation (pK_a), and conformational flexibility can provide valuable input in developing co-crystals. Low molecular weight APIs with highly symmetrical structure possessing strong non-bonded interactions are good candidates for co-crystal formation. Selection of co-formers is based on hydrogen bonding rules (Etter, 1990), probable molecular recognition events (Fábián, 2009), and the toxicological profile of the co-former. A few examples of co-formers include saccharin, nicotinamide, and acetic acid (Childs et al., 2008; Di Profio et al., 2011; Zhang et al., 2013). Because of toxicological and regulatory requirements, co-formers with known safety profiles are preferred. Further, there are some examples of co-crystals of two active pharmaceutical ingredients in literature as well (Arenas-García et al., 2010; Vishweshwar et al., 2005).

After the selection of various co-formers, methods such as melt-crystallization, grinding and re-crystallization from different solvent systems can be employed to develop co-crystals (Childs et al., 2008; Dhumal et al., 2010; Friščic et al., 2006; Kojima et al., 2010; Lu et al., 2008). The choice of solvent systems depends on the solubility of the API and co-former. Similar to salt and polymorph screening, high-throughput screening methods are usually designed to screen solvents and co-formers.

Similar to salts, the solubility of the co-crystal is governed by the solubility product of the API combined with the co-former. The solubility of the co-crystal may or may not be greater than that of the API. Practically speaking, co-formers with higher solubility tend to increase co-crystal solubility. Furthermore, co-crystals may improve other physicochemical characteristics including hygroscopicity and mechanical properties. The physical and chemical properties of a co-crystal need to be investigated in the same manner as any other solid form in order to determine developability into a marketed dosage form. Physicochemical properties, such as crystallinity, melting point, solubility, dissolution, and stability, are important when moving a new compound, such as a co-crystal, through early development.

5.3.2.3 Identifying the right salt or co-crystal

A quick and effective way to identify the need for salt or co-crystal form profiling is to perform a pharmacokinetic study in rodents comparing a solution and a suspension formulation of the free acid or base form of an NCE. If the suspension studies demonstrate approximately a two-fold reduction in systemic exposure as compared to the solution studies, such a compound could benefit from salt/co-crystal profiling. As previously mentioned, salts and co-crystals can improve the dissolution rate (kinetics), thus it is important to understand the impact improved kinetic solubility could have on biopharmaceutical performance. In particular, understanding the interplay between pH solubility, the region of GI absorption and GI transit time should be considered. For example, consider a scenario where improving the dissolution rate in the upper GI tract, thereby increasing dissolved concentrations, results in precipitation of a free form in the lower GI tract upon pH shift. This would not necessarily improve biopharmaceutical performance or bioavailability. These parameters can be very well characterized by a two-stage dissolution evaluation (Fotaki & Vertzoni, 2010) using a miniaturized dissolution apparatus (Berger et al., 2007; Zheng et al., 2012). Combination of in vivo data results and in vitro two-stage dissolution data should be considered when selecting the appropriate salt or co-crystal.

Once the salt or co-crystal form is selected, further PK analysis is typically performed to confirm the selection. The analysis typically compares the oral bioavailability (%F) of oral suspensions of the selected salt or co-crystal forms with suspensions of the free form at a high dose. Then, dose linearity studies are performed to demonstrate the likelihood of achieving enough exposure for safety assessment of the NCE. The solid form that gives dose-linear exposure which is scalable in its manufacture and physicochemically stable is selected for further development.

5.3.3 Particle Size Reduction

A very common and typically easy way to assess dissolution rate enhancement for poorly soluble molecules is through micronization. In this process, larger particles are reduced through milling to a mean particle size typically in the range of 1–10 microns (Englund & Johansson, 1981; Hargrove et al., 1989; Nimmerfall & Rosenthaler, 1980; Shastri et al., 1980; Watari et al., 1983). The goal is to increase the dissolution rate by enlarging the surface area of the powder. Micronization is a commonly applied technology for BCS Class 2 drugs (low solubility/high permeability) (Amidon et al., 1995; Lennernas & Abrahamsson, 2005; Yu et al., 2002). Because their good permeability, accelerating the dissolution rate of BCS class 2 drugs often improves oral bioavailability. One of the first examples of size reduction to micron scale is griseofulvin (oral anti-fungal agent) (Kraml et al., 1962), where a four-fold reduction of particle size to 1 μm doubled the absorbed amount of drug in humans. However, particle size reduction into the low micrometer range is sometimes insufficient. For example, compounds with very low solubility can never achieve a dissolution rate to drive sufficiently high bioavailability. The problem could be due to the hydrophobic nature of the molecule, high melting point ($> 250°C$) or high enthalpy (> 100 J/gm) of the molecule. Although very small particles give rise to high surface area, eventually dissolution is still required to achieve absorption though the GI tract.

Further particle size reduction to < 500 nanometer (nm) range can offer non-linear improvement in kinetic solubility/dissolution rates (Jia et al., 2002; Liversidge & Conzentino, 1995; Liversidge & Cundy, 1995). The size reduction can also result in increased bioavailability, provided the limited absorption of the drug is due to dissolution rate and not permeability or metabolism. Nano-sizing of hydrophobic drugs generally involves the production of drug nanocrystals through either chemical precipitation or disintegration precipitation method is a bottom-up approach and disintegration method is a top-down approach. Elan's NanoCrystal® wet milling technology and SkyePharma's Dissocubes® high-pressure homogenization technology enable production of nanoparticles via the top-down approach. The advances in milling technologies have made it possible to produce drug particles in the range of 100 to 200 nm in a reproducible manner (Chan & Kwok, 2011; Van Eerdenbrugh et al., 2008). In fact, today there are a number of drug products in the market based on particle size reduction to nano scale through milling techniques. The first pharmaceutical application of this technology was performed by wet milling of danazol, a neutral compound (Liversidge & Cundy, 1995). The relative bioavailability of danazol in dogs increased more than 10 times when the particle size was reduced from 10 μm to an average of 84.9 nm, which was measured using a sedimentation field flow fractionator. The particles varied in size from 26 to 340 nm. A summary of oral tablets and capsules that employ nano-sized active is provided in Table 5.8.

Table 5.8   Marketed oral products using nanocrystalline API

Product name	Technology	Uses	Manufacturer
Emend™	Nanocrystalline aprepitant	Antiemetic	Elan Drug Delivery (King of Prussia, PA, USA), Merck & Co. (Whitehouse Station, NJ, USA)
Megace ES™	Nanocrystalline megesterol acetate	Eating disorders	Elan Drug Delivery, Par Pharmaceutical Companies (Woodcliff Lake, NJ, USA)
Rapamune™	Nanocrystalline sirolimus	Immunosuppressant	Elan Drug Delivery, Wyeth Pharmaceuticals (Collegeville, PA, USA)
Tricor™	Nanocrystalline fenofibrate	Lipid regulation	Elan Drug Delivery, Abbott (Abbott Park, IL, USA)
Triglide™	Nanocrystalline fenofibrate	Lipid regulation	SkyePharma, First Horizon Pharmaceuticals (Alpharetta, GA, USA)

5.3.4 Amorphous Form and Solid Dispersion

Amorphous materials are non-equilibrium systems and have short-range molecular order, i.e., they do not pack as a crystalline form would. Typically, these regions of short-range order extend to only a few molecular layers (Angell, 1995; Lubchenko & Wolynes, 2007). The apparent solubility enhancement of amorphous-based drug formulations has been extensively studied (Abu-Diak et al., 2011; Gupta et al., 2004; Kai et al., 1996; Law et al., 2004; Prasad et al., 2014). The enthalpy, entropy, and free energy of amorphous solids are high relative to the corresponding crystalline solid due to the disordered structure of the amorphous form. The higher free energy results in enhanced apparent solubility and dissolution rate.

A model developed by Parks et al. (Parks et al., 1927) can be used to predict the relative solubilities of the amorphous and crystalline forms of drug based on the difference in their free energies (ΔG). The model shows that the solubility ratio ($S_{amorphous}/S_{crystalline}$) of the two forms at a certain temperature is directly related to the free energy difference (ΔG) between these two forms (Eq. 5.3).

$$\Delta G = RT \ln \frac{S_{amorphous}}{S_{crystalline}} \tag{5.3}$$

where
R = gas constant
T = temperature
S = solubility

To estimate the free energy difference between the amorphous and crystalline forms, the Hoffman equation (Eq. 5.4) can be applied, if the melting temperature (T_m) and heat of fusion (ΔH_f) of the crystalline form are known.

$$\Delta G = \frac{\Delta H_f (T_m - T) T}{\Delta T_m^2}$$

The Hoffman equation provides a reasonable estimate of the free energy difference between amorphous and crystalline drug (Abu-Diak et al., 2011; Marsac et al., 2006).

If a higher apparent solubility is maintained for a sufficient time in the GI tract, enhanced bioavailability can be obtained. However, the higher free energy and higher molecular mobility of the amorphous form can result in physical instability in the solid state, namely form conversion to lower energy crystalline form(s) and higher chemical degradation rates. Thus, an amorphous form can offer solubility advantages over the crystalline form, but it can also suffer from lower stability, which could limit its use in formulation development.

Another way to improve the oral bioavailability of poorly water-soluble drugs is through the use of solid dispersion. Solid dispersion is a comparatively old technique; Sekiguchi and Obi were the first to describe a solid dispersion in 1961 (Sekiguchi & Obi, 1961). They found that the formulation of eutectic mixtures can improve the rate of drug release and, consequently, the bioavailability of poorly water-soluble drugs. Then, in 1971, Chiou and Riegelman defined solid dispersion as the dispersion of one or more APIs in an inert carrier or matrix in the solid state (Chiou & Riegelman, 1971). Solid dispersion can be used for structurally different compounds with diverse physicochemical properties.

Solid dispersions can create high GI drug concentrations by increasing the apparent solubility and dissolution rate of poorly water-soluble drugs. Various factors such as a reduction in particle size, improved wetting, and reduced agglomeration contribute to an increase in apparent solubility and dissolution rate which translates into higher bioavailability.

More recently, the importance of combining the amorphous form of drugs with various polymers in formulating amorphous solid dispersions has been realized and efforts have focused on dispersing amorphous APIs in polymer systems. The difficulty in formulating amorphous systems is that they are thermodynamically unstable and tend to revert to the more thermodynamically stable crystalline form upon storage. Thus, the primary challenge in the development of amorphous drug-based formulations is assuring the long-term physical stability of the formulation. Stability issues can be overcome by developing amorphous solid dispersions. In amorphous solid dispersions, polymers have been added to stabilize the amorphous drugs, however, the stabilizing mechanism is not fully understood and is likely an interplay of multiple mechanisms (Chiou & Riegelman, 1971; Serajuddin, 1999).

Form conversion of the drug from crystalline to amorphous can be achieved during preparation of the amorphous solid dispersions. There are two main factors for physical stabilization of the amorphous solid dispersions: (i) interactions between drug and polymer molecules and (ii) molecular mobility. In general, inhibition of crystallization is affected by more than one factor. The stabilizing ability of a polymer has been ascribed to various phenomena such as the polymer's ability to decrease molecular mobility, increase the glass transition temperature, disrupt drug–drug molecular interactions and stabilize drug–polymer interactions. Moreover, the stability of amorphous dispersions also depends on the intrinsic crystallization tendency of the pure amorphous drug (Bhugra & Pikal, 2008). Various mechanisms through which polymers help in stabilizing amorphous drug are discussed below.

Drug-polymer interaction plays an important role in stabilization of amorphous solid dispersions. The literature confirms the importance of interactions such as hydrogen bonding, hydrophobic interactions and in some cases ionic interactions in stabilizing solid dispersions. Taylor and Zografi showed the presence of intermolecular hydrogen bonding interactions in solid dispersions of indomethacin and polyvinylpyrrolidone (PVP). Polymers that can form such interactions were shown to have stabilization effects even if they had no effect on the Tg of the system, or when the anti-plasticizing effects of the polymers were low (at low polymer concentration) (Taylor & Zografi, 1997). Also, the importance of hydrogen bonding has been highlighted in stabilizing the amorphous phase through disruption of drug-drug interactions and the formation of drug–polymer interactions (Chauhan et al., 2013; Tobyn et al., 2009; Vasanthavada et al., 2005).

Another important effect is the decrease in molecular mobility. The addition of a polymer decreases the molecular mobility of the drug molecule and therefore decreases the crystallization tendency. The T_g is also related to the mobility of a material and the relative strength of adhesive and cohesive interactions. The addition of a polymer with high T_g increases the glass transition of the system and may help in the stabilization of an amorphous solid dispersion (Aso et al., 2004; Zhou et al., 2007).

Also, the extent of drug–polymer miscibility influences the stability of the amorphous solid dispersions. Systems containing drug–polymer mixtures are considered miscible if they have a single T_g. More than one glass transition represents phase separation which results in a less stable system. The formation of single or multiple phases depends upon the initial concentration of drug and the nature of the drug–polymer interaction. Drug–polymer miscibility can provide an insight into the physical stability of drug–polymer mixtures and can be measured by various techniques including Flory–Huggins interaction parameter, solubility parameters, T_g measurements by DSC, computational analysis of X-ray diffraction (XRD) data, solid state nuclear magnetic resonance

(NMR) spectroscopy and atomic force microscopy (AFM) (Meng et al., 2015; Qian et al., 2010; Rumondor et al., 2009; Schachter et al., 2004; Yoo et al., 2009).

The dissolution of amorphous solid dispersions has been reported to yield higher solution concentrations than that of the crystalline form or the pure amorphous form (Bikiaris, 2011; Chauhan et al., 2014; Hancock & Parks, 2000). Due to the high energy state of amorphous drugs, amorphous solid dispersions have higher apparent solubility compared to crystalline forms. Furthermore, the presence of hydrophilic polymers in close contact with the drug molecules also increases the solubility by enhancing the wettability and maximizing the apparent surface area of the compound. When both the amorphous solid and the supersaturated solution generated via dissolution are stabilized against crystallization, maximum supersaturation can be obtained (Brouwers et al., 2009). The amorphous form can provide a high-fold enhancement in solubility which leads to a significant improvement in oral bioavailability.

There are two main methods for preparing solid dispersions, melting and solvent evaporation. Quench cooling and melt extrusion involves melting the drug and polymers together followed by cooling to generate a solid dispersion. In hot melt extrusion, the drug-carrier mix is simultaneously melted, homogenized and then extruded from a machine. Then, the mixture is molded into various forms such as tablets, granules, pellets, sheets, sticks or powder. Various techniques such as rotary evaporation, spray drying, lyophilization, and use of supercritical fluids rely on evaporating the solvent from the solution containing the drug and polymeric carriers to produce the amorphous solid dispersion.

Solid dispersions have defined physical properties (e.g., glass transition temperature) and such physical properties depend on the properties of the API, polymeric carriers and interactions between them. Table 5.9 summarizes various techniques for characterization of amorphous solid dispersions. Recently, it has been shown that preparation methods can also impact the physicochemical properties of solid dispersions. For example, hydrophilic polymers such as polyvinylpyrrolidone (PVP), polyethyleneglycols (PEG), polymethacrylates and cellulose derivatives, are among the most common polymers used to develop solid dispersions. Many

Table 5.9 Techniques used for material characterization of solid dispersions

Qualitative	Quantitative
Hot plate microscopy	Differential scanning calorimetry
Polarization microscopy	Solid state NMR
Scanning electron microscopy	X-ray diffraction
AFM	Fourier transform infrared spectroscopy (FTIR)
	Raman spectroscopy

researchers have shown that surfactants can be further added to avoid drug re-crystallization and enhance solubility (Dave et al., 2012; Ghebremeskel et al., 2007; Fan Meng et al., 2015).

The release of drug is controlled by the dissolution of the hydrophilic carrier and can be manipulated by polymer properties. The properties include the type of polymer and its molecular weight, particle porosity, and wettability.

With increased research, better understanding of solid dispersions, and successful advancement of these formulations in development pipelines, the number of drugs formulated as a solid dispersion is expected to increase in the coming years. Commercial examples of amorphous solid dispersions include quinapril hydrochloride (Accupril®), zafirlukast (Accolate®), and nelfinavir mesylate (Viracept®). Novartis's soft gelatine capsule Gris-PEG® is based on a solid dispersion of griseofulvin in PEG 8000. Fujisawa's Prograf® is a solid amorphous dispersion of tacrolimus in hydroxypropylmethylcellulose (HPMC). Cesamet®, a capsule product by Eli Lilly, is based on a solid dispersion of nabilone in povidone. An example of an amorphous dispersion as a soft gelatin capsules is Fortovase® by Roche USA, which contains saquinavir mesylate suspended in glycerides. Kaletra®, a product of Abbott Laboratories, contains the protease inhibitors lopinavir/ritonavir and the formulation is prepared by melt extrusion. Other examples of solid dispersions for oral administration are nimodipine (Nimotop®) by Bayer, nilvadipine (Nivadil®) by Astellas, troglitazone (Rezulin®) by Pfizer, rosuvastatin calcium (Crestor®) by Astrazeneca, etravirine, (Intelence®) by Tibotec, and everolimus (Certican®/Zortress®) by Novartis and most recently

Telaprevir (Incivek®) and Ivacaftor (Kalydeco®) by Vertex (Brough & Williams, 2013; Kawabata et al., 2011).

The fact that almost 40% of the new chemical entities are poorly water soluble (Fahr & Liu, 2007) implies that studies with solid dispersions will continue, and more drug compounds formulated as solid dispersions will reach the pharmaceutical market in the future.

5.3.5 Lipid-Based Formulations

Lipid-based drug delivery systems represent a wide variety of formulations composed of lipophilic, amphiphilic and hydrophilic excipients that are able to solubilize poorly water-soluble and lipophilic drugs. Lipid-based delivery systems range from simple oil solutions to complex mixtures of oils, surfactants, co-surfactants and co-solvents. The latter mixtures are typically self-dispersing systems often referred to as self-emulsifying drug delivery systems (SEDDS) or self-micro emulsifying drug delivery systems (SMEDDS).

5.3.5.1 Oil/lipid solutions

One of the earliest and simplest approaches to increase the bioavailability of poorly water-soluble drugs is a solution of drug in lipid/oil. The advantages of an oil/lipid solution are in the system's relative simplicity and ability to dissolve highly lipophilic drugs (Hauss, 2007). Generally, various digestible lipids such as soybean oil, corn oil, or olive oil are used for such formulations. Non-digestible oils such as mineral oil are not preferred as they tend to reduce drug absorption by retaining part of the co-administered drug in the undigested formulation. The formulations are prepared by dissolving the drug in oil and formulating in the form of soft or hard gelatin capsules. The drug release and absorption from oil/lipid solutions is influenced by the GI lipid digestion that promotes emulsification; an essential step for both drug release and absorption (Palin & Wilson, 1984; Pouton & Porter, 2008).

5.3.5.2 Emulsions

An emulsion is a colloidal system that contains either oil-in-water (o/w) or water-in-oil (w/o) particles stabilized by surfactants in interfacial phases. The use of oil-in-water emulsion is more prevalent as a formulation strategy for poorly water-soluble drug. An example of a commercial emulsion formulation is Diprivan®, which is an anesthetic (IV bolus and infusion). Emulsions are also used broadly in nutritional supplements such as Intralipid®, Nutralipid®, Liposyn® and Lipofundin® (Li & Zhao, 2007).

According to Danielsson and Lindman, a microemulsion, is a single optically isotropic system that is thermodynamically stable. The key differences between emulsions and microemulsions are that microemulsions are transparent and are thermodynamically stable (Lindman & Danielson, 1981). Microemulsion transparency is due to small droplet size, typically less than 100 nm. Such small droplet size produces weak scattering of visible light when compared with that of coarse droplets of normal emulsions (1–10 μm). Another important difference between microemulsions and emulsions is that microemulsions form spontaneously and require limited or no mechanical work or energy to form whereas energy inputs are essential for the preparation of emulsions. The stability of a microemulsion is an important feature from a formulation perspective. Normal emulsions age by coalescence of droplets and by Ostwald ripening (transfer of material from small droplets to large droplets). The process leads to a decrease of the interfacial area and subsequent decrease of the free energy of the system. In microemulsions, the interfacial tension is sufficiently low to compensate for the dispersion entropy, therefore the systems are thermodynamically stable. Thus, microemulsions offer several potential advantages as a result of their solubilization capacity, optical transparency, high stability and simplicity of manufacturing (Kreuter, 1994).

Microemulsions can be used to formulate hydrophilic, lipophilic and amphiphilic drugs. Water-insoluble lipophilic drugs can be incorporated into the disperse oil phase and/or hydrophobic tail region of the surfactant and the hydrophilic drug can be incorporated into the disperse water-in-oil aqueous phase. Formulation of oil-in-water microemulsions is preferred for water-insoluble drugs instead of water-in-oil microemulsions because the droplet structure of oil-in-water microemulsions is often retained upon dilution by aqueous biological fluid, whereas for water-in-oil microemulsions, phase separation can occur due to increase in droplet size on dilution with GI fluids. This can lead to precipitation of drug upon dosing.

5.3.5.3 Self-emulsifying systems

Among the more recent advances in lipid dosage form, self-emulsifying drug delivery systems (SEDDS) and self-microemulsifying drug delivery systems (SMEDDS) have gained popularity. These are isotropic mixtures of oils, surfactants, solvents and co-solvents/surfactants. SEDDS typically produce emulsions with a droplet size between 100 and 300 nm while SMEDDS form transparent microemulsions with a droplet size of less than 50 nm (Neslihan Gursoy & Benita, 2004). SEDDS/SMEDDS are best suited for dissolution-rate limited BCS class 2 drugs, where the drug can be introduced to the GI tract in a ready-to-absorb form. Thus, these systems may offer an improvement in the rate and extent of absorption and result in a more reproducible pharmacokinetics profile. The formulations usually consist of oil, surfactants, co-surfactants and co-solvents in a certain ratio such that a thermodynamically stable microemulsion can be formed upon exposure to water. When compared with emulsions, which are sensitive and metastable dispersed forms, SEDDS are more physically stable formulations that are easy to manufacture (Hauss, 2013). SEDDS can be orally administered in soft or hard gelatin capsules and form fine, relatively stable oil-in-water emulsions upon aqueous dilution owing to the gentle agitation of the GI fluids in GI tract. When dosed, providing mild agitation by GI motility and dilution in aqueous media such as GI fluids, these systems form fine oil-in-water emulsions or microemulsions (SMEDDS) (Hauss, 2013).

The first step in formulating SEDDS/SMEDDS is understanding the solubility of the drug substance in different lipophilic, amphiphilic, and hydrophilic excipients. Phase diagrams with blends of excipients, consisting of a combination of a lipophilic (oil), co-solvent and surfactant/co-surfactants are then created to determine the design space which can produce microemulsions. Those ratios of excipients are further studied for their phase behavior and thermodynamic stability. The oil or lipid affects the SEDDS emulsification characteristics. It also affects the drug bioavailability through enhancing the drug uptake via the intestinal lymphatic system, thus enhancing the total amount of drug absorption from the GI tract (Holm et al., 2002; Lindmark et al., 1995). In the preparation of SEDDS, both long and medium chain triglyceride oils with different degrees of saturation have been used. Various unmodified edible oils such as soybean oil, corn oil, etc. offer logical choices for the formulation of SEDDS because of their safety and digestibility. However, their use is limited to highly lipophilic potent drugs because of their inability to dissolve large amounts of drugs and their relative difficulty in efficient self-emulsification. Modified or hydrolyzed vegetable oils and medium chain triglycerides (MCTs) are preferred in the self-emulsifying formulations because of their higher fluidity, better solubility properties and self-emulsification ability (Constantinides, 1995). Recently, the advent of novel semi-synthetic medium chain derivatives which are defined as amphiphilic compounds with surfactant properties offer more choices to formulators besides MCTs and hydrolyzed vegetable oils (Neslihan Gursoy & Benita, 2004). Various formulation factors which affect the self-emulsifying efficiency and oral absorption of the drug compound from the SEDDS include surfactant concentration, oil/surfactant ratio, polarity of the emulsion, droplet size and charge.

Various non-ionic surfactants such as the polysorbates (e.g., Tween® 80) and polyoxyls (e.g., Cremophor® EL), covering the Hydrophilic-Lypophilic Balance (HLB) range from 2 to 18, are used in the formulation of SEDDS/SMEDDS. The importance of surfactants in SEDDS lies in the fact that surfactants are essential for the immediate formation of oil-in-water droplets and rapid spreading of the formulation in the aqueous environment, thus providing good self-emulsification. As surfactants are amphiphilic molecules, they are able to dissolve and solubilize large amounts of lipophilic drug. The solubilization of drug in the GI tract is important to prevent drug precipitation and maintenance of the drug in a soluble form for effective absorption. Due to the relatively low toxicity and emergence of several successful marketed products such as Neoral®, Sandimmune®, Kaletra®, Aptivus®, Hectoral®, containing non-ionic surfactants, there is strong precedence for use of multiple surfactant systems in humans. In fact, the amount of surfactant that can be used is often limited by the compatibility between surfactant and capsule shell (Strickley, 2007). High concentrations of surfactant may cause brittleness of hard and soft gelatin capsules due to their dehydrating effects on capsule gelatin (Porter et al., 2008). Extremes in surfactant concentration may irritate the GI tract, so the safety of the surfactant should be carefully monitored and may still prove to be an important limiting factor.

Various materials such as Transcutol®, PEG 400, propylene glycol, ethanol and glycerol are used as co-solvents in the formulation of SEDDS/SMEDDS (Porter & Charman, 2001b; Prasad et al., 2013). Co-solvents are used to increase the solvent capacity of the formulation for drugs. A second reason for inclusion of co-solvents is to aid dispersion of systems, which contain a high proportion of water-soluble surfactants. There are practical limits on the concentrations of co-solvents which can be used depending on miscibility of co-solvents with oil components, potential incompatibilities of low molecular weight co-solvents with capsule shells and the possible risk of drug precipitation associated with the use of high concentrations of co-solvents upon dilution, as co-solvents tend to lose their solvent capacity rapidly following aqueous dilution (Cole et al., 2008).

Finally, lipid-soluble antioxidants such as α-tocopherol, β-carotene, butylated hydroxytoluene (BHT), butylated hydroxyanisole (BHA) or propyl gallate can be incorporated into the formulation to protect the ingredients, especially oil from being oxidized (Pouton & Porter, 2008).

Important characterization studies of SEDDS include ease of emulsification, droplet size distribution, in vitro dispersion and digestion tests, and surface charge determination.

The primary means of self-emulsification assessment is qualitative visual evaluation with a binary ranking system of either good or poor. There is no practical advantage in developing a precise method to examine the ease of emulsification if a simple visual observation can be used to confirm that dispersion is sufficiently rapid. Generally well-formulated SEDDS or SMEDDS are dispersed within seconds under gentle stirring. Turbidity measurements can be carried out to determine the rapidity of equilibrium reached by the dispersion and the reproducibility of the process. It also relates to the particle size of the self-emulsified system (Gursoy et al., 2003).

The determination of the droplet size distribution of the emulsion formed after dilution of SEDDS is an important factor in self-emulsification performance, as it affects the rate and extent of drug release. Droplet size analysis can be carried out using techniques such as laser diffraction, Coulter counter, photon correlation spectroscopy, or ultrasonic spectroscopy. Photon correlation spectroscopy (PCS) is a useful method for determination of droplet size especially when the emulsion properties do not change upon infinite aqueous dilution. However, microscopic techniques should be employed at relatively low dilutions for accurate droplet size evaluation (Kathe et al., 2014; Neslihan Gursoy & Benita, 2004).

In vitro dispersion and digestion tests are crucial for the formulation optimization. These methods can be used to predict the fate of drug in the GI tract. The test is of essential importance because the formulation can lose its solvent capacity when it comes in contact with the GI medium. Thus, this testing can help in predicting in vivo drug precipitation. Currently, no official pharmacopeoial method is available for in vitro dispersion testing; however, dispersion testing can be carried out by using a standard dissolution apparatus and assessing the drug release similar to solid dosage form dissolution using USP apparatus 2 (Griffin et al., 2014).

In vitro digestion/lipolysis testing of lipidic formulations is also important as it can be used to predict the fate of a drug formulation in the intestinal lumen prior to absorption. Its importance is further enhanced considering the fact that digestion/lipolysis of lipid components in a formulation greatly influence the release and absorption of drug. In vitro digestion methods have been established and described by many laboratories in recent years (Anby et al., 2014; Cuine et al., 2008; Dahan & Hoffman, 2007; Devraj et al., 2014), although the details of the methods used are varied among laboratories. As in vitro digestion/lipolysis is becoming a well-accepted characterization method for lipid formulation, research is on-going to establish a standard method (Porter & Charman, 2001a). Since 2012, the Lipid Formulation Classification System (LFCS) Consortium has published a series of articles to standardize the in vitro digestions tests for lipid-based formulations (Williams et al., 2012a; 2012b; 2013; 2014).

The characterization of surface charge may be done by determining the zeta potential of droplets. The charge of the oil droplets in conventional SEDDS is negative due to the presence of free fatty acids (Porter & Charman, 2001a; Pouton & Porter, 2008). However, incorporation of a cationic lipid, such as oleylamine may yield cationic SEDDS. Such formulations have shown to undergo electrostatic interaction with the Caco-2 monolayer and the mucosal surface of the everted rat intestine (Gershanik et al., 1998). The magnitude of droplet surface charge relates to the stability of the emulsions. Zeta potential is used as a measure of surface charge with higher absolute value of zeta potential correlating with higher emulsion stability.

The mechanism by which self-emulsification occurs is not well understood. In the case of conventional emulsion, the free energy of an emulsion formulation is a direct function of the energy required to create a new surface between the oil and water phases. In the case of self-emulsifying systems, it has been suggested that self-emulsification takes place when the entropy change favoring dispersion is greater than the energy required to increase the surface area of the dispersion. Thus, emulsification occurs spontaneously with SEDDS because the free energy required to form the emulsion is either very low positive, or it is negative (Neslihan Gursoy & Benita, 2004).

The performance of lipid-based delivery systems depends on the digestion of lipidic excipients along with the particle size of the initial dispersion. SEDDS enhances the bioavailability of drugs via various mechanisms. Some of the prominent mechanisms include alteration in gastric transit time, increase in effective luminal drug solubility, stimulation of intestinal lymphatic transport, and changes in the biochemical and physical barrier function of the GI tract. The high content of lipid leads to the increase in gastric transit time of the drug, thereby slowing delivery to the absorption site and increasing the time available for dissolution (Porter & Charman, 2001a). This modulation of GI transit time accounts for an extended release effect sometimes observed with SEDDS. The presence of lipids in the GI tract stimulates an increase in the secretion of bile salts and endogenous biliary lipids including phospholipid and cholesterol, leading to the formation of their intestinal mixed micelles and an increase in the solubilization capacity of the GI tract. This increased solubilization of drug in the GI tract increases the concentration available for absorption. The presence of lipid in the formulation enhances lymphatic drug transport, thus directly or indirectly increasing the bioavailability via a reduction in first-pass metabolism (Dintaman & Silverman, 1999). It is clear that certain lipids and surfactants may attenuate the activity of intestinal efflux transporters, for example by inhibiting the p-glycoprotein efflux pump. Lipids and surfactants may also reduce the extent of enterocyte-based metabolism (Benet & Cummins, 2001). Various combinations of lipids, lipid digestion products, and surfactants have been shown to have permeability enhancing properties, i.e., they help in increasing the membrane fluidity, thus facilitating transcellular absorption (Tang et al., 2007).

5.3.6 Controlled-Release Formulations

Controlled release may be defined as a technique in which active chemicals are made available to specified targets at a rate and duration designed to accomplish an intended effect (Kydonieus, 1980). Controlled release drug delivery currently involves control of either the time course (temporal) or location (spatial) parameters of drug delivery. While control of the time course of the drug delivery is the more classical approach, site specific or targeted delivery to specific physiological sites or cellular targets (e.g., to cancer cells, hepatocytes or across blood brain barrier) are documented (Berner & Dinh, 1992). The ultimate goal of controlled release dosage forms is to release drug(s) in a manner that produces the maximum simultaneous safety, effectiveness and reliability. The ideal attributes of such controlled release drug delivery systems are (Kumar & Banker, 1996):

(1) capability of controlled-delivery rates to accommodate the PK target of various drugs (flexible programming)
(2) capability of precise control of a constant delivery rate (precise programming)

(3) not highly sensitive to physiological variables

  (a) gastric motility and emptying, pH, fluid volume, and content of gut
  (b) state of fasting and type of food present
  (c) physiological position and activity of subject
  (d) individual variability
  (e) disease state

(4) drug stability is maintained

(5) controlled mechanism adds little mass to the dosage form

Poor solubility of the drug substance influences drug release from controlled release dosage forms. In order for a drug to be absorbed, it must first dissolve in the physiological conditions in the GI tract. The aqueous solubility of a drug influences its dissolution rate, which in turn establishes its concentration in solution providing the driving force for diffusion across membranes (Longer & Robinson, 1990). A drug with very low solubility (<0.01 mg/mL) and a slow dissolution rate will exhibit dissolution-limited absorption and will give inherently sustained blood levels (Jantzen & Robinson, 1996). Such compounds usually suffer oral bioavailability problems because of limited GI transit time of the undissolved drug particles and limited solubility at the site of absorption. The lower limit for the solubility of the drug to be formulated in a controlled/sustained release system has been reported to be 0.1 mg/mL (Fincher, 1968). Diffusional systems are a poor choice for slightly soluble drugs since the driving force for diffusion is the drug concentration in solution, which will be low. Osmotic delivery has been utilized widely for poorly soluble drugs (e.g., controlled release of prednisolone (Krögel & Bodmeier, 1999) and chlorpromazine (Okimoto et al., 1999)) using controlled porosity osmotic pumps. In osmotic tablets the solubilizer can be used as a core excipient (cyclodextrins).

  In addition to solubility, other important parameter to consider while formulating controlled release dosage forms are the dissociation constant of the molecule (pK_a) as well as understanding the charged state of the molecule at its targeted location of release and absorption. Recall that the uncharged form of a drug species is preferentially absorbed in a passive manner through biological membranes. Hence, it is important to note the relationship between the pK_a of the drug and absorptive environment. Delivery systems that are dependent on dissolution or diffusion will likewise be dependent on the solubility of the drug in physiological media. Considering controlled-release delivery systems must function in an environment of changing pH, the stomach being acidic and the small intestine more neutral, the effect of the pH on the release process must be defined (Jantzen & Robinson, 1996). There are many products on the market with controlled/sustained/extended release dosage forms. The most important pre-requisite for any drug is to have a short half-life (less than 4 hours and low dose (<100 mg)). Controlled release products on the market with poor solubility at physiological pH include morphine sulfate (Avinza®) by Pfizer, morphine sulfate and naltraxone (Embeda®) by Pfizer, oxymorphone hydrochloride (Opana ER®) by Endo Pharmaceuticals, oxycodone hydrochloride (Oxy Contin®) and hydromorphone hydrochloride (Palladone®) by Purdue Pharma, palliperidone (Invega®), tapentadol (Nucynta®) by Janssen Pharmaceuticals, and carbamazepine (Tegretol®), and darifenacin (Enablex)® by Novartis Pharmaceuticals. Though there are few technologies available to deliver poorly soluble drugs via controlled release, it is very important to understand the dose requirement. For a poorly soluble drug to be a good candidate for controlled release, it typically should be a lower dose with a higher potency.

5.4 Summary/Concluding Remarks

Poor solubility of new chemical entities can be overcome by a number of enabling technologies. It is very important to build the right physicochemical and biopharmaceutical properties into a molecule during the hit-to-lead and lead optimization process. However, often times the nature of the target constrains the ability to advance highly soluble compounds. Enabling such NCEs is critical to advancing the development of important therapies. Active engagement of the pharmaceutical scientists and the utilization of appropriate formulations for in vivo studies, while maintaining a clear line of sight to commercial development are essential to the success of any discovery program. For poorly soluble compounds, the importance of identifying the critical principles required to achieve predictable and robust bioperformance is intensified, thus justifying a need for understanding the validity of the target in humans as soon as possible.

Acknowledgments

The authors would like to sincerely thank Derek Jackson for his immense help in completion of this book chapter.

References

Abu-Diak, O. A., Jones, D. S., & Andrews, G. P. (2011). An investigation into the dissolution properties of celecoxib melt extrudates: understanding the role of polymer type and concentration in stabilizing supersaturated drug concentrations. *Mol. Pharm.,* **8**(4), 1362–1371.

Amidon, G. L., Lennernäs, H., Shah, V. P., & Crison, J. R. (1995). A theoretical basis for a biopharmaceutic drug classification: the correlation of in vitro drug product dissolution and in vivo bioavailability. *Pharm. Res.,* **12**(3), 413–420.

Anby, M. U., Nguyen, T. H., Yeap, Y. Y., Feeney, O. M., Williams, H. D., Benameur, H., et al. (2014). An in vitro digestion test that reflects rat intestinal conditions to probe the importance of formulation digestion vs first pass metabolism in Danazol bioavailability from lipid based formulations. *Mol. Pharm.,* **11**(11), 4069–4083.

Angell, C. A. (1995). Formation of glasses from liquids and biopolymers. *Science,* **267**(5206), 1924–1935.

Arenas-García, J. I., Herrera-Ruiz, D., Mondragón-Vásquez, K., Morales-Rojas, H., & Höpfl, H. (2010). Co-crystals of active pharmaceutical ingredients: acetazolamide. *Cryst. Growth Des.,* **10**(8), 3732–3742.

Artursson, P., & Tavelin, S. (2003). Caco-2 and emerging alternatives for prediction of intestinal drug transport: a general overview. *Drug Bioavailability: Estimation of Solubility, Permeability, Absorption and Bioavailability,* 72–89.

Aso, Y., Yoshioka, S., & Kojima, S. (2004). Molecular mobility-based estimation of the crystallization rates of amorphous nifedipine and phenobarbital in poly (vinylpyrrolidone) solid dispersions. *J. Pharm. Sci.*, **93**(2), 384–391.

Avdeef, A. (2003). *Charge State Absorption and Drug Development* (pp. 22-41): John Wiley & Sons.

Avdeef, A. (2007). Solubility of sparingly-soluble ionizable drugs. *Adv. Drug Deliv. Rev.*, **59**(7), 568–590.

Avdeef, A. (2012). *Absorption and Drug Development: Solubility, Permeability, and Charge State.* John Wiley & Sons.

Balimane, P. V., Han, Y.-H., & Chong, S. (2006). Current industrial practices of assessing permeability and P-glycoprotein interaction. *AAPS J.*, **8**(1), E1–E13.

Bauer, J., Spanton, S., Henry, R., Quick, J., Dziki, W., Porter, W., & Morris, J. (2001). Ritonavir: an extraordinary example of conformational polymorphism. *Pharm. Res.*, **18**(6), 859–866.

Benet, L. Z., & Cummins, C. L. (2001). The drug efflux–metabolism alliance: biochemical aspects. *Adv. Drug Deliv. Rev.*, **50**, S3–S11.

Berger, C. M., Tsinman, O., Voloboy, D., Lipp, D., Stones, S., & Avdeef, A. (2007). Technical note: miniaturized intrinsic dissolution rate (Mini-IDR (TM)) measurement of griseofulvin and carbamazepine. *Dissolut. Technol.*, **14**(4), 39–41.

Berner, B., & Dinh, S. (1992). Fundamental concepts in controlled release. *Treatise on Controlled Drug Delivery. Fundamentals, Optimization, Application*, 1–35.

Bhugra, C., & Pikal, M. J. (2008). Role of thermodynamic, molecular, and kinetic factors in crystallization from the amorphous state. *J. Pharm. Sci.*, **97**(4), 1329–1349.

Bikiaris, D. N. (2011). Solid dispersions, part I: recent evolutions and future opportunities in manufacturing methods for dissolution rate enhancement of poorly water-soluble drugs. *Expert Opin. Drug Deliv.*, **8**(11), 1501–1519.

Bourdet, D. L., & Thakker, D. R. (2006). Saturable absorptive transport of the hydrophilic organic cation ranitidine in Caco-2 cells: role of pH-dependent organic cation uptake system and P-glycoprotein. *Pharm. Res.*, **23**(6), 1165–1177.

Box, K., Bevan, C., Comer, J., Hill, A., Allen, R., & Reynolds, D. (2003). High-throughput measurement of p K a values in a mixed-buffer linear pH gradient system. *Anal. Chem.*, **75**(4), 883–892.

Brough, C., & Williams, R. O., 3rd. (2013). Amorphous solid dispersions and nano-crystal technologies for poorly water-soluble drug delivery. *Int. J. Pharm.*, **453**(1), 157–166.

Brouwers, J., Brewster, M. E., & Augustijns, P. (2009). Supersaturating drug delivery systems: the answer to solubility-limited oral bioavailability? *J. Pharm. Sci.*, **98**(8), 2549–2572.

Byrn, S. R., Pfeiffer, R. R., & Stowell, J. G. (1999). *Solid-State Chemistry of Drugs*. SSCI Inc.

Carrier, R. L., Miller, L. A., & Ahmed, I. (2007). The utility of cyclodextrins for enhancing oral bioavailability. *J. Control. Release*, **123**(2), 78–99.

Challa, R., Ahuja, A., Ali, J., & Khar, R. K. (2005). Cyclodextrins in drug delivery: an updated review. *AAPS PharmSciTech*, **6**(2), E329–E357.

Chan, H.-K., & Kwok, P. C. L. (2011). Production methods for nanodrug particles using the bottom-up approach. *Adv. Drug Deliv. Rev.*, **63**(6), 406–416.

Charman, S. A., Perry, C. S., Chiu, F. C., McIntosh, K. A., Prankerd, R. J., & Charman, W. N. (2006). Alteration of the intravenous pharmacokinetics of a synthetic ozonide antimalarial in the presence of a modified cyclodextrin. *J. Pharm. Sci.*, **95**(2), 256–267.

Chauhan, H., Hui-Gu, C., & Atef, E. (2013). Correlating the behavior of polymers in solution as precipitation inhibitor to its amorphous stabilization ability in solid dispersions. *J. Pharma. Sci.*, **102**(6), 1924–1935.

Chauhan, H., Kuldipkumar, A., Barder, T., Medek, A., Gu, C.-H., & Atef, E. (2014). Correlation of inhibitory effects of polymers on indomethacin precipitation in solution and amorphous solid crystallization based on molecular interaction. *Pharm. Res.*, **31**(2), 500–515.

Childs, S. L., Rodriguez-Hornedo, N., Reddy, L. S., Jayasankar, A., Maheshwari, C., McCausland, L., et al. (2008). Screening strategies based on solubility and solution composition generate pharmaceutically acceptable cocrystals of carbamazepine. *CrystEngComm*, **10**(7), 856–864.

Childs, S. L., Stahly, G. P., & Park, A. (2007). The salt-cocrystal continuum:? the influence of crystal structure on ionization state. *Mol. Pharm.*, **4**(3), 323–338.

Chiou, W. L., & Riegelman, S. (1971). Pharmaceutical applications of solid dispersion systems. *J. Pharm. Sci.*, **60**(9), 1281–1302.

Cleveland, J., Benko, M., Gluck, S., & Walbroehl, Y. (1993). Automated pKa determination at low solute concentrations by capillary electrophoresis. *J. Chromatogr. A*, **652**(2), 301–308.

Cole, E. T., Cadé, D., & Benameur, H. (2008). Challenges and opportunities in the encapsulation of liquid and semi-solid formulations into capsules for oral administration. *Adv. Drug Deliv. Rev.,* **60**(6), 747–756.

Constantinides, P. P. (1995). Lipid microemulsions for improving drug dissolution and oral absorption: physical and biopharmaceutical aspects. *Pharm. Res.,* **12**(11), 1561–1572.

Cuine, J. F., McEvoy, C. L., Charman, W. N., Pouton, C. W., Edwards, G. A., Benameur, H., & Porter, C. J. (2008). Evaluation of the impact of surfactant digestion on the bioavailability of danazol after oral administration of lipidic self-emulsifying formulations to dogs. *J. Pharm. Sci.,* **97**(2), 995–1012.

Dahan, A., & Hoffman, A. (2007). The effect of different lipid based formulations on the oral absorption of lipophilic drugs: the ability of *in vitro* lipolysis and consecutive *ex vivo* intestinal permeability data to predict *in vivo* bioavailability in rats. *Eur. J. Pharm. Biopharm.,* **67**(1), 96–105.

Dave, R. H., Patel, A. D., Donahue, E., & Patel, H. H. (2012). To evaluate the effect of addition of an anionic surfactant on solid dispersion using model drug indomethacin. *Drug. Dev. Ind. Pharm.,* **38**(8), 930–939.

Dearden, J. C. (2012). Prediction of physicochemical properties. *Methods Mol. Biol.,* **929**, 93–138.

Devraj, R., Williams, H. D., Warren, D. B., Porter, C. J., & Pouton, C. W. (2014). Choice of nonionic surfactant used to formulate type IIIA self-emulsifying drug delivery systems and the physicochemical properties of the drug have a pronounced influence on the degree of drug supersaturation that develops during in vitro digestion. *J. Pharm. Sci.,* **103**(4), 1050–1063.

Dhumal, R. S., Kelly, A. L., York, P., Coates, P. D., & Paradkar, A. (2010). Cocrystalization and simultaneous agglomeration using hot melt extrusion. *Pharm. Res.,* **27**(12), 2725–2733.

Di Profio, G., Grosso, V., Caridi, A., Caliandro, R., Guagliardi, A., Chita, G., et al. (2011). Direct production of carbamazepine-saccharin cocrystals from water/ethanol solvent mixtures by membrane-based crystallization technology. *CrystEngComm,* **13**(19), 5670–5673.

Dintaman, J. M., & Silverman, J. A. (1999). Inhibition of P-glycoprotein by D-a-tocopheryl polyethylene glycol 1000 succinate (TPGS). *Pharm. Res.,* **16**(10), 1550–1556.

Egan, W. J., Merz, K. M., & Baldwin, J. J. (2000). Prediction of drug absorption using multivariate statistics. *J. Med. Chem.,* **43**(21), 3867–3877.

Elder, D. P., Holm, R., & Diego, H. L. D. (2013). Use of pharmaceutical salts and cocrystals to address the issue of poor solubility. *Int. J. Pharm.*, **453**(1), 88–100.

Englund, D., & Johansson, E. (1981). Oral versus vaginal absorption in oestradiol in postmenopausal women. Effects of different particles sizes. *Upsala J. Med. Sci.*, **86**(3), 297–307.

Etter, M. C. (1990). Encoding and decoding hydrogen-bond patterns of organic compounds. *Acc. Chem. Res.*, **23**(4), 120–126.

Fábián, L. (2009). Cambridge structural database analysis of molecular complementarity in cocrystals. *Cryst. Growth Des.*, **9**(3), 1436–1443.

Fahr, A., & Liu, X. (2007). Drug delivery strategies for poorly water-soluble drugs. *Expert Opin. Drug Deliv.*, **4**(4), 403–416.

Faller, B., & Ertl, P. (2007). Computational approaches to determine drug solubility. *Adv. Drug Deliv. Rev.*, **59**(7), 533–545.

FDA (April 2013). Guidance for industry: regulatory classification of pharmaceutical co-crystals (cited June 30, 2015) http://www.fda.gov/downloads/Drugs/Guidances/UCM281764.pdf.

Fincher, J. H. (1968). Particle size of drugs and its relationship to absorption and activity. *J. Pharm. Sci.*, **57**(11), 1825–1835.

Fotaki, N., & Vertzoni, M. (2010). Biorelevant dissolution methods and their applications in in vitro-in vivo correlations for oral formulations. *Open Drug Deliv. J.*, **4**(2), 2–13.

Friščic, T., Trask, A. V., Jones, W., & Motherwell, W. D. S. (2006). Screening for inclusion compounds and systematic construction of three-component solids by liquid-assisted grinding. *Angew. Chem. Int. Ed.*, **45**(45), 7546–7550.

Galia, E., Nicolaides, E., Hörter, D., Löbenberg, R., Reppas, C., & Dressman, J. (1998). Evaluation of various dissolution media for predicting in vivo performance of class I and II drugs. *Pharm. Res.*, **15**(5), 698–705.

Gershanik, T., Benzeno, S., & Benita, S. (1998). Interaction of a self-emulsifying lipid drug delivery system with the everted rat intestinal mucosa as a function of droplet size and surface charge. *Pharm. Res.*, **15**(6), 863–869.

Ghebremeskel, A. N., Vemavarapu, C., & Lodaya, M. (2007). Use of surfactants as plasticizers in preparing solid dispersions of poorly soluble API: Selection of polymer–surfactant combinations using solubility parameters and testing the processability. *Int. J. Pharm.*, **328**(2), 119–129.

Glomme, A., März, J., & Dressman, J. (2005). Comparison of a miniaturized shake-flask solubility method with automated potentiometric acid/base titrations and calculated solubilities. *J. Pharm. Sci.,* **94**(1), 1–16.

Griffin, B. T., Kuentz, M., Vertzoni, M., Kostewicz, E. S., Fei, Y., Faisal, W., et al. (2014). Comparison of in vitro tests at various levels of complexity for the prediction of in vivo performance of lipid-based formulations: case studies with fenofibrate. *Eur. J. Pharm. Biopharm.,* **86**(3), 427–437.

Gupta, P., Kakumanu, V. K., & Bansal, A. K. (2004). Stability and solubility of celecoxib-PVP amorphous dispersions: a molecular perspective. *Pharm. Res.,* **21**(10), 1762–1769.

Gursoy, N., Garrigue, J. S., Razafindratsita, A., Lambert, G., & Benita, S. (2003). Excipient effects on in vitro cytotoxicity of a novel paclitaxel self-emulsifying drug delivery system. *J. Pharm. Sci.,* **92**(12), 2411–2418.

Hämäläinen, M., & Frostell-Karlsson, A. (2004). Predicting the intestinal absorption potential of hits and leads. *Drug Discovery Today: Technol.,* **1**(4), 397–405.

Hancock, B. C., & Parks, M. (2000). What is the true solubility advantage for amorphous pharmaceuticals? *Pharm. Res.,* **17**(4), 397–404.

Hancock, B. C., & Zografi, G. (1997). Characteristics and significance of the amorphous state in pharmaceutical systems. *J. Pharm. Sci.,* **86**(1), 1–12.

Hargrove, J. T., Maxson, W. S., & Wentz, A. C. (1989). Absorption of oral progesterone is influenced by vehicle and particle size. *Am. J. Obstetrics Gynecol.,* **161**(4), 948–951.

Hariharan, M., Ganorkar, L. D., Amidon, G. E., Cavallo, A., Gatti, P., Hageman, M. J., et al. (2003). Reducing the time to develop and manufacture formulations for first oral dose in humans. *Pharm. Technol.,* **27**(10), 68–84.

Hauss, D. J. (2007). Oral lipid-based formulations. *Adv. Drug Deliv. Rev.,* **59**(7), 667–676.

Hauss, D. J. (2013). *Oral Lipid-Based Formulations: Enhancing the Bioavailability of Poorly Water-Soluble Drugs.* CRC Press.

Hilgers, A. R., Smith, D. P., Biermacher, J. J., Day, J. S., Jensen, J. L., Sims, S. M., et al. (2003). Predicting oral absorption of drugs: a case study with a novel class of antimicrobial agents. *Pharm. Res.,* **20**(8), 1149–1155.

Holm, R., Porter, C. J., Müllertz, A., Kristensen, H. G., & Charman, W. N. (2002). Structured triglyceride vehicles for oral delivery of halofantrine: examination of intestinal lymphatic transport and bioavailability in conscious rats. *Pharm. Res.,* **19**(9), 1354–1361.

Hurst, S., Loi, C. M., Brodfuehrer, J., & El-Kattan, A. (2007). Impact of physiological, physicochemical and biopharmaceutical factors in absorption and metabolism mechanisms on the drug oral bioavailability of rats and humans. *Expert Opin. Drug Metab. Toxicol.,* **3**(4), 469–489.

Ingels, F., Deferme, S., Destexhe, E., Oth, M., Van den Mooter, G., & Augustijns, P. (2002). Simulated intestinal fluid as transport medium in the Caco-2 cell culture model. *Int. J. Pharm.,* **232**(1), 183–192.

Ishihama, Y., Nakamura, M., Miwa, T., Kajima, T., & Asakawa, N. (2002). A rapid method for pKa determination of drugs using pressure-assisted capillary electrophoresis with photodiode array detection in drug discovery. *J. Pharm. Sci.,* **91**(4), 933–942.

Jantarat, C., Sirathanarun, P., Ratanapongsai, S., Watcharakan, P., Sunyapong, S., & Wadu, A. (2014). Curcumin-hydroxypropyl-ß-cyclodextrin inclusion complex preparation methods: effect of common solvent evaporation, freeze drying, and pH shift on solubility and stability of curcumin. *Trop. J. Pharm. Res.,* **13**(8), 1215–1223.

Jantratid, E., Janssen, N., Reppas, C., & Dressman, J. B. (2008). Dissolution media simulating conditions in the proximal human gastrointestinal tract: an update. *Pharm. Res.,* **25**(7), 1663–1676.

Jantzen, G. M., & Robinson, J. R. (1996). Sustained-and controlled-release drug delivery systems. *Drugs Pharm. Sci.,* **72**, 575–610.

Jelfs, S., Ertl, P., & Selzer, P. (2007). Estimation of pKa for druglike compounds using semiempirical and information-based descriptors. *J. Chem. Inf. Model,* **47**(2), 450–459.

Jia, L., Wong, H., Cerna, C., & Weitman, S. D. (2002). Effect of nanonization on absorption of 301029: ex vivo and in vivo pharmacokinetic correlations determined by liquid chromatography/mass spectrometry. *Pharm. Res.,* **19**(8), 1091–1096.

Jullian, C., Cifuentes, C., Alfaro, M., Miranda, S., Barriga, G., & Olea-Azar, C. (2010). Spectroscopic characterization of the inclusion complexes of luteolin with native and derivatized beta-cyclodextrin. *Bioorg. Med. Chem.,* **18**(14), 5025–5031.

Kai, T., Akiyama, Y., Nomura, S., & Sato, M. (1996). Oral absorption improvement of poorly soluble drug using solid dispersion technique. *Chem. Pharm. Bull.,* **44**(3), 568–571.

Kathe, N., Henriksen, B., & Chauhan, H. (2014). Physicochemical characterization techniques for solid lipid nanoparticles: principles and limitations. *Drug Dev. Ind. Pharm.,* **40**(12), 1565–1575.

Kawabata, Y., Wada, K., Nakatani, M., Yamada, S., & Onoue, S. (2011). Formulation design for poorly water-soluble drugs based on bio-pharmaceutics classification system: basic approaches and practical applications. *Int. J. Pharm.,* **420**(1), 1–10.

Kawakami, K., Oda, N., Miyoshi, K., Funaki, T., & Ida, Y. (2006). Solubilization behavior of a poorly soluble drug under combined use of surfactants and cosolvents. *Eur. J. Pharm. Sci.,* **28**(1–2), 7–14.

Kerns, E. H., Di, L., Petusky, S., Farris, M., Ley, R., & Jupp, P. (2004). Combined application of parallel artificial membrane permeability assay and Caco-2 permeability assays in drug discovery. *J. Pharm. Sci.,* **93**(6), 1440–1453.

Kibbey, C. E., Poole, S. K., Robinson, B., Jackson, J. D., & Durham, D. (2001). An integrated process for measuring the physicochemical properties of drug candidates in a preclinical discovery environment. *J. Pharm. Sci.,* **90**(8), 1164–1175.

Kleiber, M. (1947). Body size and metabolic rate. *Physiol. Rev.,* **27**(4), 511–541.

Kojima, T., Tsutsumi, S., Yamamoto, K., Ikeda, Y., & Moriwaki, T. (2010). High-throughput cocrystal slurry screening by use of in situ Raman microscopy and multi-well plate. *Int. Pharm. Sci.,* **399**(1-2), 52–59.

Kostewicz, E. S., Aarons, L., Bergstrand, M., Bolger, M. B., Galetin, A., Hatley, O., et al. (2014a). PBPK models for the prediction of in vivo performance of oral dosage forms. *Eur. J. Pharm. Sci.,* **57**, 300–321.

Kostewicz, E. S., Abrahamsson, B., Brewster, M., Brouwers, J., Butler, J., Carlert, S., et al. (2014b). In vitro models for the prediction of in vivo performance of oral dosage forms. *Eur. J. Pharm. Sci.,* **57**, 342–366.

Kraml, M., Dubuc, J., & Gaudry, R. (1962). Gastrointestinal absorption of griseofulvin. II. Influence of particle size in man. *Antibiot. Chemother.,* **12**, 239.

Kreuter, J. (1994). *Colloidal Drug Delivery Systems* (Vol. 66): CRC Press.

Krögel, I., & Bodmeier, R. (1999). Floating or pulsatile drug delivery systems based on coated effervescent cores. *Int. J. Pharm.,* **187**(2), 175–184.

Ku, M. S. (2008). Use of the biopharmaceutical classification system in early drug development. *AAPS J.,* **10**(1), 208–212.

Kuentz, M., Nick, S., Parrott, N., & Röthlisberger, D. (2006). A strategy for preclinical formulation development using GastroPlus™ as pharma-cokinetic simulation tool and a statistical screening design applied to a dog study. *Eur. J. Pharm. Sci.,* **27**(1), 91–99.

Kumar, L., Amin, A., & Bansal, A. K. (2008). Preparation and characterization of salt forms of enalapril. *Pharm. Dev. Technol.,* **13**(5), 345–357.

Kumar, V., & Banker, G. (1996). Target-oriented drug delivery systems. *Drugs Pharm. Sci.,* **72**, 611–680.

Kurkov, S. V., & Loftsson, T. (2013). Cyclodextrins. *Int. J. Pharm.,* **453**(1), 167–180.

Kurkov, S. V., Madden, D. E., Carr, D., & Loftsson, T. (2012). The effect of parenterally administered cyclodextrins on the pharmacokinetics of coadministered drugs. *J. Pharm. Sci.,* **101**(12), 4402–4408.

Kwan, K. C. (1997). Oral bioavailability and first-pass effects. *Drug Metab. Dispos.,* **25**(12), 1329–1336.

Kydonieus, A. F. (1980). *Fundamental Concepts of Controlled Release.* CRC Press: Boca Raton, FL.

Law, D., Schmitt, E. A., Marsh, K. C., Everitt, E. A., Wang, W., Fort, J. J., et al. (2004). Ritonavir-PEG 8000 amorphous solid dispersions: in vitro and in vivo evaluations. *J. Pharm. Sci.,* **93**(3), 563–570.

Lawrence, M. J. (1994). Surfactant systems: their use in drug delivery. *Chem. Soc. Rev.,* **23**(6), 417–424.

Lennernas, H., & Abrahamsson, B. (2005). The use of biopharmaceutic classification of drugs in drug discovery and development: current status and future extension. *J. Pharm. Pharmacol.,* **57**(3), 273–285.

Li, P., & Zhao, L. (2007). Developing early formulations: practice and perspective. *Int. J. Pharm.,* **341**(1), 1–19.

Lind, M. L., Jacobsen, J., Holm, R., & Mullertz, A. (2007). Development of simulated intestinal fluids containing nutrients as transport media in the Caco-2 cell culture model: assessment of cell viability, monolayer integrity and transport of a poorly aqueous soluble drug and a substrate of efflux mechanisms. *Eur. J. Pharm. Sci.,* **32**(4–5), 261–270.

Lindman, B., & Danielson, I. (1981). The definition of microemulsion. *Colloid Surf.,* **3**, 391–392.

Lindmark, T., Nikkilä, T., & Artursson, P. (1995). Mechanisms of absorption enhancement by medium chain fatty acids in intestinal epithelial Caco-2 cell monolayers. *J. Pharm. Exp. Ther.,* **275**(2), 958–964.

Lipinski, C. A., Lombardo, F., Dominy, B. W., & Feeney, P. J. (2001). Experimental and computational approaches to estimate solubility and permeability in drug discovery and development settings. *Adv. Drug Deliv. Rev.,* **46**(1–3), 3–26.

Liversidge, G. G., & Conzentino, P. (1995). Drug particle size reduction for decreasing gastric irritancy and enhancing absorption of naproxen in rats. *Int. J. Pharm.,* **125**(2), 309–313.

Liversidge, G. G., & Cundy, K. C. (1995). Particle size reduction for improvement of oral bioavailability of hydrophobic drugs: I. Absolute oral bioavailability of nanocrystalline danazol in beagle dogs. *Int. J. Pharm.,* **125**(1), 91–97.

Loftsson, T., & Brewster, M. E. (1996). Pharmaceutical applications of cyclodextrins. 1. Drug solubilization and stabilization. *J. Pharm. Sci.,* **85**(10), 1017–1025.

Loftsson, T., & Brewster, M. E. (2010). Pharmaceutical applications of cyclodextrins: basic science and product development. *J. Pharm. Pharmacol.,* **62**(11), 1607–1621.

Lombardo, F., Shalaeva, M. Y., Tupper, K. A., Gao, F., & Abraham, M. H. (2000). ElogPoct: a tool for lipophilicity determination in drug discovery. *J. Med. Chem.,* **43**(15), 2922–2928.

Longer, M. A., & Robinson, J. R. (1990). Sustained-release drug delivery systems (pp. 1676–1693): Mack Publishing Company: Easton, PA, USA.

Lu, E., Rodriguez-Hornedo, N., & Suryanarayanan, R. (2008). A rapid thermal method for cocrystal screening. *CrystEngComm,* **10**(6), 665–668.

Lubchenko, V., & Wolynes, P. G. (2007). Theory of structural glasses and supercooled liquids. *Annu. Rev. Phys. Chem.,* **58**, 235–266.

Maag, H. (2012). Overcoming poor permeability: the role of prodrugs for oral drug delivery. *Drug Discovery Today: Technol.,* **9**(2), e121–e130.

Maher, S., & Brayden, D. J. (2012). Overcoming poor permeability: translating permeation enhancers for oral peptide delivery. *Drug Discovery Today: Technol.,* **9**(2), e113–e119.

Marsac, P. J., Konno, H., & Taylor, L. S. (2006). A comparison of the physical stability of amorphous felodipine and nifedipine systems. *Pharm. Res.,* **23**(10), 2306–2316.

Martignoni, M., Groothuis, G. M., & de Kanter, R. (2006). Species differences between mouse, rat, dog, monkey and human CYP-mediated drug metabolism, inhibition and induction. *Expert Opin. Drug Metab. Toxicol.,* **2**(6), 875–894.

Mathias, N. R., Xu, Y., Patel, D., Grass, M., Caldwell, B., Jager, C., et al. (2013). Assessing the risk of pH-dependent absorption for new molecular entities: a novel in vitro dissolution test, physicochemical analysis, and risk assessment strategy. *Mol. Pharm.,* **10**(11), 4063–4073.

Meng, F., Dave, V., & Chauhan, H. (2015). Qualitative and quantitative methods to determine miscibility in amorphous drug-polymer systems. *Eur. J. Pharm. Sci.,* **77**, 106–111.

Meng, F., Gala, U., & Chauhan, H. (2015). Classification of solid dispersions: correlation to (i) stability and solubility (ii) preparation and characterization techniques. *Drug Dev. Ind. Pharm.,* **41**(9), 1401–1415.

Merkus, H. (2009). Overview of size characterization techniques. *Particle Size Measurements* (vol. 17, pp. 137–194). Springer, Netherlands.

Neslihan Gursoy, R., & Benita, S. (2004). Self-emulsifying drug delivery systems (SEDDS) for improved oral delivery of lipophilic drugs. *Biomed. Pharmacother.,* **58**(3), 173–182.

Nimmerfall, F., & Rosenthaler, J. (1980). Dependence of area under the curve on proquazone particle size and in vitro dissolution rate. *J. Pharm. Sci.,* **69**(5), 605–607.

Okimoto, K., Ohike, A., Ibuki, R., Aoki, O., Ohnishi, N., Irie, T., et al. (1999). Design and evaluation of an osmotic pump tablet (OPT) for chlorpromazine using (SBE)7m-beta-CD. *Pharm. Res.,* **16**(4), 549–554.

Palin, K. J., & Wilson, C. G. (1984). The effect of different oils on the absorption of probucol in the rat. *J. Pharm. Pharmacol.,* **36**(9), 641–643.

Parks, G. S., Huffman, H. M., & Cattoir, F. R. (1927). Studies on glass. II: The transition between the glassy and liquid states in the case of glucose. *J. Phys. Chem.,* **32**(9), 1366–1379.

Pole, D. L. (2008). Physical and biological considerations for the use of nonaqueous solvents in oral bioavailability enhancement. *J. Pharm. Sci.,* **97**(3), 1071–1088.

Porter, C. J., & Charman, W. N. (2001a). In vitro assessment of oral lipid based formulations. *Adv. Drug Deliv. Rev.,* **50**, S127–S147.

Porter, C. J., & Charman, W. N. (2001b). Intestinal lymphatic drug transport: an update. *Adv. Drug Deliv. Rev.,* **50**(1), 61–80.

Porter, C. J., Pouton, C. W., Cuine, J. F., & Charman, W. N. (2008). Enhancing intestinal drug solubilisation using lipid-based delivery systems. *Adv. Drug Deliv. Rev.,* **60**(6), 673–691.

Pouton, C. W., & Porter, C. J. (2008). Formulation of lipid-based delivery systems for oral administration: materials, methods and strategies. *Adv. Drug Deliv. Rev.,* **60**(6), 625–637.

Prasad, D., Chauhan, H., & Atef, E. (2013). Studying the effect of lipid chain length on the precipitation of a poorly water soluble drug from self-emulsifying drug delivery system on dispersion into aqueous medium. *J. Pharm. Pharmacol.,* **65**(8), 1134–1144.

Prasad, D., Chauhan, H., & Atef, E. (2014). Amorphous stabilization and dissolution enhancement of amorphous ternary solid dispersions: combination of polymers showing drug–polymer interaction for synergistic effects. *J. Pharm. Sci.,* **103**(11), 3511–3523.

Prueksaritanont, T., DeLuna, P., Gorham, L. M., Ma, B., Cohn, D., Pang, J., et al. (1998). In vitro and in vivo evaluations of intestinal barriers for the zwitterion L-767,679 and its carboxyl ester prodrug L-775,318: roles of efflux and metabolism. *Drug Metab. Dispos.,* **26**(6), 520–527.

Qian, F., Huang, J., & Hussain, M. A. (2010). Drug-polymer solubility and miscibility: Stability consideration and practical challenges in amorphous solid dispersion development. *J. Pharm. Sci.,* **99**(7), 2941–2947.

Rubino, J. T. (2006). Cosolvents and cosolvency. *Encyclopedia of Pharmaceutical Technology,* 3rd ed., Marcel Dekker: New York, pp. 806–819.

Rumondor, A. C., Stanford, L. A., & Taylor, L. S. (2009). Effects of polymer type and storage relative humidity on the kinetics of felodipine crystallization from amorphous solid dispersions. *Pharm. Res.,* **26**(12), 2599–2606.

Rupp, M., Korner, R., & V Tetko, I. (2011). Predicting the pKa of small molecules. *Comb. Chem. High Throughput Screening,* **14**(5), 307–327.

Sanghvi, T., Katstra, J., Quinn, B. P., Thomas, H., & Hurter, P. (2010). Formulation development of amorphous dispersions. *Pharmaceutical Sciences Encyclopedia.* John Wiley & Sons.

Sarma, B., Nath, N. K., Bhogala, B. R., & Nangia, A. (2009). Synthon competition and cooperation in molecular salts of hydroxybenzoic acids and aminopyridines. *Cryst. Growth Des.,* **9**(3), 1546–1557.

Schachter, D. M., Xiong, J., & Tirol, G. C. (2004). Solid state NMR perspective of drug-polymer solid solutions: a model system based on poly(ethylene oxide). *Int. J. Pharm.,* **281**(1–2), 89–101.

Schultheiss, N., & Newman, A. (2009). Pharmaceutical Cocrystals and Their Physicochemical Properties. *Cryst. Growth Des.,* **9**(6), 2950–2967.

Sekiguchi, K., & Obi, N. (1961). Studies on absorption of eutectic mixture. I. A comparison of the behavior of eutectic mixture of sulfathiazole and that of ordinary sulfathiazole in man. *Chem. Pharma. Bull.,* **9**(11), 866–872.

Serajuddin, A. (1999). Solid dispersion of poorly water-soluble drugs: early promises, subsequent problems, and recent breakthroughs. *J. Pharm. Sci.,* **88**(10), 1058–1066.

Shastri, S., Mroszczak, E., Prichard, R. K., Parekh, P., Nguyen, T. H., Hennessey, D. R., & Schiltz, R. (1980). Relationship among particle size distribution, dissolution profile, plasma values, and anthelmintic efficacy of oxfendazole. *Am. J. Vet. Res.,* **41**(12), 2095–2101.

Shiotani, K., Uehata, K., Irie, T., Uekama, K., Thompson, D. O., & Stella, V. J. (1995). Differential effects of sulfate and sulfobutyl ether of beta-cyclodextrin on erythrocyte membranes in vitro. *Pharm. Res.,* **12**(1), 78–84.

Singh, R., Bharti, N., Madan, J., & Hiremath, S. (2010). Characterization of cyclodextrin inclusion complexes: a review. *J. Pharm. Sci. Technol.,* **2**(3), 171–183.

Sliwoski, G., Kothiwale, S., Meiler, J., & Lowe, E. W., Jr. (2014). Computational methods in drug discovery. *Pharmacol. Rev.,* **66**(1), 334–395.

Spiegel, A. J., & Noseworthy, M. M. (1963). Use of nonaqueous solvents in parenteral products. *J. Pharm. Sci.,* **52**(10), 917–927.

Stella, V. J., & He, Q. (2008). Cyclodextrins. *Toxicol. Pathol.,* **36**(1), 30–42.

Strickley, R. G. (2007). Currently marketed oral lipid-based dosage forms: drug products and excipients. *Oral Lipid-Based Formulations* (pp. 1–32).

Tang, J.-l., Sun, J., & He, Z.-G. (2007). Self-emulsifying drug delivery systems: strategy for improving oral delivery of poorly soluble drugs. *Curr. Drug Ther.,* **2**(1), 85–93.

Taylor, L., & Zografi, G. (1997). Spectroscopic characterization of interactions between PVP and Indomethacin in amorphous molecular dispersions *Pharm. Res.,* **14**(12), 1691–1698.

Thomas, V. H., Bhattachar, S., Hitchingham, L., Zocharski, P., Naath, M., Surendran, N., et al. (2006). The road map to oral bioavailability: an industrial perspective. *Expert Opin Drug Metab. Toxicol.,* **2**(4), 591–608.

Tobyn, M., Brown, J., Dennis, A. B., Fakes, M., Gao, Q., Gamble, J., et al. (2009). Amorphous drug-PVP dispersions: application of theoretical, thermal and spectroscopic analytical techniques to the study of a molecule with intermolecular bonds in both the crystalline and pure amorphous state. *J. Pharm. Sci.,* **98**(9), 3456–3468.

Tong, W. Q., & Whitesell, G. (1998). In situ salt screening–a useful technique for discovery support and preformulation studies. *Pharm. Dev. Technol.,* **3**(2), 215–223.

Uekama, K., Hirayama, F., & Irie, T. (1998). Cyclodextrin drug carrier systems. *Chem. Rev.,* **98**(5), 2045–2076.

Van Eerdenbrugh, B., Van den Mooter, G., & Augustijns, P. (2008). Top-down production of drug nanocrystals: nanosuspension stabilization, miniaturization and transformation into solid products. *Int. J. Pharm.,* **364**(1), 64–75.

Varma, M. V., Ashokraj, Y., Dey, C. S., & Panchagnula, R. (2003). P-glycoprotein inhibitors and their screening: a perspective from bioavailability enhancement. *Pharmacol. Res.,* **48**(4), 347–359.

Varma, M. V., Sateesh, K., & Panchagnula, R. (2005). Functional role of P-glycoprotein in limiting intestinal absorption of drugs: contribution of passive permeability to P-glycoprotein mediated efflux transport. *Mol. Pharm.,* **2**(1), 12–21.

Vasanthavada, M., Tong, W.-Q., Joshi, Y., & Kislalioglu, M. S. (2005). Phase behavior of amorphous molecular dispersions. II: Role of hydrogen bonding in solid solubility and phase separation kinetics. *Pharm. Res.,* **22**(3), 440–448.

Vertzoni, M., Dressman, J., Butler, J., Hempenstall, J., & Reppas, C. (2005). Simulation of fasting gastric conditions and its importance for the in vivo dissolution of lipophilic compounds. *Eur. J. Pharm. Biopharm.,* **60**(3), 413–417.

Vishweshwar, P., McMahon, J. A., Peterson, M. L., Hickey, M. B., Shattock, T. R., & Zaworotko, M. J. (2005). Crystal engineering of pharmaceutical co-crystals from polymorphic active pharmaceutical ingredients. *Chem. Commun.* (36), 4601–4603.

Wan, H., & Ulander, J. (2006). High-throughput pKa screening and prediction amenable for ADME profiling. *Expert Opin. Drug Metab. Toxicol.,* **2**(1), 139–155.

Wang, J., & Faller, B. (2007). Progress in bioanalytics and automation robotics for ADME screening. *Compre. Med. Chem.,* **5**, 341–356.

Wang, J., Urban, L., & Bojanic, D. (2007). Maximising use of in vitro ADMET tools to predict in vivo bioavailability and safety.

Watari, N., Funaki, T., Aizawa, K., & Kaneniwa, N. (1983). Nonlinear assessment of nitrofurantoin bioavailability in rabbits. *J. Pharmacokinet. Biopharm.,* **11**(5), 529–545.

Williams, H. D., Anby, M. U., Sassene, P., Kleberg, K., Bakala-N'Goma, J.-C., Calderone, M., et al. (2012a). Toward the establishment of standardized in vitro tests for lipid-based formulations. 2. The effect of bile salt concentration and drug loading on the performance of type I, II, IIIA, IIIB, and IV formulations during in vitro digestion. *Mol. Pharm.,* **9**(11), 3286–3300.

Williams, H. D., Sassene, P., Kleberg, K., Bakala-N'Goma, J. C., Calderone, M., Jannin, V., et al. (2012b). Toward the establishment of standardized in vitro tests for lipid-based formulations, part 1: method parameterization and comparison of in vitro digestion profiles across a range of representative formulations. *J. Pharm. Sci.,* **101**(9), 3360–3380.

Williams, H. D., Sassene, P., Kleberg, K., Calderone, M., Igonin, A., Jule, E., et al. (2013). Toward the establishment of standardized in vitro tests for lipid-based formulations, part 3: understanding supersaturation versus precipitation potential during the in vitro digestion of Type I, II, IIIA, IIIB and IV lipid-based formulations. *Pharm. Res.,* **30**(12), 3059–3076.

Williams, H. D., Sassene, P., Kleberg, K., Calderone, M., Igonin, A., Jule, E., et al. (2014). Toward the establishment of standardized in vitro tests for lipid-based formulations, part 4: proposing a new lipid formulation performance classification system. *J. Pharm. Sci.,* **103**(8), 2441–2455.

Williams, R. O. (2000). Solubility and Solubilization in Aqueous Media By Samuel H. Yalkowsky (University of Arizona). Oxford University Press: New York. 1999. xvi + 464 pp. $165. ISBN 0-8412-3576-7. *J. Am. Chem. Soc.,* **122**(40), 9882–9882.

Wohnsland, F., & Faller, B. (2001). High-throughput permeability pH profile and high-throughput alkane/water log$P$ with artificial membranes. *J. Med. Chem.,* **44**(6), 923–930.

Wolynes, P. G., & Lubchenko, V. (2012). *Structural Glasses and Supercooled Liquids: Theory, Experiment, and Applications.* John Wiley & Sons.

Yoo, S. U., Krill, S. L., Wang, Z., & Telang, C. (2009). Miscibility/stability considerations in binary solid dispersion systems composed of functional excipients towards the design of multi-component amorphous systems. *J. Pharm. Sci.,* **98**(12), 4711–4723.

Yu, L., Amidon, G., Polli, J., Zhao, H., Mehta, M., Conner, D., et al. (2002). Biopharmaceutics classification system: the scientific basis for biowaiver extensions. *Pharm. Res.,* **19**(7), 921–925.

Zhang, S.-W., Harasimowicz, M. T., de Villiers, M. M., & Yu, L. (2013). Cocrystals of Nicotinamide and (R)-Mandelic Acid in Many Ratios with Anomalous Formation Properties. *J. Am. Chem. Soc.,* **135**(50), 18981–18989.

Zheng, W., Jain, A., Papoutsakis, D., Dannenfelser, R. M., Panicucci, R., & Garad, S. (2012). Selection of oral bioavailability enhancing formulations during drug discovery. *Drug Dev. Ind. Pharm.,* **38**(2), 235–247.

Zhou, D., Grant, D. J., Zhang, G. G., Law, D., & Schmitt, E. A. (2007). A calorimetric investigation of thermodynamic and molecular mobility contributions to the physical stability of two pharmaceutical glasses. *J. Pharm. Sci.,* **96**(1), 71–83.

Chapter 6

A Staged Approach to Pharmaceutical Dissolution Testing

Gregory K. Webster, Xi Shao, and Paul D. Curry, Jr.

AbbVie Inc., Global Research and Development, 1 N. Waukegan Rd.,
North Chicago, IL 60064, USA

gregory.webster@abbvie.com

6.1 Introduction

The development of a meaningful dissolution procedure for drug products at different development stages has been a consistent challenge to the pharmaceutical industry.[1,2] The objective of dissolution testing, in general, varies during the development stages of a dosage form.[3] During preclinical development, dissolution plays an important role of selecting appropriate drug substance and dosage form for the toxicology formulation development to ensure adequate bioavailability. The primary objective of dissolution testing at phase 1 is to characterize the active pharmaceutical ingredient (API), ensure that phase 1 formulation does not release faster than the formulation used in the toxicology assessment, develop first-in-human (FIH) formulations, and establish the preliminary

Poorly Soluble Drugs: Dissolution and Drug Release
Edited by Gregory K. Webster, J. Derek Jackson, and Robert G. Bell
Copyright © 2017 Pan Stanford Publishing Pte. Ltd.
ISBN 978-981-4745-45-1 (Hardcover), 978-981-4745-46-8 (eBook)
www.panstanford.com

dissolution mechanism. During phases 2 and 3, the objective shifts towards developing an understanding of the impact of key formulation/process parameters on the dissolution profile and an in vitro/in vivo correlation or relationship (IVIVC/IVIVR), if possible. At product registration and beyond, the goal is to identify a quality control (QC) dissolution test method for the product that correlates to the clinical formulations. Dissolution testing is also used by regulatory agencies for granting biowaivers and for post-approval manufacturing changes. The FDA guidance for waivers of in vivo bioavailability and bioequivalence studies ("Waiver of in vivo Bioavailability and Bioequivalence Studies for Immediate-Release Solid Oral Dosage Forms Based on a Biopharmaceutics Classification System") employs dissolution testing to demonstrate rapid dissolution of immediate-release solid oral dosage forms so that a biowaiver can be granted.[4] Continual evolution of these objectives during the drug product life cycle may require different concentrations and, therefore, require changes in the analytical finish.

This chapter reviews the dissolution testing considered at the stages of drug development and the collaborations among the many functional departments to deliver a meaningful dissolution method. Business concerns and individual companies may vary in specific approaches; nonetheless, the overall strategy is often consistent at each development stage. Because of the highly regulated and controlled environment of drug development, there are a range of considerations to be addressed for dissolution testing during drug development to ascertain adequacy and accuracy of data submitted to the regulatory agencies. Some considerations are:

- What dissolution testing should be conducted to ensure effective drug development?
- What criteria should be utilized to decide which formulations are likely commercial winners and losers using dissolution testing?
- What is an efficient experimental design to avoid wasted laboratory resources and unnecessary expenditure on projects?

These concerns will be addressed in the context of the broad strategic challenges faced in developing new drugs.

Typically, development projects fail to adhere to a single development process as individual companies tend to employ their own drug delivery platforms. Generalized key activities undertaken by functional departments conducted in each development phase will be described. The first section describes the "terrain" of drug development, the phases of drug development, and the key objectives of these phases conducted by different functional departments: the Chemistry Manufacturing Control (CMC) project team, Formulation Leads, Dissolution Leads, Clinical Pharmacokinetics (PK) group, and Clinical supply team. The titles and specific groups of individual business concerns may vary, but the general deliverables remain the same.

6.2 Dissolution at Each Stage of Drug Development

Development of dissolution methods and testing typically follows the drug development paradigm.

6.2.1 Pilot Formulation Development

The initial pilot formulation exists up to the FIH/phase 1a clinical trials. These formulations may be tested in animal studies to get an estimation of how the formulations may perform in humans.[5] Additionally, the formulations may be environmentally stressed to evaluate if a potential impact during stability testing exists.

The FIH stage illustrated in Fig. 6.1 is intended to establish the safety of the drug in humans and to gather initial human pharmacokinetic data for the drug and formulation. At the beginning of a new FIH project, the properties of the API must first be determined and aligned to the target market profile for the drug product. Some questions that should be considered are:

- Is the drug going to be crystalline or amorphous?
- What is the compound's pK_a and salt form?
- What is the permeability of this compound and its absorption pathway?

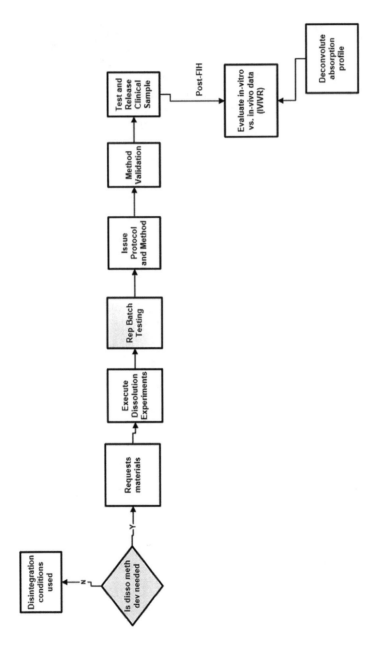

Figure 6.1 FIH schematic.

A pH solubility profile of the API needs to be established as well as its specific solubility in common dissolution buffers and surfactants (if needed). A preliminary assignment to a BCS (Biopharmaceutics Classification System) class is needed.[6]

6.2.1.1 CMC project team

The CMC project team is organized and sets the preliminary plans to formulate the new API to begin FIH studies. The project team sets its initial scheduling for formulation and methods development to meet the requirements of the early prototype of the desired target drug profile. While the goal of FIH is to establish safety, these clinical studies are the first time the drug will be manufactured and tested under GMP requirements and the trials yield the first data on how the drug will behave in human subjects.

6.2.1.2 Formulation lead

With the information provide above, the formulation group can begin their preliminary methods of formulating the API.[7]

6.2.1.3 Dissolution lead

Concurrently in this stage of development time, the dissolution lead should be identifying the complexity of the development project. For example, will this work use a simple buffer system and sink solubility, or will solubility be aided through the use of surfactants? At this early stage and with a highly soluble drug, perhaps disintegration testing replaces dissolution as the path forward. However, disintegration testing would not be appropriate for poorly soluble drugs.

Generally, materials such as excipients and drug product samples are needed for the preliminary dissolution work. If exploratory formulation (e.g., API in capsule and simple formulated capsule/tablet) is intended to be used at this stage, the dissolution lab will minimally require 4 to 6 dosage units of the highest dose to begin method development. At this stage of development, the dissolution method is solely intended to establish that the formulation provides full release *in-vitro*. When a suitable method is developed, the laboratory will execute a life-cycle appropriate level of validation in order to test and release clinical supplies. (Chapter 18 of this book details the process for dissolution method development.) The most important aspect of this stage is the establishment of a representative dissolution profile of the specific FIH formulation.

6.2.1.4 PK group

Once the FIH clinical data becomes available, the group responsible for pharmacokinetics needs to establish the drug absorption profile for each formulation. This preliminary C_{max} and T_{max} will be the basis for the next stage of development. This information also provides valuable insight into the changes needed in the dissolution method to make it representative of the observed in vivo data.

6.2.1.5 Samples needed

For the early clinical lots, sample requirements are relatively low. As a minimum, our experience suggests the development team set aside 24 dosage units of each representative lot for each stability pull point in case investigations are needed and an additional 24 dosage units of each clinical lot for release testing. In addition, the development team should retain a minimum of 100 dosage units of each clinical lot for future dissolution method development investigations, as well as to analyze using the proposed commercial dissolution method.

6.2.2 Development Formulations: Phase 1B/2A

During phase 1B/2A, the goal is to verify the acute toxicity dose proportionality and biopharmaceutical proprieties of preliminary market formulation(s). Combined with phase 1B studies, phase 2A initiates pilot clinical trials not only to understand the safety of a potential drug, but also provide an early evaluation of potential efficacy and dose strength need in selected populations of patients. Objectives may focus on dose response, type of patient, frequency of dosing, or numerous other characteristics of safety and efficacy. Dissolution strategies illustrated in Fig. 6.2 can be useful in bridging the FIH/phase 1a clinical data and formulation development in

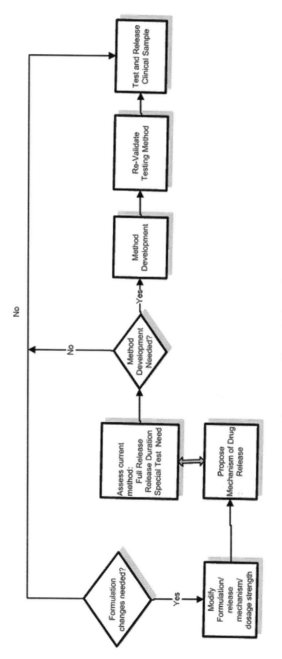

Figure 6.2 Phase 1B/2A schematic.

phase 1B/2A, understanding the drug release mechanism, building the knowledge foundation for the manufacturing design space and target product profile, as well as monitoring the stability performance of formulation prototypes.

6.2.2.1 Project team

The results from FIH studies are evaluated to compare the dissolution method and absorption profiles. The project team, as a whole, will review whether the clinical absorption profile of the FIH formulation study is representative of the target profile that the business is targeting. If not, new formulation studies and development samples are needed.

6.2.2.2 Formulation lead

If the clinical drug absorption profile is misaligned with the desired target profile, a change in formulation is needed. The formulation lead needs to evaluate whether the project should modify the formulation, change the release mechanism from what was used at FIH, or simply change the dosage strength. Most importantly, the formulation team works to propose and narrow the target mechanism of drug release, based on dissolution and clinical data, for the formulation design going forward.

6.2.2.3 Dissolution lead

The work of the dissolution lead at this point is twofold. First, the project team needs to evaluate whether the in vitro method is consistent with the clinical absorption profile. If not, the dissolution method needs to be optimized to correlate with the C_{max} and T_{max} of the preliminary drug absorption profiles. Such alignment with the absorption profile will lead to a more clinically relevant dissolution method in the long run.

Secondly, the dissolution lead must work with the formulation lead to experimentally evaluate any new formulation changes resulting from the evaluation of the first clinical trial. Some researchers at this point would argue that the dissolution lead should continue to use the initial FIH dissolution method to evaluate the phase 1B/2A samples in order to maintain consistency in results. Others argue that if the dissolution and in vivo results do not correlate the laboratory should evaluate changes to the method and then, in subsequent in vivo studies, compare how much closer the updated method has gotten. In either case, the team must maintain sufficient development samples of these individual formulations so that the final method can bridge to earlier dissolution method results. A systematic dissolution method development process should be maintained.

During phase 1B/2A formulation optimization, such as new excipients and release mechanisms, may be evaluated. The dissolution team evaluates and assists this effort through continuing dissolution method development and testing. The dissolution lead should be constantly looking to ensure full release in various dissolution media and evaluating the ability of the method to discriminate critical process control parameters. As possible, this is all done with hope of maintaining consistency to the C_{max} and T_{max} from the clinical drug absorption profile!

This is the point in dissolution method development where the potential to deliver a clinically relevant in vitro method is evaluated. Every change in formulation or manufacturing process requires re-evaluation of whether the change affects the discrimination capability of the dissolution method and the impact upon in vitro release. As such, true clinical relevance evaluation of the in vitro method should take place at later development stages (after proof of principle) with the "final" formulation and release mechanism. The greatest challenge of this phase of dissolution method development is to understand the relationship of method variability, method reproducibility, and drug-product variability so as to be able to determine whether sources of error are coming from the sample or from the method.

With each new formulation going into clinical studies, the project team must evaluate impact on the method validation and suitability of the specification as they change. At this early stage of development, the business may use a general release specification of "report results" for the dissolution profile or propose acceptance criteria to ensure there are no gross changes in the dissolution profile.

6.2.2.4 PK group

The primary goal for the PK group at this stage is maintaining the drug absorption profiles for the project and correlating them to formulation development goals, as well as evaluating how each formulation changes the C_{max}, T_{max}, AUC, and duration time from the clinical data. Evaluation of the rank order in the deconvolution absorption profile is critical in assisting formulation and dissolution method development at this point in the project studies. This group is now exploring how the critical manufacturing attributes (CMA's) and critical process parameters (CPP's) affect the in vitro and in vivo data.

6.2.2.5 Samples needed

Sample requirements at this stage of development remains at a minimum of 24 dosage units for each formulation being developed and 100 dosage units for each formulation used in the clinic.

6.2.3 Development Formulations: Phase 2B

Pharmaceutical phase 2B development is significant because the financial investment at this stage of development becomes substantial. To the business, the drug looks like it has real market potential and the studies are now being designed to build a path to approval. At this development stage, manufacturing control strategies are being implemented, optimized, and finalized. Formulation strategies and analytical methods must be implemented, finalized, and validated for use in the final stages of development. Analytical and formulation considerations such as whether the methods and processes are robust, confirming the formulation is delivering the drug as targeted and processes are under control are evaluated.

6.2.3.1 Project team

Phase 2B is the development stage to finalize and update every setting needed in the analytical methods and formulation manufacturing process (Fig. 6.3). The next development stage requires significant financial investment by the sponsor. Any future changes

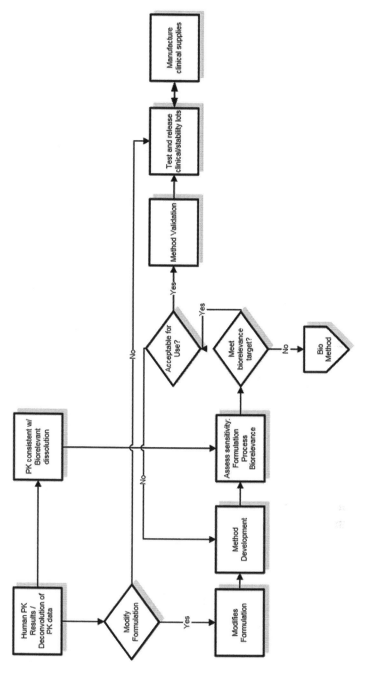

Figure 6.3   Phase 2B schematic.

will come at a significant financial and timing cost. The goal is to have methods and processes finalized by the end of the phase 2B development stage. Thus, barring any unforeseen testing or clinical results, this stage should be the final time in which the team will modify the formulation. The final decision needs to be made whether the project can support and pursue an in vitro/in vivo correlation (IVIVC) as part of the regulatory filings. At this stage, adequate data to evaluate the solubility, pharmacokinetic, and permeability attributes of drug substance is available. The dissolution profiles of dosage forms should be available to evaluate whether biowaiver is appropriate and can be proposed.

6.2.3.2 Formulation lead

The formulation lead will be updating and finalizing every setting for the formulation manufacturing process. In addition, efficacy and toxicity data of candidate drug product obtained from phase 2A studies will be used as a crucial standard to determine the necessity of formulation optimization in phase 2B stage. This should be the final time to modify the formulation and ultimately, towards the end of this process, to decide whether to pursue an IVIVC for the formulation. However, IVIVC is only possible if formulations using the same release mechanism have shown a difference in the in vivo pharmacokinetic profile.

6.2.3.3 Dissolution lead

The work of the dissolution lead continues to evaluate whether the in vitro method is consistent with the clinical drug absorption profile and discriminates critical process attributes. Further method optimization can proceed, as warranted. The analytical laboratory is focused on release and stability testing for clinical lots.

Phase 2B studies are focused on defining the dosage form, strength(s) and related acceptance criteria, which are heavily reliant upon the efficacy and toxicity data collected in the phase 2A clinical studies. Dissolution method development continues in parallel with formulation optimization and manufacturing process characterization. The studies eventually lead to the final dosage form that will be evaluated in phase 3 trials. Further method optimization needed as a result of any formulation change proceeds as warranted. Nevertheless, the evaluation of whether the in vitro method is consistent with the clinical drug absorption profile and can discriminate critical process attributes will be addressed at this stage. A correlation between design space and the quality target product profile will also be demonstrated.

The establishment of IVIVC or IVIVR dissolution method is critical at this stage of development not only due to the urgency of building a link between the final formulation and efficacy, but also for guiding the quality control strategy needed in the coming phase 3 pivotal studies. During the continuous development of IVIVC/IVIVR, dissolution data always serves as the most significant set of input utilized in the modeling tools (Gastroplus, PDx-IVIVC, or WinNonlin), in comparing pharmacokinetic/pharmacodynamics properties, API properties, and dosage form information. Therefore, the dissolution method used to build the IVIVC or IVIVR should be capable of reflecting a clinical relevance, while also discriminate formulation and process changes.

6.2.3.4 PK group

The primary goal the PK group continues to be maintaining the project absorption profiles and their correlation to formulation development. The pharmacokinetic and pharmacodynamic properties of the drug product should also be provided in order to support the potential development of an IVIVC/IVIVR. The drug absorption mechanism and window should be available for formulation optimization at this stage.

6.2.3.5 Samples needed

No changes from phase 1B/2A.

6.2.4 Final Formulation Development: Phase 3

Phase 3 is the final stage of validation for the new drug. At this point, the processes and methods are final. It is now up to the manufacturing concern to show they are in control of the process and the resulting product repeatedly meets product specifications and stability. Three consecutive registration lots are produced and tested for documentation of this manufacturing control as well as for final clinical testing. This is also the time to potentially demonstrate the suitability of the dissolution method to regulatory authorities.

6.2.4.1 Project team

At this last major stage for the development project, the project team should be working within the proposed clinical dose strengths of the product. Working with the marketing team, the coating of the product should be finalized as well as any embossing. The project team should have a clear understanding of any potential stability constraints on the expiration dating for the product. The dissolution method should be finalized and sensitive enough to monitor sample stability and process parameter changes. The formulation and the process are finalized entering this stage of development and the critical quality attributes are identified from formulation and bench process development studies. The dissolution method should be finalized and the dissolution specifications are proposed by the team for the regulatory submission.

6.2.4.2 Formulation lead

The formulator will be addressing any requests to modify the formulation in terms of strength, coating, or stability attributes. With each modification the corresponding dissolution method must examined the samples need to be provided to ensure relevant discrimination is maintained. The formulator leads the effort to identify the critical process parameters in the manufacturing process and provides samples to determine if the dissolution method is selective for such parameters. Phase 3 requires the manufacturing of significant quantities of clinical supplies. A proper design of experiments (DOE) that includes an examination of the manufacturing process is needed in order to establish the control space for quality by design (QbD) manufacturing of the product, which is encouraged. Things to consider are drug substance particle size distribution, granule particle size, compression force, tapped/bulk density, levels of release and binding excipients, effect of coating material, etc., relevant to the type of release mechanism specific to the final formulation.

6.2.4.3 Dissolution lead

The analytical groups at this stage work to evaluate the stability shelf-life of the drug product based on the phase 2B clinical materials and support formulation development with the phase 2B dissolution method (Fig. 6.4). Ideally, the formulation should be finalized and process parameters tuned to determine the critical process parameters. With any modification to the formulation or manufacture process, the dissolution method must also be re-evaluated to ensure that it remains sensitive to the critical process parameters. The analytical labs test and release the lots for the final clinical studies and initiate stability registration investigations. Dissolution studies will be conducted by the analytical group to bridge phase 2 and phase 3 clinical supplies and commercial lots and, if necessary, bioequivalence studies will be conducted by the PK group. The dissolution method is used to support any process controls and identify relevant control space. Clinically relevant release and stability specifications should be targeted.

6.2.4.4 PK group

The main task of the PK group is to evaluate the in vivo performance of the drug products. This group provides the link between the in vivo performance and the formulation, and between the in vivo results to in vitro testing as well. Evaluation of these links is paramount to establishing clinically relevant specifications for the final product. Additionally, this group should be looking to confirm whether different formulations (i.e., changes in the composition of release-controlling excipients) behave differently in the body. This can provide justification of the range of the acceptance criteria proposed.

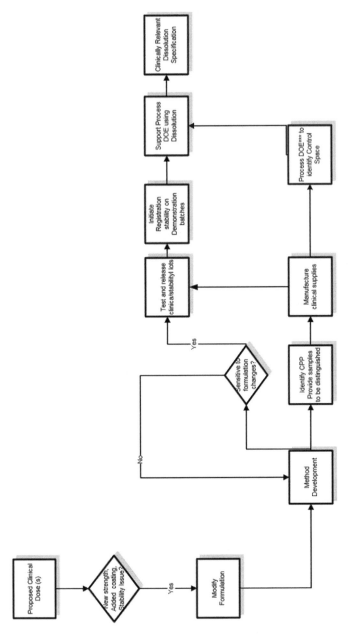

Figure 6.4 Phase 3 schematic.

6.2.4.5 Samples needed

The phase 3 stage of development has significant sampling needs: the project needs retained samples from formulations illustrating differing drug release behavior, stress samples from storage and stability, as well as formulation and process development samples. Again, our experience is that dissolution studies of these samples generally require 24 dosage units of each representative condition. The method validation requires 100 dosage units for each strength, which may vary depending on the specific validation protocol. Additional samples to establish statistical equivalency may be needed.

6.2.5 In vitro/in vivo Correlation

IVIVC (in vitro/in vivo correlation) is discussed extensively in later chapters. However, since biorelevance is often discussed during the stages of development, an overview is presented here. It is important to note that biorelevance and clinical relevance is often misunderstood. Oftentimes, the term biorelevance dissolution is simply referring to dissolution studies in media simulating the gastrointestinal tract. Clinical relevancy, which is much more important to the pharmaceutical scientist, is a dissolution method that is capable of predicting or correlating to clinical results. The process and requirements to establish an in vitro/in vivo correlation for a dissolution method are outlined in the FDA guidance on the subject.[9]

The requirements for IVIVC include the following: (a) all IVIVC lots must release drug by the same mechanism, (b) all IVIVC lots much have the same excipients, (c) all drug release testing must be comprised of $n \geq 12$ number samples, (d) formulations used in IVIVC evaluations should be different by f2 test calculations, and (e) external validation is required for any low therapeutic index drug. Additionally, the in vivo tests of the products must demonstrate a difference in the pharmacokinetic profiles. If such a difference cannot be demonstrated, the data can be used to support wider acceptance criteria for the chosen product.

6.2.5.1 Project team

Generally, drug candidates from BCS Class I are expected to establish an IVIVC for the final dissolution method. Not all formulations and dosage forms have the potential for establishing IVIVC.[8] The project team should review if the criteria for the dosage forms being developed meet the IVIVC criteria, especially for other BCS classes of drugs.

6.2.5.2 Formulation lead

The Formulation lead should establish the mechanism of release for the drug product formulation. This is critical for IVIVC development for all the subsequent samples need to use this mechanism of release as well. The IVIVC studies require dosage forms that release faster and slower than the current target profile.

6.2.5.3 Dissolution lead

As illustrated in Fig. 6.5, the dissolution lead begins establishing an IVIVC by performing in vitro/in vivo relationship (IVIVR) studies to determine biorelevance of the release method. This method is needed to characterize trial formulations. If the results of the faster and slower releasing formulations illustrate a difference by the f2 test, then the team can proceed with manufacture of additional clinical supplies for the IVIVC studies. Obtaining similar results by f2 testing illustrates the potential need for further dissolution method development. Once an IVIVC method is obtained, the method should be validated and used to test all phase 3 clinical supplies to support external validation.

6.2.5.4 PK group

The main task of the PK group is to evaluate the human PK results from clinical testing and produce deconvoluted dissolution profiles for the dissolution lead to design a method to meet. The PK group also evaluates the in vitro results against the actual clinical in vivo performance of the drug products.

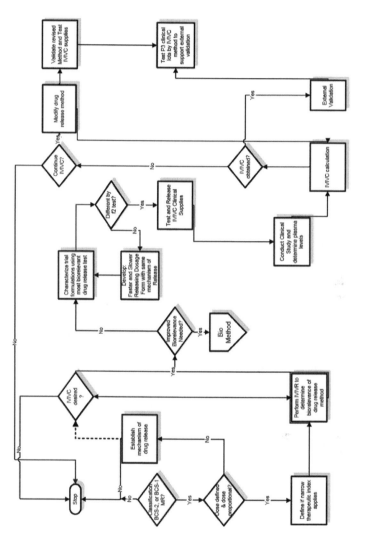

Figure 6.5 IVIVC schematic.

The PK group needs help to determine if the API is classified as BCS class 1–2 (generally, class 3 and 4 drugs cannot obtain an IVIVIC). With the dosage definition, the PK group needs to evaluate if the drug is dose proportional and if a narrow therapeutic index applies. The PK group helps design clinical studies that are sufficiently powered for the IVIC trial requirements. The trials need to determine plasma levels of the drug and use these results to develop the IVIVC calculation and ultimately in external validation of the prediction model.

6.2.5.5 Samples needed

Representative samples from phase 2 (if applicable) and phase 3 development studies are required as well as samples of the trial formulations for release characterization.

6.3 Conclusion

The process of dissolution testing has been described as it develops in relation to a drug's journey through the drug product development process. This chapter illustrates the function of dissolution testing at each drug development stage, as well as the essential collaborations among the sponsor's functional groups needed to deliver meaningful dissolution methodology and the results from early development to the post launch surveillance of product quality. In early phases of product development, dissolution serves as a tool for API screening, formulation selection and optimization, understanding the drug-release mechanism and critical manufacture parameter identification. During later stages of product development, dissolution testing is demanded as a sensitive and a reliable predictor of stability, ensuring batch-to-batch consistency and helping to establish bioequivalence between formulations. With the increasing utilization of the BCS classification system and introduction of new regulatory QbD concepts, more innovative and science-based approaches need to be assessed in order to ensure the consistency of dissolution development for the oral dosage forms. With a greater focus on achieving IVIVC/R in the pharmaceutical industry, more cross functional efforts and rational dissolution activities need to be emphasized in order to improve dissolution testing to meet the demands for rational drug development, meaningful acceptance-criteria setting and strong regulatory justifications.

Acknowledgments

The authors would like to thank Jian-Hwa Han, Anagha Vaidya, and Susan George of AbbVie for their insightful discussions.

Author Disclosure

AbbVie provided no financial support outside of Drs. Webster, Shao, and Curry being employees of AbbVie. AbbVie participated in the writing, reviewing, and approving the publication. The presentation contains no proprietary AbbVie data.

References

1. Lipsky, M. S., and Sharp, L. K. *J. Am. Board Fam. Med.*, 2001, 14(5), 362–367.
2. Crowther, J. B., Lauwers, W., Adusumalli, S., and Shenbagamurthi, P. The laboratory analyst's role in the drug development process, in *Analytical Chemistry in a GMP Environment*, J. M. Miller and J. B. Crowther, eds. (Wiley, New York, 2000), Chapter 1, p. 3.
3. Hawley, M. Stage appropriate dissolution methods in formulation development. *Am. Pharm. Rev.*, 2013, 16, 1–8.
4. FDA Guidance for Industry Waiver of In Vivo Bioavailability and Bioequivalence Studies for Immediate-Release Solid Oral Dosage Forms Based on a Biopharmaceutics Classification System (CDER, FDA).
5. ICH Topic E8: General considerations for clinical trials.
6. Benet, L. Z. *J. Pharm. Sci.*, 2013, 102(1), 34–42.
7. Guidelines for Industry: CGMP for Phase 1 Investigational Drugs, July 2008.

8. Validation of Analytical Procedures: Text and Methodology, Q2(R1), International Conference on Harmonization of Technical Requirements for Registration of Pharmaceuticals for Human Use, November 1996.
9. Guidelines for Industry: Extended Release Oral Dosage Forms Development, Evaluation, and Application of in vitro/in vivo Correlations, September 1997.

Chapter 7

Development and Application of in vitro Two-Phase Dissolution Method for Poorly Water-Soluble Drugs

Ping Gao, Yi Shi, and Jonathan M. Miller

GPRD, AbbVie, 1 North Waukegan Road, North Chicago, IL 60064, USA
ping.gao@abbvie.com

7.1 Introduction

Dissolution of drugs from their oral dosage forms in the gastrointestinal (GI) tract is a prerequisite to the absorption process. Examination of dissolution attributes of a drug from its formulation is of critical importance in the development of oral drug products. It is highly desirable that the in vitro dissolution test for drug products closely mimics the in vivo situation in preclinical species and human subjects in order to assess its bioperformance.[1-3] Biorelevant media such as fasted simulated gastric fluids (FaSGF), fasted simulated small intestine fluids (FaSSIF), fed simulated small intestine fluids (FeSSIF), and bicarbonate buffers have been widely used to simulate conditions in the GI tract for in vitro dissolution

Poorly Soluble Drugs: Dissolution and Drug Release
Edited by Gregory K. Webster, J. Derek Jackson, and Robert G. Bell
Copyright © 2017 Pan Stanford Publishing Pte. Ltd.
ISBN 978-981-4745-45-1 (Hardcover), 978-981-4745-46-8 (eBook)
www.panstanford.com

and solubility assessments,[4–5] especially for Biopharmaceutical Classification System (BCS) type II and IV drugs. Biorelevant dissolution test methods with appropriate simulated media and hydrodynamics are useful during early stages of drug development for identifying the key issues associated with the biopharmaceutical performance of the drug (i.e., solubility limited- or dissolution-limited absorption, food effect, precipitation in the small intestine, etc.) and assessing manufacturability of the drug product with its impact on bioperformance. Application of biorelevant dissolution tests can effectively facilitate formulation selection and optimization during drug product development.

One of the most important yet also most difficult tasks in the biopharmaceutical evaluation of a drug is to establish a quantitative or qualitative relationship between in vitro dissolution profiles and in vivo pharmacokinetic performance.[1–5] Commonly referred to as in vitro–in vivo relationship (IVIVR) or in vitro–in vivo correlation (IVIVC), this approach usually not only enables rapid development of drug product with optimal bioperformance attributes but also minimizes animal experimentation, and bioavailability studies in human subjects.

Understanding of the human GI physiology and identification of key formulation attributes are prerequisites to the rational development of biorelevant drug release tests. Dissolution test devices enabling a realistic simulation of the mechanical as well as physicochemical factors experienced by dosage forms during the GI transit can be realized in many ways. The two-phase dissolution test method involving sequential "dissolution-partition" processes in two different phases was first introduced for evaluating drug formulations in 1967.[6] This test is designed to evaluate the dissolution of a drug from its dosage form into an aqueous phase and subsequent partitioning into a water immiscible organic phase. In this way, the two-phase dissolution test method mimics both the dissolution and membrane permeation processes simultaneously occurring in vivo during oral absorption. The presence of the "absorptive phase" by the two-phase dissolution test offers a distinct advantage as compared to other dissolution tests involving only an aqueous phase. This chapter will summarize the most recent developments in experimental methods, theoretical models and applications of the two-phase dissolution test method in drug formulation development with emphasis on IVIVR.

7.2 Development and Applications of Two-Phase Dissolution Test

7.2.1 Overview of Two-Phase Dissolution-Partition Method

Two-phase dissolution test was first reported by Niebergall and coworkers, who used an upper organic phase (i.e., octanol) with the presence of a lower phase of an aqueous medium.[6] This novel in vitro test method utilized media comprising immiscible aqueous and organic phases. Following dissolution in the aqueous phase, the dissolved drug rapidly partitioned into the organic phase due to its high lipophilicity. The dissolution and partition occurred simultaneously and the drug was continuously transferred into the organic phase which maintained a sink condition. Since 1967, further research has been reported to develop and apply this novel methodology, with major focus on characterization of simultaneous drug dissolution and partitioning kinetics of drug products.[6–31] The major feature that differentiates the two-phase dissolution system from the single aqueous phase dissolution system is the presence of the organic phase, resulting in simultaneous partitioning of the drug from the aqueous phase to the organic phase. The presence of the organic phase with sink capacity fundamentally impacts the dissolution kinetics in the aqueous phase due to continuous and rapid removal of the drug from the aqueous phase to the organic phase. The drug concentration–time profile observed in the aqueous phase is essentially dictated by the dynamic balance between the drug dissolution rate and partitioning rate. Therefore, the drug–concentration–time profile observed in the aqueous phase may be more representative of the dynamic dissolution-absorption processes that occur in the GI tract. In contrast, the drug concentration–time profile observed from a single phase dissolution system is accumulative, approaching a complete release under sink conditions. Even when a biorelevant dissolution media is employed and the test is conducted under non-sink conditions, the drug dissolution–time profile of a poorly soluble drug observed in such single phase aqueous media is not representative of the in vivo situation due to the lack of partitioning kinetics.

During the last decade, great efforts have been made to understand the supersaturation of poorly soluble weak acidic and basic drugs in the GI tract and its relevance to oral absorption.[32–38] It is well understood that generating a highly supersaturated state of the drug in vivo may result in spontaneous precipitation. The kinetics and mechanism(s) of nucleation and growth of a drug are known to be highly sensitive to the degree of supersaturation, as well as the presence of certain excipients (e.g., water soluble cellulosic polymers) which can inhibit or retard precipitation and thus stabilize the supersaturated state. It has been widely recognized that the competition among the kinetics of dissolution, partitioning and precipitation jointly dictates the amount of free drug in the aqueous phase available for absorption through the GI membrane. In vitro two-phase dissolution tests are highly suitable to characterize the interplay among the drug dissolution rate, the degree of supersaturation, and the precipitation kinetics from different formulations.[26,30,31] In addition, theoretical models describing both the kinetic processes of dissolution and partition have been developed and reported in the literature.[6–8,11,21,22,29]

In general, the two-phase dissolution test method offers three major advantages over single aqueous phase based dissolution tests with respect to establishing IVIVR First, a biorelevant aqueous medium under non-sink condition can be utilized for poorly water soluble drugs and the dissolution rate observed in such medium reflects the amount of drug available for partitioning into the organic phase. In contrast, the organic phase possessing a sink-condition for the drug of interest can be obtained at the same time, permitting rapid partition of the drug into this phase. For BCS II/IV drugs, the coexistence of a non-sink condition associated with dissolution and a sink condition associated with absorption is well recognized in GI tract and the two phase dissolution system is designed to mimic these features.

Second, the dissolution and partition of drug between the two phases may mimic the processes of drug dissolution and absorption in the GI tract. Unlike conventional single phase dissolution tests the two-phase dissolution test usually reveals a steady state without accumulation of the dissolved drug in the aqueous phase due to rapid removal into the organic phase through partition (assuming a high log P which is generally the case for a poorly soluble, lipophilic drug).

In principle, the presence of the "absorptive phase" by the two-phase dissolution test provides a distinct advantage as compared to other dissolution tests. When the organic phase, the "absorptive environment" is present, the dissolution profile in the aqueous phase should more closely represent the drug dissolution in vivo. Concentration–time profiles of the drug observed in the aqueous phase reflect a combined effect of both dissolution and partitioning, and thus would differ from those observed from single phase dissolution tests.[3,26,30,31] In particular, the two-phase dissolution test is highly desirable to evaluate a supersaturated state of a drug when the kinetic processes of dissolution, permeation and precipitation occur simultaneously.

Third, a rapid and reliable quantitation of drug concentrations in the organic phase from the two-phase dissolution test provides a significant advantage in experimental setting. Due to the presence of drug particles (i.e., undissolved and/or precipitated drug) in the aqueous medium under non-sink conditions, reliable and accurate determination of free drug concentrations is technical challenging and none trivial. However, determination of the drug concentration in the organic phase would be a simple analytical task. This is because the partition process effectively acts as an analytical "filter" to prevent drug particles including micelles from moving into the organic phase. The drug concentration in the organic phase can be readily determined with the use of calibrated UV probes.[24,26] Such drug concentrations observed in the organic phase provide indirect measurement of the drug concentration in the aqueous phase and can be further utilized for establishing IVIVR As the free drug concentration in the aqueous phase is the sole driving force for partitioning, the drug–concentration profiles observed in the organic phase is directly proportional to the drug concentration in the aqueous phase. Often, the drug concentration observed in

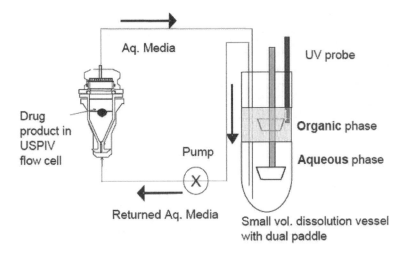

Figure 7.1 Schematic of an in vitro two-phase dissolution test system.

the organic phase via the two-phase dissolution system would be a meaningful surrogate for in vivo bioperformance of drug products.

7.2.2 Experimental

7.2.2.1 Apparatus

A schematic of the two-phase dissolution system currently employed in our laboratory is shown in Fig. 7.1 and a photograph of this system with all associated instruments is shown in Fig. 7.2. This system consists of both USP 2 and USP 4 apparatuses. The drug product for testing is loaded into the flowcell of USP 4 apparatus, permitting accommodation of a variety of dosage forms (e.g. tablet, capsule, solution, suspension).[24,26] Both aqueous and organic phases are contained in the vessel of USP 2 apparatus. The aqueous phase is circulated in a close loop between the flowcell and the USP 2 vessel by the USP 4 pump at a typical rate of 5 to 20 mL/min. A dual paddle resides in the middle of both phases in the USP 2 vessel. The paddle speed is typically set at 50 or 75 rpm in order to achieve sufficient hydrodynamics in both phases while keeping the water–octanol interface undisturbed (avoiding

Figure 7.2 An in vitro two-phase dissolution test system combining USP 2 and USP 4 apparatuses.

the formation of octanol-in-water emulsions in the aqueous phase).

7.2.2.2 Selection of aqueous media

Because weakly acidic or basic drugs possess pH-dependent solubility due to ionization, it is desirable to apply the two-phase dissolution system with "two compartments" containing aqueous media of two different pH values.[25,31] In our laboratory, this is typically achieved as described below. The first compartment, or the flowcell of USP 2 apparatus, contains simulated gastric fluids (SGF, pH 1–2). The drug product is placed in the flowcell for a predetermined time (typically 15–30 min) prior to initiate circulation of the aqueous medium with the USP 2 vessel. Upon circulation, the SGF in flowcell is rapidly mixed with the simulated intestinal fluid (SIF) in the USP 2 vessel. The volume and buffer capacity of SIF are chosen to rapidly reach an equilibrium pH of desired pH (e.g., 5.5–7.0) in the USP 2 vessel. The fluid volume in the UPS 4 flowcell is about 12 mL. Depending on the size of the USP 2 vessel and desired experimental conditions, the fluid volume in USP 2 vessel ranges between 50 and 250 mL.

7.2.2.3 Selection of organic phase

Octanol is the preferred choice for the organic phase in most studies associated with the two-phase dissolution system. This is based on its desirable physicochemical properties as summarized below:[6, 8, 23]

(1) practically insoluble in water (solubility of octanol in water: 0.05 g/100 mL) and immiscible with water,
(2) lower density than water (specific gravity 0.825 at 20°C),
(3) low volatility, and
(4) commonly used in determination of the partition coefficient (K_{ow}) of drug substances. It generally provides sufficient solubility of poorly soluble drugs with high log P.

Several other organic solvents and their mixtures, such as nonanol, chloroform, and octanol/nonanol/cyclohexane, have also been used for two-phase dissolution testing.[23, 24]

7.2.3 Case Studies with Emphasis on IVIVR

Grundy et al.[17, 18] has investigated both single-phase and two-phase dissolution systems to determine the release profiles of nifedipine from extended release (ER) formulations of two strengths with emphasis on their correlation with absorption profiles observed in human subjects.

Nifedipine ER formulations showed relatively linear fractional absorption–time plots between 6 and 24 hours. According to the authors, the conventional single phase dissolution test accurately measured drug release, but failed to account for drug dissolution from suspended particles released from ER tablets. This may explain the discrepancies found with the IVIVC of these nifedipine ER formulations. Release profiles of nifedipine ER formulations from the two-phase dissolution test were compared to the in vivo PK data. Deconvolution of the in vivo PK data was performed using either model-dependent (Wagner–Nelson) or model-independent (DeMons) methods. The deconvoluted profiles are indicative of release rates that are considerably more non-linear than the zero-order drug release rates resulting from the single-phase dissolution method.[18] In contrast, a significantly improved IVIVC was demonstrated with the two-phase dissolution test, yielding a level A correlation (r^2 = 0.99 in all cases). Improved IVIVC with the two-phase dissolution system was attributed to that this test simultaneously simulated dissolution of nifedipine suspension particles in the GI tract as well as absorption from the gut lumen. The dissolution of nifedipene from released suspension particles is considered to be the rate-limiting factor for absorption as evidenced by comparison between the PK data and in vitro release profiles.

To our best knowledge, Vangani et al.[24] reported for the first time a two-phase dissolution system combining a USP 4 apparatus with USP 2 apparatus. The authors applied this system to evaluate a variety of dosage forms (e.g., tablets, capsules, suspensions, etc.) of several poorly water-soluble drugs, demonstrating a high flexibility of USP 4 flowcell for such applications. Evaluation of enabling formulations of AMG 517, a BCS II drug with extremely low aqueous solubility, revealed a rank order relationship between the in vitro release and the in vivo absorption profiles. In addition, this test method successfully discriminated between the bioequivalent and non-bioequivalent marketed drug products of lovastatin and griseofulvin.

Heigoldt et al.[25] reported their exploration of a two-phase dissolution test with pH alternation in aqueous media to evaluate multiple modified release (MR) formulations of two BCS II weakly basic drugs, dipyridamole and BIMT 17. The two-phase dissolution test permitted maintenance of quasi sink conditions in the aqueous phase due to removal of dissolved drugs through partitioning into the organic phase. Various formulations of dipyridamole and BIMT 17 were evaluated using the two-phase dissolution test and their profiles were compared with in vivo PK data. While the ranking of release profiles of dipyridamole from MR formulations in the single phase aqueous media at a constant pH was inconsistent with their in vivo bioperformance, the release profiles obtained from the two-phase dissolution method with pH-alternation followed the rank order of in vivo exposures in healthy human subjects. Similarly, the two-phase dissolution method with pH alternation was able to differentiate MR formulations of BIMT 17 derived from different formulation concepts and achieve qualitative IVIVR. In comparison, the authors concluded that conventional single phase dissolution tests with pH-alternation failed to establish an IVIVR among these MR formulations and the dissolution test results without pH-alternation were misleading.

7.2.4 Case Studies Assessing Supersaturation and IVIVR

With an increasing number of poorly water-soluble drugs in current drug discovery pipelines, the concept of controlling supersaturation as an effective formulation approach for enhancing bioavailability is gaining momentum.[32–38] The key concept of this formulation approach is to design a formulation to yield and sustain intra-luminal concentrations of the drug significantly higher than its thermodynamic equilibrium solubility (i.e., supersaturation) and thus to improve the intestinal absorption. It has been widely recognized that a dynamic balance among the kinetic processes of drug dissolution and precipitation dictates the amount of free drug available for partitioning into the GI membrane and, therefore, its bioavailability. In vitro characterization methods have been designed to understand the interplay among the drug dissolution rate, the degree of supersaturation, and precipitation kinetics and provide valuable insight for formulation development and optimization.

While it is possible to characterize dissolution, supersaturation, and precipitation phenomena with the use of single phase dissolution tests, these experiments may not be adequate to assess the overall effect of these simultaneous kinetic processes upon drug absorption in the GI tract. It is worth noting that the two-phase dissolution test is especially suitable to study the supersturated state of formulations. This test not only reveals the dissolution kinetics in the aqueous phase, but also simultaneous partitioning kinetics of the drug into the organic phase in a dynamic manner. In principle, this should provide a better opportunity to simulate the in vivo situation in which there is constant removal of the dissolved drug as a result of intestinal absorption with its rate depending on the degree of supersaturation in the aqueous luminal phase.

7.2.4.1 Felodipine

Worthen et al.[30] investigated amorphous solid dispersions (ASDs) of felodipine (FLD, a neutral drug) with the use of several polymers including Eudragit-EPO, Eudragit-L-100-55, AQOAT-AS-LF, Pharmacoat-603, and Kollidon-VA-64. This study was intended to evaluate the effect of polymeric excipients upon sustaining a supersaturated state of FLD and identify the most effective polymer for drug product. The time dependent profiles of dissolution, supersaturation, and precipitation of the ASDs were obtained using both single and two-phase dissolution methods.

The percent of FLD vs. time profiles from various ASDs in both aqueous and organic phases are shown in Figs. 7.3a,b.[30] Essentially no drug was dissolved in the aqueous phase from the ASD containing Eudragit-EPO (Fig. 7.3a). As a result, this ASD showed the lowest amount of drug partitioning into the organic phase as compared to

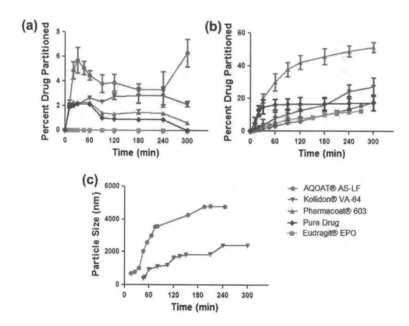

Figure 7.3 Partitioning profiles of FLD in the aqueous phase (a) and organic phase (b) and crystal growth profiles of precipitated drug in the aqueous phase (c) during two-phase dissolution of FLD ASDs.

the other ASDs and the pure crystalline drug over a period of 5 hours (Fig. 7.3b).

The ASD containing Kollidon-VA-64 showed higher concentrations of FLD in the organic phase as compared to ASDs containing AQOAT-AS-LF and Eudragit-EPO (Fig. 7.3b), indicating superior performance due to inhibition of nucleation and crystal growth of FLD in the aqueous phase. Among all ASDs tested, Pharmacoat-603 appeared to be the most effective polymer to yield the highest concentrations of FLD in the organic phase (Fig. 7.3b). No precipitation of FLD was observed in the aqueous phase during the two-phase dissolution of the ASD containing Pharmacoat-603, although FLD concentrations in the aqueous phase were lower than those of the ASDs containing AQOAT-AS-LF and Kollidon-VA-64 (Fig. 7.3a). These results suggest that the ASD containing Pharmacoat-603 achieved the highest concentration of FLD in the organic phase due to optimal dissolution rate of the ASD.

7.2.4.2 Celecoxib

Celecoxib (CEB), a selective cyclooxygenase-2 (COX-2) inhibitor, is a marketed drug that is used for the treatment of osteoarthritis, rheumatoid arthritis and acute pain. CEB is a weak acid with a pKa of 11.1 and is characterized as a BCS II drug due to its low aqueous solubility ($\sim$5 µg/mL) and high permeability.[26] We explored both single-phase and two-phase dissolution tests to characterize three CEB formulations including the commercial Celebrex® capsule, a solution formulation (containing cosolvent and surfactant) and a supersaturatable self-emulsifying drug delivery system (S-SEDDS).[26] As these formulations possessed distinctly different release kinetics, we examined the dissolution profiles with relevance to the oral exposure observed in human subjects.

CEB concentration profiles from the three formulations observed in the single phase dissolution test under non-sink conditions are shown in Fig. 7.4. The Celebrex® capsule showed a consistently low CEB concentration of $\sim$5 µg/mL during the 2 hour test, comparable to its equilibrium solubility. CEB concentrations from the solution and S-SEDDS formulations quickly reached $\sim$0.085 mg/mL around 10 min, and slightly decreased afterwards indicative

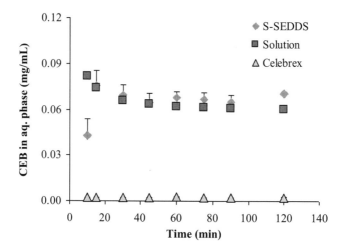

Figure 7.4 CEB release profiles from the three formulations obtained from the *single-phase dissolution test* under non-sink conditions.

of a supersaturated state of CEB. Meanwhile, fine precipitates from both formulations were observed in the dissolution medium.

CEB concentration profiles from the three formulations in the aqueous and octanol phases obtained from a two-phase dissolution test are shown in Figs. 7.5 and 7.6. Celebrex® capsules yielded continuously low CEB concentrations in the aqueous medium during the 2 hour test, which was similar to that observed in the single phase dissolution test under non-sink conditions (Fig. 7.4). Consistently, a low CEB concentration (~0.05 mg/mL) in the octanol was observed from Celebrex® capsules at $t = 2$ hours. The CEB concentration-time profile in the aqueous phase from Celebrex® capsules appeared unaffected by the presence of octanol. This was indicative of a steady state between dissolution and partitioning processes and the presence of octanol did not alter the CEB concentration profile in the aqueous phase.

With the use of two-phase dissolution test, the solution formulation generated high CEB concentrations in the aqueous phase (Fig. 7.5), similar to those observed in the single phase dissolution test (Fig. 7.4). However, this formulation yielded a CEB concentration of ~0.07 mg/mL in octanol at $t = 2$ hours, only slightly higher than that of Celebrex® capsules.

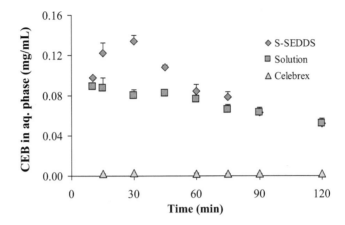

Figure 7.5 CEB release profiles from the three formulations in the *aqueous phase* obtained from the two-phase dissolution test.

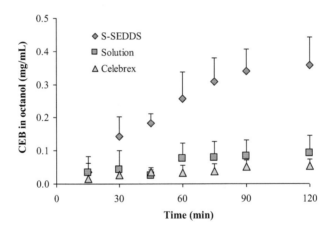

Figure 7.6 CEB release profiles from the three formulations in the *octanol phase* obtained from the two-phase dissolution test.

In contrast, the S-SEDDS formulation showed noticeable higher CEB concentrations in the aqueous medium (Fig. 7.5), especially during the time period of 10 to 60 min compared to those observed in the single-phase dissolution test under non-sink conditions (Fig. 7.4). This clearly indicates that a higher degree of supersaturation of CEB in the aqueous phase was observed due to the presence of the octanol phase. The concentration of CEB in octanol from the S-SEDDS formulation was ~0.35 mg/mL at $t = 2$ hours, significantly higher than those of the solution formulation and Celebrex® capsules (Fig. 7.6).

Although Celebrex® capsule and the solution formulation showed highly different drug release profiles in the aqueous phase from both single phase and two-phase dissolution tests, these two formulations yielded comparable CEB concentrations in octanol. The solution and S-SEDDS formulations in the aqueous phase showed similar concentration-time profiles from the single phase dissolution test (Fig. 7.4) and comparable concentration-time profiles in the aqueous phase (Fig. 7.5) from the two-phase dissolution test. However, they yielded significantly different concentrations in the octanol phase (Fig. 7.6). These observations indicate that the apparent drug concentrations observed in the aqueous phase regardless of the presence of octanol did not represent the true free drug concentration. Consequently, the apparent CEB concentrations observed in the aqueous phase from the three formulations exhibited no relevance to the oral bioavailability in human subjects. A rank-order correlation was obtained between CEB concentration in the octanol phase from the three CEB formulations and their corresponding in vivo exposures (e.g., AUC and C_{max}) in human subjects.

7.2.4.3 BIXX

BIXX is a weakly basic drug of BCS II type and its related formulations were evaluated using a small scale two-phase dissolution test with pH alternation in aqueous media from pH 2.2 to pH 5.5 (referred as "pH-shift" in the paper) by Wagner et al.[31] As shown in Fig. 7.7, Formulation A showed a rapid decrease of the concentration in the aqueous phase after the pH shift at 60 min, suggesting a short-lived supersaturation in the aqueous media of pH 5.5. Formulations B and C containing pH-modifying agents showed slightly delayed precipitation and Formulation D exhibited more stable supersaturation. PXRD analysis indicated the precipitates were amorphous in all cases.[31]

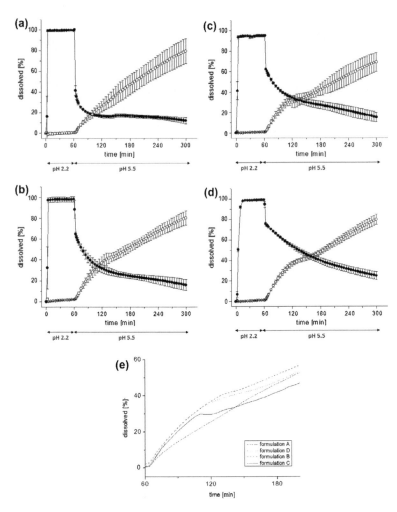

Figure 7.7 Concentration profiles of BIXX obtained from the two-phase dissolution test: pH profile: $t = 0$ until $t = 60$ min, pH 2.2; at 60 min, change to pH 5.5. $n = 3$, mean ±SD, 100% = 27.5 mg (salt). (a–d): solid circles: amount in aqueous phase; hollow circle: amount partitioned in the octanol phase. (a) Formulation A. (b) Formulation B. (c) Formulation C. (d) Formulation D. (e) Mean "absolute" amount dissolved in the lipophilic phase of all four formulations between $t = 60$ min and $t = 200$ min (100% = 18.3 mg (base)). For more details, see Ref. 31.

Formulation A exhibited a consistent partition rate after pH-shift at 60 min as evidenced by its linear BIXX concentration-time profile in octanol while concentration-time profiles of Formulations B, C and D exhibited two different partition rate (Fig. 7.7). A more rapid partition into the octanol phase during the time interval between 60 to 120 min for Formulations B-D was apparent as compared to that after $t > 120$ min. The authors concluded that the rate of partitioning of BIXX from the aqueous phase into the octanol phase was dependent on three competitive processes including supersaturation, precipitation and re-dissolution of precipitates. The initial higher partition rate observed for Formulations B, C, and D was driven by a high degree of supersaturation of BIXX. The second (slower) partitioning rate was presumably associated with the re-dissolution of the amorphous precipitates. All three Formulations B, C, and D exhibited an identical partition rate into the octanol phase. It is worth noting that Formulation A without supersaturating stabilizing excipients showed an almost constant partition rate attributed to the re-dissolution of drug precipitates.

A level A correlation as plotted in Fig. 7.8 was obtained revealing a linear dependency ($R^2 = 0.95$) between the relative fraction of the drug absorbed in dogs vs. relative fraction of the drug in octanol. For comparison, the authors pointed out that the single phase dissolution profiles were not predictive of the in vivo bioperformance; the ranking of formulations was not in accordance with the in vivo PK results.

These cases described above indicate that the in vitro drug concentration observed in the organic phase may serve as a surrogate for correlation with the pharmacokinetic observations in preclinical species and/or human subjects. Thus, the two-phase dissolution test methods should provide an excellent opportunity to establish IVIVR or IVIVC for immediate and extended release formulations of BCS II/IV drugs.

Further, these cases reveal the critical importance of yielding and maintaining a supersaturated state in order to improve bioavailability of BCS II and IV drugs. The strategy for achieving optimal absorption should focus on yielding and maintaining a supersaturated state by (i) controlling the dissolution rate to obtain a supersaturated state, (ii) utilizing effective precipitation inhibitors

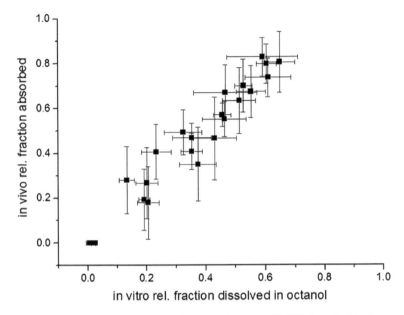

Figure 7.8   Level A correlation of relative fraction of BIXX absorbed in dogs and relative fraction partitioned in octanol at $t = 3$ hours. $R^2 = 0.95$. For more details, see Ref. 31.

to maintain the supersaturation state with minimal precipitation, and (iii) controlling the degree of supersaturation with sufficiently long duration. As demonstrated by the case studies above, the two-phase dissolution test plays an invaluable role in assessing the supersaturated state of formulations with different approaches (e.g., ASD, S-SEDDS, solution) and forecasting in vivo bioperformance.

7.3 Theoretical Modeling of Drug Dissolution and Transfer between Two Phases

Since 1967, several publications described mathematical models of mass transport of the solute from the aqueous phase to the organic phase of a two-phase system.[6–9,13,14,21,22,29] Among all theoretical models reported, we select two models to be highlighted below.

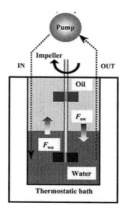

Figure 7.9 Experimental apparatus to evaluate the Grassi model.[21]

7.3.1 Grassi Model

In 2002, Grassi et al. published a comprehensive mathematical model describing the mass transport kinetics of a drug substance between the aqueous phase and the organic phase in a two-phase dissolution apparatus.[21,22] A schematic diagram of the two-phase dissolution apparatus is shown in Fig. 7.9. Both the aqueous and organic media were contained in a vessel and agitated by a single shaft fitted with two impellers. The major assumptions of the model included pseudo-steady-state conditions, a dilute aqueous solution, and diffusion-controlled transport through the water–organic interfaces. Input parameters can be measured experimentally or estimated a priori.

The mathematical description of the kinetics was described using four rate equations involving a steady-state differential rate equation for aqueous drug concentration as a function of rate constants from the aqueous to the organic (k_{wo}) phase and from the organic to aqueous (k_{ow}) phases. The solutions for drug concentration in the aqueous phase (C_w) and in the organic phase (C_o) as a function of time were resolved as described in Ref. 21.

The agreement between the apparent partition coefficient (P) of piroxicam obtained by means of the kinetic partition and equilibrium measurements confirmed the reliability of this model. Partitioning of drug from the aqueous to the organic media was monitored as a function of time until the equilibrium concentration of drug in each phase was reached. This model allows the determination of the rate constants, k_{ow} and k_{wo}. Since the model accommodates both the kinetics and the equilibrium characteristics of the experimental data, the drug concentration–time profile in the organic phase may be predicted via mass balance.[21]

The authors had to select proper equations based upon the values of defined model parameters that were a function of both experimental and fitted parameters (e.g., k_{ow} and k_{wo}), which cannot be determined a priori. Therefore, the Grassi model assumes different expressions depending on the parameter values. It is usually not a trivial task to know a priori which model expression should be suitable. In addition, as the Grassi model assumes that the drug initially is completely dissolved in a solution which is well mixed into the aqueous phase of the two-phase system as the input, a major limitation of this model (as the authors pointed out) is that it could not describe the partitioning profile when a drug of low solubility precipitates in the aqueous phase.[21]

Grassi et al.[22] further advanced the model to take into consideration of the drug dissolution kinetics from solid state into the aqueous phase and simultaneous partitioning into the organic phase. A simple experimental approach with the use of a rotating disk containing a solid layer of a drug substance was utilized since the dissolution rate from this system was well described by the Levich theory. This advancement permits the incorporation of the drug dissolution rate in the aqueous phase and its effect on partitioning kinetics into the organic phase, analogous to the fate of solid dosage forms in vivo.[22] The authors conducted experimental verification using a model drug (piroxicam) in two aqueous buffers with different pH values. Piroxicam concentration-time profiles observed were in agreement with the predicted profiles in the aqueous phase with appropriate experimental parameters. However, the drug-concentration profile in the organic phase was not reported and discussed.

7.3.2 Amidon Model

In 2011, Amidon et al.[29] applied a mechanistic approach to understand the drug transport phenomenon associated with the two-phase dissolution system. They performed a mass transport analysis of the kinetics of partitioning of drugs (dissolved in solution) from the aqueous to the organic phase. The kinetics of drug partitioning from the aqueous phase to the organic phase in a two-phase dissolution system was described based on simultaneous chemical equilibria and mass transfer of dissolved drug through the water–organic interfacial barrier. The drug transport was assumed to be controlled by diffusion through a hydrodynamically controlled or "stagnant" diffusion layer on each side of the aqueous–organic interface. A steady diffusion across the thin interface was assumed to predict the total flux of drug across the two diffusion layers in series.

An advantage of the newly developed Amidon model over the Grassi model is that all model parameters associated with the former can be either measured or estimated a priori. For instance, the model parameters (e.g., M_T, V_a, V_o, and A_I) are defined by the experimental setup.[29] The apparent partition coefficient, K_{ap}, can be measured using established methods or can be estimated using molecular descriptors.[29]

As it is difficult to capture the range of physiological conditions affecting dissolution and absorption, it is important that the chosen apparatus encompasses the most important factors for the particular drug product of interest. If the key physiological scaling parameters (A_I/V_a, M_T/V_a, and $V_a/(K_{ap}V_0)$) for the two-phase system described above are properly designed, and a biorelevant aqueous buffer is used, it is reasonable to expect similar saturation conditions occurred between the in vitro aqueous medium and the intestinal lumen in vivo and, therefore, to expect an in vitro partitioning rate that is similar to the in vivo absorption rate of a drug substance.

Amidon et al. revealed that the in vitro partitioning rate coefficient, k_p (equal to $(A_I/V_a)*P_I$), reflects the rate of drug partitioning into the organic phase. Therefore, one approach to establish physiological relevance is to keep the in vitro k_p equal to the expected in vivo absorption rate coefficient, k_a, as described below.

$$k_p = \left(\frac{A_I}{V_a} P_I \right)_{\text{in vitro}} = k_a \left(\frac{A}{V} P_{\text{eff}} \right)_{\text{in vivo}}$$

This approach assumes first-order absorption kinetics and a relatively high fraction absorbed in vivo (F_a). On the basis of a known or estimated k_a and observed or estimated P_l, one can adjust A_l/V_a so that k_p and k_a are similar or equal. In Ref. 29, the authors presented case studies to demonstrate how a two-phase system could be set up to mimic in vivo absorption rates of ibuprofen, piroxicam, and nimesulide (all BCS II weak acid drugs) with known in vivo k_a values in human subjects.[29] The authors demonstrated the effectiveness of the model in predicting experimental results by the use of apparatuses of different geometry.

Since the in vitro A_l/V_a is dictated by the diameter and geometry of the vessel, options for this parameter are limited to the use of standard, hemispherical vessels in experimental setting. The authors showed the minimum and maximum A_l/V_a achieved in a 1000 mL USP 2 vessel and a 100 mL vessel of similar proportions.[29] These estimates are based on practical constraints such as maintaining a minimum aqueous volume to achieve a reasonable liquid height. The authors pointed out that because a limited range of A_l/V_a practically available and inability to fully control P_l, the desired k_p may not be achievable in all cases. Table 6 in Ref. 29 provides estimated ranges for k_a and k_p for BCS II drugs.

It is worth noting that a major limitation of the Amidon model is that it can only deal with the drug in solution and could not take into consideration of the drug dissolution kinetics from solid state into simulation at the present time.

7.4 Conclusions

Biorelevant in vitro dissolution tests are an essential tool for drug development. As we demonstrated in this chapter, the two-phase dissolution tests, containing both aqueous and organic phases are designed to be more physiologically relevant than conventional single-phase dissolution tests with the involvement of the "absorptive environment." Accordingly, two-phase dissolution systems are intended to simulate both kinetic processes of drug dissolution and partitioning into the GI tract. Case studies summarized in this book chapter illustrate that this two-phase dissolution test methods enable better opportunity for establishing IVIVR or IVIVC of several drugs from their immediate or extended dosage forms as compared to conventional single-phase dissolution methods. In particular, the two-phase dissolution test may be extremely useful in assessing the metastable supersaturated state of BCS II drugs in a dynamic manner, including the duration and degree of supersaturation of poorly soluble drugs from variable formulation approaches, apparent effect of precipitation inhibition by polymeric excipients, and an overall effect upon drug partition. In addition, current theoretical models of the two-phase dissolution systems are reviewed. These models are far from perfection and desired to be further advanced to account for dissolution kinetics of drug substances under non-sink conditions and simultaneous partitioning kinetics responding to the resulting aqueous concentration for BCS II/IV drugs. In summary, it is our opinion and recommendation that the two-phase dissolution test methods should be broadly applied to characterize BCS II/IV drug products and further advanced to become a better in vitro tool for achieving IVIVR or IVIVC.

References

1. Sjögren, E., et al. In vivo methods for drug absorption: comparative physiologies, model selection, correlations with in vitro methods (IVIVC), and applications for formulation/API/excipient characterization including food effects. *Eur. J. Pharm. Sci.*, **57**, 99–151 (2014).

2. Tsume, Y., Mudie, D. M., Langguth, P., Amidon, G. E., Amidon, G. L. The Biopharmaceutics Classification System: subclasses for in vivo predictive dissolution (IPD) methodology and IVIVC. *Eur. J. Pharm. Sci.*, **57**, 342–368 (2014).

3. Augustijns, P., et al. In vitro models for the prediction of in vivo performance of oral dosage forms. *Eur. J. Pharm. Sci.*, **57**, 152–163 (2014).

4. Koziolek, M., Garbacz, G., Neumann, M., Weitschies, W. Simulating the postprandial stomach: biorelevant test methods for the estimation of intragastric drug dissolution. *Mol. Pharm.*, **10**, 2211–2221 (2013).

5. Reppas, C., Vertzoni, M. Biorelevant in-vitro performance testing of orally administered dosage forms. *J. Pharm. Pharmacol.*, **64**, 919–930 (2012).

6. Niebergall, P. J., Pastil, M. Y., Sugita, E. T. Simultaneous determination of dissolution and partitioning rates in vitro. *J. Pharm. Sci.*, **56**(8), 943–947 (1967).

7. Gibaldi, M., Feldman, S. Establishment of sink conditions in dissolution rate determinations: theoretical considerations and application to nondisintegegrating dosage forms. *J. Pharm. Sci.*, **56**(10), 1238–1242 (1967).

8. Niebergall, P. J., et al. Dissolution rates under sink conditions. *J. Pharm. Sci.*, **60**(10), 1575–1576 (1971).

9. Takayama, K., Nambu, N., Nagai, T. Analysis of interfacial of indomethacin following dissolution of indomethacin/ polyvinylpyrrolidone coprecipitates. *Chem. Pharm. Bull.*, **29**(9), 2718–2721 (1981).

10. Stead, S. A., et al. Ibuprofen tablets: dissolution and bioavailability studies. *Int. J. Pharm.*, **14**, 59–72 (1983).

11. Fini, A., et al. Three phase dissolution partition of some non-steroidal anti-inflammatory drugs. *Acta Pharm. Technol.*, **32**(2), 86–8 (1986).

12. Chaudhary, R. S., et al. Dissolution system for nifedipine sustained release formulations. *Drug Dev. Ind. Pharm.*, **20**(7), 1267–1274 (1994).

13. Kinget, R., et al. In vitro assessment of drug release from semi-solid lipid matrices. *Eur. J. Pharm. Sci.*, **3**, 105–111 (1995).

14. Hoa, N. T., et al. Design and evaluation of two-phase partition-dissolution method and its use in evaluating artemisinin tablets. *J. Pharm. Sci.*, **85**(10), 1060–1063 (1996).

15. Ngo, T. H., Vertommen, J., Kinget, R. Formulation of artemisinin tablets. *Int. J. Pharm.*, **146**, 271–274 (1997).

16. Hg, T. H., et al. Bioavailability of different artemisinin tablet formulations in rabbit plasma-correlation with results obtained by an in vitro dissolution. *J. Pharm. Biom. Anal.*, **16**, 185–189 (1997).

17. Grundy, J. S., et al. Studies on dissolution testing of the nifedipine gastrointestinal therapeutic system. I. Description of a two-phase in vitro dissolution test. *J. Control. Release*, **48**, 1–8 (1997).

18. Gundy, J. S., et al. Studies on dissolution testing of the nifedipine gastrointestinal therapeutic system. II. Improved in vitro-in vivo correlation using a two-phase dissolution test. *J. Control. Release*, **48**, 9–17 (1997).

19. Pillay, V., Fassihi, R. Evaluation and comparison of dissolution data derived from different modified release dosage forms: an alternative method. *J. Control. Release*, **55**, 45–55 (1998).

20. Pillay, V., et al. A new method for dissolution studies of lipid-filled capsules employing nifedipine as a model drug. *Pharm. Res.*, **16**(2), 333–337 (1999).

21. Grassi, M., et al. Modeling partitioning of sparingly soluble drugs in a two-phase liquid system. *Int. J. Pharm.*, **239**, 157–169 (2002).

22. Grassi, M., Grassi, G., Lapasin, R., Colombo, I. Drug dissolution and partition, chap. 5 in *Understanding Drug Release and Absorption Mechanisms: A physical and Mathematical Approach*. by CRC Press, 2007.

23. Gabriels, M., et al. Design of a dissolution system for the evaluation of the release rate characteristics of artemether and dihydroartemisinin from tablets. *Int. J. Pharm.*, **274**, 245–260 (2004).

24. Vangani, S., et al. Dissolution of poorly water-soluble drugs in biphasic media using USP 4 and fiber optic system. *Clin. Res. Regul. Aff.*, **26**(1–2), 8–19 (2009).

25. Heigoldt, U., Sommer, F., Daniels, R., Wagner, K.-G. Predicting in vivo absorption behavior of oral modified release dosage forms containing pH-dependent poorly soluble drugs using a novel pH-adjusted biphasic in vitro dissolution test. *Eur. J. Pharm. Biopharm.*, **76**, 105–111 (2010).

26. Shi, Y., Ping, G., Yuchuan, G., Haili, P. Application of a biphasic test for characterization of in vitro drug release of immediate release formulations of celecoxib and its relevance to in vivo absorption. *Mol. Pharm.*, **7**(5), 1458–1465 (2010).

27. Phillips, D. J., Pygall, S. R., Cooper, V. B., Mann, J. C. Overcoming sink limitations in dissolution testing: a review of traditional methods and the potential utility of biphasic systems. *J. Pharm. Pharmacol.*, **64**(11), 1549–1559 (2012).

28. Phillips, D. Toward biorelevant dissolution: application of a biphasic dissolution model as a discriminating tool for HPMC matrices containing a model BCS class II drug. *Dissolut. Technol.*, **19**(1), 25 (2012).

29. Mudie, D. M., Shi, Y., Ping, H., Gao, P., Amidon, G. L., Amidon, G. E. Mechanistic analysis of solute transport in an in vitro physiological two-phase dissolution apparatus. *Biopharm. Drug Dispos.*, **33**(7), 378–402 (2012).

30. Sarode, A. L., Wang, P., Obara, S., Worthen, D. R. Supersaturation, nucleation, and crystal growth during single- and biphasic dissolution of amorphous solid dispersions: polymer effects and implications for oral bioavailability enhancement of poorly water soluble drugs. *Eur. J. Pharm. Biopharm.*, **86**, 351–360 (2014).

31. Frank, K. J., Locher, K., Zecevic, D. E., Fleth, J., Wagner, K. G. In vivo predictive mini-scale dissolution for weak bases: advantages of pH-shift in combination with an absorptive compartment. *Eur. J. Pharm. Sci.*, **61**, 32–39 (2014).

32. Gao, P., Morozowich, W. Development of supersaturatable SEDDS (S-SEDDS) formulations for improving the oral absorption of poorly soluble drugs. *Expert Opin. Drug Deliv.*, **3**, 97–110 (2005).

33. P. Gao, A. Akrami, F. Alvarez, J. Hu, L.Li, C. Ma, S. Surapaneni. Characterization and optimization of AMG 517 supersaturatable self-emulsifying drug delivery system (S-SEDDS) for improved oral absorption. *J. Pharm. Sci.*, **98**(2), 516–628 (2009).

34. Brouwers, J., Brewster, M., Augustijns, P. Supersaturating drug delivery systems: the answer to solubility-limited oral bioavailability. *J. Pharm. Sci.*, **98**(8), 2549–2572. (2009).

35. Bevernage, J., Forier, T., Brouwers, J., Tack, J., Annaert, P., Augustijns, P. Excipient-mediated supersaturation stabilization in human intestinal fluids. *Mol. Pharm.*, **8**, 564–570. (2011).

36. Gao, P., Shi, Y. Characterization of supersaturatable formulations for improved absorption of poorly soluble drugs. *AAPS J.*, **14**,(4): 703–713 (2012).

37. Bevernage, J., Brouwers, J., Annaert, P., Augustijns, P. Drug precipitation-permeation interplay: supersaturation in an absorptive environment. *Eur. J. Pharm. Biopharm.*, **82**(2), 424–428 (2012).

38. Gao, P. Amorphous pharmaceutical solids: characterization, stabilization, and development of marketable formulations of poorly soluble drugs with improved oral absorption. *Mol. Pharm.*, **5**(6) 903–904 (2008).

Chapter 8

The Use of Apparatus 3 in Dissolution Testing of Poorly Soluble Drug Formulations

G. Bryan Crist

Agilent Technologies, Inc., 2500 Regency Parkway, Cary, NC 27518, USA
bryan.crist@agilent.com

8.1 Background of the Reciprocating Cylinder Apparatus

The development of the reciprocating cylinder apparatus arose from a need for an alternative drug release apparatus capable of providing pharmacokinetic and mechanical conditions that more closely represented the various regions throughout the gastrointestinal tract. While the traditional paddle and basket apparatus offered a convenient means to evaluate most oral drug formulations at single and multiple pHs and over long periods, it was difficult to change pH during the test and changes in agitation rates during the in vitro test are seldom noted as well.

Poorly Soluble Drugs: Dissolution and Drug Release
Edited by Gregory K. Webster, J. Derek Jackson, and Robert G. Bell
Copyright © 2017 Pan Stanford Publishing Pte. Ltd.
ISBN 978-981-4745-45-1 (Hardcover), 978-981-4745-46-8 (eBook)
www.panstanford.com

Traditional dissolution testing is conducted at a single speed which is justified to be low enough for adequate discrimination in the test to provide batch-to-batch uniformity, consistency, and detection of potentially bioinequivalent lots from bioequivalent ones. At lower speeds however, cone formation may become an issue due to drug product in the core remaining unexposed to media, which may lead to inaccurate drug release profiling. Changing pH also proves to be a challenge and typically demands a considerable amount of time for the analyst.

The needs for an alternative approach to determine drug product release rate arose during a Federation Internationale Pharmaceutique (FIP) drug release conference in 1980. Analytical development chemists initially utilized the rotating bottle apparatus that was introduced into the National Formulary in 1965 as the Timed-Release Tablets and Capsules – In Vitro Test Procedure. This predated the paddle and basket apparatus and was considered at the time to provide suitable in vitro test criteria to assure product uniformity of most timed-release tablets and capsules.[1] The rotating bottle provided an excellent and reproducible environment to determine the rate of drug release by changing media composition and agitation rate and allowing profile time points for extended periods of time with virtually no evaporative loss. Yet, it was quite labor intensive to physically remove the apparatus from its circulator, remove the vessels, remove/replace media, reattach the vessels, and return the apparatus to the water bath. These challenges led to the collaborative development of the reciprocating cylinder apparatus under the direction of Arnold Beckett of the University of London, Ian Borst of Health Canada, and Sydney Ugwu with VanKel Industries. The reciprocating cylinder apparatus combined the mixing and alternative flow capabilities of the original rotating bottle apparatus as well as the disintegration apparatus to expose drug product to various media contained in 300 mL vessels without analyst intervention during the test period. Studies indicated that the reciprocating cylinder, also referred to as the "Bio-Dis" for biorelevant dissolution, was comparable with the rotating bottle method under various experimental conditions and was found to be a suitable alternative to the rotating bottle method.[2] The reciprocating cylinder apparatus also provided programmable time,

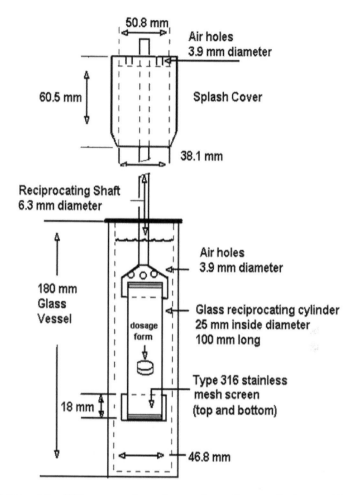

Figure 8.1 USP apparatus 3: reciprocating cylinder with a non-disintegrating dosage form.

agitation rate and media change operations, which were necessary during dissolution method development of a product to study in vitro–in vivo correlation.[3]

The reciprocating cylinder apparatus (Fig. 8.1) was introduced in the USP in 1990 and is contained in Physical Test Chapter <711> Dissolution and is now harmonized with the European Pharmacopeia in 2.9.3 Dissolution Test for Solid Dosage Forms.[4,5]

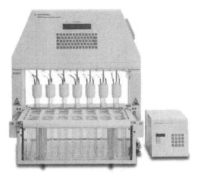

Figure 8.2   USP apparatus 3: reciprocating cylinder apparatus.

8.2 Operational Overview

The reciprocating cylinder apparatus shown in Fig. 8.2 consists of six cylinders, each containing a dosage form. The reciprocating cylinder is made of a glass inner tube that is threaded on each end. Upper and lower caps containing screens are attached to the upper and lower portion of the tube to retain the dosage form. The cylinder is attached to shafts that allow the cylinder to move up and down in the glass vessel at a programmed rate referred to in dips per minute (DPM). The most common agitation rates range from 5 to 40 DPM. The vessels are also made of glass and are arranged in a series of trays that allow the vessels to be removed or added easily. Similar to other dissolution apparatus, six dosage forms are tested at a time and a seventh position is also provided for an optional control sample, blank solution, or standard solution. The seventh position is also useful for providing automated collection of blank and reference standard solutions used for the various media during the test since standard solutions must be prepared in the same media that represents each row.

The 300 mL flat-bottom vessels, or outer tubes, are arranged on the vessel plates in six rows primarily to represent multiple regions of the GI tract: gastric, duodenum, jejunum, proximal ileum, and distal ileum. The extra row may be used to represent an area of the jejunum with two rows of the same media if a majority of the drug is released in this region or if sink conditions become challenging

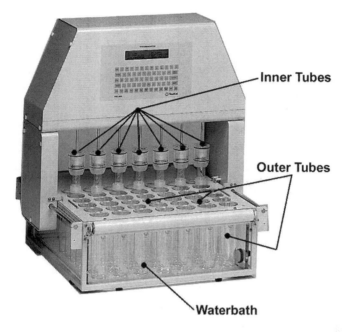

Figure 8.3 USP apparatus 3: reciprocating cylinders (inner tubes) and vessels (outer tubes).

in a specific region. The versatile apparatus allows a maximum of up to 42 vessels for testing six dosage forms during a single run. Additionally, a poorly soluble immediate release drug for example, may be tested with up to 1,800 mL (6 × 300 mL) of media, which is twice the capacity of a traditional dissolution test performed in 900 mL to study the drug at a single pH representing gastric or intestinal fluids.

Screens for the reciprocating cylinders are chosen based on their ability to retain the dosage form and allow the maximum amount of media to move through the double-ended chamber. Unlike the traditional harmonized USP apparatus 1 basket method, which utilizes a 40-mesh basket with specific wire diameter and aperture characteristics, the harmonized reciprocating cylinder apparatus does not contain specifications and tolerances for the screen mesh, micrometer size, or composition. Screens are produced in stainless steel or polypropylene and range from 20 to 400 mesh. Larger mesh

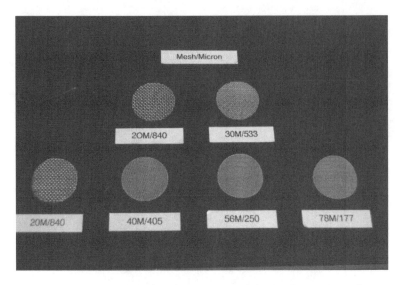

Figure 8.4 Typical stainless steel (upper row) and polypropylene (lower row) screens used for the top and bottom end caps.

screens may be used to contain and transport the undissolved or un-disintegrated formulation from row to row. However, there is a tradeoff with very fine mesh screens because they will restrict the flow of media through the cylinder and may compromise filling and draining the cylinder once the screen is wet. Media flow through fine mesh screens may be improved by omitting the use of an upper screen and using a level of media at which the upper cap is not submerged. This will improve the movement of media through the lower screen.

Once the dosage form is introduced and attached to the reciprocating shaft, the method may be programmed to specify the number of rows the product will be introduced to, the dip rate, time in each row, time that the cylinder will pause once lowered to allow the cylinder to fill, and how long the tube will drain before moving to the next row. The reciprocating motion of the cylinder allows the drug to be freely exposed to media at all times. Based on the specific density of the drug formulation, the particles, pellets, beads, and/or tablets should float freely through the liquid column on the

Figure 8.5   A set of reciprocating cylinders conforming to USP apparatus 3.

downstroke. On the upstroke, the product will contact the bottom screen only to be suspended again on the next downstroke. This motion overcomes the coning issues associated with USP apparatus 2 paddle where the agitation is minimized at the center of rotation on the bottom of the vessel.

Media considerations vary widely with the application and examples of biorelevant solutions will be discussed later in this chapter. Claims have been made that for certain non-disintegrating dosage forms, deaeration has negligible effect on the dissolution rate because the absence or presence of bubbles does not alter the movement of the dosage form and the hydrodynamics of the system sufficiently to affect their dissolution rates.[3] However, poorly deaerated media has caused bead formulations to aggregate with air bubbles, which could restrict the entire surface of the bead being exposed to the media, as shown in Fig. 8.6. Similar to most dissolution methods, the effects deaeration may cause are product specific and the impact must be studied on a case-by-case basis during dissolution method development.

Figure 8.6   Bead aggregates due to effects of air bubbles in the media.

8.3 Typical Applications for Poorly Soluble Compounds

The poorly soluble compounds addressed in this book cover many categories from immediate, delayed, and sustained release drugs, to targeted drug delivery systems and the testing of microspheres. While apparatus 3 was designed primarily for modified release formulations tested between 1 and 24 hours, it has shown versatility for testing a variety of drug delivery systems.

The reciprocating cylinder apparatus has successfully shown discrimination for poorly soluble compounds used for immediate release products due to its aggressive mixing characteristics. A study on the utilization of the apparatus for immediate release products demonstrated that when operated at the extreme low end of the agitation range, such as 5 DPM, hydrodynamic conditions equivalent to USP apparatus 2 at 50 rpm were achieved when compared with the f_2 similarity test.[6]

Although not poorly soluble, theophylline has been well documented regarding its modified release formulations and profiles to document profiles corresponding to the pH-dependent release due to the coating used for the formulation. Table 8.1 shows typical percentages of theophylline in which coefficients of variation obtained were less than 5%.[2]

Table 8.1 Percentages of theophylline dissolved in vitro under various conditions

Condition/ in vitro %	1 hour pH 1.2	2 hour pH 2.5	3.5 hour pH 4.5	5 hour pH 7.0	7 hour pH 7.2
Form A	28.8	50.1	71.7	82.4	91.5
Form B	13.0	32.2	71.7	82.4	91.5
Form C	25.9	36.5	45.6	74.0	96.2

Nifedipine ER tablets are one of the best compendial examples of a USP method for the dissolution of a poorly soluble compound with the reciprocating cylinder apparatus. The tablets are retained in cylinders containing 20-mesh polypropylene screen on the bottom cap and reciprocated at 20 DPM in 250 mL of media throughout the run with one minute drip times between rows. The test consists of reciprocating for the first hour in simulated gastric fluid without enzyme containing 3% polysorbate 80, pH 1.2; followed by four consecutive rows of 0.01 M sodium phosphate buffer containing 3% polysorbate 80, pH 6.8. The remaining sample intervals are at 2, 8, 12, and 24 hours and the results are calculated cumulatively:

At 1 hour: NMT 5% dissolved

$$D = (r_u/r_s) \times (C_s/L) \times V \times 100 = D_1$$

At 2 hours: 0%–10% dissolved

$$D_2 = D_1 + D$$

At 8 hours: 25%–60% dissolved

$$D_8 = D_2 + D$$

At 12 hours: 45%–80% dissolved

$$D_{12} = D_8 + D$$

At 24 hours: NLT 80%

$$D_{24} = D_{12} + D$$

r_u = response of sample solution
r_s = response of standard solution
C_s = concentration of standard solution (mg/mL)

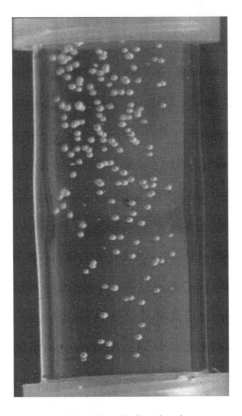

Figure 8.7 Suspended theophylline beads in apparatus 3.

$L = $ label claim (mg/tablet)
$V = $ volume of media (250 mL)[7]

A similar test for oxybutynin chloride ER tablets is contained in the USP with similar media although the non-ionic surfactant polysorbate 80 has not been added. The method allows the characterization of limited release in the simulated gastric fluid for the acid stage medium and then profiles the full release through the 24 hours in simulated intestinal fluid for rows 2–4 representing the buffer stage.[7]

The advantage of the multiple rows of similar media in this case allows the product to be exposed to media in excess of 900 mL, which is ideal for poorly soluble products. Surfactants are an obvious first choice for poorly soluble compounds in traditional dissolution apparatus 1 and 2. However, the addition of surfactants to apparatus 3 with moderate to high dip rates tends to cause excessive foaming in the vessel. If excessive foaming is encountered, simethicone antifoam may be utilized in very small amounts to control foaming. The effect of the reduction of foaming, due to the addition of simethicone in synthetic gastric media containing surface-active surfactants, causes the film surface to rupture which allowed air to escape and foaming to be minimized.[8]

Workshops held by FIP and AAPS to outline guidelines for the dissolution and drug release testing of novel dosage forms indicated that the reciprocating cylinder apparatus may be suitable for testing chewable tablets. The addition of glass beads would be required to provide more intensive agitation to the in vitro dissolution test.[9]

Chewable tablets often contain highly soluble API, but they are very difficult to disintegrate and solubilize in traditional dissolution apparatus. For this reason, chewable formulations share a need for additional mechanical forces similar to poorly soluble compounds. Chewable tablets may be tested in the reciprocating cylinder apparatus with the addition of plastic beads, or possibly glass beads, to provide an intense contact with the dosage form through reciprocation and break down the chewable tablet relatively quickly with mechanical shearing. Figure 8.8 shows chewable tablets reciprocated with beads of various sizes at 30 DPM.

A reciprocating cylinder dissolution method was developed using beads for testing hydrophilic matrix tablets to predict in vivo performance, which enabled the development of level A IVIVC for several formulations. It was noted that high mechanical forces in vitro were necessary to provide a satisfactory correlation with the in vivo data. For the bead-based method, synthetic polymeric bead material from 1 to 8 mm with a density of about 1.1 g/cm were used, which allowed good interaction with the tablet. It was noted that glass beads are generally unsuitable because the usually stay on the bottom of the cylinder and do not exert any mechanical forces on the tablets even when agitating with a high dip rate. This testing summarized that the use of apparatus 3 methods with beads of proper size and density supply additional mechanical stress that is advantageous for evaluating matrix tablets where erosion

Figure 8.8 Chewable aspirin tablets in the presence of 1.5 mm beads (left) and 6 mm beads (right).

is involved in the release mechanism. Additionally, beads may be useful in future studies since robust matrix formulations that are bioequivalent to the reference product could be planned during early stages of the development.[10]

Poorly soluble acyclovir has also been successfully tested with the reciprocating cylinder to overcome the effects of coning and better mimic the changes in physiochemical conditions and mechanical forces experienced by products in the gastrointestinal tract. In a study with the FDA, 5 acyclovir generic products were tested along with the innovator product and compared between USP apparatus 2 paddle and USP apparatus 3. Figure 8.9 illustrates that both apparatus produce similarly rapid release rates.[7]

A study conducted to evaluate the effects of hypromellose viscosity on ketoprofen release from matrix tablets yielded that the reciprocating cylinder method was the most appropriate for evaluating the dissolution characteristics of hydrophilic matrices. The results showed a direct relationship of the influence of medium pH with the ketoprofen release rate. The study was conducted with 250 mL of fasted media with an agitation of 8 DPM. Drug release

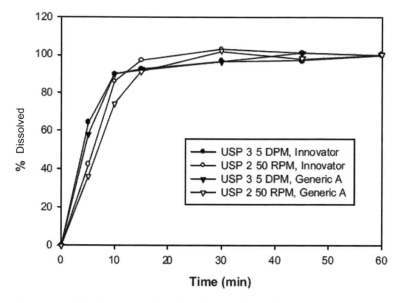

Figure 8.9 Dissolution profiles of acyclovir tablets of innovator and generic products.

conditions consisted of pH 1.2 for 1 hour, pH 4.5 for 0.5 hour, pH 6.0 for 2.5 hours, pH 6.8 for 6 hours, and pH 7.2 for 2 hours and the tubes were allowed to drain for 1 minute prior to moving to the next row. This suggested that the Bio-Dis provided a better approximation of the dissolution in vivo performance of hypromellose matrix tablet containing drugs with pH-dependent solubility.[11]

Apparatus 3 has been highly successful characterizing drug release at specific sites within the GI tract where high concentrations may be necessary to treat chronic intestinal inflammation such as Inflammatory Bowel Disease (IBD). A sudden release of the drug in the stomach, for instance, would deplete the drug due to absorption in the duodenum, leaving insufficient therapeutic levels in the lower small intestine and colon. Studies were conducted on several mesalazine products to evaluate the enteric coating properties of IBD medications, in the fasted state with typical mean transit times. The tests were conducted at 10 DPM with 220 ml of the various media and transit times described in Table 8.2.[12]

Table 8.2 Dissolution media and transit times used in the mesalazine study[12]

GI segment	Transit time	Medium	pH value
Stomach	120 min	Simulated gastric fluid USP 24 sine pepsin (SGFsp)	1.2
Duodenum	10 min	Phosphate buffer Ph Eur. 1997	6.0
Jejunum	120 min	Simulated intestinal fluid USP 24 sine pancreatin (SIFsp)	6.8
Proximal ileum	30 min	Phosphate buffer Ph Eur. 1997	7.2
Distal ileum	30 min	Simulated intestinal fluid USP 23 sine pancreatin (SIFsp)	7.5

The reciprocating cylinder apparatus in this case illustrated drug controlled release behavior as it passes through the GI tract and it provides a discriminating method that may be utilized to compare formulations batch to batch. This drug release method is not confined to differentiating between formulations but could also be used to examine the effects of inter- and intraindividual variations in the gastrointestinal composition and transit of both multiple and single unit dosage forms. Therefore, the method should not only be applicable to compare marketed products but also for the development of new modified release formulations.[12]

8.3.1 Modifications to the Reciprocating Cylinder

Two non-compendial modifications have been developed over the years to provide solutions for analytical development: the wide lower cap and the 100 mL vessel. The wide lower cap (Fig. 8.10) has been used for highly disintegrating formulations with particles that fall through the lower screen only to experience limited mixing with the regular cap. The wide lower cap provides greater agitation due to the piston action between the cylinder and vessel wall.[3]

The 100 mL version is a non-compendial version of the USP apparatus 3 300 mL system. The miniaturization of the apparatus has been used for highly potent low dose systems that require small-volume dissolution and is shown in Fig. 8.11.

Each of these modifications may have limited application for poorly soluble formulations because the wide cap is intended

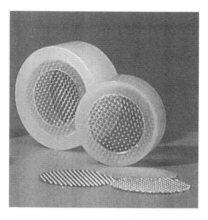

Figure 8.10 Non-compendial wide cap (left) compared with compendial lower cap (right).

Figure 8.11 Non-compendial reciprocating inner tube for the 100 mL vessel (left) compared to compendial reciprocating inner tube (right).

for agitating particles within a single row to ensure complete dissolution and insufficient sink conditions would probably exist for many poorly soluble compounds with a total volume less than 300 mL. The 100 mL vessel, while beneficial for small-volume requirements for low dose products, may also challenge sink conditions for poorly soluble formulations.

8.4 Reciprocating Cylinder Apparatus Qualification

In addition to conducting routine physical parameter measurements, USP apparatus 3 has historically benefited from a USP Performance Verification Test with USP Chlorpheniramine Maleate ER tablets. However, effective February 1, 2012, USP removed the requirement for apparatus 3 PVT Apparatus Suitability section of

General Chapter <711> Dissolution based on unavailability of the product. Although USP remains convinced that a PVT is a critical element in the qualification of in vitro performance test equipment, a suitable replacement for the apparatus 3 PVT has not been found at the time of this writing.[13]

In lieu of the PVT, an alternative procedure based on enhanced mechanical qualification approach was proposed at the USP Challenges in Dissolution Workshop held in June of 2012.[14] The enhanced mechanical approach mirrors the procedures recommended in FDA guidance for the qualification of dissolution apparatus 1 and 2 by controlling variability in dissolution testing through more comprehensive and stringent measurement of physical parameters.[15] The proposed focused on five main areas; certification of components, documentation of preventative maintenance, mechanical qualification parameters, operational checks, and laboratory controls.

Component certification. Components may be certified by direct measurement or possibly obtaining documentation from a supplier that verifies each specification outlined in Fig. 8.12 has been measured, documented and is found to be within the tolerances outlined in USP <711> Dissolution.

Preventative maintenance (PM). PM should be performed on a prescribed frequency and the instrument should be evaluated to lubricate, repair or replace worn items. Internal components, power and communication cables should be evaluated for wear. The water bath, heater/circulator, drive components, vessels, inner tubes, screens, sampling lines, evaporation covers, and splash covers should all be evaluated for wear and replaced if necessary.

Verification of physical parameters. The operational parameters must be routinely measured and documented for vessel temperature, dip rate, stroke distance, and condition of screens. Automated systems also need to be evaluated for timing and volume accuracy.

Operational checks. Before each run, the analyst is responsible to ensure the integrity of the apparatus:

- Reciprocating cylinders and vessels are free from residue, scratched, and cracks.

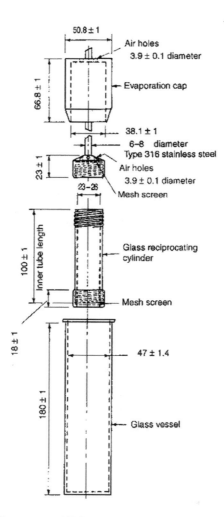

Figure 8.12 ICH harmonized USP apparatus 3.[16]

- Screens are of the required dimension and are not damaged, frayed, misshapen, film coated, or corroded.
- Upper and lower caps are clean and free from residue and the vent holes in the upper cap are unobstructed.
- Vessel temperature of media must be measured and documented at 37.0°C.

- Evaporation covers (2) for the vessels are installed with the proper tension to retract and move freely during the test.

Laboratory controls. The laboratory is required to control sources of variation that will affect any dissolution and drug release apparatus in terms of dissolved gasses, vibration, and component dimensions.

8.5 Summary

In conclusion, apparatus 3 reciprocating cylinder was originally designed for testing extended release products and exposing drug formulation to pharmacokinetic and mechanical properties similar to those found in the GI tract. The instrument has proven useful for poorly soluble compounds, chewable formulation, immediate release, delayed release, and numerous modified release products. At the time of this writing, adaptions to the reciprocating cylinder to contain microspheres within a dialysis membrane are under evaluation so this may provide additional benefits to biorelevant drug release profiling from micro- and nanoparticles.

Because of the apparatus' ability to characterize the release profile of early drug formulation candidates, the apparatus has the potential to provide knowledge-based assessments of formulations developed through quality by design (QbD) and further characterization of the products design space for post approval manufacturing.

In the words of the inventor of the Bio-Dis, Dr. Beckett, "The USP 3 should be considered the first line apparatus in product development of controlled release preparations, because of its usefulness and convenience in exposing products to mechanical as well as a variety of physicochemical conditions which may influence the release of product in the GI tract."[3]

References

1. National Formulary XIV, Timed-Release Tablets and Capsules – In Vitro Test Procedure; 1975; 985–986.
2. B. Esbelin, E. Beyssac, J. M. Aiache, G. K. Shiu, and J. P. Skelly (1991). A new method of dissolution in vitro, the "Bio-Dis" apparatus: comparison

with the rotating bottle method and in vitro-in vivo correlations. *J. Pharm. Sci.*, **80**(10):991–994.

3. I. Borst, S. Ugwu, and A. H. Beckett (1997). New and extended applications for USP drug release apparatus 3. *Dissolut. Technol.*, **4**(1):11–15.

4. U.S. Pharmacopeia XXXVI/National Formulary XXXI, through Second Supplement, <711> Dissolution, U.S. Pharmacopeial Convention, Rockville Maryland, USA, 2013.

5. European Pharmacopoeia 8. Methods of Analysis 2.9.3 Dissolution test for solid dosage forms. The European Directorate for the Quality of Medicines and HealthCare, 8th ed., 2014

6. L. X. Yu, J. T. Wang, and A. S. Hussain (2002). Evaluation of USP apparatus 3 for dissolution testing of immediate-release products. *AAPS PharmSci*, **4**(1):E1.

7. US Pharmacopeia XXXVI, 2013.

8. L. Brecevic, I. Bosan-Kilibarda, F. Strajnar (1994). Mechanism of antifoaming action of simethicone. *J. Appl. Toxicol.*, **14**(3):207–211.

9. M. Siewert, J. Dressman, C. K. Brown, and V. P. Shah (2003). FIP/AAPS guidelines to dissolution/in vitro release testing of novel/special dosage forms. *AAPS PharmSciTech*, **4**(1):E7.

10. U. Klancar, B. Markun, S. Baumgartner, and I. Legen (2013). A novel beads-based dissolution method for the *in vitro* evaluation of extended release hpmc matrix tablets and the correlation with the *in vivo* data. *AAPS J.*, **15**(1):267–277.

11. B. R. Pezzini and H. G. Ferraz (2009). Bio-Dis and the paddle dissolution apparatuses applied to the release characterization of ketoprofen from hypromellose matrices. *AAPS PharmSciTech*, **10**(3):763.

12. S. Klein, M. W. Rudolph, and J. B. Dressman (2002). Drug release characteristics of different mesalazine products using USP apparatus 3 to simulate passage through the GI tract. *Dissolut. Technol.*, **9**(4):6–12.

13. S. Paul. USP Notification Memo to USP customers about the discontinuance of chlorpheniramine maleate ER tablets, December 20, 2011.

14. B. Crist, Qualification of USP Apparatus 3. USP Workshop Challenges in Dissolution, USP Headquarters, Rockville, MD, June 11–12, 2012.

15. US Food and Drug Administration, Guidance for Industry. The Use of Mechanical Calibration of Dissolution Apparatus 1 and 2 – Current Good Manufacturing Practice (CGMP), January 2010.

16. U.S. Pharmacopeia XXXVI/National Formulary XXXI, through Second Supplement, <711> Dissolution, U.S. Pharmacopeial Convention, Rockville MD, USA, 2013.

Chapter 9

Use of Apparatus 4 in Dissolution Testing, Including Sparingly and Poorly Soluble Drugs

Rajan Jog,[a] Geoffrey N. Grove,[b] and Diane J. Burgess[a]

[a]*University of Connecticut, Department of Pharmaceutical Sciences, 69 N. Eagleville Rd. U3092, Storrs, CT 06269, USA*
[b]*Sotax Corporation, 2400 Computer Drive, Westborough, MA 01581, USA*
geoffrey.grove@sotax.com

9.1 Introduction

The earliest recorded mention of a flow-through cell dissolution apparatus which later entered the USP (United States Pharmacopeia) and EP (European Pharmacopoeia) as apparatus 4, and the JP (Japanese Pharmacopeia) as apparatus 3, was in 1957, pre-dating the basket and paddle apparatus. This earliest dissolution system, contained in essence the same components present in modern-day systems: a flow cell, a reservoir, and a pump (Fig. 9.1).[1] Slightly more than a decade later in 1968, Pernarowski published an article on a continuous flow dissolution apparatus which, like the previous system, contained a flow cell, a reservoir, and a pump. However,

Poorly Soluble Drugs: Dissolution and Drug Release
Edited by Gregory K. Webster, J. Derek Jackson, and Robert G. Bell
Copyright © 2017 Pan Stanford Publishing Pte. Ltd.
ISBN 978-981-4745-45-1 (Hardcover), 978-981-4745-46-8 (eBook)
www.panstanford.com

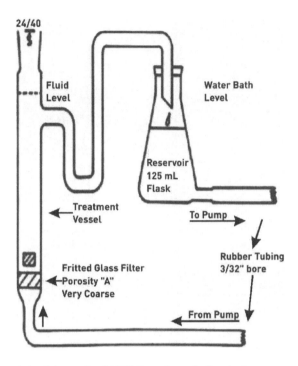

Figure 9.1   Schematic of 1957 flow-through dissolution apparatus.[1]

in this later system, the "flow cell" consisted of a 1 L flask and a rotating basket which later evolved into the modern apparatus 1 (Fig. 9.2).[2]

Not long thereafter, in the early 1970s, Dr. F. Langenbucher, then at Ciba-Geigy, conceptualized and made drawings for what would become the prototype version of a commercial flow-through system, and in 1973 he commissioned a small Swiss engineering company (Sotax) to build the prototype system to his design. Sotax continued development based on this prototype until a commercial system was finally released in 1976 (see Fig. 9.3).[3]

In 1977, in his article titled "Dissolution Rate Testing with the Column Method: Methodology and Results," Dr. Langenbucher presented the system which would become the modern EP/USP 4 and JP 3. In this article he made the case that this system permitted reproducible performance among different laboratories and, that

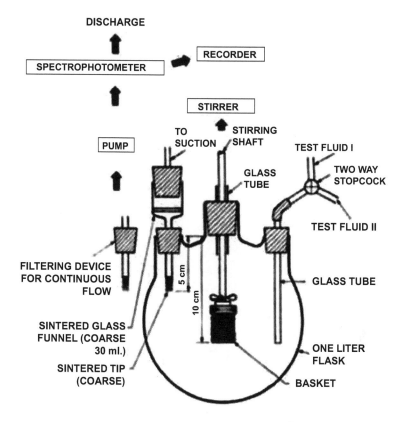

DISCHARGE

SPECTROPHOTOMETER

RECORDER

STIRRER

PUMP

TO SUCTION

STIRRING SHAFT

GLASS TUBE

TEST FLUID I

TWO WAY STOPCOCK

TEST FLUID II

FILTERING DEVICE FOR CONTINUOUS FLOW

5 cm

10 cm

GLASS TUBE

SINTERED GLASS FUNNEL (COARSE 30 ml.)

SINTERED TIP (COARSE)

ONE LITER FLASK

BASKET

Figure 9.2 Pernarowski apparatus.[2]

it could be used for a variety of dosage forms, including powders, granules, tablets, coated tablets, and capsules. He also made a case for a correlation with in vivo data and suggested that this new system be regarded as an alternative to the existing dissolution apparatus types.[4]

In support of his recommendation, and based on precedent, Dr. Langenbucher also noted that looking back at some of the original FDA work, and citing a 1957 letter by E. B. Vliet from the FDA, and an article by D. J. Campbell. The original column method (also known as the flow-through method) had been favored by the FDA for testing release from time-release preparations, although the USP did not choose to pursue it at that time.[5]

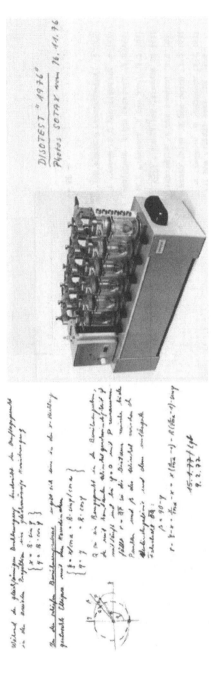

Figure 9.3 Langenbucher's flow-through system.[3]

Despite these early efforts by Dr. Langenbucher, it was not until 1990 that the USP and EP introduced the flow-through cell as apparatus 4, and later in 1996 when the JP introduced the same system as apparatus 3.

9.2 Pharmacopeial Considerations

The first compendial methods included drawings and specifications for the flow cells to be used. However, although apparatus 4 entered into both the USP and EP at roughly the same time, different flow cells were included in these respective pharmacopeia. Today several different flow cells are listed in both pharmacopeias, but the chapters are still not completely harmonized. In the main dissolution chapter of USP (chapter <711>) the 22.6 mm and 12 mm cells appear while in the main chapter of the EP, the 12 mm and the two-chambered lipidic cells appear. The USP has in recent years added the two chambered lipidic cell in chapters other than <711> (i.e., chapter <2040>), and it is not unreasonable to assume that as new flow cell types are designed and adopted by the industry, the pharmacopeias will expand their listings.

It should be noted that in USP chapter <1092>, it is mentioned that "apparatus 4 may have utility for soft gelatin capsules, bead products, suppositories, or **poorly soluble drugs**." Additionally, at the time of this writing, a revision to chapter <1092> has been communicated through the USP's Pharmacopeial Forum. In this suggested revision, the stimuli article proposes that "Apparatus 4 (flow-through cell) may offer advantages for modified-release dosage forms that contain active ingredients with limited solubility".[6] The stimuli article goes on to state that "Apparatus 4 may have utility for soft gelatin capsules, bead products, suppositories, or poorly soluble drugs." A later suggested version of this text also included immediate released products with limited solubility, and expanded the forms to include injectable depot dosage forms as well as suspension-type extended-release dosage forms for oral, parenteral or ocular application.

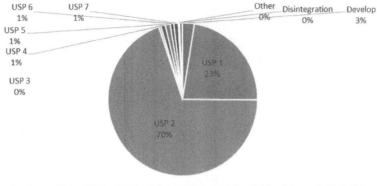

■ Develop ■ USP 1 ■ USP 2 ■ USP 3 ■ USP 4 ■ USP 5 ■ USP 6 ■ USP 7 ■ Other ■ Disintegration

Figure 9.4 Percentage of use of differing USP apparatus types in FDA registered dissolution methods.

9.3 FDA Considerations

The FDA maintains a database currently containing more than 1000 dissolution methods, aptly named the "FDA Dissolution Methods Database." This database includes methods which refer to the USP as well as dissolution methods presently recommended by the FDA Division of Bioequivalence, Office of Generic Drugs. An analysis of the methods in this database shows that while USP apparatus 1 and 2 hold a commanding lead in the number of recommended methods, perhaps somewhat due to their early recognition in the compendia, that there is roughly an equal division of recommendations for the remaining apparatus types (see Fig. 9.4). Note: The category "Develop" indicates that a dissolution method should be developed, but that there is not yet an existing method listed. The category "Disintegration" indicates that a disintegration test may be used instead of a dissolution test.

It may also be worth noting that in recent years the FDA has recommended for a variety of products (parenteral, implants, microparticles, suspensions, liposome formulations, rectal and vaginal suppositories, etc.) that companies should "[d]evelop an in vivo release method using USP IV (Flow-Through Cell), and, if applicable, apparatus 2 (Paddle) or any other appropriate method." In their online database, the majority of the times the FDA recommends that a dissolution method should be developed, the formulation in question is an injectable suspension, but there is also a recommendation for a tablet and capsule formulation as well.

9.4 System Configuration

The configuration of an apparatus 4 system at its simplest requires a choice of whether the system will be operated in the open or closed mode, the analytical finish, and the flow cell type. Of particular importance when dealing with poorly soluble compounds is the choice of whether to operate the system in the open or closed mode. When operated in the closed mode (see Fig. 9.4), the dissolution media circulates through a reservoir. In practice, closed systems can be operated with volumes as low as 50 mL without modification; however, apparatus 4 can be operated with reservoir sizes far exceeding the capacities of apparatus 1 and 2. This is of particular utility when working with poorly soluble compounds.

When operated in the open mode (see Fig. 9.5), fresh media continuously passes through the samples in the flow cells. Methods can either use a single media, or, with the use of the media selector valve, the media can be changed throughout the run, to cover a range of media types for the product, or to mimic pH changes which happen in vivo.

The open mode configuration, like the closed loop configuration, can easily accommodate volumes much larger than is possible in compendial apparatus 1 or 2. A good example of this from the FDA's dissolution methods database is medroxy progesterone acetate. This

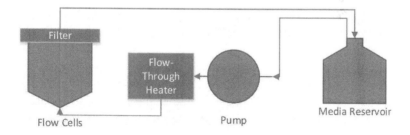

Figure 9.5 Closed-loop mode schematic.

test is run with a pump setting of 17 mL/min, and the final sampling point is taken at 90 minutes, for a total volume of 1530 mL.

When using apparatus 1 or 2, it may be necessary to use high concentrations of surfactants in order to work with poorly (water) soluble compounds. However, both the closed and open loop configurations of apparatus 4 allow the use of much larger media volumes. Therefore, it may be possible by increasing media volume, to develop an apparatus 4 method with lower surfactant concentration. In some cases achieving conditions which may allow correlation with in vivo results.

The two most common choices for analytical finish are UV and LC. If the drug and drug concentration permit UV detection, both the open and closed loop configurations can be used with a UV equipped with a cell changer and UV flow cells for online detection. For LC detection, samples are typically collected and analyzed offline. In the closed loop mode, samples are collected in capped LC vials using an in-line autosampler, whereas in the open loop mode, samples are collected using a fraction collector.

9.4.1 Flow Cell Selection and Design

Including the compendial cells, there are a wide variety of flow cell types and cell design choices. The principal choice of flow cell type generally depends on the product dosage form. While tablets and capsules and other solid dosage forms may fit equally well inside several of the flow cells listed in the compendia; suspensions, emulsions, and semisolid dosage forms may require inserts. Nanoparticle formulations, which, would require filtration at such a fine level that pump back pressure could become an issue can be sequestered in a space contained behind a membrane, much like in a Franz cell, with flow passing along the membrane surface.

One of the most basic compendial cell design choices is illustrated in the following figures. In the first figure (Fig. 9.6) a dosage form is shown in a 22.6 mm compendial cell with a ruby bead at the bottom. The ruby bead acts as a check valve and also has an effect on the fluid flow dynamics. In the second figure (Fig. 9.7), glass beads have been added to the cell and the dosage form is resting

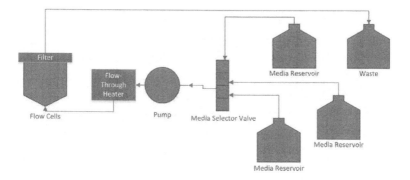

Figure 9.6 Open loop mode schematic.

Figure 9.7 No glass beads.   **Figure 9.8** With glass beads.

on top of the bead bed. In the third figure (Fig. 9.8) in addition to the bead bed, a tablet holder is inserted in the cell to hold the dosage form above the bead bed. A fourth case (not shown), would be to have the dosage form held by a tablet holder, but without the bead bed below. In each of these cases, the fluid flow that the dosage form experiences is different, and some, but not necessarily all dosage forms may be sensitive to these changes. When developing a method, it is best to choose the simplest cell design which provides reproducible results.

When working with powders, a compendial cell design could resemble that shown in Fig. 9.9, or, to prevent the powder from floating in the cell, a "Sandwich" technique is sometimes used

Figure 9.9 With glass beads and tablet holder.

Figure 9.10 Powder layered on glass beads.

Figure 9.11 Powder in glass bead "sandwich."

Figure 9.12 Layering of a Semisolid product.

(Fig. 9.10). These same layering or sandwiching techniques can also be used with suspensions and creams.

While USP <711> mentions glass beads "of about 1 mm diameter," there are a variety of sizes available. It has been shown that the size of the beads can affect some products, so bead size should be noted during method development.[7]

Other cell inserts for example to support drug eluting contact lenses (Fig. 9.13), or to contain semisolids behind a membrane as in the Franz cell design (Fig. 9.14), add a great deal of versatility to

Figure 9.13 Sandwich technique with semi-solid product.

Figure 9.14 Contact lens holder.

Figure 9.15 Semisolid insert.

Figure 9.16 Lipidic flow cell.

the existing compendial 22.6 mm cell, making it perhaps the most versatile of flow cells available.

Beyond these simple inserts, however, there are also a range of cells designed with particular products in mind. For example, the compendial lipidic cell (Fig. 9.15) has a two chambered flow path meant to capture and wash a lipidic layer with the dissolution media. Originally developed for suppository formulations, this cell was first adopted in the EP and later entered the USP. The principal is somewhat like a classical separatory funnel, where active ingredients can be extracted from the lipidic layer into the aqueous

Figure 9.17 Powder flow cell.   **Figure 9.18** Small-volume flow cell.

Figure 9.19 Stent cell (shown with filter cap).

solution. This is not dissolution in the traditional sense, where a solid is dissolving into a media, but rather a partitioning of the active ingredients from the lipidic phase to the aqueous phase.

For work with powdered or granular substances including powder formulations, powder mixtures prior to tableting, etc., the "powder" cell was developed (Fig. 9.16). This cell appears in the EP under 2.9.43. Apparent dissolution is not currently harmonized across all pharmacopeias, though it has been suggested that this should be seriously considered. An apparent dissolution experiment can capture useful information regarding changes that may occurring during interim steps in processing prior to the final tableting, encapsulation, etc.

When working with dosages, such as ocular implants, where the method, or the in vivo environment suggests that a low volume dissolution would be appropriate, a low volume cell, which allows the system to be run at a total working volume under 25 mL may be used (Fig. 9.17). Working at lower volumes may also be useful when dealing with low dose products. Cells have also been designed to be loaded in a way which mimics in vivo implantation procedures. For example, a stent cell (Fig. 9.18), designed with a single inner diameter tube, such that a stent can be expanded into the cylinder, mimicking in vivo stent expansion, and facilitating the majority of aqueous flow to pass through the inside of the stent, as occurs in vivo.

9.4.2 Pump Selection

Historically, the compendial flow rates were 4, 8, and 16 mL/min, with a pulsation rate of 120 pulses per minute. However, a recent change allows a pump "without pulsation" to be used.[8] Producing a pump that is completely free of pulsation is expensive, but this description of the pumping system is probably being loosely interpreted to include peristaltic pumps, which have somewhat less pulsation than piston pumps.

When considering the linear flow rate that the sample experiences, there is a simple relation to the cell diameter that should be considered (Table 9.1). The addition of glass beads and/or inserts, and the size of the dosage form itself are also important.

Table 9.1 Linear velocity relative to pump rate

	Flow rate (mL/min)	2	4	8	16	32
Cell diameter (mm)	Cell area (cm^2)	Linear velocity (cm/min)				
12.0	1.1	1.8	3.5	7.1	14.1	28.3
22.6	4.0	0.5	1.0	2.0	4.0	8.0

There has been some debate in the literature regarding the way in which the presence or absence of glass beads may change the flow properties in the cell relative to laminar and turbulent flow.[9-11] At the very least, it can be said with some certainty that the addition of glass beads decreases the cross-sectional cell area through which the flow travels, and would therefore serve to increase the linear velocity of the media as it passes through and exits the bead bed. If the sample is placed in the bed, or at the surface, then it would also experience this increased linear flow velocity.

9.4.3 Flow-Through Cell and Dissolution Testing of Different Dosage Forms

As described above, apparatus 4 has been used for a number of dosage forms (such as microspheres, capsules, lozenges, suppositories, extended release tablets, drug-resin complexes, nanoparticles, etc.). In addition, USP apparatus 4 has been used for poorly soluble drugs. Dissolution method development for compounds with low solubility in solid dosage forms is challenging.[12,13] However, the fluid dynamics in the flow-through cell simulates that in the gastrointestinal tract (GIT). In order to maintain sink conditions throughout the experiment, large media volumes are required for poorly soluble drugs and this may result in drug concentrations that are below the limit of detection at the early stages of dissolution testing. This may require advanced analytical techniques such as LC-MS, LC-NMR, etc., to detect the analyte at such a low detection limit.

The following Tables 9.2a-9.2h describe a list of dosage forms that have been analyzed using apparatus 4 along with the instrumental design, experimental conditions and significant findings relevant to the use of apparatus 4.

Table 9.2a Dissolution testing of capsules using apparatus 4

Title [year]	Instrumental design	Drug	Experimental conditions	Significant findings relevant to the use of apparatus 4	Refs.
A dissolution method for hard and soft gelatin capsules containing testosterone undecanoate in oleic acid [1986]	Langenbucher flow-through cell (Sotax, type CE-1)	Testosterone undecanoate	**Media:** 0.1M HCl **Filter:** nitrocellulose filter (Millipore, type AA, 0.8 μm) **Flow rate:** 5 mL/min **Temperature:** 37 ± 0.5°C **UV:** 380 nm	→ The method used a flow-through dissolution cell and a dissolution medium the composition of which was optimized for both its capacity to dissolve the gelatin capsule wall completely and to give homogeneous filtrates of drug and oily excipient for convenient analysis of the collected fractions.	[14]
A collaborative study of the in vitro dissolution of acetylsalicylic acid gastro-resistant capsules comparing the flow-through cell (FTC) method with the USP paddle method [1997]	Sotax CE6 dissolution-testing apparatus (Sotax AG, Switzerland) with six cells and the CY6D piston pump	Aspirin	**Media:** 0.1 M HCl mixed with 0.2M Na_3PO_4 **Flow rate:** 8, 12 and 16 mL/min ± 5% **Temperature:** 37°C **Glass bead:** 1 mm **UV:** 296 nm	→ The FTC method was found to be more convenient to use for changing the pH from pH 1.2 to 6.8. → The effect of different dissolution media ionic strengths on the dissolution rate results was reported.	[15]

(Contd.)

Table 9.2a *(Contd.)*

Title [year]	Instrumental design	Drug	Experimental conditions	Significant findings relevant to the use of apparatus 4	Refs.
A comparison of dissolution testing on lipid soft gelatin capsules using USP apparatus 2 and apparatus 4 **[2005]**	Sotax CE7 Dissolution system equipped with soft gel capsule flow-through cells	Free base of secondary amine	**Media:** 0.01N HCl, 0.01N HCL/0.25% Tween 80, 0.1% acetic acid/0.25% Tween 80, 0.01N HCl/0.5% Tween 80 **Filter:** Glass microfiber **Flow rate:** 16 mL/min **Temperature:** 37°C **UV:** 273 nm	→ Apparatus 2 resulted in a faster dissolution rate, however apparatus 4 was better able to discriminate between the different formulations. → Apparatus 4 method is more suitable to assess excipient and/or process changes that may affect the release rate of the drug from the dosage form during formulation development. → 0.01 N HCl / Tween 80 medium resulted in the most complete drug release.	[16]

Table 9.2b Dissolution testing of tablets using apparatus 4

Title [year]	Drug	Instrumental design	Experimental conditions	Significant findings relevant to the use of apparatus 4	Refs.
A collaborative in vivo dissolution study: comparing the flow-through method with the USP paddle method using USP prednisone calibrator tablets [1989]	Prednisone	Flow-through apparatus [Sotax AG, Switzerland] with 12.0 and 22.6 mm diameter cells	**Media:** deaerated water **Filter:** 15 – 25 mm **Flow rate:** 12 mm cell – 9 & 16 mL/min 22.6 mm cell – 32 mL/min **Temperature:** 37°C **Glass beads:** 1 & 4 mm **UV:** 242 nm	→ For the flow-through method, the linear flow rate of the dissolution medium in the cells is a fundamental parameter to consider. → The flow-through method may be valuable as an alternative dissolution method capable of generating reproducible results for disintegrating tablets, if the flow rate is properly selected.	[17]
A flow-through dissolution approach to in vitro/in vivo correlation of adinazolam release from sustained release formulation [1989]	Adinazolam	Flow-through system, which comprises a Disotest CE6 thermostatted flow-cell unit, a Disopump CY6 (Sotax AG, Switzerland) and a Dissoette fraction collector (Copley, UK)	**Media:** pH 1.2 simulated gastric fluid without enzymes (USP XXI) and phosphate buffers at pH 4, pH 7 and pH 9 **Flow rate:** closed loop – 8 or 15 mL/min, open loop – 8 mL/min **Temperature:** 37 ± 0.1°C **UV:** 220 nm	→ The open-loop flow-through dissolution differential profiles provide more information about the release characteristics of the sustained release formulations than the closed-loop cumulative profiles. → The similarity of the open-loop differential profiles to the in vivo concentration-time profiles appears to allow marginally better in vitro/in vivo correlation than the cumulative closed-loop profiles.	[18]

(Contd.)

Table 9.2b (*Contd.*)

Title [year]	Instrumental design	Drug	Experimental conditions	Significant findings relevant to the use of apparatus 4	Refs.
Evaluation of the Flow-through cell dissolution apparatus: effects of flow rate, glass beads and tablet position on drug release from different types of tablets **[1994]**	Flow-through cell dissolution apparatus (Erweka) with cell of an internal diameter of 22.6 mm	Water soluble drug A & water sparingly soluble drug B	**Media:** distilled water **Filter:** glass microfiber (GF/D grade) **Flow rate:** 7–21 mL/min **Temperature:** 37°C **Glass beads:** 1 mm	→ Drug release increases with increasing flow rate. → Flow rate appreciably impacts drug release from erodible tablets when glass beads are not utilized in the cells due to turbulent flow causing the erosion rate to change. → For erodible tablets, the horizontal position results in faster drug release, possibly due to more surface area exposed to turbulent flow. → For coated matrix tablets, the vertical position results in faster drug release possibly due to change in the coating breakage rate under turbulent flow conditions.	[19]

| Application of flow-through dissolution method for evaluation of oral formulations of nifedipine [1994] | Flow-through dissolution apparatus (SOTAX Dissotest) – Leap Technologies (NC, USA) | Nifedipine | **Media:** water or phosphate buffer (0.05 or 0.1M; pH 7.4) with/out Tween
Flow rate: 12.5 mL/min
Temperature: 37°C
UV: 254 nm | → In comparison to beaker type systems, flow-through dissolution apparatus offers a potentially better alternative for the characterization and comparison of in vitro drug release from different types of formulations.
→ As large volumes of dissolution media can be employed, addition of solubility enhancers such as Tween may not be necessary, hence the dissolution profiles obtained using the flow-through system are likely to be more representative of the effect of formulations on the drug-release characteristic in vivo. | [20] |

(Contd.)

Table 9.2b (*Contd.*)

Title [year]	Drug	Instrumental design	Experimental conditions	Significant findings relevant to the use of apparatus 4	Refs.
A flow-through dissolution method for a two component drug formulation where the actives have markedly differing solubility properties **[1998]**	Atovaquone	Flow-through dissolution system – Sotax equipment: a CE6 Dissotest, a CY7 piston pump, an MS36 media switcher controlled by a PD29 Technical Interface	**Media:** 0.1M sodium hydroxide **Filter:** Whatman **Flow rate:** 16 mL/min **Temperature:** 37°C **Glass beads:** 1 mm **UV:** 487 nm	→ A flow-through dissolution test was developed for a two component anti-malarial tablet formulation as an alternative to a test using the EP/USP paddle apparatus (which was shown to provide unsatisfactory release data for the low solubility component in the formulation).	[21]
Robustness testing, using experimental design, of a flow-through dissolution method for a product where the actives have markedly differing solubility properties **[2000]**	Atovaquone	Flow-through apparatus – Sotax (Basingstoke, UK); a CE6 Dissotest, a CY7 piston pump and an MSV-6 media switcher with 22.6 mm internal diameter cell	**Media:** 0.1M sodium hydroxide **Filter:** Whatman Maidstone, UK) GF:F, GF:B, GF:C, GF:A and GF:D **Flow rate:** 16 mL/min **Temperature:** 37°C **Glass beads:** 1 mm **UV:** 487 nm	→ The robustness of a flow-through dissolution method for atovaquone in a two component antimalarial tablet formulation was assessed using DoE (design of experiments) study. → Results demonstrate that the flow-through dissolution method may be considered robust to changes in all the main parameters (dissolution media, peristaltic pump speed and flow rate) evaluated at sample times of 30 min and above.	[22]

Studies of floating dosage forms of furosemide: in vitro and in vivo evaluations of bilayer tablet formulations [2000]	Continuous flow-through cell method	Furosemide	**Media:** artificial gastric fluid of pH 1.2 without enzymes & 0.02% polysorbate 20 **Flow rate:** 9 mL/min **Temperature:** 37°C **UV:** 274 nm	→ The in vitro–in vivo correlation obtained indicates that the flow-through cell method used for the in vitro dissolution rate is an ideal method for testing of such dosage forms.	[23]
In vitro–in vivo correlation and comparative bioavailability of vincamine in prolonged-release preparations [2000]	Small flow-through cell (Dissotest CE-6, connected with a piston pump Y 7; Sotax AG, Basel, Switzerland)	Vincamine	**Media:** simulated gastric fluid without enzyme (pH 1.2 and 4.5) and simulated intestinal fluid without enzyme (pH 6.9 and 7.5) **Filter:** 0.45 μm **Flow rate:** 8 ± 0.2 mL/min **Temperature:** 37°C **UV:** 267 nm	→ Good correlation was obtained between fraction absorbed in vivo and fraction dissolved in vitro using the flow-through cell in the open mode.	[24]

(Contd.)

Table 9.2b (*Contd.*)

Title [year]	Instrumental design	Drug	Experimental conditions	Significant findings relevant to the use of apparatus 4	Refs.
Predicting dissolution via hydrodynamics: salicylic acid tablets in flow-through cell dissolution [2000]	Flow-through cell (USP apparatus 4)	Salicylic acid	**Media:** 0.05M phosphate buffer (pH 7.4) **Filter:** 3 μm **Temperature:** 37.5 ± 0.5°C **UV:** 29 nm	→ Dissolution results obtained using USP apparatus 4 demonstrated that the dissolution rate was approximately 52 mL/min with tablets in the horizontal position in the 12 mm cells, the same could not be achieved with tablets in the vertical orientation in 12 mm cells, or in any orientation in the 22.6 mm cells at the maximum velocity available on the pump. → The dissolution rate of approximately 52 mL/min obtained using USP 4 was not achieved in USP 2 at 100 rpm.	[25]
Performance of USP calibrator tablets in flow-through cell apparatus [2002]	Dissolution tester Dissotest CE-6, Sotax AG, Basel; Dissolution flow-through cell for tablets and capsules with a diameter of 22.6 and 12 mm	Salicylic acid, Prednisone	**Flow rate:** 8 & 16 mL/min **Temperature:** 37°C **Filter:** Glass microfibre, Whatman GF/F **Glass beads:** 5 mm	→ The USP calibrator tablets: salicylic acid and prednisone were shown to be appropriate for use as an apparatus suitability test for the flow-through cell dissolution method — USP apparatus 4.	[26]

Study	Apparatus/Method	Drug	Conditions	Observations	Ref.
Comparison of dissolution profiles for albendazole tablets using USP apparatus 2 and 4 **[2003]**	USP apparatus 4 (flow-through cell method), Sotax CH4123 automatized dissolution tester – open system	Albendazole	**Media:** 0.1N HCl **Filter:** Whatman **Flow rate:** 16 mL/min **Temperature:** $37 \pm 0.5°C$ **Glass beads:** 0.75–1 mm **UV:** 291 nm	→ Significant differences were found between albendazole dissolution profiles for generic and reference products, using either USP apparatus 2 or 4. → USP apparatus 4 demonstrated a greater sensitivity and discriminative capability than USP apparatus 2 in detecting differences in the dissolution behavior of the albendazole products studied.	[27]
In vitro–in vivo correlation for wet-milled tablet of poorly water-soluble cilostazol **[2008]**	Flow-through cell dissolution tester (PT-DZ7, Pharma Test Apparatebou GmbH, Hainburg, Germany)	Cilostazol	**Media:** 900 mL of 0.30% SLS, 4.0% Polysorbate 80 or water **Filter:** 0.3 μm, Gamma 12 filter, Grade 03, Whatman **Temperature:** 37°C **UV:** 257 nm	→ The compendial dissolution method using apparatus 2 did not correlate well with the bioavailability enhancement of the wet-milled tablet. → An IVIVC was obtained for dissolution rates determined using apparatus 4 in the closed-loop mode with a large amount of aqueous SLS solution (at a concentration below the critical micelle concentration) for both the fasted and fed states.	[28]

(Contd.)

Table 9.2b *(Contd.)*

Title [year]	Instrumental design	Drug	Experimental conditions	Significant findings relevant to the use of apparatus 4	Refs.
Comparison of three dissolution apparatuses for testing calcium phosphate pellets used as ibuprofen delivery systems [2009]	USP apparatus 4 automated system CE 7smart (Sotax, Basel, Switzerland)	Ibuprofen	**Media:** phosphate buffer solution (pH 7.48) **Flow rate:** 8 mL/min **Temperature:** 37°C **Glass beads:** 1 mm **UV:** 264 nm	→ The paddle apparatus and the flow-through cell were preferred. Dissolution testing in the reciprocating cylinder required a shorter time and smaller media volume compared to the paddle and flow-through methods. However the reciprocating cylinder method was less discriminating, due to the specific design and the motion of the apparatus, which resulted in an undesirable granule disintegration. → The compendial flow-through cell was determined to be more appropriate to develop a dissolution testing method for bone implantable materials used as drug delivery systems.	[29]

Comparison of flow-through cell and paddle methods for testing vaginal tablets containing a poorly water-soluble drug [2013]	Erweka Flow-Through Cell Dissolution Tester, type DFZ 720 (Heusenstamm, Germany) equipped with 22.6 mm diameter cells – open loop	Clotrimazole	**Media:** 0.1M HCl or acetate buffer pH 5.2 + 1% SDS **Flow rate:** 4, 8 or 16 mL/min **Temperature:** 37 ± 0.5°C **UV:** 210 nm	→ Faster release was obtained using the flow-through cell method with 0.1 M HCl (85.9 % in 10 min) when compared to acetate buffer pH 5.2/1 % SDS (>80 % of the drug released in 30 min). → The slower dissolution obtained with acetate buffer (pH 5.2) /1 % SDS was useful in distinguishing between different formulations. → The paddle method resulted in faster dissolution rates, however the flow-through cell method was more reproducible.	[30]
Comparative in vitro dissolution study of carbamazepine immediate-release products using the USP paddle method and the flow-through cell system [2014]	Sotax CE6, Sotax AG, Switzerland with 22.6 mm cells (i.d.) and a piston pump (Sotax CY7–50, Sotax AG, Switzerland) – open system	Carbamazepine	**Media:** 1.0% sodium lauryl sulfate aqueous solution **Filter:** 0.45 μm nitrocellulose membranes **Flow rate:** 16 mL/min **Temperature:** 37 ± 0.5°C **UV:** 285 nm	→ The flow-through cell method has greater discriminating ability than the USP paddle method to identify significant differences between rate and extent of dissolution of carbamazepine immediate release tablets.	[31]

Table 9.2c Dissolution testing of drug moiety/investigational compounds using apparatus 4

Title [year]	Instrumental design	Drug	Experimental conditions	Significant findings relevant to the use of apparatus 4	Refs.
A collaborative study of the in vitro dissolution of phenacetin crystals comparing the flow-through method with the USP Paddle method [**1991**]	Flow-through apparatus [Sotax AG, Switzerland]	Phenacetin	**Media:** deaerated water **Flow rate:** 16 and 32 mL/min ±5% **Temperature:** 37 ± 0.5°C **Glass bead:** 1 mm **UV:** 244 nm	→ The flow-through cell apparatus proved less dependent on the hydrodynamic intensity compared to the USP Paddle method. → A faster drug release rate was observed in case of the flow-through cell apparatus, (which operates under more efficient sink conditions since, fresh solvent is constantly supplied), compared to the USP Paddle method. → The flow-through method provided more efficient wetting of the crystals and was less dependent upon sample size, when compared to the USP Paddle method.	[32]

Dissolution testing of a poorly soluble compound using the flow-through cell dissolution apparatus [2002]	USP 4, Erweka Instruments Inc., with 12 mm diameter cell	PD198306	Media: 25 mM pH 9 sodium phosphate solution with 0.5% SLS Flow rate: 4 or 8 mL/min Temperature: 37°C UV: 280 nm	→ It was determined that a better dissolution profile in terms of rate and extent of dissolution was obtained when drug powder was suspended prior to adding to the flow-through cell.	[33]
Assessment of oral bioavailability enhancing approaches for SB-247083 using flow-through cell dissolution testing as one of the screens [2003]	Flow-through cell dissolution apparatus (Sotax AG CH-4008 Basel) with internal diameter of 22.6 mm	SB-247083 – ETA-selective endothelin receptor antagonist	Media: pH 1.2 to 6.8 dissolution medium Flow rate: 4 mL/min Temperature: 37 ± 1°C Glass beads: 1 mm UV: 286 nm	→ Flow-through cell dissolution testing was a useful predictor of the oral bioavailability of various formulations of SB-247083.	[34]

(Contd.)

Table 9.2c *(Contd.)*

Title [year]	Instrumental design	Drug	Experimental conditions	Significant findings relevant to the use of apparatus 4	Refs.
In vitro/in vivo correlations for a poorly soluble drug, danazol, using the flow-through dissolution method with biorelevant dissolution media [2005]	USP apparatus 4 – Sotax Dissotest CE70, Sotax, Basel, Switzerland with 22.6 mm internal diameter	Danazol	**Media:** fed state – Bile salts, phospholipids, fatty acids & monoglycerides (pH – 5.5) fasted state – Bile salts, phospholipids, fatty acids & monoglycerides (pH – 6.8) **Filter:** fiber glass wool **Flow rate:** 8, 16 or 32 mL/min **Temperature:** 37 ± 0.1°C **Glass beads:** 1 mm **UV:** 285 nm	→ In vitro/in vivo correlations of danazol were obtained under fed and fasted conditions using the flow-through dissolution method.	[35]

An in vitro dissolution study of ibuprofen using a flow-through cell **[2007]**	Sotax CE70 flow-through cell (12 mm & 22.6 mm diameter)	Ibuprofen	**Media:** HCl, HCl/NaCl, HCl/NaCl/0.5% Cetrimide **Filter:** Metal & Glass fiber **Flow rate:** 8 or 16 mL/min **Temperature:** 37°C **UV:** 221 nm	→ The dissolution rate for the 12 mm cell was higher compared to the 22.6 mm cell, since the smaller diameter cell resulted in a higher linear velocity (cm/min) of the medium, which in turn increased the dissolution rate. → A faster dissolution rate was obtained when the media flow rate was increased, since faster flow past the drug particles results in a thinner boundary layer and a faster dissolution rate. → Addition of surfactant increased the dissolution rate drastically, since surfactants both increase the wetting of the drug particles as well as form micelles that aid particle dissolution.	[36]

(Contd.)

Table 9.2c *(Contd.)*

Title [year]	Instrumental design	Drug	Experimental conditions	Significant findings relevant to the use of apparatus 4	Refs.
Dynamic dissolution testing to establish in vitro/in vivo correlations for montelukast sodium, a poorly soluble drug [2008]	Custom made flow-through cells, Scientific Glass Blowing Services, Chemistry Dept., University of Alberta with 22.6 mm internal diameter	Montelukast sodium	**Media:** SGF, 0.01 M HCl, consisting of 2 g/l NaCl and 0.1% w/v SLS, pH 2.0 and Biorelevant Dissolution Media (FaSSIF pH 5.0, 6.5, 7.5) consisting of 3.75 mM sodium taurocholate and 0.75 mM lecithin **Filter:** fiber glass wool **Flow rate:** 3.3 (SGF) & 5.8 (FaSSIF) mL/min **Temperature:** 37 ± 0.5°C **Glass beads:** 1 & 4 mm **UV:** 389 nm	→ Flow-through dissolution testing followed by dynamic pH change protocols are able to mimic the environmental changes that an orally administered drug usually encounters in the GI tract. → Simulations using the flow-through dissolution data were able to define the entire in vivo profile of montelukast sodium and an in vitro/in vivo correlation was established.	[37]

Dissolution of poorly water-soluble drugs in biphasic media using USP 4 and fiber optic system **[2009]**	USP 4, Sotax USP 4 – CE 70 system with CP-7 piston pump, equipped with 22.6 mm diameter cells	AMG 517 (Investigational compound – Amgen)	**Media:** phosphate buffer pH 6.8 **Flow rate:** 0 to 35 mL/min **Temperature:** 37°C **UV:** 290 nm	→ The dissolution experimental parameters were optimized *viz.* filter position, flow rate, glass beads and dimensions of USP cells for flow-through cell apparatus 4.	[38]
Flow-through cell method and IVIVR for poorly soluble drugs **[2011]**	USP apparatus 4 – open mode	Diclofenac, Meclofenamic acid	**Media:** Simulated Gastric Fluid sine pepsin (SGFsp) & Sodium-Simulated Intestinal Fluid sine pancreatin (So-SIFsp) **Filter:** glass microfiber **Flow rate:** 2 mL/min **Temperature:** 37°C **UV:** 214 nm	→ The flow-through cell represents an easy and economical method of obtaining highly standardized dissolution data for different salts of a new API.	[39]

Table 9.2d Dissolution testing of suppositions apparatus 4

Title [year]	Instrumental design	Drug	Experimental conditions	Significant findings relevant to the use of apparatus 4	Refs.
Suppository dissolution utilizing USP apparatus 4 [1996]	Sotax CE6, Dissotest – closed system with internal diameter of 22 mm	Acetaminophen	**Media:** USP purified water **Flow rate:** 20 ± 0.5 mL/min **Temperature:** $37.5 \pm 0.5°C$ **Glass beads:** 1 & 4 mm	→ Flow-through cell apparatus may be suitable for quality control monitoring of finished product.	[40]
In vitro release of ketoprofen suppositories using the USP basket and the flow-through cell dissolution methods [2014]	Automated USP apparatus 4 (Dissotest CE- 6; Sotax AG, Basel, Switzerland) coupled to a UV/Vis spectrophotometer (Perkin Elmer Lambda 10; Norwalk CT, USA) and a piston pump (Sotax CY7-50; Sotax AG, Basel, Switzerland)	Ketoprofen	**Media:** phosphate buffer pH 8 and 1% sodium lauryl sulfate aqueous solution **Filter:** 0.45 μm nitrocellulose membranes **Flow rate:** 16, 24 & 32 mL/min **Temperature:** 37 $\pm 0.5°C$ **UV:** 260 nm	→ The flow-through cell method proved to be a suitable method to evaluate the dissolution performance of ketoprofen.	[41]

Table 9.2e Dissolution testing of drug-resin complexes using apparatus 4

Title [year]	Instrumental design	Drug	Experimental conditions	Significant findings relevant to the use of apparatus 4	Refs.
The effect of loading solution and dissolution media on release of diclofenac from ion exchange resins [2002]	Flow-through cell – apparatus 4 – open type	Diclofenac sodium	**Media:** SGF (simulated gastric fluid) **Flow rate:** 7 mL/min **Temperature:** 37°C **UV:** 275 nm	→ Diclofenac-resin complexes did not release their drug content in simulated gastric fluid but released it in simulated intestinal fluid independent of exposure time under acidic conditions using apparatus IV.	[42]
Comparison of dissolution profiles for sustained release resinates of BCS class I drugs using USP apparatus 2 and 4: a technical note [2008]	Electrolab Ltd., reservoir with vertically positioned flow-through cell and water bath	Ciprofloxacin HCl, Ofloxacin, Verapamil HCl, Diltiazem HCl	**Media:** pH 1.2 buffer **Filter:** glass fiber **Flow rate:** 4 mL/min **Temperature:** 37 ± 0.5°C **Resin:** Indion® 244 (cation exchange resin)	→ 80% release in 8 h for verapamil HCl resinates and more than 90% release in 12 h for ciprofloxacin HCl, ofloxacin and diltiazem HCl resonates using USP apparatus 4 was observed. → Drug-resin complexation and drug release from the resinate are equilibrium processes. In the case of USP apparatus 2 equilibrium was achieved prior to complete release of drug, while in the case of USP apparatus 4, the resinate was continuously exposed to fresh medium (thus maintaining sink conditions) which facilitated the continuous release of actives from the resinate obviating equilibrium attainment, thereby simulating in vivo conditions.	[43]

Table 9.2f Dissolution testing of microspheres using apparatus 4

Title [year]	Instrumental design	Drug	Experimental conditions	Significant findings relevant to the use of apparatus 4	Refs.
Application of USP apparatus 4 and *in situ* fiber optic analysis to microsphere release testing [2005]	USP apparatus 4 -Sotax CE7 smart with CY 7 piston pump, Sotax, Horsham, PA with 12 or 22.6 mm diameter cells – closed mode	Dexamethasone	**Media:** 250 mL PBS (pH 7.4) with 0.1% sodium azide **Filter:** 0.45 mm fiberglass **Flow rate:** 4-35 mL/min **Temperature:** 37°C	→ The release profiles of the drug was compared using USP apparatus 4 and sample-and-separate methods, the initial burst release and lag phase (7 days) were similar in both methods, however the total cumulative release determined was 16% higher when investigated using USP apparatus 4 (30 days) compared to the sample-and-separate method. → Comprehensive characterization of the burst release phase was possible using fiber optic – UV monitoring with USP apparatus 4, since multiple data points are collected over a short period of time.	[44]

Elevated temperature accelerated release testing of PLGA microspheres **[2006]**	USP apparatus 4 –Sotax CE7 smart with CY 7 piston pump, Sotax, Horsham, PA with 12 or 22.6 mm diameter cells – closed mode	Dexamethasone	**Media:** 250 mL 0.1M PBS with 0.1% sodium azide **Filter:** 0.45 μm fiber glass **Flow rate:** 20 mL/min **Temperature:** 37, 45, 53, 60 and 70°C **Glass beads:** 1 mm **UV:** 242 nm	→ Drug release from four different PLGA microsphere formulations (Mw -5K, 25K, 28K and 70K) was evaluated under real-time (37°C) and accelerated release testing conditions of elevated temperature (45, 53, 60 and 70°C) as well as increase in flow rate (4–35 ml/min) using USP apparatus 4. → Formulations composed of low Mw (5 K) PLGA exhibited diffusion-controlled kinetics in real-time, whereas, formulations composed of higher Mw PLGA (25 K, 28 K and 70 K) followed erosion-controlled kinetics at 37°C.	[45]

(Contd.)

Table 9.2f

Title [year]	Instrumental design	Drug	Experimental conditions	Significant findings relevant to the use of apparatus 4	Refs.
Effect of acidic pH on PLGA microsphere degradation and release [2007]	USP apparatus 4 –Sotax CE7 smart with CY 7 piston pump, Sotax, Horsham, PA with 12 or 22.6 mm diameter cells – closed mode	Dexamethasone	**Media:** 250 mL 0.1M PBS (pH 2.4 or 7.4) with 0.1% sodium azide **Filter:** 0.45 μm fiber glass **Flow rate:** 20 mL/min **Temperature:** 37°C **Glass beads:** 1 mm **UV:** 242 nm	→ The mechanism of PLGA degradation changed from inside-out at pH 7.4 to outside-in at pH 2.4, according to the morphological studies reported, and this affected the release rates of these microspheres, which followed erosion-controlled kinetics.	[46]
In situ fiber optic method for long-term in vitro release testing of microspheres [2008]	USP apparatus 4 – Sotax CE7 smart with CY 7 piston pump, Sotax, Horsham, PA with 12 or 22.6 mm diameter cells – closed mode	Cefazolin sodium	**Media:** 250 mL 0.1M PBS (pH 7.4) with 0.1% sodium azide **Filter:** 0.45 mm glass microfiber **Flow rate:** 16 mL/min **Temperature:** 37°C **Glass beads:** 1 mm **UV:** 270 nm & 288 nm	USP apparatus 4 method in conjunction with fiber optic UV probes was more appropriate for monitoring cumulative drug release from PLGA microspheres compared to USP apparatus 2, due to improved reproducibility of the data.	[47]

A novel USP apparatus 4 based release testing method for dispersed systems [2010]	USP apparatus 4 – Sotax CE7 smart with CY 7 piston pump, Sotax, Horsham, PA with 12 or 22.6 mm diameter cells – closed mode	Dexamethasone	Media: Hepes buffer Flow rate: 8 and 16 mL/min Temperature: 37°C Glass beads: 1, 4 mm	→ The dialysis adapter USP apparatus 4 method had greater discriminatory ability compared to dialysis and reverse dialysis sac methods for in vitro release testing of liposomes and other dispersed system formulations.	[48]
USP apparatus 4 method for in vitro release testing of protein loaded Microspheres [2011]	USP apparatus 4 – Sotax CE7 smart with CY 7 piston pump, Sotax, Horsham, PA with 12 or 22.6 mm diameter cells – closed mode	Bovine serum albumin (BSA) protein loaded PLGA	Media: 40 mL 0.05M PBS (pH 7.4) with/out 0.01% SDS Filter: 0.45 μm regenerated cellulose Flow rate: 8 mL/min Temperature: 37 ± 0.1°C Glass beads: 1 mm	→ Highly porous microsphere structure that resulted in high burst release was due to the presence of buffer salts in the internal aqueous phase. → The modified USP apparatus 4 method had superior reproducibility and was easier to use when compared to the sample and separate method for in vitro release testing of protein (BSA) loaded PLGA microspheres.	[49]

(Contd.)

Table 9.2f *(Contd.)*

Title [year]	Instrumental design	Drug	Experimental conditions	Significant findings relevant to the use of apparatus 4	Refs.
Validation of USP apparatus 4 method for microsphere in vitro release testing using Risperdal® Consta® [2011]	USP apparatus 4 –Sotax CE7 smart with CY 7 piston pump, Sotax, Horsham, PA with 12 or 22.6 mm diameter cells – closed mode	Risperidone	**Media:** 250 mL 0.05 M phosphate buffer saline pH 7.4 with 0.1% sodium azide **Filter:** 0.45 μm regenerated cellulose **Flow rate:** 8 or 16 mL/min **Temperature:** 37°C **Glass beads:** 1, 2.4–2.9mm	→ Modified USP apparatus 4 method was validated for robustness and reproducibility and was shown to be an appropriate method for in vitro release testing of risperidone microspheres.	[50]
Accelerated in vitro release testing of implantable PLGA microsphere/PVA hydrogel composite coatings [2012]	USP apparatus 4 –Sotax CE7 smart with CY 7 piston pump, Sotax, Horsham, PA with 12 or 22.6 mm diameter cells – closed mode	Dexamethasone	**Media:** 40 mL 0.1M PBS (pH 7.4) with 0.1% sodium azide **Flow rate:** 8 mL/min **Temperature:** 37°C	→ USP apparatus 4 was feasible for in vitro release testing of drug loaded PLGA microsphere/PVA hydrogel composite coatings and it had good discriminatory ability for the various formulations investigated.	[51]

Table 9.2g Dissolution testing of nanoparticles using apparatus 4

Title [year]	Instrumental design	Drug	Experimental conditions	Significant findings relevant to the use of apparatus 4	Refs.
What is a suitable dissolution method for drug nanoparticles? [2008]	USP 4 apparatus – flow-through cell – closed loop with 25 mm internal diameter cell	Cefuroxime axetil	**Media:** 0.1M HCl containing 0.1% w/v sodium dodecyl sulfate **Flow rate:** 1.6 mL/min **Temperature:** 37 ± 0.5°C **UV:** 278 nm	→ Flow-through cells have been shown to be suitable for dissolution analysis and performance evaluation of drug nanoparticles.	[52]
Using USP I and USP IV for discriminating dissolution rates of nano- and microparticle-loaded pharmaceutical strip-films [2012]	Flow-through cell dissolution apparatus (USP 4, Sotax, Switzerland) – closed loop configuration with 22.6 mm internal diameter cell	Griseofulvin	**Media:** 0.54% sodium dodecyl sulfate **Flow rate:** 4, 8 and 16 mL/min **Temperature:** 37 ± 0.5°C **Glass bead:** 6 mm **UV:** 251 nm	→ The dissolution behavior of strip-films containing griseofulvin microparticles was compared to the dissolution behavior of films containing griseofulvin nanoparticles. → The flow-through cell method had better discriminatory capabilities than the USP apparatus 1, which would be beneficial in the optimization of poorly soluble drugs incorporated into strip-film dosage forms.	[53]

Table 9.2h Dissolution testing of lozenges using apparatus 4

Title [year]	Instrumental design	Drug	Experimental conditions	Significant findings relevant to the use of apparatus 4	Refs.
The application of modified flow-through cell apparatus for the assessment of chlorhexidine hydrochloride release from lozenges containing sorbitol [**2009**]	In house flow-through cell – highest diameter of the biconical device – 50 mm	Chlorhexidine dihydrochloride	**Media:** water **Flow rate:** 1.9 mL/min **Temperature:** 37°C **Glass balls:** 5 mm **UV:** 254.5 nm	→ In the flow-through cell, the release of drug from the lozenges followed first-order kinetics.	[54]

9.5 Conclusions

A review of the literature (see Table 9.2) reveals a wide variety of sample dosage forms tested, with attention to flow rate, positioning, and design. Some general trends are the noted discriminating capability surpassing that of other techniques[29] and the ability to use large volumes of fresh media in the open-loop system configuration.[20,32]

For poorly soluble drugs[33,35,38,39,42,53] it is worth noting that in addition to the ability to use large volumes, significant levels of surfactant (i.e., 1% SLS) were also successfully used. Cell setup and preparation also plays a role, as in the case of poorly soluble nanoparticles delivered via strip film.[33]

With poorly soluble compounds, formulators may choose to take advantage of a wider range of micro- or nanoparticle formulations, suspensions, liposomes, and other formulation types to facilitate the delivery and bioavailability of the poorly soluble drug. Based on a review of the current literature, the USP apparatus 4 type has demonstrated its suitability for method development for a wide range of dosage forms including poorly soluble drugs.

References

1. F. Langenbucher, D. Benz, W. Kurth, H. Moller, and M. Otz, Standardized flow-cell method as an alternative to existing pharmacopoeial dissolution testing, *Pharm. Ind.*, 1989; 51(11): 1276–1281.

2. M. Pernarowski, W. Woo, and R. Searl, Continuous flow apparatus for the determination of the dissolution characteristics of tablets and capsules, *J. Pharm. Sci.*, 1968; 57(8): 1419–1421.

3. Sotax Corporation.

4. F. Langenbucher and H. Rettig, Dissolution rate testing with the column method: methodology and results, *Drug Dev. Ind. Pharm.*, 1977; 3(3): 241–263.

5. D. J. Campbell and J. G. Theivagt, *Stand.*, 1958; 26: 73.

6. Stimuli Article PF 31(6) In-Process Revision<1092> The Dissolution Procedure: Development and Validation.

7. N. Fotaki, O. Oluwasanmi, Y. Rampal, F. Baxevanis, and G. Grove, Effect of glass bead size on drug dissolution in the flow through cell apparatus, In: *AAPS Annual Meeting*, 2014, San Diego.

8. USP <711> Dissolution, The United States Pharmacopeial Convention, 2011.

9. M. Kakhi, Mathematical modeling of the fluid dynamics in the flow-through cell, *Int. J. Pharm.*, 2009; 376(1–2): 22–40.

10. G. Shiko, L. F. Gladden, A. J. Sederman, P. C. Connolly, and J. M. Butler, MRI studies of the hydrodynamics in a USP 4 dissolution testing cell, *J. Pharm. Sci.*, 2011; 100: 976–991.

11. D. M. D'Arcy, B. Liu, T. Persoons, and O. I. Corrigan, Hydrodynamic complexity induced by the pulsing flow field in USP dissolution apparatus 4, *Dissolut. Technol.*, 2011; 6–13.

12. B. R. Rohrs, Dissolution method development for poorly water soluble compounds, *Dissolut. Technol.*, 2001; 1–5.

13. S.A. Qureshi, G. Caille, R. Brien, G. Piccirilli, V. Yu, and I. J. Mcgilveray, Application of flow-through dissolution method for the evaluation of oral formulations of Nifedipine, *Drug Dev. Ind. Pharm.*, 1994; 20(11): 1869–1882.

14. S. E. Neisingh, A. P. Sam, and H. d. Nijs, A dissolution method for hard and soft gelatin capsules containing testosterone undecanoate in oleic acid, *Drug Dev. Ind. Pharm.*, 1986; 12(5): 651–663.

15. K. Gjellan, A. B. Magnusson, R. Ahlgren, K. Callmerd, D. F. Christensen, U. Espmarker, L. Jacobsen, K. Jarring, G. Lundin, G. Nilsson, and J. O. Waltersson, A collaborative study of the in vitro dissolution of acetylsalicylic acid gastro-resistant capsules comparing the flow-through cell method with the USP paddle method, *Int. J. Pharm.*, 1997; 151(1): 81–90.

16. J. Hu, A. Kyad, V. Ku, P. Zhou, and N. Cauchon, A comparison of dissolution testing on lipid soft gelatin capsules using USP apparatus 2 and apparatus 4, *Dissolut. Technol.*, 2005; 6–9.

17. B. Wennergren, L. J. Lindberg, M. Nicklasson, G. Nilsson, G. Nyberg, R. Ahlgren, C. Persson, and B. Palm, A collaborative in vitro dissolution study: comparing the flow-through method with the USP paddle method using USP prednisone calibrator tablets, *Int. J. Pharm.*, 1989; 53(35–41).

18. J. G. Philips, Y. Chen, and I. N. Wakeling, A flow through dissolution approach to *in vivo-in vitro* correlation of Adinazolam release from sustained release formulation, *Drug Dev. Ind. Pharm.*, 1989; 15(14–16): 2177–2195.

19. G. H. Zhang, W. A. Vadino, T. T. Yan, W. P. Cho, and I. A. Chaudry, Evaluation of the flow-through cell dissolution apparatus: effects of flow rate, glass beads and tablet position on drug release from differnet type of tablets, *Drug Dev. Ind. Pharm.*, 1994; 20(13): 2063–2078.

20. S. A. Qureshi, G. Caille, R. Brien, G. Piccirilli, V. Yu, and I. J. McGilverayl, Application of flow-through dissolution method for the evaluation of oral formulation of Nifedipine, *Drug Dev. Ind. Pharm.*, 1994; 20(1): 1869–1882.

21. W. C. G. Butler and S. R. Bateman, A flow-through dissolution method for a two component drug formulation where the actives have markedly differing solubility properties, *Int. J. Pharm.*, 1998; 173(1): 211–219.

22. M. S. Bloomfield and W. C. Butler, Robustness testing, using experimental design, of a flow-through dissolution method for a product where the actives have markedly differing solubility properties, *Int. J. Pharm.*, 2000; 206(1–2): 55–61.

23. N. Ozdemir, S. Ordu, and Y. Ozkan, Studies of floating dosage forms of furosemide: in vitro and in vivo evaluations of bilayer tablet formulations, *Drug Dev. Ind. Pharm.*, 2000; 26(8): 857–866.

24. L. H. Emar, B. S. El-Menshawi, and M. Y. Estefan, In vitro-in vivo correlation and comparative bioavailablity of vincamine in prolonged-release preparations, *Drug Dev. Ind. Pharm.*, 2000; 26(3): 243–251.

25. S. R. Cammarn and A. Sakr, Predicting dissolution via hydrodynamics: salicylic acid tablets in flow through cell dissolution, *Int. J. Pharm.*, 2000; 201(2): 199–209.

26. N. Bielen, Performance of USP calibrator tablets in flow-through cell apparatus, *Int. J. Pharm.*, 2002; 233(1–2): 123–129.

27. M. H. y. d. l. Pena, Y. V. Alvarado, A. M. D. Ramirez, and A. R. C. Arroyo, Comparison of dissolution profiles for albendazole tablets using USP apparatus 2 and 4, *Drug Dev. Ind. Pharm.*, 2003; 29(7): 777–784.

28. J. Jinno, N. Kamada, M. Miyake, K. Yamada, T. Mukai, M. Odomi, H. Toguchi, G. G. Liversidge, K. Higaki, and T. Kimura, In vitro-in vivo correlation for wet-milled tablet of poorly water-soluble cilostazol, *J. Control. Release*, 2008; 130(1): 29–37.

29. E. Chevalier, M. Viana, A. Artaud, L. Chomette, S. Haddouchi, G. Devidts and D. Chulia, Comparison of three dissolution apparatuses for testing calcium phosphate pellets used as ibuprofen delivery systems, *AAPS PharmSciTech*, 2009; 10(2): 597–605.

30. S. Emilia and W. Katarzyna, Comparison of flow-through cell and paddle methods for testing vaginal tablets containing a poorly water-soluble drug, *Trop. J. Pharm. Res.*, 2013; 12(1): 39–44.

31. J. R. Medina, D. K. Salazar, M. Hurtado, A. R. Corte's, and A. M. D. Ramirez, Comparative in vitro dissolution study of carbamazepine immediate-release products using the USP paddles method and the flow-through cell system, *Saudi Pharm. J.*, 2014; 22 141–147.

32. M. Nicklasson, A. Orbe, J. Lindberg, B. Borga, A. B. Magnusson, G. Nilsson, R. Ahlgren, and L. Jacobsen, A collaborative study of the in vitro dissolution of phenacetin crystals comparing the flow through method with the USP Paddle method, *Int. J. Pharm.*, 1991; 69: 255–264.

33. S. N. Bhattachar, J. A. Wesley, A. Fioritto, P. J. Martin, and S. R. Babu, Dissolution testing of a poorly soluble compound using the flow-through cell dissolution apparatus, *Int. J. Pharm.*, 2002; 236(1–2): 135–143.

34. C. Y. Perng, A. S. Kearny, N. R. Palepu, B. R. Smith, and L. M. Azzarano, Assessment of oral bioavailability enhancing approaches for SB-247083 using flow-through cell dissolution testing as one of the screens, *Int. J. Pharm.*, 2003; 250(1): 147–156.

35. V. H. Sunesen, H. G. Kristensen, B. L. Pedersen, and A. Mullertz, In vivo in vitro correlations for a poorly soluble drug, danazol, using the flow-through dissolution method with biorelevant dissolution media, *European J. Pharm. Sci.*, 2005; 24(4): 305–313.

36. S. Kallquist, An in vitro dissolution study of Ibuprofen using a flow-through cell; Department of Chemical Engineering, Lund Institute of Technology, 2007.

37. A. Okumu, M. DiMaso, and R. Lobenberg, Dynamic dissolution testing to establish in vitro/in vivo correlations for montelukast sodium, *Pharm. Res.*, 2008; 25(12): 2778–2785.

38. S. Vangani, X. Li, P. Zhou, M. A. D. Barrio, R. Chiu, N. Cauchon, P. Gao, C. Medina, and B. Jasti, Dissolution of poorly water-soluble drugs in biphasic media using USP 4 and fiber optic system, *Clin. Res. Regul. Aff.*, 2009; 26(1–2): 8–19.

39. C. Wahling, C. Schroter, and A. Hanefeld, Flow-through cell method and IVIVR for poorly soluble drugs, *Dissolut. Technol.*, 2011.

40. R. Dunn, H. Reimers, L. Ward, and J. Chapman, Suppository dissolution utilizing USP apparatus 4, *Dissolut. Technol.*, 1996; 18–19.

41. J. R. Medina, A. R. Padilla, M. Hurtado, A. R. Cortes, and A. M. D. Ramirez, In vitro release of ketoprofen suppositories using the USP basket and the flow-through cell dissolution methods, *Pak. J. Pharm. Sci.*, 2014; 27(3): 453–458.

42. F. Atyabi, M. Koochak, and R. Dinarvand, The effect of loading solution and dissolution media on release of Diclofenac from ion exchange resins, *DARU J. Pharm. Sci.*, 2002; 10(1): 17–22.

43. N. B. Prabhu, A. S. Marathe, S. K. Jain, P. P. Singh, K. Sawant, L. Rao and P. D. Amin, Comparison of dissolution profiles for sustained release resinates of BCS class I drugs using USP apparatus 2 and 4: a technical note, *AAPS PharmSciTech*, 2008; 9(3): 769–773.

44. B. S. Zolnik and D. J. Burgess, Application of USP apparatus 4 and in situ fiber optic analysis to microsphere release testing, *Dissolut. Technol.*, 2005; 11–14.

45. B. S. Zolnik, P. E. Leary, and D. J. Burgess, Elevated temperature accelerated release testing of PLGA microspheres, *J. Control. Release*, 2006; 112(3): 293–300.

46. B. S. Zolnik and D. J. Burgess, Effect of acidic pH on PLGA microsphere degradation and release, *J. Control. Release*, 2007; 122(3): 338–344.

47. J. M. Voisine, B. S. Zolnik, and D. J. Burgess, In situ fiber optic method for long-term in vitro release testing of microspheres, *Int. J. Pharm.*, 2008; 356(1–2): 206–211.

48. U. Bhardwaj and D. J. Burgess, A novel USP apparatus 4 based release testing method for dispersed systems, *Int. J. Pharm.*, 2010; 388(1–2): 287–294.

49. A. Rawat and D. J. Burgess, USP apparatus 4 method for in vitro release testing of protein loaded microspheres, *Int. J. Pharm.*, 2011; 409(1–2): 178–184.

50. A. Rawat, E. Stippler, V. P. Shah, and D. J. Burgess, Validation of USP apparatus 4 method for microsphere in vitro release testing using Risperdal Consta, *Int. J. Pharm.*, 2011; 420(2): 198–205.

51. J. Shen and D. J. Burgess, Accelerated in vitro release testing of implantable PLGA microsphere/PVA hydrogel composite coatings, *Int. J. Pharm.*, 2012; 422(1–2): 341–348.

52. D. Heng, D. J. Cutler, H. K. Chan, J. Yun, and J. A. Raper, What is a suitable dissolution method for drug nanoparticles?, *Pharm. Res.*, 2008; 25(7): 1696–1701.

53. L. Sievens-Figueroa, N. Pandya, A. Bhakay, G. Keyvan, B. Michniak-Kohn, E. Bilgili, and R. N. Dave, Using USP I and USP IV for discriminating

dissolution rates of nano- and microparticle-loaded pharmaceutical strip-films, *AAPS PharmSciTech*, 2012; 13(4): 1473–1482.

54. W. Musial and J. B. Mielck, The application of modified flow-through cell apparatus for the assessment of chlorhexidine dihydrochloride release from lozenges containing sorbitol, *AAPS PharmSciTech*, 2009; 10(3): 1048–1057.

Chapter 10

Dissolution of Nanoparticle Drug Formulations

John Bullock

John A. Bullock Consulting, LLC, West Chester, PA, USA
jabullock91@gmail.com

10.1 Introduction

The technical development and subsequent advancement of clinical studies and ultimately commercial applications of various types of nanoparticle drug formulations have accelerated over the last couple of decades. These nanoparticle formulations have been pursued as enabling technologies to achieve various modes of enhanced drug delivery of therapeutic and diagnostic agents, including the improved oral delivery of poorly water-soluble drugs in terms of increased bioavailability, onset of action, and reduction of food effects,[1–3] improved targeted delivery of therapeutic agents,[2–5] the reduction of toxicity, and enhancement in the efficacy of certain agents[2,3] and general improvements in dosing convenience and compliance.[1,2] In the field of drug delivery, the term *nanoparticles* encompasses a broad assortment of different types of nano-sized

Poorly Soluble Drugs: Dissolution and Drug Release
Edited by Gregory K. Webster, J. Derek Jackson, and Robert G. Bell
Copyright © 2017 Pan Stanford Publishing Pte. Ltd.
ISBN 978-981-4745-45-1 (Hardcover), 978-981-4745-46-8 (eBook)
www.panstanford.com

particle structural motifs. Broadly speaking, the most common nanoparticle structures used in oral drug delivery can be grouped into two different categories: (i) pure drug nanoparticles consisting of essentially 100% drug (with or without the addition of surface stabilizers) in either a crystalline (nanocrystals)[1,3] or noncrystalline state[6,7] and (ii) an assortment of different nano-sized structures in which the drug is encapsulated or dispersed in a solid, semisolid, or liquid state within a formulation matrix. In this latter category are included polymeric nanoparticles,[8] nanoemulsions,[9] liposomes,[10] and solid lipid nanoparticles,[11] among others. Additional types of nanostructured drug particles include dendrimers, quantum dots, and various types of metal-based colloidal nanoparticles to which a drug may be anchored via covalent or noncovalent mechanisms[12] and which are outside the scope of this review.

In keeping with the primary focus of this reference volume on orally delivered poorly water-soluble drugs in a solid dosage format, the subject matter of this chapter will predominately focus on dissolution testing of solid nanoparticle dosage forms which are manufactured with pure drug nanoparticles (also referred to as drug nanocrystals) with lesser discussion of matrix type nanoparticles. Regardless of which type of drug nanoparticles are discussed in this chapter, in general the fundamental aspects of the dissolution techniques to be covered could in many cases be equally applied to both pure drug nanoparticles and matrix type nanoparticle formulations.

As of 2013 there was as yet no universally agreed-upon definition for the size range that constitutes a nanoparticle in the context of drug delivery. A draft* FDA guidance on the application of nanotechnology in FDA-regulated products[13] states that FDA will consider a product to contain nanomaterials if "an engineered material or end product has at least one dimension in the nanoscale range (approximately 1 nm to 100 nm)." However, the draft guidance goes on to consider other properties that may differentiate a material above this size range from conventionally scaled materials

---

Note: This draft FDA guidance was finalized in June 2014 but the basic tenants regarding particle size of a drug nanoparticle remain consistent between the draft and final guidance.

and further states that "In the absence of a bright line as to where an upper limit should be set, the agency considers that an upper bound of one micrometer (i.e., 1,000 nm) would serve as a reasonable parameter for screening materials with dimensions beyond the nanoscale range for further examination." In two separate nanoparticle drug delivery reference texts published in the last 10 years a size below 300 nm has been suggested as a reasonable and practically achievable goal to target for drug nanoparticles in one case,[14] while in another text drug nanocrystals were defined as having a mean diameter below 1000 nm.[15] In the drug delivery patent literature an upper average size of about 400 nm has been used in one case,[16] while in another a size limit of 200 nm was claimed.[17] For the purposes of the current discussion, no upper (or lower) size limit will be specified to define drug nanoparticles although the specific types of nanoparticle products that will be addressed in this chapter typically have volume average (median or mean) particle sizes in the range from about 100 nm up to about 1000 nm. However, it will be stressed in a subsequent section provided below that appropriate consideration of the actual size range of the drug nanoparticles under evaluation is important for the proper selection and optimization of the dissolution testing procedure.

Pure drug nanoparticles can be produced by either top-down or bottom-up techniques. Top-down technologies include size reduction of conventional size drug particles via media milling[1, 18] or high shear homogenization.[19] Bottom-up techniques involve controlled crystallization/precipitation to produce drug nanoparticles from the solution state.[6, 7, 20] A detailed discussion of the technologies used to produce drug nanoparticles is beyond the scope of this chapter. The interested reader is referred to several reviews that have recently been published that compare the merits of the two approaches.[19, 21, 22] In all cases the ultimate goal is to decrease the average drug particle size into the nano-scale range and thereby increase the dissolution rate and potentially enhance other physical properties of the drug. Several theoretical models exist to describe the enhanced dissolution rate attained with decreased drug particle size. The following section provides a brief review of the theoretical underpinnings controlling dissolution of drug particles in general as well as certain considerations more important for nano-size particles. The purpose of this review is to provide the reader with sufficient theoretical understanding of the potentially relevant nanoparticle solubility and dissolution phenomena to aid in developing and troubleshooting suitable dissolution methods as well as to accurately interpret the resulting dissolution data.

10.2 Theoretical Considerations for the Dissolution of Nanoparticles

The drug dissolution process from a solid oral dosage form (e.g., tablet or capsule) can be divided into a number of steps, some of which will vary in mechanism from one type of dosage form to another. As described by Siepmann and Siepmann[23] in this context the terms "drug dissolution" and "drug release" from a dosage form are not synonymous. As an example, these authors discussed the situation in which a drug present as solid particles formulated within a polymeric matrix tablet undergoes drug release and dissolution in a USP apparatus 2 system. Several complex phenomena are involved in the overall "drug release" out of this matrix tablet, only one of which is actual dissolution of the primary drug particles. Among the additional phenomena are the absorption of water into the tablet, partial dissolution of drug molecules within the tablet matrix, and transport of dissolved drug and drug particles out of the tablet. Additionally, the polymer matrix may itself swell and eventually dissolve over time which then releases drug particles into the medium. Other types of solid dosage forms may entail different steps that contribute to the overall drug release process such as a disintegration or erosion process to free up larger granules containing multiple drug particles which eventually devolve into primary drug particles. Thus the drug release rate from the solid oral dosage form is controlled by multiple properties encompassing both the components of formulation and the solid dosage form manufacturing process.

Most theoretical equations used to elucidate or model drug dissolution rates only consider the "drug dissolution" step, which starts from the primary drug particles, in the overall drug release process. Spiepmann and Speipmann described this drug dissolution process from drug particles which occurs in a well-stirred bulk fluid (typical of a compendial dissolution apparatus) to encompass five sequential steps: (1) wetting of the drug particle surface, (2) breakdown of solid state bonds in the drug particle, (3) solvation of the individual drug molecules at the solid-liquid interface, (4) diffusion of the solvated drug molecules through a liquid boundary layer surrounding the drug particle into the bulk liquid, and (5) convection of dissolved drug within the well-stirred bulk fluid. Very often the diffusion through the liquid boundary layer (also referred to as stagnant diffusion layer) is much slower than the other processes and determines the overall drug dissolution rate.

The dissolution process occurring from primary drug particles released from a solid dosage form is dependent on the drug's physical properties (including particle size) and the physicochemical properties of the dissolution medium in the case of so called "reactive media".[24] In the diffusion layer model of drug dissolution the release of drug molecules at the drug solid–liquid interface (steps 1–3) is assumed to take place rapidly and so the overall dissolution rate for the released drug particles is controlled by the diffusion of molecules through the stagnant diffusion layer. The following discussion highlights the most important physical properties controlling the dissolution rate of drug nanoparticles.

10.2.1 Nanoparticle Solubility as a Function of Size

The saturation solubility of a drug crystal/particle increases as a function of decreasing particle size as described by the Ostwald–Freundlich equation[25]:

$$c_{s,r} = c_{s,\infty} \exp \left(\frac{2\gamma M}{r \rho R T} \right) \qquad (10.1)$$

where $c_{s,r}$ is the saturation solubility of particle of diameter r, $c_{s,\infty}$ is the saturation solubility of an infinitely large particle, γ is the interfacial surface tension between the particle and the medium, M is the compound molecular weight, ρ is the particle density, R is the universal gas constant, and T the temperature. The increase in solubility can be related to the increased curvature of a particle with decreasing radius and a corresponding increased dissolution pressure. As the particle size is decreased below 50 nm and especially below 10 nm a significant increase in saturation solubility is predicted.[25] However, only a modest solubility increase with decreasing nanoparticle size would be predicted in the range from about 1000 nm down to 100 nm which encompasses the average particle size of typical drug nanoparticle preparations. This was confirmed in a study that experimentally measured the saturation solubility of a series of nano-milled drugs in the size range from about 150–400 nm using a light-scattering technique.[26] The enhancement in solubility compared to larger micron sized drug particles was only in the range of 0–15% in line with what would be predicted based on the Ostwald–Freundlich equation (see also [61] in Section 10.8.6). Other examples have appeared in the literature claiming much more significant solubility enhancements for drug nanoparticles in this size range. However, in such cases the greater solubility enhancements may have been due to changes in drug solid state properties as discussed in Section 10.2.5.

10.2.2 Nanoparticle Dissolution Kinetics as a Function of Surface Area

Perhaps the most widely cited equation describing dissolution kinetics of a pharmaceutical solid is that of Noyes–Whitney with modifications of Nernst and Brunner[23]:

$$\frac{dm}{dt} = \frac{A\,D}{h}\,(C_s - C) \qquad (10.2)$$

where dm/dt is the rate of dissolution, A is the total exposed surface area of the particles, D is the diffusion coefficient of the dissolved molecules, h is the thickness of the stagnant diffusion boundary layer surrounding the dissolving particles, C_s is the saturation solubility in the medium, and C is the analyte solubility in the bulk medium at time t. The dissolution rate in this model is seen to be proportional to surface area, which increases with decreasing particle size, and the diffusion coefficient and inversely proportional to the stagnant diffusion boundary thickness.

10.2.3 Stagnant Diffusion Layer Thickness as a Function of Nanoparticle Size

In the Noyes–Whitney–Nernst–Brunner model the diffusion layer thickness is treated as constant throughout the dissolution process and independent of the starting particle size. However, according to the Prandtl boundary layer equation for flow passing a flat surface,[27] the diffusion layer thickness is dependent on the particle size as given in Eq. 10.3.

$$h_{\mathrm{H}} = k \left(\frac{\sqrt{L}}{\sqrt{V}} \right) \qquad (10.3)$$

where L is the length of the surface in the direction of flow, V is the relative velocity of the flowing liquid against a flat surface, and k is a constant. It has been reported that for solid particles dispersed in a liquid medium under agitation that both L and V decrease with decreasing particle size. Although the effects of these terms on the diffusion layer thickness counteract each other, it was shown that the net effect was a thinner diffusion layer and increasing dissolution rate with decreasing particle size.[27]

Recently, a quasi steady-state model (QSM) was developed to accurately predict the ratio of the diffusion layer thickness to particle radius (Sherwood number) as a constant plus a correction that depends on the degree of confinement.[28] The QSM provides a mathematical expression for the nondimensional diffusion thickness (the Sherwood number) as a function of the container radius or volume ratio.

Based on the previous discussion (and disregarding potential solid state changes), improvement in the dissolution rate of nanoparticle dosage forms is seen to be driven by three primary attributes of the nanoparticles; the increased dissolution kinetics resulting from the increased surface area of the small particles, the increased saturation solubility of the nanoparticles resulting from the increased particle curvature with decreasing particle size, and the decreasing thickness of the stagnant diffusion layer accompanying a decreasing particle diameter. To put this in perspective, the relative importance of these three properties for typically sized drug nanoparticles is approximately qualitatively

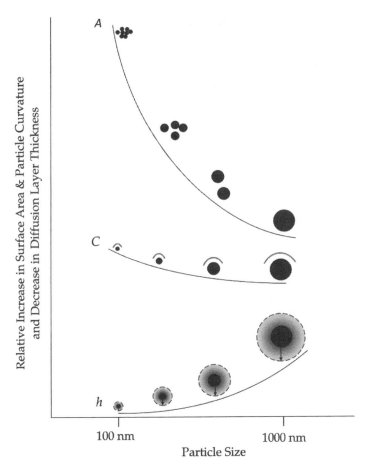

Figure 10.1 Fictive qualitative depiction of the relative change in diffusion layer thickness (h), particle curvature (C), and total exposed surface area (A) that occurs with decreasing particle size.

depicted in Fig. 10.1 which displays a fictive representation of the relative changes in these properties with particle size in the range of a typical nanoparticle dosage form (100–1000 nm). The dissolution kinetics increase with the increasing surface area, increasing saturation solubility and decreasing diffusion layer thickness that simultaneously accompanies a decreasing particle size, with the effect from increasing surface area predominating in this size range.

10.2.4 Advanced Models of Nanoparticle Dissolution Kinetics

In addition to not accounting for an initial particle size-dependent diffusion layer thickness, the Noyes–Whitney–Nernst–Brunner model does not account for the time-dependent change in either the diffusion layer thickness or particle radius as the particle size decreases during the dissolution process. Additionally, real life samples are polydisperse so that a model is needed that accounts for a distribution of different sizes of the drug particles.

Several more advanced models have been developed and the model of Johnson and coworkers[29] is provided below:

$$\frac{dX_{s_i}}{dt} = -\frac{3DX_{0_i}^{\frac{1}{3}}X_{s_i}^{\frac{2}{3}}}{\rho h_i r_{0_i}}\left(C_{s_i} - \frac{X_{d_T}}{V}\right) \tag{10.4}$$

where X_{s_i} is the mass of solid drug in particle size fraction i at any time t, X_{0_i} is the initial mass of solid drug in particle size fraction i, D is the drug diffusion coefficient, ρ is the true density of the drug, h_i is the particle size-dependent diffusion layer thickness in particle size fraction i at any time, r_{0_i} is the initial radius of particles in particle size fraction i, r_i is the radius of particles in particle size fraction i at any time, C_{s_i} is the particle size-dependent solubility of drug in particle size fraction i at any time, X_{d_T} is the total mass of drug in solution from all particle size fractions at any time, and V is the dissolution volume. This model provides a computational tool for exploring theoretical implications and explaining the behavior of nanoparticles, in particular with respect to simulating Ostwald ripening of nanoparticles as described in Section 10.2.6.

10.2.5 Solid State (Crystalline, Amorphous, Disordered) Influences on Dissolution Properties

The production of drug nanoparticles frequently involves the use of high energy manufacturing processes, particularly with respect to top-down media milling processes. As a result there exists the potential to produce drug nanoparticles in higher energy states as a result of either polymorphic changes, production of fully amorphous drug particles, various forms of surface amorphization of drug particles or the general development of less ordered, higher energy drug solid states.[30,31] Although sometimes produced intentionally, these different types of higher energy state nanoparticles are often unintended. Importantly, these higher energy state nanoparticles can possess dramatically different apparent solubility and dissolution kinetics compared with crystalline nanoparticles.

The topic of mechanically (e.g., milling) induced disordering of drug crystals is a complex subject that continues to evolve. Mosharraf et al. developed a conceptual model for describing the effects of a disordered surface layer of varying thickness and continuity on the apparent solubility of a drug substance.[31] Their model proposed that the solubility of the disordered surface of the particles appeared to be the rate-limiting factor during the initial dissolution phase, while the solubility of the crystalline core was the rate-limiting factor during the final slower phase. They developed the schematic reproduced in Fig. 10.2 to depict

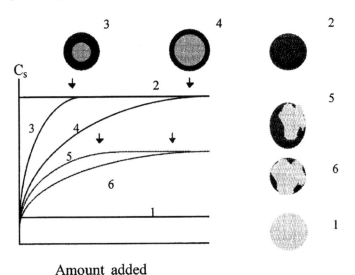

Figure 10.2 Schematic illustration of some possible plateau levels obtained for apparent solubility of drug as a function of the amount added and degree of disorder. 1: totally ordered; 2: totally disordered; 3 and 4: crystalline core surrounded by a thick or thin disordered layer, respectively; 5 and 6: the surface is not continuously disordered. Cs: apparent solubility. Reproduced from Ref. 31 with permission from Elsevier.

different types of solubility properties that can occur when the compound is either totally ordered (crystalline), totally disordered (amorphous), or when the particles are composed of different regions of disordered and ordered material. Two solubility plateaus exist for the totally crystalline (1) and totally amorphous (2) substances that are independent of the amount of material added to the solvent. However, when the crystalline and amorphous phases coexist the apparent equilibrium solubility is dependent on the amount of material added to the solvent. In cases 3 and 4 the crystalline core is surrounded by a thick and thin layer of amorphous phase, respectively. When the degree of disorder is much higher than is required to form a saturated metastable solution, the apparent solubility would be determined by the disordered phase and would approach that of the totally amorphous phase. In cases 5 and 6 the surface has exposed areas of both crystalline and amorphous phase. The resulting solution was claimed to be in apparent equilibrium with both the crystalline and amorphous phases and would plateau at an intermediate level between that of the totally crystalline and amorphous phase. This conceptual model by Mosharraf et al. was subsequently refined and expanded to consider the solubility behavior of partially crystalline drugs in both a one-state model (disordering by a continuous and homogenous increase in disorder throughout a particle progressing from a totally crystalline state to a totally amorphous state) and a two-state model (heterogeneous physical mixture of different amounts of amorphous or crystalline regions residing within a solid particle).[32] It was confirmed that the apparent solubility level in these partially crystalline systems is a complex interfacial phenomenon which depends on the amount, reactivity, and solid-state structure of the exposed solid surfaces in equilibrium with solution and that different apparent solubility plateaus for a substance can exist. It is stressed that in the presence of a disordered phase that the apparent solubility is metastable and would gradually decrease to the solubility of the crystalline phase. On the basis of these considerations it is clear that it is important to characterize the solid state properties of the drug nanoparticle formulation and understand the potential impact on solubility and dissolution properties in order to be able to interpret unusual or unexpected solubility properties or dissolution profiles.

10.2.6 Supersaturation Potential and Impact on Dissolution Profile

Supersaturation occurs when the instantaneous kinetic solubility of a material exceeds its thermodynamic equilibrium solubility in the dissolution medium. For drug nanoparticles, supersaturation during dissolution testing can occur due to at least two different phenomena: (i) the presence of higher energy solid state form nanoparticles discussed in Section 10.2.5 or (ii) when the potential for Ostwald ripening exists. Ostwald ripening describes the growth of larger particles in a suspension in the presence of faster dissolving smaller particles.[33] In both cases if the dissolution test is conducted in media that is not at true sink conditions (defined as a drug concentration 3–10× below the saturation solubility) there exists the possibility for the dissolution profile to show an initial fast rate that may plateau before eventually drifting downward until the dissolved drug concentration equals the saturation solubility.

Johnson[29] used the model presented in Section 10.2.4 to simulate the potential impact of Ostwald ripening during a dissolution experiment conducted on a formulation containing nanoparticles. The simulation modeled a situation in which two different size fractions of drug particles were present in a sample, one with an initial size of 200 nm and the other with an initial size of 2 μm. The total mass of drug and the dissolution volume in the simulation were such that supersaturation relative to the larger particles was encountered during the dissolution process. Initially, the smaller particles dissolve much faster due to their higher surface area. As these smaller particles dissolve their solubility increases exponentially according to the Ostwald–Freundlich equation. Eventually a situation is reached in which the concentration of drug in solution is supersaturated relative to the solubility of the larger particles which then begin to grow. After the smaller particles are completely dissolved, the dissolution profile reaches a maximum value after which precipitation of the larger particles begins and the concentration of dissolved drug begins to fall. This situation is more likely to be encountered when the dissolution experiment is not performed under true sink conditions.

In situations in which partially or totally amorphous nanoparticles are obtained, a dissolution profile similar to that obtained above accompanying an Ostwald ripening phenomenon may be encountered. This is due to the initial faster dissolution rate and higher solubility of the amorphous component of the nanoparticles which achieves a solubility level above that of the saturation solubility of the crystalline component of the nanoparticle.[34]

10.2.7 Summary

The relevant properties of a nanoparticle drug and its formulation that control overall dissolution rate encompass properties of the primary drug nanoparticles as well as the formulation components and the manufacturing process used to prepare the finished dosage form. It is stressed here that proper characterization of all of the properties controlling the overall drug release and dissolution process from the dosage form are important, including the particle size distribution and solid state properties of the nanoparticles, in order to develop an appropriate dissolution method and accurately interpret the resulting dissolution data.

10.3 General Considerations for Dissolution Methods for Nanoparticle Formulations

Two of the more important and distinguishing characteristics of dissolution methods for nanoparticle formulations include (i) the challenges encountered in appropriately processing samples containing small nano-size drug particles and (ii) the potential for much faster dissolution rates compared to formulations manufactured with conventional sized drug particles. The small size of the drug nanoparticles challenges the ability of the selected analytical technique used to quantify dissolution samples to be able to distinguish dissolved drug from undissolved drug nanoparticles. The fast dissolution rates can create practical limitations and constraints in the ability to process samples in a timely manner in order to accurately gauge dissolution rates. These issues tend to be most important when the dosage form is a liquid dispersion formulation where the drug is already well dispersed as primary drug nanoparticles. In such cases dissolution rates on the order of 1–2 minutes may be encountered.[35] Often though, the liquid nanoparticle dispersion is processed into solid dosage format (tablet or capsule). In such cases the dosage form must first undergo disintegration or other drug release phenomena of the dosage form before presenting the drug nanoparticles to the bulk medium for subsequent dissolution. In this case the overall measured dissolution rates are prolonged compared to the aqueous nanoparticle dispersion. Additionally, although the goal is to produce a solid dosage form that readily releases drug particles as suitably dispersed primary nanoparticles when exposed to aqueous media, often this is not achieved and the drug nanoparticles are released from the dosage form in an aggregated state in which case the effective drug particle size is much larger.[36–38] Based on these and other considerations it is difficult to define a universal approach for the dissolution testing of nanoparticle formulations. What is required is a rational approach to developing a suitable method that considers the unique properties of each drug and its formulation such as size of the primary nanoparticles, state of nanoparticle aggregation, drug solubility, mechanism of drug release from the dosage (erosion versus disintegration) and overall rate of the dissolution process.

10.4 Media Considerations for Dissolution Testing of Nanoparticles

In principle, the important considerations with respect to the selection of a suitable medium for the dissolution testing of nanoparticle formulations are the same as those that would be important for non-nanoparticulate formulations of poorly water-soluble drugs. In general, a medium that achieves sink conditions is preferred from a method robustness perspective, especially for routine quality control testing. However, it has been reported by several groups that for nanoparticle suspensions the use of non-sink dissolution conditions can provide more discrimination of rates and better prediction of in vivo performance.[39,53,54] In many cases the dissolution procedure used for initial formulation screening and optimization will be different than the method used for routine quality control and stability testing. In general, dissolution test conditions should be chosen based on the method's discriminatory capability, method ruggedness, stability of the analyte in the dissolution medium and relevance to in vivo performance as provided in USP guidance.[40]

10.4.1 Conventional Compendial Media

Conventional compendia media include simulated gastric or intestinal fluid USP (with or without enzymes), dilute HCl, and aqueous buffers in the physiological pH range of 1–7.5. For poorly water-soluble drugs the addition of low levels of surfactants such as sodium lauryl sulfate or various types of polysorbates are often required as a wetting agent when they present at levels below the critical micelle concentration (CMC) or to sufficiently enhance the solubility of the drug when present at levels above the CMC. In terms of establishing the discriminatory capability of a dissolution method for nanoparticle formulations an often used benchmark is the ability to discriminate dissolution rates for similar or identical formulations made with different sizes of drug nanoparticles. This is most conveniently evaluated using aqueous dispersions of the same nanoparticulate formulation prepared with slightly different average particle sizes. For example, it would be expected that dissolution rates for samples with D_{50} values differing by a factor of $2\times$ (e.g., 200 nm versus 400 nm versus 800 nm) could be discriminated in an optimal method. Shekunov et al.[41] demonstrated discrimination of dissolution rates for two different sized nanoparticle suspensions of griseofulvin (760 and 978 nm) along with a micronized sample with a particle size of 5.9 μm when conducted in a pH 7.4 phosphate buffer using USP apparatus 2.

10.4.2 Surfactant-Based Media

Synthetic and semisynthetic surfactants are a mainstay in the dissolution testing of poorly water-soluble drugs. The more commonly used surfactants include the ionic surfactants sodium lauryl sulfate (SLS or SDS) and cetyltrimethyl ammonium bromide (CTAB), the nonionic surfactants polyoxyethylene sorbitan monolaurate (polysorbate 20) or monooleate (polysorbate 80) and polyoxyethylene lauryl ether (Brij 35) and the zwitterionic surfactant lauryldimethylamine N-oxide (LDAO). Although effective in solubilizing poorly water-soluble drugs, many surfactants tend to be very aggressive in enhancing solubility and dissolution rates when used at levels above their CMC and so the amounts used in the dissolution medium are typically kept to a minimum level needed for achieving sink conditions. In this author's experience the ionic surfactants SLS and CTAB tend to be more aggressive and often less discriminating for dissolution rates of nanoparticle formulations compared with the non-ionic polymeric surfactants such as polysorbates. In general, the non-ionic surfactants have often been considered to be more biologically relevant than the ionic surfactants.[42] However, one of the challenges with using the non-ionic polymeric surfactants such as the polysorbates and Brij 35 is the variation in the composition of these surfactants among different suppliers or across a number of batches from the same supplier which can result in variations in drug dissolution rates. In that respect the single chemical entity surfactants such as SLS are often preferred for better method robustness for routine quality control applications.

Although these surfactant-based media have sometimes been considered not to be biorelevant there are examples in the literature in which these surfactants have produced dissolution profiles comparable to more complex biorelevant media. Lehto et al. reported on the use of both SLS and polysorbate 80 containing dissolution media with a USP apparatus IV system that produced similar dissolution profiles to the more complex and costly fasted state simulated intestinal fluid (FaSSIF) media system (refer to Section 10.4.4) for three poorly water-soluble BCS class II drugs when the levels of these surfactants were properly optimized.[43] For two of the three model drugs the optimal surfactant levels providing the best discrimination did not achieve sink conditions. Importantly, these authors concluded that the use of lower, more biorelevant levels of these two surfactants, than is typically found in USP compendia monographs, should find utility in terms of predicting in vivo absorption as well as for more meaningful quality control testing.

Numerous applications in which surfactant-based media have been successfully employed to develop discriminating dissolution methods for nanoparticle dosage forms have been reported. As examples, Talekar et al. described a dissolution method employing 1% polysorbate 20 in pH 7.4 phosphate buffered saline medium at 50 rpm using USP apparatus 2 that was useful to characterize nanoparticle and microparticle suspensions of an investigational molecule.[44] And Quinn et al.[45] used a medium containing 0.1% polysorbate 80 in 0.01 M HCl at 75 rpm with a USP apparatus 2 to discriminate between the dissolution rates of a nanoparticle suspension and un-milled suspension formulation of a poorly water-soluble BCS class II molecule. The nanoparticle formulation demonstrated near complete dissolution within 5 minutes, whereas the unmilled preparation showed only 6% dissolved after 1 hour. The improved in vitro dissolution properties for the nanoparticle preparation translated in the reduction of the food effect (fed-fasted AUC ratio) in a beagle dog food effect model from 6.2 for the unmilled formulation to 1.3 for the nanoparticle formulation.

10.4.3 Two-Stage Media

A common two-stage dissolution procedure entails an initial assessment in acidic media (0.01–0.1 M HCl) followed by a second stage in a neutral pH buffered medium (typically pH 6.8 phosphate buffer). Although classically applied as a test for delayed release dosage forms,[46] this technique is useful to screen formulations of drugs that demonstrate pH-dependent solubility, in particular weak bases.[47] Depending on its pK_a value a weakly basic drug will typically demonstrate higher solubility under acidic conditions but lower or very poor water solubility at neutral pH values. By evaluating the dissolution properties with a two-stage acidic to neutral pH media setup one can check for potential supersaturation and precipitation events that may be encountered in vivo during the transition of the dosage form through the gastrointestinal tract. For a rapidly dissolving, weakly basic drug nanoparticle dosage form, the advantage of the faster dissolving nanoparticles can be lost if complete dissolution of the dosage form occurs in the acidic stomach environment but subsequent precipitation occurs after transitioning to the neutral pH environment of the small intestine. This type of two-stage dissolution test can screen for such dissolution-precipitation phenomena and also check for the impact of formulation ingredients, in particular nanoparticle surface stabilizers, to inhibit drug precipitation. An example of conducting a two-stage dissolution test was reported by Ghosh et al. for a nanosuspension of the salt of a weakly basic drug stabilized with Vitamin E TPGS along with a coarse suspension of the drug.[48] The first stage was conducted in pH 2 media using USP apparatus 1 at 100 rpm on a sample of the nanosuspension in a prefilled capsule. After one hour the media was change to pH 6.8 (refer to Fig. 10.3). Although it was not specifically claimed by the authors, these results would suggest that some level of supersaturation was possibly maintained by the nanosuspension since the amount of drug dissolved after the switch to pH 6.8 at 60 minutes stayed relatively constant out to the final time point at 120 minutes. In the absence of supersaturation the curve for the crystalline nanosuspension should have eventually reached the same percentage dissolved level attained for the micronized and non-micronized samples. It has been reported by several researchers that Vitamin A TPGS is a potent precipitation inhibitor that is effective in maintaining drug in a supersaturated state.[49]

A similar two-stage dissolution method was used by Miller et al.[6] to characterize the supersaturation features of amorphous nanoparticles of the poorly water-soluble weakly basic drug itraconazole. In one variation of the method the pH 6.8 neutral buffer stage contained 0.17 % SDS to simulate micelles in the small intestine while in another variation SDS was not present. High levels of supersaturation were achieved for the amorphous nanoparticles in the pH 1.2 acid stage which decayed rapidly in the pH 6.8 buffer stage without SDS. However, in the presence of SDS in the pH 6.8 buffer a high level of supersaturation was maintained for 2 hours. In vivo PK data obtained in a rat model for different formulations correlated with the two-stage dissolution method containing SDS, but not without the SDS micelles, indicating the SDS was a good

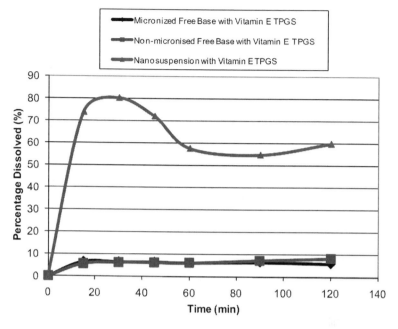

Figure 10.3 Comparison of a two-stage dissolution test with media switch at 60 min. from acid to neutral pH conducted on a weakly basic API nanosuspension vs. coarse drug (micronized and nonmicronized) each containing vitamin E TPGS. Reproduced from Ref. 48 with permission from Elsevier.

mimic for bile salt micelles in the intestine for this nanoparticle formulation.

10.4.4 Biorelevant Media

The use of so-called biorelevant media to develop more predictive in vitro dissolution methods in terms of forecasting dosage form in vivo performance is well established in the pharmaceutical industry. The topic of biorelevant media is covered in more detail in other sections of this volume. In brief, biorelevant media are constructed to possess physicochemical properties such as pH, ionic strength, surface tension, bile salt and lecithin content that are more representative of fluids of the gastrointestinal tract.[50–52] Standardized media compositions representing both the fasted and fed state in both the gastric environment (FaSSGF and FeSSGF) and in the intestinal environment (FaSSIF and FeSSIF) have been reported.[50–52]

Several research groups have reported on the successful use of biorelevant media for the dissolution testing of nanoparticle drug formulations. As an example, Shono et al.[53] studied the use of biorelevant dissolution tests in conjunction with in silico modeling based on STELLA® software to forecast the in vivo performance of a commercial nanoparticle capsule and investigational micronized oral formulation of aprepitant in the pre- and post-prandial states. A USP apparatus 2 system was used with 500 mL of media at 50 rpm. In addition to using biorelevant simulated media (FaSSGF, FeSSGF, FaSSIF and FeSSIF), USP compendia simulated gastric and intestinal fluids without enzymes (SGF_{SP} and SGF_{SP}) were used for comparison. The results of the study showed that dissolution testing in biorelevant gastric and intestinal media, coupled with in silico modeling, can forecast the human pharmacokinetics of both the micronized and nanoparticulate dosage forms of aprepitant in the fasted and fed states. It was also demonstrated that the in silico simulations using biorelevant media produced superior results compared to those obtained using the compendia simulated media. It is noteworthy that the dissolution tests were conducted under non-sink conditions with the maximum amount of drug dissolved ranging from a few percent up to about 40% depending on the drug particle size and biorelevant media composition.

In a similar study, Juenemann et al.[54] used biorelevant dissolution in conjunction with in silico modeling to predict the in vivo performance of a commercial nanoparticle tablet formulation and micronized capsule formulation of fenofibrate. It was demonstrated that the use of biorelevant media (FaSSGF, FaSSIF, FeSSIF, FaSSIF-V2 and FeSSIF-V2) in combination with in silico modeling with STELLA® software enabled in vitro-in silico-in vivo correlation for both micronized and nanoparticle formulations of fenofibrate in the fed and fasted state. Importantly, these authors highlighted the criticality of using an appropriately small pore size filter to ensure accurate dissolution results for the nanoparticle formulation (see Section 10.6.2 for a further discussion of this point).

10.4.5 Sink versus Non-Sink Conditions

As previously stated, sink conditions are generally preferred for dissolution methods for improved method robustness and to avoid artifacts due to variations in sample concentrations or precipitation phenomena. However, as highlighted in Section 10.4.4, examples exist in the literature in which non-sink conditions were found to be more discriminating as well as more prognostic in terms of gauging in vivo performance of nanoparticle dosage forms. Liu et al. experimentally studied and mathematically modeled the dissolution rate of indomethacin nanoparticle suspensions prepared with different particle sizes in the range from 340 to 1300 nm.[39] A mathematical model based on the so-called "shrinking-core model" was established to estimate the dissolution profiles and clarify the dissolution mechanism in both sink and non-sink conditions for either monomodal or bimodal nanoparticle size distributions. The dissolution experiments were performed with a USP apparatus 2 setup with either a pH 1.2 dilute HCl or pH 5.0 phthalate buffer. Based on the results, these researchers concluded that when the sample amount in the dissolution experiment is at or close to the saturation solubility of the drug in the medium, the slowest dissolution rate and the best discrimination in the profiles for the different particle sizes is obtained for either monomodal or bimodal distributions. However, using either true sink conditions or an amount of drug well in excess of the saturation solubility in the medium, the dissolution rates increase and little to no discrimination between the profiles for the different sized preparations was observed. Consistent with the experimental results, the mathematical simulations predicted the slowest dissolution rate when the experiment is conducted at the drug's solubility limit in the medium. When conducted under sink conditions the rate increases according to the Noyes-Whitney-Nernst-Brunner model (Eq. 10.2) due to the steeper concentration gradient (C_s-C). When the dissolution experiment is conducted at a level above the saturation solubility, a portion of the smaller particles in the sample dissolves quickly and produces a dissolution medium that is saturated with drug in a short period of time.

Based on these results and the previous discussion in Section 10.2.6 regarding the potential impact of supersaturation/precipitation events, the use of non-sink conditions may be better suited to supporting early formulation screening and optimization in terms of dosage form in vivo performance. Once an optimal dosage form is developed it may be advantageous to revert to a method that employs sink conditions for routine quality control applications in order to enhance method robustness.

10.5 Instrumentation and Sampling Technique Considerations

The types of dissolution apparatus reported in the literature to conduct testing on nanoparticle formulations spans everything from a customized small-volume, dynamic drug release apparatus coupled on-line with liquid chromatography[55] to the use of standard compendia apparatus setups.[56–58] It has been often stated that proliferation of new dissolution apparatuses is to be discouraged when testing novel dosage forms such as nanoparticle formulations[59] unless they offer clear advantages to existing dissolution technologies. In most cases nanoparticle dosage forms are amenable to testing with a compendia apparatus with the use of USP apparatus 1 (baskets),[56] 2 (paddles)[57] and 4 (flow-through)[58] commonly cited setups with apparatus 2 applications predominating. In one report[60] the authors compared the use of USP apparatus 1, 2, and 4 for the dissolution testing of cefuroxime axetil nanoparticle preparation (results were also obtained using a dialysis method which will be described in Section 10.7). The nanoparticle preparation was obtained using a bottom-up precipitation process leading to "chain-like" aggregates of nanoparticles after drying. In each case the dissolution medium was 0.1 M HCL containing 0.1% SLS. The authors claimed that the flow-through apparatus was unequivocally the most robust method for the dissolution analysis of drug nanoparticles producing good discrimination between the dissolution rates obtained for the nanoparticle preparation and the unprocessed drug. However, the authors noted that significant wetting problems were encountered with the drug preparations (which did not contain surface stabilizers) and the powders were aggregated and tended to float on the dissolution medium surface with the basket and paddle apparatus. In the flow-through apparatus the powders were held in place in the cell which minimized any wetting or floating problems. Thus, the advantage observed with the flow-through apparatus in this application may have been an artifact of the aggregation and poor wetting phenomena for the formulation. In this author's experience and as will be documented below with references from other groups, USP apparatus 1 and 2 setups have also been successfully used for the dissolution testing of nanoparticle dosage forms.

As is the case for formulations made with conventional sized drug particles, for the quantification of the amount of drug dissolved in the media the standard techniques of direct UV analysis or HPLC with UV analysis tend to predominate for testing nanoparticle formulations. Complicating the quantification of the amount of solubilized drug released is the need to ensure adequate differentiation of molecularly solubilized drug from the very small un-dissolved nanoparticles. As documented by different groups[61–63] in most cases the methods used to process and test the dissolution samples of the nanoparticle formulations can be broadly categorized as either (a) sample and separate via mechanical means (filtration or centrifugation), (b) membrane diffusion separation or (c) in situ techniques that rely on a variety of different spectroscopic or physical measurements to directly quantify the amount of dissolved drug in the presence of undissolved drug particles and excipients. Each technique can have its advantages and disadvantages and as of yet there is no universally optimal technique that has proven to be broadly applicable for all nanoparticle formulations.

10.6 Sample, Separate, and Analyze

In the sample, separate and analyze technique for dissolution testing of nanoparticle formulations, an aliquot of the sample is removed from the dissolution vessel at appropriate time points and the dissolved drug is mechanically separated from undissolved nanoparticlulate drug via filtration, centrifugation or a combination of centrifugation with filtration. In many cases, the processed sample is further diluted/dissolved in a solvent prior to quantitation by direct UV or HPLC with UV detection. To ensure the true dissolution rate is measured which is unbiased by undissolved or partially dissolved nanoparticles in the processed sample, it is important the pore size of the membrane filter and the filtration processing time are appropriately optimized or the centrifugation force and time sufficiently optimized.

For dosage forms manufactured with conventional sized drug particles, dissolution samples are processed with a membrane filter with an average pore size in the range from 0.45 μm up to 70 μm as described in the USP[40] with 0.45 μm filters tending to predominate. With nanoparticle formulations prepared with average drug particles as low as 100–400 nm the use of a typical 0.45 μm pore size filter can be drawn into question since a significant portion of the initial nanoparticle population would reside below the nominal size of the filter pores, although the use of 0.45 μm filters for nanoparticle formulations has been reported in numerous publications.[64–66] The use of a 0.45 μm filter may be justified in some cases based on the actual size distribution of the drug nanoparticles, their state of aggregation and the mechanism of drug release. Several examples exist in the literature in which smaller pore size filters have been used for nanoparticle formulations with sizes around 0.1 μm frequently used [see for example 67] while in some cases even smaller sized filters below 0.1 μm (typically 0.02 μm) are used.[45,54,62,68–72] In such cases the pore size of the filter is presumably selected based on the average initial size of the nanoparticles. Even with 0.1 μm filters, the low end of a typical drug nanoparticle size distribution may reside below the nominal pore size of the filters.

It is emphasized at this point that the dissolution process for all sizes of drug particles is a continuous process spanning the initial dissolution from the surface of the primary drug particle (nano- or micron-sized) to the completion of the drug particle dissolution process at which point all drug molecules are in a molecularly dissolved state. Thus, even if the membrane pore size is below the average size of the initial particles constituting the formulation there

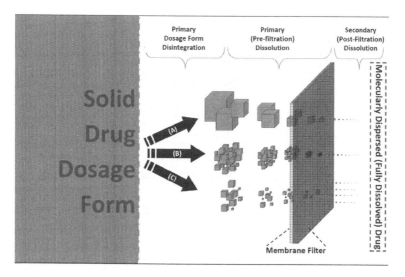

Figure 10.4 Depiction of pre- and postfiltration dissolution from a solid dosage form composed of (a) conventional micrometer-sized drug particles, (b) aggregated drug nanoparticles, and (c) redispersing drug nanoparticles.

will usually be some post-filtration dissolution of the continuously dissolving drug particles taking place until the dissolution process is complete (unless the pore size is on the order of molecular dimensions). This is demonstrated in the hypothetical examples provided in Fig. 10.4. This depicts the dissolution process from three different formulations processed using a small pore size membrane filter: (a) a conventional micrometer-sized formulation, (b) a poorly redispersing (e.g., aggregating) nanoparticle formulation, and (c) a well-redispersed nanoparticle formulation.

In this figure the pore size of the filter is taken to be just below the initial size of the primary drug nanoparticles. However, some degree of secondary post-filtration dissolution occurs with all three formulations since the nominal filter pore size is still larger than the size of the completely dissolved molecule, even with very small pore size filters (20 nm) with the largest percentage of post-filtration dissolution occurring with the well redispersed nanoparticles. Therefore, what is required is a rational development protocol to ensure that the separation process is appropriate for each particular nanoparticle preparation. Some suggestions are provided below.

10.6.1 Filter Membrane Material Selection

There are several considerations with regard to the selection of an appropriate filter membrane to process nanoparticle dissolution samples. Foremost is the need for an appropriate pore size to discriminate between dissolved and undissolved drug nanoparticles. Next is the need for an inert surface that ensures adequate recovery of dissolved drug during the filtration process. Finally, the pore structure of the filter membrane needs to not only be selective for removing undissolved nanoparticles but also, must avoid potential blockage and increased back pressure which can come from nanoparticles penetrating and blocking filter pore through paths. In this regard, Anopore® membrane filters have shown promise to address all three concerns. Anopore® filters are produced with a high-purity inorganic aluminum oxide matrix that is manufactured electrochemically producing membranes possessing a non-deformable, honeycomb pore structure.[73] In addition to an inert alumina matrix the pore size distribution is very uniform and the membranes possess a high degree of pore density and low degree of pore structure tortuosity (no lateral crossovers between pores), especially compared to typical fibrous, tortuous path membranes filters. These properties make the Anopore® membranes well suited for filtering nanoparticle dissolution samples, in particular with respect to minimizing blockage and high back pressure during filtration. Additionally, the Anopore® membranes come in three different nominal pore sizes (20 nm, 100 nm, and 200 nm) which are particularly well suited for processing typical sized drug nanoparticle samples.

10.6.2 Filter Pore Size Selection

A preliminary step to establish a filtration approach to processing nanoparticle dissolution samples is to confirm the pore size necessary to sufficiently remove undissolved nanoparticles from dissolved drug. In this author's experience and as documented by other researchers,[54] this can be achieved by selecting a series of filters with pore sizes spanning a range from well below the average primary particle size of the drug nanoparticles to a pore size above this average size. The reason for selecting filters with pore sizes above that of the nanoparticle sample is that a smaller pore size filter may not always be necessary in the context of accurately measuring nanoparticle dissolution rates. This relates to the mechanism by which the dosage form disintegrates or erodes and ultimately releases drug particles to the dissolution medium in addition to the dissolution rate for the primary drug nanoparticles. If the mechanism of release does not result in the formation of primary drug nanoparticles then the size of the filter pores does not necessarily need to be as small as the drug nanoparticles. Also, if the rate of dissolution of nanoparticles released from the dosage form is extremely fast ($<<1$ min), then it may not be necessary to use a small pore size filter for routine quality control applications. Finally, if the dosage form does not adequately redisperse to primary nanoparticles (see Section 10.9.3) then the size of the filter pores may not be as critical to the success of the method.

One approach to select a suitable filter pore size is to determine the extent of dissolution of the nanoparticle dosage form by processing the same sample using a series of decreasing pore size filters. A target level of about 50% dissolved in the experiment is defined (or another suitable target level corresponding to partial dissolution of the sample) and a series of samples are simultaneously taken at the prescribed time and immediately processed using the different pore size filters. As a control, a pre-dissolved solution of the drug in the dissolution medium representing about 100% of the target drug level is also tested to check for drug adsorption to the filter. The samples are analyzed for % drug recovery and a plot is constructed of the percent drug recovered versus the filter pore size. An example of the output that one might obtain is depicted in Fig. 10.5. Ideally, the control will demonstrate near 100 % recovery using with each different filter. The nanoparticle dissolution samples will typically show decreasing recovery with decreasing filter pore size until a threshold filter pore size is reached that sufficiently retains the undissolved nanoparticles. In Fig. 10.5 this occurs at the 0.1 µm pore size and reducing the pore size to 0.02 µm produces no further

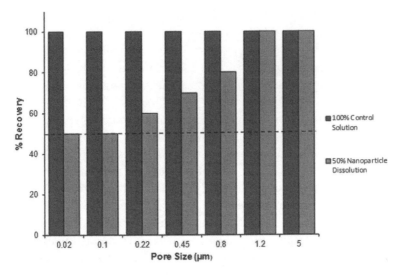

Figure 10.5 Recovery of drug solution (100% control solution) and retention of drug nanoparticles (50% nanoparticle dissolution) as a function of filter pore size.

retention of undissolved nanoparticles so the 0.1 μm filter can be selected.

An alternative approach to screen for the proper filter pore size involves the use of light scattering particle size analyzers (dynamic or static light scattering) to check the filtered dissolution samples for the presence of undissolved drug nanoparticles. With this approach dissolution samples from the nanoparticle formulation are processed through the selected filter(s) and immediately subjected to particle size analysis. If the particle size analyzer registers a significant signal for undissolved drug nanoparticles then a smaller pore size filter may be required. In cases where a significant level of undissolved nanoparticle are detected then the subsequent rate of further dissolution of these undissolved or partially dissolved drug nanoparticles can be estimated as described in Section 10.8.6. If the filtered sample contains a small amount of partially dissolved smaller nanoparticles that very rapidly dissolve (<<1 min) then their presence may not be important to adequately measuring the overall dissolution rate of the formulation. Dynamic light scattering instruments when used for this type of analysis have the advantage of requiring smaller sample volumes (<1 mL) and the ability to detect particles well below 10 nm, but analysis times (generally 1-2 min) are typically longer that for static light scattering.

A growing number of applications in which 0.02 μm (20 nm) Anopore® filters have been successfully used to process dissolution samples for different types of nanoparticle drug formulations have appeared in the literature in recent years.[41,45,54,61,68–72] There thus appears to be the potential for the broad applicability of these filters for use in testing nanoparticle dissolution samples using a sample and separate approach. When the results of a filter screening study indicate that a very small pore size filter (0.02–0.1 μm) is necessary and the amount of suspended solids, which can include undissolved drug nanoparticles and excipients, in the dissolution medium is high there may be a propensity for the filter membrane pores to completely or partially plug or exhibit high back pressure. In such situations a possible remedy is to use a larger pore size (1–10 μm) clarifying filter in series with and preceding the smaller pore size filter to reduce the undissolved solids burden on the surface of the smaller pore size filter and thereby reduce back pressure.

10.6.3 Filtration Timing

Another step in the sample and separate development protocol is to pull dissolution samples and filter them after different delay periods spanning several minutes. This allows for the characterization of any meaningful dissolution that is occurring post sampling but prior to filtration. In addition, the pulled samples are analyzed after different delay times post filtration. This is to check for the potential of a supersaturation event in which case precipitation may occur in the filtered samples as they cool to room temperature and prior to completing the quantitative analysis. This can be especially important when testing samples under non-sink conditions and in particular when working at drug concentration levels in the dissolution vessel that exceed the equilibrium solubility in the medium at 37°C.

10.6.4 Centrifugation

The use of centrifugation to process dissolution samples for a nanoparticle suspension[74] and drug nanoparticles embedded in microparticles of mannitol[75] has been reported. In both cases samples were centrifuged at 13,000 rpm for 8 minutes. For each of these applications complete release of the drug nanoparticles occurred within 3 minutes. The use of an 8 minute centrifugation procedure to test these rapidly dissolving samples can be questioned, although in one case[74] the authors claimed that the centrifugation procedure was demonstrated to be superior to using filtration. Centrifugation in conjunction with ultrafiltration has also been reported to separate released drug from a liposome-encapsulate ciprofloxacin formulation.[76] Drug release samples were taken and diluted 1:1 with buffer and added to a Nanosep® ultrafiltration device with 10 k or 30 k molecular weight cut offs and centrifuged at 10,000 rpm to separate released drug.

10.7 Membrane Diffusion (Dialysis) Analysis

Numerous examples have been published describing the use of dialysis procedures for determining the in vitro release from nanoparticle dosage forms.[77–81] In particular, this approach has been applied to a variety of matrix type drug nanoparticles such as biodegradable polymeric nanoparticles,[77] liposomal nanoparticles,[78] solid lipid nanoparticles[79] and drug-linked gelatin nanoparticles[80] in addition to pure drug nanoparticle preparations.[81]

In the dialysis technique, the sample of nanoparticles is placed into the release (donor) medium separated from the bulk (receiver) medium by a dialysis membrane that can have a range of different molecular weight cutoffs (typically 5–20 kDa). As drug dissolves it diffuses through the membrane into the larger volume bulk receiver medium whereas the intact nanoparticles are retained in the donor compartment. The dialysis membranes can come in a number of different configurations including dialysis bags, dialysis cells and other custom dialysis adapters. At predetermined time points samples are taken from the bulk receiver medium and assayed for dissolved drug by typical UV or chromatographic methods.

The dialysis technique demonstrates versatility to characterize the in vitro release rate of numerous types of nanoparticulate drug formulations. This approach has several advantages and disadvantages which were recently summarized by Modi et al.[82] On the one hand the technique precludes the need to mechanically separate free drug from the nanoparticles. It was pointed out by these researchers that the forces applied to separate free from nanoparticulate drug using sample and separate techniques can disturb the equilibrium controlling drug release. Additionally, incomplete separation can lead to errors in the measurement of drug release. On the other hand, these authors point out that the appearance of free dissolved drug in the "sink" receiver compartment is the result of both release/diffusion of free drug from the nanoparticles and diffusion of free drug across the dialysis membrane. In this regard the measured release of free drug is the net result of drug transport across two barriers in series, release and dissolution from the nanoparticles and diffusion through the dialysis membrane, and so the measured release rate my not reflect the true drug release rate. This will particularly be the case for rapidly dissolving or releasing nanoparticle preparations such as nanocrystal dispersions. In this regard the dynamic dialysis technique may be more appropriate for controlled release nanoparticle preparations in which case the diffusion through the dialysis membrane is not rate-limiting to the overall measured drug release rate. To overcome some of the shortcomings of the dynamic dialysis technique Modi et al.[82] presented mathematical models and experimental simulations that demonstrated the utility of these models to extract reliable release rates in cases where drug binding effects and/or dialysis membrane transport may be partially contributing to the apparent drug release kinetics from the nanoparticles.

Reports have surfaced in recent years in which different research teams have integrated dialysis based dissolution methods directly into standard compendia dissolution apparatus setups for characterizing the dissolution properties of various types of nanoparticle based formulations. Abdel-Mottaleb and Lamprecht[83] developed a glass basket dialysis method by modifying a USP apparatus 1 (baskets) setup. The standard basket assembly was replaced by a custom glass cylinder that is closed at the base by a 12–14 KDa cutoff cellulose dialysis membrane. The release rates of various types of colloidal drug loaded nanoparticles were measured including liposomes, polymeric nanoparticles and lipid nanoparticles. Samples of the different nanoparticle formulations were added to the custom dialysis cells which were attached to the shafts of the dissolution apparatus and rotated at 50 rpm. Drug release was conducted with a pH 7.4 phosphate buffer maintained at 37°C. At predetermined time points released drug was measured by on-line UV analysis. Comparisons were made to release rates conducted with traditional dialysis bags made with the same membrane material that were immersed in the same dissolution medium held in a container placed in a shaking water bath. The release rate results obtained with the custom USP 1-dialysis setup were more discriminating for the different formulations and better reproducibility was obtained compared to the standard dialysis bag method.

Bhardwaj and Burgess described a novel USP apparatus 4 dialysis adapter for drug release from drug loaded liposomes, drug nanosuspensions and other types of nanoparticle formulations.[84] These authors designed a dialysis adapter that can be used with a 22.6-mm USP apparatus 4 sample cell. The cell adapter is a hollow framed cylinder with a Teflon top and base. Fittings are provided to hold a dialysis membrane over the frame. The adapter cell with dialysis membrane fits the 22.6-mm USP apparatus 4 cell dimensions. The nanoparticle formulations are added to the dialysis adapter which is sealed. The USP apparatus 4 can then be operated in both the normal open and closed configurations and the flow rate varied as required. This custom adapter is now commercially available from Sotax Inc. The optimization and feasibility of the novel USP apparatus 4 dialysis adapter was demonstrated with solution, suspension and nanoparticle liposome formulations of dexamethasone and the results compared to those obtained with standard dialysis sac and reverse dialysis sac methods. The novel USP apparatus 4 dialysis method was claimed to provide better discrimination in release rates and less variation that the other methods.

Another facile manner to incorporate a dialysis membrane into a USP apparatus 2 (paddles) system was reported by Kumar et al.[85] These researchers utilized commercially available Float-A-Lyzer dialysis cells from Spectrum Labs which they secured to the cover on the USP 2 apparatus by gluing the Float-A-Lyzer cap to a plastic rod. The rod was then fixed to the dissolution vessel cover in such a way that the dialysis membrane was suspended below the surface of the dissolution medium (900 mL) in the vessel. Dissolution rates were conducted on spray-dried formulations made from crystalline nanoparticle suspensions of naproxen and indomethacin with average sizes of about 250 nm which were performed under sink conditions in a 50 mM, pH 6.8 phosphate buffer. The spray-dried formulations were re-suspended in 1 mL of distilled water and placed into the dialysis cells (cutoff 1000 kD) and the dissolution rates measured at a paddle speed of 50 rpm. At selected time points samples were removed from the dissolution vessel (receiving vessel) and analyzed directly by HPLC. The authors were able to discriminate among dissolution rates for different spray-dried preparations of each drug based on the degree of aggregation in the formulation, which was correlated with the drug-to-stabilizer ratio. In this author's experience, the Float-A-Lyzer units can be directly used by allowing the units to freely float at the surface of the dissolution medium using a USP apparatus 2 setup. In particular these dialysis units are effective for controlled release nanoparticle formulations.

10.8 In situ Analysis

A variety of different in situ analysis techniques (spectroscopic and physical measurement based) have been combined with conventional and/or novel dissolution apparatus or have been proposed as a means to measure the dissolution rate of nanoparticle formulations. Each of the different in situ measurement techniques has its own advantages and limitations but in all cases they possess the ability to measure very fast dissolution rates ($\sim$1 min). In this regard these in situ techniques are most beneficial when measuring the dissolution rates of very rapidly dissolving pure drug nanoparticle dispersions. The following is a summary of several different approaches that have been reported.

10.8.1 In situ Fiber Optic UV Analysis

In situ fiber optic UV analysis has gained general acceptance over the last 10–15 years within the pharmaceutical industry for the automated analysis of dissolution samples. In addition to eliminating the need for manual sampling and subsequent processing (e.g., filtration and dilution) of the dissolution samples, the real time sample analysis provides the opportunity to measure very rapid dissolution rates. Several publications are available highlighting the advantages as well as some of the potential challenges with using fiber optic UV analysis in conjunction with dissolution testing of conventional dosage forms, in particular with respect to scattering arising from particulates in the dissolution test medium.[86–88]

The application of fiber optic UV analysis in conjunction with dissolution testing of nanoparticle formulations has been critically evaluated by Van Eerdenbrugh et al.[89] These authors studied the application of fiber optic UV for the dissolution testing of 300 nm felodipine nanosuspensions as well as larger sized felodipine suspensions up to 400 μm. Mie theory calculations were used to model the phenomena underpinning the dissolution testing of the different sized felodipine suspensions. Notably, the authors reported that undissolved nanoparticles of felodipine, in addition to scattering light, also absorb light in a manner similar to free felodipine molecules in solution. This makes accurate assessment of dissolved drug in the presence of undissolved nanoparticles difficult even when typical second derivative corrections to address scattering are applied. The authors cautioned the use of in situ fiber optic UV measurements for dissolution testing in the presence of undissolved nanoparticles, whether they are intentionally present as a nanoparticle formulation, if they appear as a result of dissolution of larger micron sized particles or if they are present due to precipitation of supersaturating formulations.

Tsinman et al.[90] also investigated the use of in situ fiber optic UV for the dissolution testing of nanoparticle formulations. The authors applied a special ZIM (Zero Intercept Method) analysis of the second derivative UV spectrum to purportedly overcome the problems of the complex effects of simultaneous scattering and absorption of UV radiation by nanoparticles. In ZIM, the zero-crossing point of the derivative spectrum is used to separate the contributions of dissolved drug and drug nanoparticles in the overlapping UV spectrum. These authors obtained confirmation of the validity of this approach based on the good agreement obtained independently using a dialysis technique for a naproxen nanoparticle suspension formulation.

10.8.2 In situ Potentiometric Analysis

A prototype potentiometric sensor instrument was reported for the dissolution testing of four different basic drugs with different physicochemical properties.[91] The results were compared to off-line HPLC analysis and in situ fiber optic UV analysis. The potentiometric sensors demonstrated good response times for fast releasing formulations and no interference with undissolved particles or air bubbles as compared with the fiber optic UV analysis. Although the authors did not specifically present data for nanosuspensions they proposed that the potentiometric sensors may have advantages for dissolution testing of fast dissolving nanosuspensions compared to existing techniques. Disadvantages noted for the potentiometric technique included the need to precondition a new set of sensors for each drug and poor performance when using media containing certain surfactants such as polysorbate 20.

10.8.3 In situ Turbidimetric Analysis

An in situ turbibimetric technique has been described for measuring the rapid dissolution rates of 300 nm nanocrystal suspensions of the poorly water-soluble drugs danazol and itraconazole in aqueous dissolution media containing surfactant micelles.[92] Dissolution half-lives as short as a few seconds could be determined with this approach without the need to withdraw samples or to filter the particles. A mass transfer model was developed to analyze the data to determine the particle size distribution and dissolution rate in terms of two steps: interfacial reaction, consisting of micelle uptake and desorption, followed by diffusion of the drug-loaded micelles. The model provided a fundamental understanding of the roles of interfacial reaction and diffusion as a function of particle size.

10.8.4 In situ Electrochemical Analysis

An electrochemical technique for the in situ monitoring of the dissolution kinetics from drug loaded liposomal nanoparticles has been described.[93] The method was illustrated for doxorubicin-loaded liposomes using repetitive square-wave voltammetric measurements of the reduction of doxorubicin released from liposomes at a glassy-carbon electrode. The method couples high sensitivity down to 20 nM doxorubicin with high speed and stability for drug release in dissolution media (phosphate buffered saline at either pH = 7.4 or pH = 5.2) or diluted serum. Square-wave voltammetry was reported to be advantageous over potentiometric sensing and represents an attractive technique for monitoring the drug release kinetics with high temporal resolution, as it couples high sensitivity down to nanomolar (nM) range with high speed and reproducibility.

10.8.5 In situ Solution Calorimetry Analysis

The utility of using solution calorimetry to measure the dissolution rate of three model poorly water-soluble drug nanoparticle suspensions with particle sizes in the range of 211–244 nm has been reported.[94] In this technique the heat absorbed or released during the dissolution testing of nanoparticle suspensions was used to quantify and monitor the dissolution process. The dissolution apparatus setup used a 100 mL glass reaction vessel fitted with a thermistor, heater and stirring device. Samples of the drug nanoparticle or crude suspensions were added to 100 mL of simulated gastric fluid (SGF, USP), simulated intestinal fluid (SIF, USP) or a pH 10 borate buffer and stirred at 400 rpm. When the drug dissolves the solution calorimeter monitors the change in temperature which is eventually translated into heat of solution. In order to accurately convert temperature change to heat of solution for the dissolving drug nanoparticle suspensions, all heat transfers in the system must be considered, resulting in the application of a correction due to the heat capacity of the system. The technique was shown to be suitable to measure the fast dissolution rates of the nanoparticle suspensions (<1 min.) which could not be achieved using a classical filtration followed by HPLC procedure due to the inability to process these samples fast enough. A noted disadvantage of this calorimeter approach is that if there are any other heat transfers in the system or other heat transfer phenomena during the dissolution test that are not accounted for in the analysis then erroneous results will be obtained.

10.8.6 In situ Light-Scattering Analysis

Several examples exist demonstrating the utility of in situ light scattering to determine the dissolution rate of nanoparticle drug suspensions.[36,61,95] As an example, Anhalt et al.[61] described procedures to measure both the dissolution rate and saturation solubility of fenofibrate nanoparticle suspensions spanning a size range from 120 nm to 1070 nm in simulated gastric fluid containing the surfactant polysorbate 80 using dynamic light scattering. To determine the saturation solubility of the different sized nanoparticle suspensions, various dilutions of the nanoparticle suspensions in the dissolution medium were equilibrated at 25°C and samples taken at different time points and transferred to disposable polystyrene micro cuvettes. The absolute scattering intensities were measured using a dynamic light scattering particle size instrument and the values plotted against the sample concentration. The plot resulted in two linear segments; the first representing completely dissolved samples with scattering intensities equal to the dissolution medium blank and the second one representing partially dissolved samples with a slope corresponding to an increasing solid fraction of drug nanoparticles. The intercept represents the solubility of the nanoparticle suspension. Consistent with reports in the literature for other nanoparticle drugs,[26] the saturation solubility enhancements achieved with the fenofibrate nanoparticle suspensions in the size range from 120 nm to 1070 nm were only modestly higher (0–13%) compared to a conventional sized drug suspension, consistent with what would be theoretically calculated. To determine the dissolution rates of the fenofibrate nanoparticle suspensions, an initial dilution of the samples was made by adding aliquots of the suspensions to 5–10 mL of the dissolution medium and mixed using a magnetic stirrer. A final dilution of the sample was made directly into the sample cuvette and the dissolution rate measured based on absolute scattering measurements made over time which were corrected for the scattering intensity of the dissolution medium blank. This procedure was successful in discriminating dissolution rates for the different sized nanoparticle suspensions and the results correlated well with results generated with a sample and separate technique conducted in a conventional USP apparatus 2 setup using filtration with a 0.02 μm syringe filter and quantitation using HPLC analysis. It should be stressed that the in situ light scattering technique would not be applicable to nanoparticle formulations containing other insoluble excipients such as in a tablet or capsule formulation unless the scattering signal from the different types of particles could be successfully deconvoluted and separated.

10.9 Alternative in vitro Release Techniques

A number of different alternative in vitro release techniques have been reported for evaluating the performance of nanoparticle formulations. While some of these techniques are highly specialized or involve complicated setups, others have demonstrated broader applicability to a variety of nanoparticle dosage forms.

10.9.1 Asymmetric Flow Field-Flow Fractionation with Multi-Angle Light Scattering

Engel et al. developed and validated an analytical flow field-flow fractionation (AF4) procedure which they claim enables the determination of both the nanoparticle and solubilized drug released from a solid dosage form during dissolution testing in a single sample analysis.[96] The setup entails a commercial AF4 system with multi-angle light scattering (MALLS) and UV detector. Briefly, in AF4 particles are separated based on size in a flowing stream contained within a ribbon-like channel that maintains a parabolic flow profile with a cross-flow force through a semipermeable membrane. A model tablet formulation made with PLGA nanoparticles (120 and 220 nm) was used to establish the technique. For this application the PLGA particles were not loaded with drug. Dissolution of a model tablet made with 120 nm and 220 nm PLGA particles was conducted in a USP apparatus 2 with 500 mL of SIF, USP at 50 rpm. To validate the system samples were taken at various times for 2–120 minutes and directly injected into the AF4 system with no filtration and the amount of PLGA nanoparticles released from the tablets was quantified based on the MALS response. Although for this particular application the PLGA particles were not loaded with drug, it was claimed that the AF4 setup offers the possibility to measure both the release of the intact drug-loaded nanoparticles from the tablet as well as dissolved drug in the same run.

10.9.2 Dissolution/Permeation System

Buch et al.[97] reported on a combination dissolution/permeation system that was used to predict the in vivo performance of a commercial nanoparticle tablet formulation of fenofibrate in addition to a solid dispersion formulation, commercial micronized capsule formulation, lipid microparticle formulation and a micronized preparation of fenofibrate along with the un-micronized bulk drug. Biorelevant media representing the fed and fasted state (FeSSIF and FaSSIF) were used to conduct the in vitro dissolution in the donor compartment which was directly linked to a Caco-2 monolayer to measure the drug permeation into the receiver compartment. The in vitro parameters from the dissolution/permeation system reflected well the in vivo performance of the formulations when tested in rats in terms of AUC and C_{max}.

10.9.3 Nanoparticle Redispersibility

As an alternative to directly measuring the dissolution rate of nanoparticle dosage forms, it has been claimed that measuring the in vitro redispersibility properties of the primary nanoparticles from a dosage form in a biorelevant medium can be a more reliable method to gauge in vivo drug dissolution and bioavailability.[37]

The described test is conducted in aqueous biorelevant media in which the drug does not show significant solubility. Representative electrolyte solutions that were reported include dilute HCl, NaCl and phosphate buffers. It is claimed that measuring the nanoparticle redispersibility in such biorelevant media rather than the actual drug dissolution in the aggressive, surfactant-based media typically required for poorly water soluble drugs was able to provide a more reliable prediction of in vivo dosage form performance.

It was reported that the redispersibility method can be performed in a standard dissolution apparatus. Similar to a dissolution test, at selected time points samples of the redispersed dosage form are removed and the amount of redispersed active agent is quantitated by UV or HPLC analysis after removal of any larger particle size excipients or aggregated drug particles using a suitable filter. In addition, the size of the redispersed drug nanoparticles are measured using a suitable particle size analyzer and compared to the size of drug nanoparticles prior to incorporation into the dosage form. According to this report, it is expected that an optimal in vivo performing nanoparticle dosage form is one in which the size of the redispersed drug particles in the redispersibility test most closely resembles the distribution of the particles before they are incorporated into the final dosage form.

An example that was provided in this reference demonstrated that a conventional in vitro dissolution method was not able to predict the relative in vivo performance among two different nanoparticle dosage forms and a conventional non-nanoparticle dosage form of a poorly water-soluble drug while the in vitro redispersibility method was able to accurately predict which of the two nanoparticulate dosage forms would demonstrate enhanced in vivo performance in humans in terms of reduced T_{max} and increased bioavailability. These results demonstrated the importance of the formulations' redispersibility properties rather than just the size of the primary nanoparticles incorporated into the dosage form in designing a suitable nanoparticle formulation with the optimal in vivo performance.

Several other research groups have reported using a variation of this redispersibility test to assess the success of producing basic spray-dried or freeze-dried solid powder preparations made from liquid nanoparticle dispersions.[98–101] In a typical experiment a small amount of the spray-dried nanoparticle dispersion was dispersed in a small volume (1–50 mL) of water, buffer or simulated fluid and agitated with either stirring or sonication. The sample is then directly subjected to particle size analysis (static or dynamic light scattering) with no further sample processing. The resulting particle size distribution is compared to the particle size distribution for the starting liquid nanoparticle dispersion. The more closely the particle size of the redispersed powder matches that of the original dispersion the better are the redispersibility properties of the powder formulation. This mode of redispersibility test would be restricted to formulations in which the drug nanoparticles are the only insoluble component in the formulation.

10.10 Future Developments

Commensurate with the development and advancement of nanoparticle dosage formulation and manufacturing process technologies in recent years has been the development and improvement of various types of dissolution (in vitro release) technologies to characterize the performance of these dosage forms. That said, there is perhaps still room for further improvements and standardization in the approaches used to develop and perform dissolution testing of nanoparticle dosage forms. These improvements can be grouped into two different categories. The first is the establishment of a rational approach for dissolution testing performed in support of nanoparticle formulation screening and optimization and the second is in the area of routine quality control testing. Because of the wide variety of different types of nanoparticle formulation technologies being pursued and their different in vitro release/dissolution mechanisms and in vivo performance characteristics (e.g., immediate and controlled release) it is difficult to envisage a universal approach that will be broadly applicable to all nanoparticle dosage forms. Rather, what are needed are standardized protocols that can be followed to guide the researcher to develop an appropriate dissolution method for each specific type of nanoparticle formulation that takes into consideration the unique characteristics of a given formulation. To support formulation screening and optimization the wide variety of successfully applied approaches available in the contemporary literature that use either conventional or novel instrumental setups would seem to provide a good foundation of methodologies or starting points sufficient to guide the interested researcher to develop a suitable method to optimize in vivo performance. Different prognostic in silico simulation models used in conjunction with the in vitro dissolution data have been demonstrated to be beneficial in this regard. A number of different dissolution testing approaches have been demonstrated each with its own advantages and disadvantages for a specific type of nanoparticle formulation. In particular for rapidly dissolving (i.e., 1–5 min) formulations typified by drug nanoparticle aqueous dispersions, the advancement of a general purpose in situ dissolution technique would be desirable. In this respect a broadly applicable in situ fiber optic UV technique and data processing protocol would be beneficial.

What is also lacking is sufficient compendia and regulatory guidance in developing suitable methods for routine quality control applications for nanoparticle dosage forms. The currently available guidance for developing suitable dissolution methods for solid dosage forms in general is sparse [ref USP, EP and ICH] and not specifically directed toward nanoparticle dosage forms. Given the unique performance capabilities and analytical challenges of these nanoparticle dosage forms it is offered that more specific guidance is required.

References

1. Merisko-Liversidge, E., and Liversidge, G. G. (2011). Nanosizing for oral and parenteral drug delivery: a perspective on formulating poorly-water soluble compounds using wet milling technology, *Adv. Drug Deliv. Rev.*, **63**, 427–440.

2. Bosselmann, S., and Williams, R. O. (2012). Has nanotechnology led to improved therapeutic outcomes? *Drug Dev. Ind. Pharm.*, **38**, 158–170.

3. Gao, L., Liu, G., Ma, J., Wang, X., Zhou, L., and Li, X. (2012). Drug nanocrystals: in vivo performances, *J. Control. Release*, **160**, 418–430.

4. Sharma, P., and Sanjay, G. (2010). Pure drug and polymer based nanotechnologies for the improved solubility, stability, bioavailability and targeting of anti-HIV drugs, *Adv. Drug Deliv. Rev.*, **62**, 491–502.

5. Couvreur, P., and Vauthier, C. (2006). Nanotechnology: intelligent design to treat complex disease, *Pharm. Res.*, **23**, 1417–1450.

6. Miller, M. A., DiNunzio, J., Matteucci, M. E., Ludher, B. S., Williams, R. O., and Johnston, K. P. (2012), Flocculated amorphous itraconazole nanoparticles for enhanced in vitro supersatruation and in vivo bioavailability, *Drug Dev. Ind. Pharm.*, **38**, 557–570.

7. Matteucci, M. E., Brettmann, B. K., Rogers, T. L., Elder, E. J., Williams, R. O., and Johnston, K. P. (2007). Design of potent amorphous drug nanoparticles for rapid generation of highly supersaturated media, *Mol. Pharmaceutics*, **4**, 782–793.

8. Bhardwaj, V., and Kumar, M. N. V. R., (2006). *Nanoparticle Technology for Drug Delivery*, eds. Gupta, R. B., and Kompella, U. B., Chapter 9 "Polymeric nanoparticles for oral drug delivery," (Taylor and Francis Group, New York) pp. 231–271.

9. Anton, N., Benoit, J. P., and Saulnier, P. (2008). Design and production of nanoparticles formulated from nano-emulsion templates – a review, *J. Control. Release*, **128**, 185–199.

10. Sharma, A., and Sharma, U. S. (1997). Liposomes in drug delivery: progress and limitations, *Int. J. Pharm.*, **154**, 123–140.

11. Müller, R. H., Mäder, K., and Gohla, S. (2000). Solid lipid nanoparticles (SLN) for controlled drug delivery – a review of the state of the art, *Eur. J. Pharm. Biopharm.*, **50**, 161–177.

12. Moghimi, S. M., Hunter, A. C., and Murray, J. C. (2005). Nanomedicine: current status and future prospects, *FASEB J.*, **19**, 311–330.

13. U. S. FDA draft guidance: considering Whether an FDA-Regulated Product Involves the Application of Nanotechnology, available at http://www.fda.gov/regulatoryinformation/guidances/ucm257698.htm (accessed Dec. 19, 2013).

14. Gupta, R. B., (2006). *Nanoparticle Technology for Drug Delivery*, eds. Gupta, R. B., and Kompella, U. B., Chapter 1 "Fundamentals of drug nanoparticles," (Taylor and Francis Group, New York) p. 8.

15. Möschwitzer, J., and Müller, R. H. (2007). *Nanoparticulate Drug Delivery Systems*, eds. Thassu, D., Deleers, M., and pathak, Y., Chapter 5 "Drug nanocrystals – the universal formulation approach for poorly soluble drugs," (Informa Healthcare, New York) p. 73.

16. Liversidge, G. G., Cundy, K. C., Bishop, J. F., and Czekai, D. A. (1992). Surface modified drug nanoparticles, United States patent No. 5, 145, 684.

17. Payne, T., Meiser, F., Postma, A., Cammarano, R., Williams, J., McCormick, P., Dodd, A., and Caruso, F. (2009). Nanoparticle composition and methods of synthesis thereof, United States patent application 2009/0028948.

18. Peltonen, L., and Hirvonen, J. (2010). Pharmaceutical nanocrystals by nanomilling: critical process parameters, particle fracturing and stabilization methods, *J. Pharm. Pharmacol.*, **62**, 1569–1579.

19. Shegokar, R., and Müller, R. H. (2010). Nanocrystals: industrially feasible multifunctional formulation technology for poorly soluble actives, *Int. J. Pharm.*, **399**, 129–139.

20. Sinha, B., Müller, R. H., and Möschwitzer, J. P. (2013). Bottom-up approaches for preparing drug nanocrystals: formulations and factors affecting particle size, *Int. J. Pharm.*, **453**, 126–141.

21. Müller, R. H., Gohla, S., and Keck, C. M. (2011). State of the art nanocrystals – special features, production, nanotoxicology aspects and intracellular delivery, *Eur. J. Pharm. Biopharm.*, **78**, 1–9.

22. Möschwitzer, J. P. (2013). Drug nanocrystals in the commercial pharmaceutical development process, *Int. J. Pharm.*, **453**, 142–156.

23. Siepmann, J., and Siepmann, F. (2013). Mathematical modeling of drug dissolution, *Int. J. Pharm.*, **453**, 12–24.

24. Serajuddin, A. T. M. (2007). Salt formation to improve drug solubility, *Adv. Drug Deliv. Rev.*, **59**, 603–616.

25. Grassi, M., Grassi, G., Lapasin, R., and Colombo, I. (2007). *Understanding Drug Release and Absorption Mechanisms: A Physical and Mathematical Approach*, Chapter 6 *"Dissolution of crystallites: effects of size on solubility,"* (Taylor and Francis Group, Boca Raton), pp. 333–370.

26. Van Eerdenbrugh, B., Vermant, J., Martens, J. A., Froyen, L., Van Humbeeck, J., Van den Mooter, G., and Augustijns, P. (2010). Solubility increases associated with crystalline drug nanoparticles: methodologies and significance, *Mol. Pharmaceutics*, **7**, 1858–1870.

27. Mosharraf, M., and Nyström, C. (1995). The effect of particle size and shape on the surface specific dissolution rate of micronized practically insoluble drugs, *Int. J. Pharm.*, **122**, 35–47.

28. Wang, Y., Abrahamsson, B., Lindfors, L., and Brasseur, J. G. (2012). Comparison and analysis of theoretical models for diffusion-controlled dissolution, *Mol. Pharmaceutics*, **9**, 1052–1066.

29. Johnson, K. C. (2012). Comparison of methods for predicting dissolution and the theoretical implications for particle-size-dependent solubility, *J. Pharm. Sci.*, **101**, 681–689.

30. Sharma, P., Denny, W. A., and Garg, S. (2009). Effect of wet milling process on the solid state of indomethacin and simvastatin, *Int. J. Pharm.*, **380**, 40–48.

31. Mosharraf, M., Sebhatu, T., and Nyström, C. (1999). The effects of disordered structure on the solubility and dissolution rates of some hydrophilic, sparingly soluble drugs, *Int. J. Pharm.*, **177**, 29–51.

32. Mosharaf, M., and Nyström, C. (2003). Apparent solubility of drugs in partially crystalline systems, *Drug Dev. Ind. Pharm.*, **29**, 603–622.

33. Verma, S., Kumar, S., Gokhale, R., and Burgess, D. J. (2011). Physical stability of nanosuspensions: investigation of the role of stabilizers on Ostwald ripening, *Int. J. Pharm.*, **406**, 145–152.

34. Lindfors, L., Skantze, P., Skantze, U., Westergren, J., and Olsson, U. (2007). Amorphous drug nanosuspensions. 3. Particle dissolution and crystal growth, *Langmuir*, **23**, 9866–9874.

35. Quan, P., Shi, K., Piao, H., Piao, H. Liang, N., Xia, D., and Cui, F. (2012). A novel surface modified nitrendipine nanocrystals with enhancement of bioavailability and stability, *Int. J. Pharm.*, **430**, 366–371.

36. Chaubal, M. V., and Popescu, C. (2008). Conversion of nanosuspensions into dry powders by spray drying: a case study, *Pharm. Res.*, **25**, 2302–2308.

37. Cooper, E. R., Bullock, J. A., Chippari, J. R., Schaefer, J. L., Patel, R. A., Jain, R., Strasters, J., Ryde, N. P., Ruddy, S. B. (2010). In vitro methods for evaluating the in vivo effectiveness of microparticulate or nanoparticulate active agent compositions, United States patent No. **7**, pp. 695–739.

38. Heng, D., Ogawa, K., Cutler, D., Chan, H. K., Raper, J. A., Ye, L., and Yun, J. (2010). Pure drug nanoparticles in tablets: what are the dissolution limitations? *J. Nanopart. Res.*, **12**, 1743–1754.

39. Liu, P., De Wulf, O., Laru, J., Heikkilä, T., van Veen, B., Kiesvaara, J., Hirvonen, J., Peltonen, L., and Laaksonen, T. (2013). Dissolution studies of poorly soluble drug nanosuspensions in non-sink conditions, *AAPS PharmSciTech*, **14**, 748–756.

40. USP 37-NF 32, (2014). <1092> The dissolution procedure: development and Validation, United States Pharmacopeial Convention, Rockville, MD.

41. Shekunov, B. Y., Chattopadhyay, P., Seitzinger, J., and Huff, R. (2006). Nanoparticles of poorly water-soluble drugs prepared by supercritical fluid extraction of emulsions, *Pharm. Res.*, **23**, 196–204.

42. Brown, C. K., Chokshi, H. P., Nickerson, B., Reed, R. A., Rohrs, B. R., and Shah, P. A. (2004). Acceptable analytical practices for dissolution testing of poorly soluble compounds, *Pharm. Technol.*, **28**(12), 56–65.

43. Lehto, P., Kortejärvi, H., Liimatainen, A., Ojala, K., Kangas, H., Hirvonen, J., Tanninen, V. P., and Peltonon, L. (2011). Use of conventional surfactant media as surrogates for FaSSIF in simulating in vivo dissolution of BCS class II drugs, *Eur. J. Pharm. Biopharm.*, **78**, 531–538.

44. Talekar, M., Ganta, S., Amiji, M., Jamieson, S., Kendall, J., Denny, W. A., and Garg, S. (2013). Development of PIK-75 nanosuspension formulation with enhanced delivery efficiency and cytotoxicity for targeted anti-cancer therapy, *Int. J. Pharm.*, **450**, 278–289.

45. Quinn, K., Gullapalli, R. P., Merisco-Liversidge, E., Goldbach, E., Wong, A., Liversidge, G. G., Hoffman, W., Sauer, J. M., Bullock, J., and Tonn, G. (2012). A formulation strategy for gamma secretase inhibitor ELND006, a BCS class II compound: development of a nanosuspension formulation with improved oral bioavailability and reduced food effects in dogs, *J. Pharm Sci.*, **101**, 1462–1474.

46. USP 37-NF 32, (2014). <711> Dissolution, United States Pharmacopeial Convention, Rockville, MD.

47. Hsieh, Y. L., Ilevbare, G. A., Van Eerdenbrugh, B., Box, K. J., Sanchez-Felix, M. V., and Taylor, L. S. (2012). pH-induced precipitation behavior of weakly basic compounds: determination of extent and duration of supersaturation using potentiometric titration and correlation to solid state properties, *Pharm. Res.*, **29**, 2739–2753.

48. Ghosh, I., Bose, S., Vippagunta, R., and Harmon, F. (2011). Nanosuspension for improving the bioavailability of a poorly soluble drug and screening of stabilizing agents to inhibit crystal growth, *Int. J. Pharm.*, **409**, 260–268.

49. Brewster, M. E., Vandecruys, R., Peeters, J., Neeskens, P., Verreck, G., and Loftsson, T. (2008). Comparative interaction of 2-hydroxypropyl-β-cyclodextrin and sulfobutylether-β-cyclodextrin with itraconzaole: phase-solubility behavior and stabilization of supersaturated drug solutions, *Eur. J. Pharm. Sci.*, **34**, 94–103.

50. Dressman, J. B., Amidon, G. L., Reppas, C., and Shah, V. P. (1998). Dissolution testing as a prognostic tool for oral drug absorption: immediate release dosage forms, *Pharm. Res.*, **15**, 11–22.

51. Jantratid, E., Janssen, N., Reppas, C., and Dressman, J. B. (2008). Dissolution media simulating conditions in the proximal human gastrointestinal tract: an update, *Pharm. Res.*, **25**, 1663–1676.

52. Otsuka, K., Shono, Y., and Dressman, J. (2013). Coupling biorelevant dissolution methods with physiologically based pharmacokinetic modeling to forecast in-vivo performance of solid oral dosage forms, *J. Pharm. Pharmacol.*, **65**, 937–952.

53. Shono, Y., Jantratid, E., Kesisoglou, F., Reppas, C., Dressman, J. B. (2010). Forecasting in vivo oral absorption and food effect of micronized and nanosized aprepitant formulations in humans, *Eur. J. Pharm. Biopharm.*, **76**, 95–104.

54. Juenemann, D., Jantratid, E., Wagner, C., Reppas, C., Vertzoni, M., and Dressman, J. B. (2011). Biorelevant in vitro dissolution testing of products containing micronized or nanosized fenofibrate with a view to predicting plasma profiles, *Eur. J. Pharm. Biopharm.*, **77**, 257–264.

55. Helle, A., Hirsjärvi, S., Peltonen, L., Hirvonen, J., Wiedmer, S. K., and Hyötyläinen, T. (2010). Novel, dynamic on-line analytical separation system for dissolution of drugs from poly(lactic acid) nanoparticles, *J. Pharm. Biomed. Anal.*, **51**, 125–130.

56. Bose, S., Schenck, D., Ghosh, I., Hollywood, A., Maulit, E., and Ruegger, C. (2012). Application of spray granulation for conversion of a nanosuspension into a dry powder form, *Eur. J. Pharm. Sci.*, **47**, 35–43.

57. Salazar, J., Müller, R. H., and Möschwitzer, J. P. (2013). Application of the combinative particle size reduction technology H 42 to produce fast dissolving glibenclamide tablets, *Eur. J. Pharm. Sci.*, **49**, 565–577.

58. Sievens-Figueroa, L., Pandya, N., Bhakay, A., Keyvan, G., Michniak-Kohn, B., Bilgili, E., and Davé, R. N. (2012). Using USP I and USP IV for discriminating dissolution rates of nano- and microparticle-loaded pharmaceutical strip-films, *AAPS PharmSciTech,* **13**, 1473–1482.

59. Brown, C. K., Friedel, H. D., Barker, A. R., Buhse, L. F., Keitel, S., Cecil, T. L., Kraemer, J., Morris, J. M., Reppas, C., Stickelmeyer, M. P., Yomota, C., and Shah, V. P. (2011). FIP/AAPS joint workshop report: dissolution/in vitro release testing of novel/special dosage forms, *AAPS PharmSciTech,* **12**, 782–794.

60. Heng, D., Cutler, D. J., Chan, H. K., Yun, J., and Raper, J. A. (2008). What is a suitable dissolution method for drug nanoparticles, *Pharm. Res.*, **25**, 1696–1701.

61. Anhalt, K., Geissler, S., Harms, M., Weigandt, M., and Friker, G. (2012). Development of a new method to assess nanocrystal dissolution based on light scattering, *Pharm. Res.*, **29**, 2887–2901.

62. Jünemann, D., and Dressman, J. (2012). Analytical methods for dissolution testing of nanosized drugs, *J. Pharm. Pharmacol.*, **64**, 931–943.

63. Shen, J., and Burgess, D. J. (2013). In vitro dissolution testing strategies for nanoparticulate drug delivery systems: recent developments and challenges, *Drug Deliv. Transl. Res.*, **3**, 409–415.

64. Lai, F., Sinico, C., Ennas, G., Marongiu, F., Marongiu, G., and Fadda, A. M. (2009). Dicolfenac nanosuspensins: influence of preparation procedure and crystal form on drug dissolution behavior, *Int. J. Pharm.*, **373**, 124–132.

65. Lai, F., Pini, E., Angioni, G., Manca, M. L., Perricci, J., Sinico, C., and Fadda, A. M. (2011). Nanocrystals as tool to improve piroxicam dissolution rate in novel orally disintegrating tablets, *Eur. J. Pharm. Biopharm.*, **79**, 552–558.

66. Li, W., Yang, Y., Tian, Y., Xu, X., Chen, Y., Mu, L., Zhang, Y., and Fang, L. (2011). Preparation and in vitro/in vivo evaluation of revaprazan hydrochloride nanosuspension, *Int. J. Pharm.*, **408**, 157–162.

67. Xia, D., Cui, F., Piao, H., Cun, D., Piao, H., Jiang, Y., Ouyang, M., and Quan, P. (2010). Effect of crystal size on the in vitro dissolution and oral absorption of nitrendipine in rats, *Pharm. Res.*, **27**, 1965–1976.

68. Mohammadi, G., Nokhodchi, A., Barzegar-Jalali, M., Lotfipour, F., Adibkia, K., Ehyaei, N., and Valizadeh, H. (2011). Physicochemical and anti-bacterial performance characterization of clarithromycin nanoparticles as colloidal drug delivery system, *Colloids Surf. B*, **88**, 39–44.

69. Pardeike, J., Strohmeier, D. M., Schrödl, N., Voura, C., Gruber, M., Khinast, J. G., and Zimmer, A. (2011). Nanosuspensins as advanced printing ink for accurate dosing of poorly soluble drugs in personalized medicines, *Int. J. Pharm.*, **420**, 93–100.

70. Yin, S. X., Franchini, M., Chen, J., Hsieh, A., Jen, S., Lee, T., Hussain, M., and Smith, R. (2005). Bioavailability enhancement of a COX-2 inhibitor, BMS-347070, from a nanocrystalline dispersion prepared by spreay-drying, *J. Pharm. Sci.*, **94**, 1598–1607.

71. Sylvestre, J. P., Tang, M. C., Furtos, A., Leclair, G., Meunier, M., and Leroux, J. C. (2011). Nanonization of megestrol acetate by laser fragmentation in aqueous milieu, *J. Control. Release*, **149**, 273–280.

72. Khan, S., de Matas, M., Zhang, J., and Anwar, J. (2013). Nanocrystal preparation: low-energy precipitation method revisited, *Cryst. Growth Des. 13*, 2766–2777.

73. Whatman – GE Healthcare Life Sciences.

74. Liu, P., Rong, X., Laru, J., van Veen, B., Kiesvaara, J., Hirvonen, J., Laaksonen, T., and Peltonen, L. (2011). Nanosuspensions of poorly soluble drugs: preparation and development by wet milling, *Int. J. Pharm.*, **411**, 215–222.

75. Laaksonen, T., Liu, P., Rahikkala, A., Peltonen, L., Kauppinen, E. I., Hirvonen, J., Jävinen, K., and Raula, J. (2011). Intact nanoparticle indomethacin in fast-dissolving carrier particles by combined wet milling and aerosol flow reactor methods, *Pharm. Res.*, **28**, 2403–2411.

76. Cipolla, D., Wu, H., Eastman, S., Redelmeier, T., Gonda, I., and Chan, H. K. (2014). Develoment and characterization of an in vitro release assay for liposomal ciprofloxacin for inhalation, *J. Pharm. Sci.*, **103**, 314–327.

77. Muthu, M. S., Rawat, M. K., Mishra, A., and Singh, S. (2009). PLGA nanoparticle formulations of risperidone: preparation and neuropharmacological evaluation, *Nanomed-Nanotechnol.*, **5**, 323–333.

78. Shazly, G., Nawroth, T., and Langguth, P. (2008). Comparison of dialysis and dispersion methods for in vitro release determination of drugs from multilamellar liposomes, *Disolut. Technol.*, **15**, 7–10.

79. Chakraborty, S., Shukla, D., Vuddanda, P. R., Mishra, B., and Singh, S. (2010). Utilization of adsorption technique in the development of oral delivery system of lipid based nanoparticles, *Colloids Surf. B*, **81**, 563–569.

80. Leo, E., Cameroni, R., and Forni, F. (1999). Dynamic dialysis for the drug release evaluation from doxorubicin-gelatin nanoparticle conjugates, *Int. J. Pharm.*, **180**, 23–30.

81. Thombre, A. G., Caldwell, W. B., Friesen, D. T., McCray, S. B., and Sutton, S. C. (2012). Solid nanocrystalline dispersions of ziprasidone with enhanced bioavailability in the fasted state, *Mol. Pharmaceutics*, **9**, 3526–3534.

82. Modi, S., and Anderson, B. D. (2013). Determination of drug release kinetics from nanoparticles: overcoming pitfalls of the dynamic dialysis method, *Mol. Pharmaceutics*, **10**, 3076–3089.

83. Abdel-Mottaleb, M. M. A., and Lamprecht, A. (2011). Standardized in vitro drug release test for colloidal drug carriers using modified USP dissolution apparatus I, *Drug Dev. Ind. Pharm.*, **37**, 178–184.

84. Bhardwaj, U., and Burgess, D. J. (2010). A novel USP apparatus 4 based release testing method for dispersed systems, *Int. J. Pharm.*, **388**, 287–294.

85. Kumar, S., Xu, X., Gokhale, R., and Burgess, D. J. (2014). Formulation parameters of crystalline nanosuspensions on spray drying processing: a DOE approach, *Int. J. Pharm.*, **464**, 34–45.

86. Wiberg, K. H., and Hultin, U. K. (2006). Multivariate chemometric approach to fiber-optic dissolution testing, *Anal. Chem.*, **78**, 5076–5085.

87. Nir, I., Johnson, B. D., Johansson, J., and Schatz, C. (2001). Application of fiber-optic dissolution testing for actual products, *Pharm. Technol.*, 33–40.

88. Zhang, Y., Bredael, G., and Armenante, P. M. (2013). Dissolution of prednisone tablets in the presence of an arch-shaped fiber optic probe in a USP dissolution testing apparatus 2, *J. Pharm. Sci.*, **102**, 2718–2729.

89. Van Eerdenbrugh, B., Alonzo, D. E., and Taylor, L. S. (2011). Influence of particle size on the ultraviolet spectrum of particulate-containing solutions: implications for in-situ concentration monitoring using UV/Vis fiber-optic probes, *Pharm. Res.*, **28**, 1643–1652.

90. Tsinman, K., Tsinman, O., Riebesehl, B., Grandeury, A., and Juhnke, M. (2013). In situ method for monitoring free drug concentration released from nanoparticles, 2013 AAPS Annual Meeting, San Antonio, TX.

91. Peeters, K., De Maesschalck, R., Bohets, H., Vanhoutte, K., and Nagels, L. (2008). In situ dissolution testing using potentiometric sensors, *Eur. J. Pharm. Sci.*, **34**, 243–249.

92. Crisp, M. T., Tucker, C. J., Rogers, T. L., Williams, R. O., and Johnston, K. P. (2007). Turbidimetric measurement and prediction of dissolution rates of poorly soluble drug nanocrystals, *J. Control. Release*, **117**, 351–359.

93. Mora, L., Chumbimuni-Torres, K. Y., Clawson, C., Hernandez, L., Zhang, L., and Wang, J. (2009). Real-time electrochemical monitoring of drug release from therapeutic nanoparticles, *J. Control. Release*, **140**, 69–73.

94. Kayaert, P., Li, B., Jimidar, I., Rombaut, P., Ahssini, F., and Van den Mooter, G. (2010). Solution calorimetry as an alternative approach for dissolution testing of nanosuspensions, *Eur. J. Pharm. Biopharm.*, **76**, 507–513.

95. Tucker, C. J. (2004). Real time monitoring of small particle dissolution by way of light scattering, United States patent No. 6,750,966.

96. Engel, A., Plöger, M., Mulac, D., and Langer, K. (2014). Asymmetric flow field-flow fractionation (AF4) for the quantification of nanoparticle

release from tablets during dissolution testing, *Int. J. Pharm.*, **461**, 137–144.

97. Buch, P., Langguth, P., Kataoka, M., and Yamashita, S. (2009). IVIVC in oral absorption for fenofibrate immediate release tablets using a dissolution/permeation system, *J. Pharm. Sci.*, **98**, 2001–2009.

98. Niwa, T., Miura, S., and Danjo, K. (2011). Design of dry nanosuspension with highly spontaneous dispersible characteristics to develop solubilized formulation for poorly water-soluble drugs, *Pharm. Res.*, **28**, 2339–2349.

99. Bhakay, A., Azad, M., Bilgili, E., and Dave, R. (2014). Redispersible fast dissolving nanocomposite microparticles of poorly water-soluble drugs, *Int. J. Pharm.*, **461**, 367–379.

100. Bhakay, A., Dave, R., and Bilgili, E. (2103). Recovery of BCS class II drugs during aqueous redispersion of core-shell type nanocomposite particles produced via fluidized bed coating, *Powder Technol.*, **236**, 221–234.

101. Kesisoglou, F., and Mitra, A. (2012). Crystalline nanosuspensions as potential toxicology and clinical oral formulations for BCS II/IV compounds, *AAPS J.*, **14**, 677–687.

Chapter 11

Dissolution of Lipid-Based Drug Formulations

Stephen M. Cafiero

Boehringer Ingelheim Pharmaceuticals Inc., Analytical Development,
900 Old Ridgebury Road, Ridgefield, CT 06877, USA
stephen.cafiero@boehringer-ingelheim.com

11.1 Introduction

As more and more active pharmaceutical ingredients in drug development are being classified as BCS II/IV due to poor aqueous solubility,[1] there have been significant challenges to successfully formulate these sparingly soluble compounds. One approach to make these drug substances more bioavailable is to dissolve them in lipid based, or lipophilic, matrices. Because the drug substances are typically pre-dissolved in the formulation matrix, dissolution testing is not conducted with the same focus on drug substance solubility. As the drug substance solubility is not a strong test consideration, the dissolution test may be more aptly referred to as a dispersion test, meaning that the lipid-based formulation will be released in aqueous dissolution medium and form an emulsion. In fact, there are a few notable differences in dissolution testing of lipid-based

Poorly Soluble Drugs: Dissolution and Drug Release
Edited by Gregory K. Webster, J. Derek Jackson, and Robert G. Bell
Copyright © 2017 Pan Stanford Publishing Pte. Ltd.
ISBN 978-981-4745-45-1 (Hardcover), 978-981-4745-46-8 (eBook)
www.panstanford.com

formulations versus more conventional formulations that will be discussed in this chapter.

The focus of the chapter will be to describe the general nature of these types of formulations and how dissolution testing may be conducted. The availability of the drug for in vivo absorption can be enhanced by presentation of the drug as dissolved within a colloidal dispersion. Due to an industry trend of an increasing number of poorly soluble drugs formulated in lipophilic matrices, a collaboration of expert groups at USP has authored a general chapter with specific focus on dissolution testing of these formulations. Many of the aspects discussed in the general chapter USP <1094> Liquid-Filled Capsules—Dissolution Testing and Related Quality Attributes are discussed here. There are also more investigative discussions presented on topics such as a capsule rupture test, emulsion droplet size determinations and a case study to verify the residence of a drug substance in the emulsion phase by UV analysis.

11.2 Lipid-Based Formulations

Much has been published on the topic of lipid-based formulations and their benefits with regard to increasing bioavailability for poorly soluble compounds.[2–4] It is important to gain an understanding of the distinctions between lipid-based and conventional formulations prior to any discussion on the development of the dissolution test for lipid-based formulations. The content of this section will give an overview summary of lipid-based drug delivery systems (LBDDS) (a general classification), their in vivo functionality, and the behavior of lipid-based formulations when dispersed in aqueous dissolution media.

There are a number of formulation options for LBDDS. The various types include lipid suspensions, solutions, emulsions, microemulsions, mixed micelles, SEDDS (self-emulsifying drug delivery systems), SMEDDS (self micro-emulsifying drug delivery system), thixotropic vehicles, thermo-softening matrices, and liposomes.[5] Each approach has its advantages depending on the nature of the active ingredient to be formulated. Typically hard or soft gelatin shells are utilized for solid oral options. In addition to improving bioavailability, LBDDS have also shown the ability to reduce or eliminate the influence of food on the absorption of drugs.[6]

LBDDS for oral administration generally consist of a drug dissolved in a blend of two or more excipients, which may be triglycerides, partial glycerides, oils, surfactants or co-surfactants. Some examples of frequently used excipients in the formulation of LBDDS are phosphatydilcholine, α-tocopherol, glycerin, Cremophor (polyoxyl 35 castor oil), oleic acid, Capmul, propylene glycol, polysorbates, and various triglycerides (olive oil, sesame oil, hydrogenated vegetable oil, peanut oil).[7] Selection of the appropriate lipid excipient(s) not only affects drug solubility in the formulations, but also the solubility of drugs in the GI tract during lipid digestion, as well as the absorption and bioavailability of drugs.[8] However, only a limited number of studies have compared the performance of different LBDDS in vivo to in vitro.[9]

To date, there are a number of LBDDS that are approved and commercially available. A few examples are listed in Table 11.1 below.

Table 11.1 Examples of commercially available LBDDS[10,11]

Drug product	Dose	Lipid-based excipients	Indication
Kaletra® (lopinavir & ritonavir), Oral Solution	80 mL, 20 mL	Glycerin, peppermint oil, propylene glycol	HIV antiviral
Norvir® (ritonavir) capsules	100 mg	Oleic acid, Cremophor	HIV antiviral
Cyclosporine	25 mg, 100 mg	Cremephor RH 40	Immunosuppressant
Aptivus® (tipranavir), capsule	250 mg	Cremophor, Capmul MCM, propylene glycol	HIV antiviral
Rocaltrol® (Calcitriol) capsule	0.25 µg, 0.5 µg	Fractionated triglyceride of coconut oil	Calcium regulator
Accutane® (isotretinoin) capsule	10, 20, 40 mg	Beeswax, hydrogenated vegetable oil, soybean oil	Anti-acne

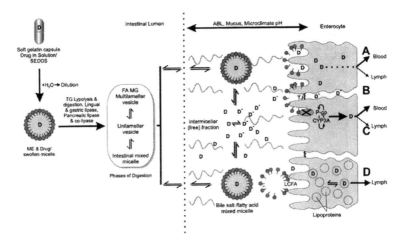

Figure 11.1 In vivo depiction of absorption of a lipid-based formulation. Reproduced with permission from O'Driscoll.[15]

The fate of drugs during transit through the GI tract may be affected by many factors. Fine emulsions are formed upon dispersing of the lipid-based solution in the GI fluids, which facilitates lipid hydrolysis by co-lipase-dependent pancreatic lipase on the oil/water interface.[8,12] The lipolytic products together with bile salts and phospholipids form micelles and other colloidal structures.[8,13] If the solubility capacity of the micelles is lower than that of the initial fine emulsion, drug compounds may precipitate in the GI tract.[8] Drug precipitation during digestion may play a significant role in bioavailability of LBDDS and studies on this topic have been studied published.[14] For water-insoluble drugs, the formulation goal is to maintain the drug substance in the lipid phase droplets after emulsification. This will increase the chances for luminal absorption. Figure 11.1 is a schematic diagram summarizing intestinal drug transport from lipid-based formulations.

The Lipid Formulation Classification System (LFCS) was described by Pouton[16] as a means to guide formulation development with regard to LBDDS. Table 11.2 summarizes the characteristics of each type of LBDDS, including component types and advantages/disadvantages.

Table 11.2 Lipid Formulation Classification System: characteristic features and advantages and disadvantages of the four essential types of lipid formulations

Formulation type	Materials	Characteristics	Advantages	Disadvantages
I	Oils without surfactants (e.g. tri-, di-and mono-glycerides)	Non-dispersing, requires digestion	GRAS status; simple; excellent capsule compatibility	Formulation has poor solvent capacity unless drug is highly lipophilic
II	Oils and water insoluble surfactants	SEDDS without water soluble components	Unlikely to lose solvent capacity on dispersion	Turbid o/w dispersion (particle size 0.25–2 µm)
IIIA	Oils, surfactants, co-solvents (both water insoluble & water soluble excipients)	SEDDS/SMEDDS with water soluble components	Clear or almost clear dispersion; drug absorption without digestion	Possible loss of solvent capacity on dispersion; less easily digested
IIIB	Oils, surfactants, co-solvents (both water insoluble & water soluble excipients)	SMEDDS with water soluble components and low oil content	Clear dispersion; drug absorption without digestion	Likely loss of solvent capacity on dispersion
IV	Water-soluble surfactants and co-solvents (no oils)	Formulation disperses typically to form a micellar solution	Formulation has good solvent capacity for many drugs	Likely loss of solvent capacity on dispersion; may not be digestible

Figure 11.2 Dissolution of a SEDDS solid oral dosage form in aqueous media.

On the basis of this work, an industrial/academic consortium was formed to promote continued research into this topic. Most notably, this consortium has published a summary[17] describing in vitro testing methods to predict the digestive power of biorelevant media and how lipid-based excipients would be digested in the gut. This titration-based technique, lipolysis, is not strictly related to dissolution but is a useful tool to determine whether or not the drug is maintained in a solubilized form to allow for intestinal absorption.

From an in vitro standpoint, lipid-based formulations are tested for product quality using standard USP apparatus described in General Chapter <711>.[18] Essentially, the lipid content will be dispersed throughout the dissolution media to, most likely, form some level of emulsion. An example of the resultant emulsified solution is shown in Fig. 11.2.

The rate of formation and stability of the emulsion will depend on the effectiveness of the formulation design. The unique characteristics of LBDDS, as compared to conventional forms, contribute to specific approaches needed to develop an appropriate dissolution platform. Section 11.4 discusses specific aspects of developing dissolution methods as the system is not entirely aqueous-based due to the introduction of a lipid-based fill to dissolution media. Prior to method development attempts, though, an understanding of manufacturing and packaging aspects of LBDDS is needed to determine the potential for problems with the dissolution method.

11.3 Manufacturing/Packaging of LBDDS and Potential for Dissolution Problems

Special consideration should be paid to the way that LBDDS are manufactured and packaged. In the case of solid oral LBDDS, the capsule fill may be enclosed by a gelatin ribbon that has the potential to cause some problems with dissolution testing.

The following are manufacturing and packaging issues related to LBDDS that may play some role in the rate of release during dissolution testing of a product. These have been summarized in a USP general chapter proposal that was published in the Pharmacopoeia Forum Volume 38 (1) in 2012.[19]

- Moisture content from the fill may migrate to the gelatin capsule shell. Moisture may also migrate in the reverse direction, from the shell to the fill material. This may cause the shell to become brittle.
- The residence of the active ingredient in the fill and potential transit to the shell should be investigated. The active ingredient could migrate from the fill solution to the shell (particularly gelatin) during exposure to routine stability conditions.
- Potential aldehyde formation in the drug product. Aldehydes are a well-documented cause of cross-linking in gelatin capsule shells.[10]
- Manufacturing process parameters related to the cooling and drying of the drug product could affect certain components of the formulation fill.
- If gelatin is used as a capsule shell, it is important to monitor exposure to heat and/or moisture. This exposure can contribute to the formation of pellicles on the inner portion (which is in contact with the capsule fill) of the

gelatin. This phenomenon is known as cross-linking and is discussed in more detail in Section 11.3.1.

- Packaging that does not appropriately control the amount of moisture ingress to the drug product. As previously mentioned, cross-linking in the capsule shell can occur as a result of exposure to moisture.
- Stability conditions chosen where the humidity is controlled will allow for effective monitoring of moisture effect on the capsule burst time via the dissolution test.

If moisture does appear to cause a slow-down of the dissolution profile, appropriate packaging and storage requirements should be implemented to better control the release rate. If a slower dissolution profile is not related to cross-linking, factors listed above could be the cause and worthy of investigation.

11.3.1 Cross-linking

According to the general chapter USP <1094>, cross-linking involves strong chemical linkages between gelatin chains and affects the thermal reversibility of the gelatin shell. It may be caused by agents in the capsule fill that react with gelatin molecules resulting in the formation of a pellicle on the internal surface of the shell. A pellicle is a thin, clear membrane of cross-linked protein surrounding the fill or the capsule that prevents the capsule fill from being released. Pellicles are difficult to confirm visually, but may manifest themselves during dissolution testing. One of the strongest and most common types of cross-linking involves the covalent bonding of the amine group of a lysine side chain of one gelatin molecule to a similar amine group on another. This is usually caused by trace amounts of reactive aldehydes. Formaldehyde, glutaraldehyde, glyoxyl, and reducing sugars are the most common cross-linking agents. These bonds are mostly irreversible. Dissolution testing of suspected cross-linked capsule shells may continue with the use of enzymes added to the media as described in USP General Chapters Dissolution <711> and USP Dissolution Procedure: Development and Validation <1092.[20] However, some regulatory agencies may require evidence that the cross-linking does not affect in vivo performance before the use of enzymes is accepted. The enzyme exposure to the gelatin shell will eventually force breakages of the peptide bonds within the gelatin protein chains, which will aide in the capsule burst. Another possible way to reduce this type of cross-linking is to chemically modify the gelatin with succinic acid groups to the lysine side chains. Additionally, it has been demonstrated that manufacture of gelatin containing some combination of glycine and citric acid will reduce cross-linking.[21] However, as previously discussed in Section 11.2.1 of this chapter, lipid-based formulations may contain ingredients that are ripe for the formation of these pellicles.

One class of capsule fill components which are known to cause cross-linking is aldehydes. It is important to understand and study aldehyde formation within the capsule fill/gelatin system and how this formation will impact subsequent pellicle formation. It is possible to spike known amounts of aldehydes into the formulation during development in order to better understand its effects on dissolution and the amount of cross-linking that may result in its presence.[22]

Besides the aforementioned reasons, cross-linking has also been known to be caused by some of the following:[10,19]

- Substances that promote decomposition of stabilizer in corn starch (hexamethylenetetramine), resulting in the formation of ammonia and formaldehyde, which cause the cross-linking reaction
- Rayon coilers that contain an aldehyde functional group
- Polyethylene glycol that may auto-oxidize to form aldehydes
- UV light, especially with high heat and humidity
- Aldehyde formation promoted by elevated temperatures

Dissolution testing demonstrates that cross-linked capsules can result in slower release of the capsule fill (and active ingredient(s)) or, in more severe cases, no release at all. It is important to understand the potential causes of cross-linking to reduce product development problems. However, the use of enzymes (where accepted) in the dissolution medium as described in the regulatory guidelines[18] can help to overcome this problem. The use of enzymes in method development is explained further in Section 11.4.4 later in this chapter.

11.4 Dissolution Development, Validation, and Testing Considerations: USP <1094> General Chapter

The subject of lipid-based formulations became the focus of a USP experts group that has authored a general chapter, USP <1094>, which summarizes the various aspects associated with manufacture, packaging, and testing of liquid-filled capsules. The following subsections will elaborate on specific issues within the USP chapter that are related to dissolution method development and validation of lipid-filled formulations. While the majority of the development recommendations are for solid oral LBDDS dosage forms, some of these principles may be applied to non-solid oral LBDDS.

11.4.1 Apparatus

Dissolution apparatus selection for lipid-based formulations may occur very similarly to most other solid oral dosage forms. USP 1 (baskets), USP 2 (paddles), USP 3 (reciprocating cylinder), and USP 4 (flow-through cell) are the apparatuses most often chosen during development for lipid-filled solid oral dosage forms. However, there are certain aspects specific to LBDDS formulations that should be considered when selecting the appropriate testing apparatus:

- Fill solution in a capsule that is less dense than water may require the use of sinkers (if paddles are used) or may require the use of baskets as the capsule may float.
- Wire coil sinkers, as specified in the USP General Chapters <711> and <1092>, are effective if paddles are used to not only keep the dosage form from floating, but they help keep capsule shells from sticking to the vessel bottom during the test. As such, they help to improve result variability from capsule to capsule.
- The lipid-based formulation may release, but not necessarily form a homogeneous mixture in the vessel. Care should be taken to ensure that the appropriate apparatus and rotation speed combination is employed.

- Certain capsule shell materials may not fully dissolve during a test. Regardless of the rotation speed chosen, this aspect of the test may be inconsequential if the materials do not interfere with the analysis.
- Gelatin material may rupture and clog the wires on a 40 mesh basket to the point where dispersion of the fill may be affected. In this case, use of a larger mesh size basket may be justifiable.
- Apparatus 3 and 4 contain wire mesh filters (similar to USP 1 baskets) that may become clogged as a gelatin capsule shell ruptures and dissolves. USP <2040>[23] specifies a USP 4 flow-through cell that is specific for lipid-filled softgels which may alleviate these problems.

In general, apparatus 1 or 2 may be used for these types of dosage forms. As specified in the USP and regulatory guidelines,[18,20] the agitation rate is typically set at 50–100 rpm for baskets and 50–75 rpm for paddles. Other rates should be appropriately justified. Nonetheless, apparatus 3 and 4 may be more appropriate for certain formulations and can be justified. In one case, USP apparatus 4 was shown to distinguish between different LBDDS formulations in a direct comparison with apparatus 2.[24] Regardless of the apparatus chosen, the hydrodynamic forces applied to the dosage form by each one will generally force the rupture of a capsule shell and dispersion of the fill.

11.4.2 Media

Media pH can be a critical method parameter selection for lipid-based formulations despite the fact that the drug substance is most likely already dissolved in the capsule fill. Although this is likely, it is still recommended to perform solubility studies of the API in media over the physiological pH range of 1.0–8.0.[25]

In addition to establishing a drug pH solubility profile during the dissolution development, the following specific issues should also be evaluated:

- Effect of the medium pH on the swelling or dissolution of the capsule based on the type of gelatin used in the product:

either Type A gelatin (where the medium pH is typically 7–9) or Type B gelatin (where the medium pH is typically 4.7–5.4).[26] Additional details to describe these are written in the USP *Gelatin* monograph.

- Effect of the medium pH on the potential adverse interactions of surfactants or enzymes/activity that are used.
- Selection of a medium pH that will be suitable for the solubility (e.g., if drug substance precipitates from an emulsion) and stability of the active ingredient and emulsion (if appropriate, see Section 4.3.3.4 of this chapter).

As part of development of any effective dissolution procedure, the medium should preferably (but not necessarily) provide sink conditions for the active ingredient(s) in an environment that ensures suitable stability and is ideally physiologically relevant for the product. Lipid-filled formulations however can contain either a matrix or drug substance (or both) that are hydrophobic and/or water-insoluble. Therefore, surfactants or enzymes may be needed as media additives. The use of organic co-solvents to achieve sink conditions is discouraged and should be employed only as a last resort with appropriate justification. The presence of organic solvents in the medium may inhibit the dissolution of a capsule shell.[19] In addition to dispersing/dissolving the matrix and active ingredient(s), the medium must neither interfere with the activity of enzymes that are used nor negatively interact with the capsule shell or formulation. Developing a suitable medium for these dosage forms may require considerable experimentation.

11.4.3 Surfactants

For some dissolution methods, surfactants may be needed as a media additive in order to achieve sink conditions for sparingly soluble drug substances. However, LBDDS are systems where the active ingredient(s) is dissolved in a lipid-based fill during formulation and manufacture. In these cases, it may be safe to assume that surfactants are not needed to improve drug solubility in media. While this may be true, surfactants, dispersing agents or solubility enhancers can still play a useful role when the formulation matrix is hydrophobic. USP <1094> states that surfactants may be used to ensure sink conditions if it is possible that the emulsion created after the LBDDS is mixed in aqueous medium is not sufficiently stable for analytical testing. Finally, surfactants may be useful as agents for systems where the lipid based formulation does not readily disperse with the recommended agitation rates for USP apparatuses.

The following are taken from USP <1094> and are examples of commonly used surfactants, dispersing agents and solubility enhancers:

- Anionic
  - Sodium dodecyl (or lauryl) sulfate (SDS/SLS)
  - Bile Salts: sodium deoxycholate, sodium cholate
- Cationic
  - Cetyltrimethylammonium bromide (CTAB)
  - Hexadecyltrimethylammonium bromide (HTAB)
  - Methylbenzethonium chloride (Hyamine)
- Non-ionic
  - Polysorbates (Tween)
  - Polyoxyethylene sorbitan esters
  - Octoxynol (Triton X100^{TM})
  - *N*, *N*-Dimethyldodecylamine-*N*-oxide
  - Brij 721^{TM}
  - Polyoxyl castor oil (Cremophor ^{TM})
  - Nonylphenol ethoxylate (Tergitol ^{TM})
  - Cyclodextrins
  - Polyoxyl 10 lauryl ether
- Zwitterionic
  - Lauryl dimethyl amine oxide (dodecyldimethylamine oxide, DDAO)
  - Lecithin

If a surfactant is needed for a LBDDS dissolution method, it may be prudent to choose one based on the ionic characteristic of the reagent. It has been shown that SDS, which is an anionic surfactant, has several hindering effects on the dissolution of some LBDDS.[27] Namely, it may interact with gelatin and artificially extend a capsule's burst time. It may also reduce the activity of enzymes added to the media if cross-linking is observed. Finally, SDS cannot be dissolved into potassium phosphate buffer, or mixed with any potassium-containing reagents, as insoluble precipitates are formed. On the other hand, cationic surfactants have the capability of forming precipitates with fatty acid-containing formulations.[19] And fatty acids may be derived from triglycerides, which may be an excipient component of LBDDS. In these cases, cationic surfactants, like CTAB, should be avoided. A more practical solution may be to add a non-ionic surfactant to the dissolution media. Polysorbates (Tween 20 or Tween 80, for example) could have sufficient solubilizing or dispersing characteristics and have been shown to not significantly affect the activity of enzymes needed to overcome cross-linking in gelatin capsule material as has been observed with SDS.[19,27] Alternatively, a pretreatment period with enzyme-containing media, followed by a later addition of the surfactant to the medium may be useful. It is recommended to utilize a pretreatment period of not more than 15 minutes to allow for the enzymatic activity to occur.[19]

As with any dissolution, the usage and choice of surfactant needs to be appropriately justified. Although sparingly soluble active ingredients are pre-dissolved in lipid-based matrices, there may be reasons to add surfactants to the media, for instance to help stabilize the emulsion in order to perform offline UV or HPLC sample analysis.

11.4.4 Enzymes

During method development of LBDDS, the addition of proteolytic enzymes such as pepsin or pancreatin may be needed when the gelatin capsule shell does not dissolve in aqueous media due to cross-linking, as discussed in Section 11.3.1 of this chapter. The use of enzymes for this purpose is documented in USP General Chapter <711> and is referred to as tier testing in FDA's Guidance for Dissolution of Immediate Release Solid Oral Dosage Forms. Tier 1 media is developed without enzymes, while tier 2 media is defined as tier 1 media with the addition of enzymes. Generally the dissolution test is first conducted in tier 1 media. If the acceptance criteria is not met due to evidence of cross-linking, USP Stage testing may be conducted in tier 2 media starting with $n = 6$ replicates. However, it is important to note that the text concerning enzyme use in USP <711> is not harmonized between the European Pharmacopoeia (Ph. Eur.) and the Japanese Pharmacopoeia (JP). Therefore, enzyme use may not be accepted by regulatory agencies outside of the US without in vivo data to demonstrate that the presence of cross-linking does not impact bioavailability of the drug. While the Ph. Eur. currently states that "[i]n specific cases, and subject to approval by the competent authority, dissolution media may contain enzymes," accompanying pharmacokinetic data can present strong evidence to verify enzymatic relevance to the dissolution test. For Japan, justification for enzyme use in the medium may prove even more difficult as the Japanese Pharmacopoeia does currently not make mention of enzymes in relation to dissolution testing.

Pellicles may be formed on the inner or outer portion of a gelatin capsule shell for a variety of reasons and the resulting cross-linking can be significant enough to prevent the capsule fill from being released during a dissolution test. As such, there may be limited or no release of the active ingredient and products may fail their stated acceptance criteria. Cross-linking can also be visually confirmed by the analyst. One may see the outer coating of the gelatin to be dissolved; however, the fill remains intact due to pellicles formed on the inner layer of the shell, nearest the capsule fill. A secondary use of enzymes in the media may be to break weaker ionic cross-links within the gelatin material.

For LBDDS with a gelatin capsule shell, dissolution method development should include testing to show that the enzymes used do not adversely interact with the lipid fill of the product. Also, as mentioned in Section 11.4.3, surfactants in the media may have some effect on enzyme activity. The combination of surfactants and enzymes in the media should be thoroughly investigated during method development.

One of the studies conducted during method development should be to force formation of cross-linking in products with gelatin capsule shells. This study is useful in determining the amount of enzyme that will be added to the media for the tier 2 dissolution

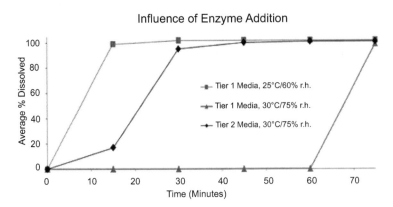

Figure 11.3 Influence of the addition of enzymes to dissolution media for a SEDDS capsule.

test. Cross-linking can be achieved in most cases by simply exposing the capsules to high humidity for an extended period of time.

The study below demonstrates this simplified approach. A dissolution method was developed in pH 6.8 phosphate buffer media for a SEDDS product encapsulated in soft gelatin material. The capsules were tested using USP 2 (paddles) rotating at 75 rpm. During stability studies, it was determined that enzyme would be needed as reduced dissolution was observed after the capsules were exposed to higher humidity. For pH 6.8 media and above, Pancreatin enzyme reagent was chosen and the amount added to the tier 1 (phosphate) medium was not more than 1750 USP protease activity units, per the directive in USP <711>. In order to calculate the amount of enzyme needed, the activity test as indicated in the USP monograph for the enzyme may be performed. However, it may be justified to obtain a certificate of analysis from the enzyme reagent vendor which should indicate its protease enzymatic activity.

Figure 11.3 above shows dissolution profiles of a SEDDS product tested in the media describe above after exposure to the following storage conditions:

- 25°C/60% relative humidity (proposed storage conditions)
- 30°C/75% relative humidity (intermediate accelerated storage conditions)

The capsules exposed to the proposed storage conditions show acceptable release, as demonstrated by the profile utilizing the square markers. These capsules were tested in dissolution media without enzymes, or tier 1 media. However, when the same batch of capsules was exposed to the intermediate accelerated conditions, testing in tier 1 media showed that the capsules did not burst until the rotation speed was increased to 250 rpm at the 60 minutes time point. These results are represented by the profile with triangular markers. When this behavior is observed, tier 1 media may be modified according to USP <711> to include either pepsin (in low pH media) or pancreatin (at media with pH 6.8 or above) to facilitate the capsule burst. The enzyme reagent is added to tier 1 media with appropriate mixing. The same batch of capsules that were exposed to intermediate accelerated conditions is retested with tier 2 (or enzyme containing) media. The results are represented by the profile with diamond markers. In this case, the enzymatic action in combination with the paddle rotation speed of 75 rpm does need time to promote capsule burst, but the dissolution is nearly complete by the 30 minutes time point.

The use of enzymes has been conditionally accepted as a way to appropriately deal with the challenge of cross-linking in gelatin material, which may be observed in capsule formulations for LBDDS. This practice has been documented in USP <711> and FDA's guidance on dissolution of immediate release solid oral dosage forms. However, despite harmonization of many dissolution practices in USP <711>, the text on the use of enzymes in dissolution media has not been agreed upon for inclusion in the Ph. Eur. and the JP.

11.4.5 Sampling

Removal of LBDDS sample aliquots from the dissolution vessel for analysis is similar to that for a typical dissolution test. Guidance for these are described in USP General Chapters <711> and <1092>. However, there are a few points to consider that are more specific when testing lipid-based formulations:

- The capsule fill material may form a thin layer on the surface of the media in the vessel after burst of the capsule shell. Depending on the thickness of the layer, it is important that the sampling cannula does not become clogged. This may impact consistent sampling volume for later time pulls.
- By the same token, the sample filter may become clogged, either from an oily layer that has not mixed appropriately mixed with the media, or possibly from undissolved gelatin. Care should be taken to ensure, as with any test, that the filter does not become clogged to the point where undissolved material is being passed through.
- For automated samplers, it is important that effective filtration is maintained so that transfer lines do not become clogged.
- Finally, if resident sampling probes are in use, the hydrodynamic effect of these probes on the release of the dosage form in this system should be evaluated as part of a robustness study.

One recommendation to circumvent issues with regard to filtration is to employ surfactants and/or enzymes as described in Sections 11.4.3 and 11.4.4, respectively. This would help to control the amount of undissolved materials that would potentially clog the filters or sampling lines in use.

11.4.6 Analysis

Typical methods of detection may be used for lipid-based formulations. HPLC by UV detection is certainly viable for compounds with UV chromophores, as are other spectrophotometric and combination (LC-MS) techniques. These techniques are listed in USP <1092> as being the most common forms of sample quantitation.

A point to consider when analyzing LBDDS dissolution samples is that emulsion formation will most likely create a cloudy suspension in the vessel after a dosage form is introduced to aqueous media. This could pose problems with accurate analysis of the sample solution if UV/vis spectrophotometric detectors are used. Light-scattering is possible if the sample solution is turbid enough to the point where the spectral baseline observed begins to drift or slope. Baseline correction techniques may often solve this problem. Also, some in situ UV systems offer second derivative algorithms for more complicated corrections.

Sampling points to obtain a dissolution profile should be established during the development phases to identify when the rate of active ingredient dissolved has reached a plateau. During development, it is recommended to acquire data at time points most likely to be used for the specification. Regulatory agencies may focus their reviews on data at earlier time points such as 10, 15 or 20 minutes for an immediate release LBDDS product as capsule burst and subsequent mixing of the fill with media are the two most likely rate-limiting parameters for this type of dissolution. Depending on the type and/or thickness of the capsule shell (especially for gelatin) and the method rotation speed, regulatory agencies could expect capsule burst to occur relatively early in the test. Therefore it is important to collect early time point dissolution profile data for representative batches (pivotal clinical and/or registration batches) prior to regulatory submission.

11.4.7 Validation

Validation of dissolution methods for LBDDS requires a similar approach to that taken for most solid oral formulations. The specific parameters to evaluate are described in detail in USP <1092> and ICH Q2 (R1)—Validation of Analytical Procedures: Text and Methodology.[28] However, there are some specific considerations to focus on for lipid-based dosage forms. These are summarized in USP <1094> and listed below.

- Drug substance solubility may need to be evaluated in dissolution media where the ionic strength and/or pH have been intentionally changed. Additionally, solubility can be checked in media containing enzymes (tier 2). This would be considered method robustness validation and may be part of a design of experiments matrix of tests.

- The relative partitioning and the potential effect of release of hydrophobic active ingredient(s) between the lipid formulation and the aqueous medium may be investigated.
- For chromatographic procedures, any potential adverse effects on the system caused by the addition of surfactant should be investigated. This may be an effect on peak shape, peak retention time and/or peak separation. Variations in the surfactant concentration, type, or interactions with buffer salts should be checked.
- HPLC column life may need to be checked as the addition of surfactant or enzymes can shorten the column usage time.
- As previously mentioned, the sampling procedure should be investigated to be sure that automation sample lines and/or filters do not become clogged.
- According to USP <1092>, the stability of the sample solution should be evaluated (if appropriate) with a minimum of at least 24 hours of stability. It is possible that the emulsion formed from the LBDDS mixture in the medium may not be stable and phase separation may occur. One approach to overcome this is to add an organic diluent after the sample has been pulled from the vessel. Another approach may be to stabilize the emulsion with a surfactant that will not introduce mixed micelles. While additional sample preparation (for HPLC or offline UV analysis) is not ideal and will add costs and reduce sample throughput, generally, these types of solutions are more stable than emulsions and can alleviate problems later in development.

Another aspect of validation may be to investigate the window of use for dissolution media where enzymes have been added (tier 2). Once mixed, the enzymatic activity may be reduced over the course of time. Tier 2 media stability may be evaluated via a side-by-side comparison of freshly prepared vs. aged media on a well characterized cross-linked batch of capsules. This may also be included as part of a robustness evaluation.

11.5 Discriminatory Capability Assessments

It is critical to assess the ability of the developed dissolution method to discriminate between results for various intentional changes to the formulation and manufacturing process in order to validate its usage as a quality control tool. Typically, intentional changes to the formulation may include drug substance particle size or morphology adjustments, differing drug substance batches, removal/addition of a functional excipient or changes to excipient levels/ratios. Examples of intentional changes to the manufacturing process include variations to solution mixing times, temperatures, forces, and equipment. The dissolution method developed for quality control use should also detect physical changes on stability over time. Ideally, the dissolution method should detect some (or all) of these differences via a dissolution profile to the point where a proper design space window for the formulation and process can be drawn. Also, the method should detect these physical and chemical changes as they affect the biopharmaceutical product performance.

All of the above assertions are applicable for lipid-based formulations. However, the following are more specific examples of focus to determine the discriminatory power of a LBDDS dissolution method:

- Capsule shell type or composition
- Changes to levels of excipient composition
- Changes to the rate of emulsion
- Changes to the droplet size of the emulsion
- Changes on stability: projected storage conditions and/or gelatin cross-linking

11.5.1 Changes to the Formulation

In general, for an immediate release solid oral dosage form the discriminatory power of the dissolution method depends on its ability to detect differences in the drug product that may potentially be clinically relevant. This aspect is slightly more challenging for LBDDS. Typically, drug substance-related physical properties do not play a role in the in vitro drug product performance as the drug

Table 11.3 Comparison of lipid-based capsule fill dissolution with various levels of tris

Time (min)	Samples (average % dissolved)			
	0% Tris	0.2% Tris	0.3% Tris	0.5% Tris
15	11	50	89	91
30	13	52	89	91
45	15	49	89	92
60	17	49	90	93

substance is dissolved within the lipid-fill capsule formulation. The lipid-based material can form an emulsion at a rate which depends on mixing of the fill solution in the aqueous medium or this mixture followed by the disintegration of a capsule shell.

While the dissolution method may not discriminate changes in drug substance related characteristics, changes in excipients or amounts of these excipients in a LBDDS formulation are capable of being detected. The following shows how the discriminatory capability of a dissolution method for a SEDDS formulation was demonstrated via changes in the excipient composition of the capsule fill.

Several bulk-fill SEDDS formulations were prepared with varying levels of the excipient trometamol (tris, or tris(hydroxymethyl) aminomethane $(HOCH_2)_3CNH_2)$. For this product, tris was added as a solubility agent for the sparingly soluble active ingredient, as well as a means to facilitate the dispersion of the capsule fill in an aqueous medium. A study was designed to determine the effect of the tris level on the dissolution profile of the capsule fill. The purpose of this experiment was to determine the level that was necessary to achieve a result of at least 70% of active ingredient dissolved in vitro after 60 minutes, which was the proposed specification limit. To do this, experimental solutions were prepared according to the fill composition at release testing, varying only the tris content. The tris content of the experimental solutions was 0%, 0.2%, 0.3%, and 0.5% of the total volume per capsule.

The dissolution results are summarized in Table 11.3 and presented graphically in Fig. 11.4. The dissolution conditions

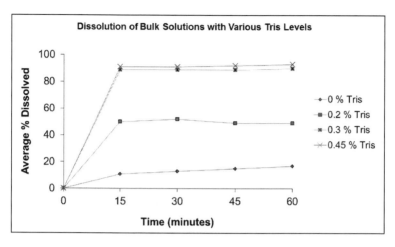

Figure 11.4 Comparison of lipid-based capsule fill dissolution with various levels of tris.

utilized were USP apparatus 2 (paddles) at 50 rpm in pH 6.8 phosphate buffer.

The dissolution profile becomes substantially lower only when the tris level dips below the 0.3% level in the bulk capsule fill. The results indicate that the in vitro dissolution profile of the drug product should remain unchanged as long as the level of tris in the lipid-based fill is no less than 0.3%.

For the same SEDDS product, experiments were performed to gauge the discriminating ability of the same dissolution method with regard to changes in a second excipient of the capsule fill: ethanol. In this case, ethanol is employed to aide in the solubility of the active ingredient as well as to improve dispersion characteristics. Experimental solutions were prepared by varying ethanol content from 2% through 7% (w/w) of the capsule fill weight. The same method conditions used for the tris study were applied.

For this change, the dissolution profiles are not significantly affected when the ethanol level is varied over the specified range. The results are summarized in Table 11.4 and presented graphically in Fig. 11.5.

The results from these two tests helped to determine the appropriate design space for the lipid-based fill. From the in vitro

Table 11.4 Changes to a SEDDS formulation: ethanol levels

Time (min)	Samples (average % dissolved)			
	7% Ethanol	**5% Ethanol**	**4% Ethanol**	**2% Ethanol**
15	85	91	77	72
30	85	92	85	78
45	89	95	88	84
60	87	94	89	87

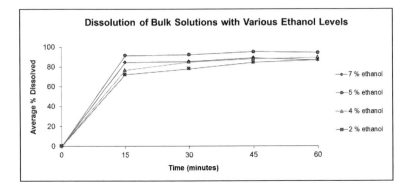

Figure 11.5 Changes to a SEDDS formulation: ethanol levels.

results calculated, it was observed that while the ethanol level could theoretically vary from 2% to 7%, the level of tris could not drop below 0.3%. These varying formulations were not tested in vivo to gauge the clinical relevance of the dissolution method; however, it was clear that the dissolution method was capable of showing differences in certain excipient levels for this lipid-based formulation.

11.5.2 Indication of Product Stability via the Dissolution Method

The dissolution method that is to be used in the quality control of a LBDDS product should be evaluated to demonstrate that it is stability-indicating. In many cases, however, the capability of the methodology may be limited to a check for capsule cross-linking. As shown in Fig. 11.3 in Section 11.4.4, capsule burst can be delayed significantly as the product is exposed to high humidity. In this case, pancreatin is added to overcome this problem that allows for capsule burst and dispersion of the fill. The chosen dissolution conditions should differentiate between capsules that have some degree of cross-linking and capsules that do not, especially if there is evidence of in vivo differences.

There may be cases where excipient levels could change in the fill as LBDDS are exposed to stability conditions. Changes in excipient levels can be evaluated and monitored as previously described in Section 11.5.1 of this chapter. In general, it may be difficult to show differences within the fill content on stability if delay of (or lack of) capsule burst slows the dissolution profile considerably.

11.6 Setting Specifications

Developing a dissolution procedure that is capable of monitoring meaningful changes in the release profile is the first step in selecting an appropriate specification for the product. These changes in the product may be due to stability storage conditions, formulation changes, or manufacturing/process alterations. Example changes to the formulation of LBDDS have previously been discussed in Section 11.5.1.

During method development, it is important to understand which product aspects are considered to be critical quality attributes (CQA) of the formulation. That understanding can be applied to determine if changes to the CQAs (intentional or not) will cause a change in the release rate using an appropriate dissolution platform. Perhaps more importantly, if a change in the product will lead to some change in in vivo performance, selection of a clinically relevant specification may be possible. Quality by design (QbD) principles can be applied to design a matrix of experiments to focus on certain product changes.

In particular, it may be advantageous to study the release mechanism of the dosage form and how it relates to the in vivo absorption characteristics of the drug. Historically, in vivo/in vitro relationships (IVIVR) for immediate release products have been difficult to achieve. However, there has been a growing expectation from regulatory authorities that sponsors should investigate these possibilities during development.

Guidelines for setting appropriate dissolution specifications for all release types have been well documented. FDA's Guidance for Dissolution Testing of Immediate Release Solid Oral Dosage Forms briefly mentions a distinction between setting specifications for highly soluble (BCS 1 and 3) and poorly soluble (BCS 2) drug products. While an LBDDS formulation strategy is employed due to a drug's poor aqueous solubility, it may be best to abide by the recommendations for highly soluble drug product as the drug substance is typically dissolved in the lipid-based fill.

In some cases, choosing acceptance criteria for encapsulated LBDDS (specifically in gelatin) may rely strictly on when the capsule shell bursts. Given the inherent variability in gelatin burst time due to cross-linking, especially in media with pH 6.8,[10,29] this could pose problems in trying to set an earlier time point. Many times, this variability can be attributed to the lot of gelatin used for encapsulation, but it may also be seen on a capsule-to-capsule basis within the same batch of gelatin. The example below in Fig. 11.6 illustrates this variability. Three clinical release batches of a LBDDS formulation were tested and the profiles obtained are graphed with

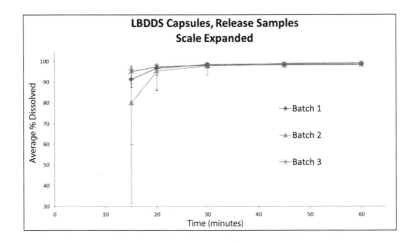

Figure 11.6 Example of burst variability for LBDDS capsules.

the error bars indicating the range of results for 12 replicates. The variability of capsule burst for batch 2 is much higher than batches 1 or 3 at the 15 minutes time point. In this case, the variability was attributed to slight differences in the gelatin batches procured for use in the encapsulation of each batch. While the use of enzymes in the dissolution medium to combat cross-linking in capsules that have been exposed to high humidity is a generally accepted practice, the use of enzymes in medium for capsules at release will likely be questioned. At the very least, one might expect capsule burst problems with tier 1 media throughout the usage life of this batch.

As such, dosage form burst and fill dispersion in the vessel are generally the rate-limiting factors to dissolution and the key parameters in setting appropriate acceptance criteria.

11.7 Dissolution-Related Testing

As discussed in the previous sections of this chapter, there is much to consider when developing a dissolution test for lipid-based formulations. However, there are other testing techniques that are related to (or including) dissolution which may provide valuable information to the support of LBDDS formulation development. The following sections will mention a few of these tests which may be helpful for a variety of reasons. For instance, it may be possible to use a capsule burst, or rupture test, as a surrogate to dissolution testing. Also evaluation of the sizes of the lipid-based droplets formed when fill is introduced to aqueous media in the dissolution vessels may be important from an absorption perspective. Finally, a test is described where UV spectroscopy can assist in the determination of whether or not a compound remains in an emulsion following capsule burst of a SEDDS formulation. These aspects are discussed below.

11.7.1 Rupture Test

As previously mentioned, the dissolution of lipid-based formulations in soft-gelatin capsules may be considered more of a dispersion test than a true dissolution. One consideration for a surrogate to the dissolution test is contained in the USP Test <2040> Disintegration and Dissolution of Dietary Supplements. In this test, the Rupture Test for Soft Shell Capsules describes conditions to apply as a quality control test for disintegration of a soft shell capsule. As is the case with any surrogate testing, a strong justification with testing correlations to dissolution should be provided for support.

The testing conditions described in USP <2040> for soft shell capsules require the use of a dissolution bath as specified in USP Chapter <711>. Toward the end of 2014, a proposal to replace pepsin with enzyme reagents exhibiting increased activity in water was being addressed as part of USP's In Process Revision procedure. The proposed conditions are as follows:[23]

- Medium: water, 500 mL
- Apparatus: 2 at 50 rpm
- Time: 15 minutes
- Procedure: Place 1 capsule in each vessel, and allow the capsule to sink to the bottom of the vessel before starting rotation of the blade. Use sinkers if the capsules float. Observe the capsules, and record the time taken for each capsule shell to rupture.
- Tolerances: The requirements are met if all of the capsules tested rupture in not more than 15 minutes. If 1 or 2 of the capsules rupture in more than 15 but not more than 30 minutes, repeat the test on 12 additional capsules; not more than 2 of the total of 18 capsules tested rupture in more than 15 but not more than 30 minutes. For soft-gelatin capsules that do not conform to the above rupture test acceptance criteria, repeat the test with the addition of papain to the Medium in the amount that results in an activity of not more than 555,000 units/L of medium or with the addition of bromelain in the amount that results in an activity of not more than 30 gelatin dissolving units (GDU)/L of medium.

The testing conditions described in USP <2040> may be worth investigating specifically for lipid-based formulations. However, there can be several problems associated with attempting to replace

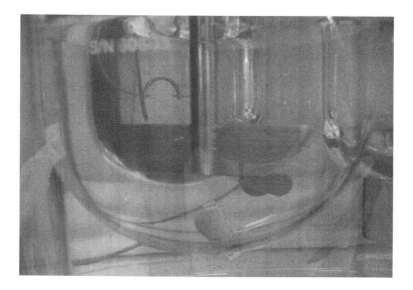

Figure 11.7   Capsule burst of a soft gelatin SEDDS formulation.

the dissolution method of a lipid-based formulation with the rupture test. From a practical, laboratory standpoint, it may be difficult to visually observe the exact point of capsule rupture. Upon very close examination (depending on the content fill of the capsule), one may begin to see a "swirling" action or "strands" of fill being released from the capsule prior to a definitive rupture point of the soft shell. The media may also begin to turn cloudy as the lipid-containing contents of the dosage form begin to disperse and emulsify. However, this may not be the case for all formulations and is highly subjective. Figure 11.7 above shows a SEDDS capsule that has begun to release active ingredient but is not definitively ruptured.

Of course it is possible to measure a solution (removed from the vessel or in situ) at the 15 minutes time point using ultraviolet spectroscopy, but this essentially defeats the purpose of replacing the dissolution test with an easier methodology as some quantitation will be required. In addition, the test language is strict on medium and agitation conditions. As such, there may be cases where a capsule shell exposed to the 50 rpm rotation speed may not rupture within 15 minutes despite the addition of papain or bromelain to break the pellicles that contribute to cross-linking.

In many cases, the dissolution test developed for lipid-based formulations may be best described as a dispersion test, with capsule rupture being the most critical of all dosage form parameters to monitor. A strong justification replete with testing comparisons and correlations will be needed if the rupture test is to replace a dissolution method in a quality control environment.

11.7.2 Droplet Size Testing

Investigation into the size of the droplets formed upon emulsification of a LBDDS is valuable to the understanding of how the formulation behaves once the fill is introduced to aqueous dissolution media. The droplet sizes produced in vitro may be indicative of the size of the droplets formed in vivo which could theoretically affect absorption. Tarr and Yalkowsky[30] were able to demonstrate that in a gut perfusion experiment, the emulsion droplet size affected the rate of absorption of cyclosporin A. Essentially absorption was more rapid from the finer of two emulsions. However, the size of the droplets upon capsule burst may not be the most important factor with regard to absorption. As the droplets move through the gut, they become susceptible to digestion and/or solubility via mixed micelles of bile salts and phospholipids. It has been reported that type IIIB formulations (see Table 11.1) generally produce the finest dispersions, due to their high content of water-soluble solubilizing agents.[4]

Experimentally, the study of LBDDS droplet sizes can be evaluated via light scattering technique or electron microscopy. These are powerful techniques that can be combined with dissolution to potentially assist in the formulation effort. The following example describes how this combination was put to use for development of a SEDDS formulation.

A study was completed to perform dissolution of a LBDDS capsule to confirm its excipient ratio functionality. The quantitative composition of the key formulation excipients (lipid and surfactant) that drive the performance characteristics of an emulsion was carefully selected based on previous in vitro experiments.

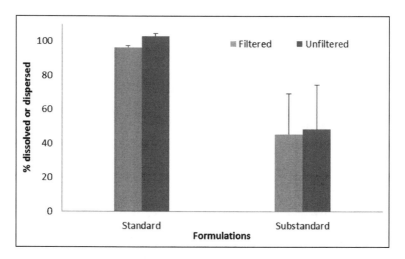

Figure 11.8 Comparison of lipid-based fill dissolution results: standard and substandard excipient ratios.

This formulation effort primarily involved monitoring turbidity of placebo formulations with different lipid/surfactant ratios. A standard capsule formulation (including excipients such as a co-solvent and antioxidant) and a capsule fill formulated with an altered (or substandard) ratio of lipid/surfactant were prepared and comparative dissolution data in 250 mL of pH 6.8 phosphate media were generated. Average results ($n = 3$) showed that roughly one half of the drug was dissolved for the substandard fill as compared to the standard formulation. These data could be confirmed visually as it was observed that the substandard fill seemed to settle or precipitate to the bottom of the dissolution vessel, whereas the standard fill did not. Finally, these solutions were filtered to compare the difference between filtered and unfiltered percent dissolved results. The results are presented graphically in Fig. 11.8.

For each formulation, it can be surmised that the amount of drug dissolved (the filtered solutions) is roughly equivalent to that of the amount that was dispersed (unfiltered). The premise here is that it is possible that some of the active ingredient may not remain in the emulsion and is subject to filtration if it is not fully dissolved.

This provides some assurance that the drug is presented to the physiological medium entirely in the form of emulsion droplets.

Actual lipid droplet sizes were then measured by laser diffraction of LBDDS capsule sample solutions prepared in both pH 2 and pH 6.8 media. In short, there were differences observed between droplet sizes of the standard and substandard formulations. Droplet sizes (X_{90}) measured in the substandard fill were approximately eight times the size of that in standard fill formulation (6 µm vs. 48 µm). Some qualitative differences in droplet size upon dissolution were apparent between these two formulations during microscopic examination of the sample in pH 2 dissolution media. Larger droplet sizes were evident when the ratio of lipid/surfactant is altered. It is acknowledged that although such measurements provide guidance about emulsion performance as a function of quantitative formulation composition, they do not necessarily represent the size of droplets obtained during dissolution. This is because the low dispersion phase volume fractions needed for laser diffraction in order to avoid interference from droplet-droplet interactions required a dilution of the samples. It is possible that this dilution could change the emulsion characteristics.

In addition, it is possible to experimentally verify that the rapid release and emulsification from the LBDDS capsule formulation is maintained throughout the shelf life of the product. In particular, since the LBDDS excipients may not be controlled via a specification, dissolution can be utilized as a means of assuring emulsion performance. For this case, the dispersed LBDDS droplet size distribution has been characterized for a drug product batch after 2 years of storage in the proposed commercial packaging versus a control. The LBDDS dispersions were prepared in 250 mL of pH 2 medium with paddles stirring at 75 RPM for 60 minutes. The droplet size was measured as a volume distribution by a laser diffraction technique, concurrent with HPLC analysis of the amount of active ingredient dispersed. The results are shown in Table 11.5.

While the average droplet size of the resultant LBDDS emulsions are approximately 1 µm in diameter, about 10% of lipid droplets are observed to be >5 µm as seen by the X_{90} results. These results seem to indicate that there is no significant change in the dispersed droplet size distribution after drug product storage, indicating that

Table 11.5 HPLC and laser diffraction results for droplet size study

	LBDDS product - control	LBDDS product - after 2 year storage
% Dispersed (by HPLC)	100% (3.5)	105% (0.5)
Droplet size by laser diffraction (μm, $n = 3$)		
X10	0.4	0.4
X50	1.1	1.2
X90	5.9	7.8

the standard excipient ratio LBDDS formulation emulsification is robust.

These studies demonstrate that it is possible to use supportive dissolution testing in combination with other analytical techniques to determine the in vitro robustness of a LBDDS formulation. These tests may provide an indication of the behavior of the formulation; however, true in vivo performance of the standard and substandard systems can only be fully evaluated via bioavailability or bioequivalence studies.

11.7.3 A Simple Test to Verify Drug Residence in an Emulsion

The work in the previous section described procedures to quantitate (in part) the active ingredient in a LBDDS formulation that is contained in the emulsion after capsule burst in aqueous media. However, there is another approach that is simpler and relies strictly on the use of a dissolution apparatus and the collection of UV spectra.

Background: A conventional dissolution test can determine if a SEDDS formulation is released from a capsule following burst. Ideally, it may also be able to discriminate between drug that is present in the lipid phase of the emulsion and drug that is not contained in the emulsion, or has precipitated. This is an important consideration, since the observed increase in bioavailability for a SEDDS formulation is brought about by the formation of the emulsion and the dissolution of the active ingredient in the lipid phase of the emulsion.[3] It is critical for the drug to be maintained in the emulsion as it makes its way through the GI tract in order to maintain efficacy as the drug will not penetrate the duodenal walls of the GI tract unless it remains solubilized. However, the environment in the dissolution bath is quite different from the GI tract, in most cases, as it is strictly an aqueous buffer environment.

It is possible that in situ UV methodology can be utilized to monitor the spectrum of an active ingredient if it happens to shift with changes in medium pH. This shift in UV spectrum can effectively measure changes in ionization state using in situ UV techniques that will allow for the determination of which phase the compound resides in: a surfactant/aqueous phase or the lipid phase of the emulsion. In many cases, the addition of surfactant to media for a LBDDS is needed for the reasons described previously in Section 11.4.3. But it should be noted that the level of surfactant in sink condition media may serve to not only solvate the drug substance, but will also potentially disperse an emulsion.

Experimental: Due to the presence of a UV chromophore of a novel compound which exhibits a shift in absorbance from acidic to neutral pH, the ionization state of a molecule can be effectively monitored. Also, information about the environment in which the molecule resides may be derived. When exposed to neutral pH conditions, or organic environments, the chromophore for the drug substance reference standard exhibits the UV spectrum shown in Fig. 11.9. This measurement was made in a 1:1 (v/v) acetonitrile:water diluent. However, when the same drug was exposed to an acidic environment, the compound was protonated and the UV chromophore shifted to a longer wavelength. The drug substance reference standard exhibits a UV spectrum as shown in Fig. 11.9. This measurement was made in a pH 2, 50 mM phosphate buffer diluent (with a small volume of acetonitrile to assist in drug solubility).

During drug development, the compound had been formulated as a SEDDS gelatin capsule. Dissolution testing was conducted with USP apparatus 2 (paddles) set to rotate at 50 rpm and using 500 ml

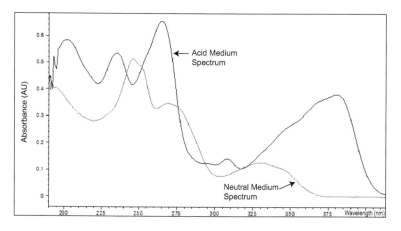

Figure 11.9 Spectra of compound in acidic and neutral pH environments.

of simulated gastric fluid (SGF) containing 2 g/L of sodium chloride. When a capsule was dissolved in an acidic dissolution medium, the UV spectrum observed using in situ monitoring was similar to the neutral medium spectrum in Fig. 11.9, or the drug substance dissolved in neutral conditions. The amount of drug present in the dissolution vessel was determined by comparing the absorbance of the solution in the vessel against a response factor of a standard solution scanned before the experiment began. Two standard solutions were utilized, one prepared in an acidic media (pH2, SGF) and one prepared in a neutral solution (1:1 (v/v) acetonitrile:water.

It could be deduced that the formulation successfully formed a microemulsion as the longer wavelength species was not observed. The compound was located inside the lipid-based phase of the emulsion, and was not exposed to the acidic media. To prove this assertion, SDS (an anionic surfactant) was added in excess to the vessel containing the dissolved capsule. After appropriate mixing, the UV spectrum observed had changed to that of the compound exposed to an acidic environment. The addition of SDS created a mixed micelle environment that allowed partitioning of the lipid and surfactant/aqueous phases to the point where the drug was out of the emulsion and into the SDS/micelle environment.

In order to utilize this type of methodology, a compound must exhibit a chromophore shift when exposed to acidic media:

a dissolution medium with a low pH should be utilized. If wavelength changes on stability are observed, this could indicate the presence of drug substance that has precipitated from the emulsion and dissolved in the surfactant media upon rupture. The effectiveness of this technique could be limited to a smaller subset of active pharmaceutical ingredients, but for compounds where this chromophore shift is evident, it can be a powerful tool to monitor the formation of microemulsions during in vitro dissolution testing.

11.8 Summary

Dissolution testing of lipid-based formulations has its own set of challenges as compared to testing of more conventional dosage forms. Some of these challenges are due in part to the fact that a poorly soluble drug substance has been dissolved in a lipid-based matrix that, when mixed with an aqueous dissolution medium, forms an emulsion. Despite this, development approaches for a suitable dissolution method are similar in certain respects. The differences in development and validation of dissolution methods for lipid-based formulations, particularly dosage forms encapsulated in gelatin, have been discussed in this chapter. Manufacture and packaging aspects of these formulations should also be considered when developing the dissolution method. These methods can be discriminating, despite the fact that drug substance properties may not impact in vitro dissolution. Additional dissolution-related testing of lipid-based formulations can prove to be quite useful in formulation development and in vitro determinations as to whether or not the drug is maintained in the emulsion after introduction of the fill to an aqueous environment.

A team under the direction of the USP Expert Committee on General Chapters: Dosage Forms has issued a general chapter, <1094>, to lend guidance in the development and validation of dissolution methods for lipid-based formulations. This is validation of the fact that there has been increasing focus on development for LBDDS (and other liquid filled capsule formulations) and it is an evolving topic of interest for industry and regulatory agencies.

Acknowledgements

Thanks to Mr. Kevin C. Bynum (Phibro Animal Health Pharmaceuticals), Dr. Edwin Gump (Boehringer Ingelheim Pharmaceuticals, Inc.), Mr. Mike Meyers (Boehringer Ingelheim Pharmaceuticals, Inc.), and Dr. Chitra Telang (Boehringer Ingelheim Pharmaceuticals, Inc.).

References

1. Lipinski, C. A. (2000). Drug-like properties and the causes of poor solubility and poor permeability, *J. Pharmacol. Toxicol. Methods*, 44(1):235–249.

2. Müllertz, A., Ogbonna, A. Ren, S., Rades. T., J. (2010). New perspectives on lipid and surfactant based drug delivery systems for oral delivery of poorly soluble drugs, *J. Pharm. Pharmacol.*, 62 (11):1622–1636.

3. Porter, C. J., Trevaskis, N. L., Charman, W. N (2007). Lipids and lipid-based formulations: optimizing the oral delivery of lipophilic drugs, *Nat. Rev. Drug Discovery*, 6:231–248.

4. Constantinides, P. (1995). Lipid microemulsions for improving drug dissolution and oral absorption: physical and biopharmaceutical aspects, *Pharm. Res.*, 12(11):1561–1572.

5. Patel, M., Patel, S., Patel, N., Patel, M. (2011). A review: novel oral lipid based formulation for poorly soluble drugs, *Int. J. Pharm. Sci. Nanotechnol.*, 3(4):1182–1192.

6. Grove, M., Müllertz, A., Pederson, G., Nielsen, J. (2007). Bioavailability of seocalcitol III. Administration of lipid-based formulations to mini-pigs in the fasted and fed state, *Eur. J. Pharm. Sci.*, 31(1):8–15.

7. Pouton, C. W. (2000). Lipid formulations for oral administration of drugs: non-emulsifying, self-emulsifying and "self-microemulsifying" drug delivery systems, *Eur. J. Pharm. Sci.*, 11(suppl. 2): S93–S98.

8. Mu, H., Holm, R., Müllertz, A. (2013). Lipid-based formulations for oral administration of poorly water-soluble drugs, *Int. J. Pharm.*, 453(1):215–224.

9. Porter, C., Pouton, C. W., Cuine, J., Charman, W. (2008). Enhancing intestinal drug solubilisation using lipid-based delivery systems, *Adv. Drug Deliv. Rev.*, 60(6): 673–691.

10. Gullapalli, R. P (2010). Review: soft gelatin capsules (softgels), *J. Pharm. Sci.*, 99(10):4107–4148.

11. Bhamidipati, S. P., Pope, S. C., Selen, A. (2009). Lipid based drug products: scientific and regulatory product quality and biopharmaceutics considerations, AAPS Workshop on Scientific and Technological Advances in the Use of Lipid-Based Drug Delivery Systems for Bioavailiability Enhancement and Tissue Targeting, Baltimore, MD.

12. Chapus, C., Sémériva, M, Bovier-Lapierre, C., Desnuelle, P. (1976). Mechanism of pancreatic lipase action. 1. Interfacial activation of pancreatic lipase, *Biochemistry*, 15(23):4980–4987.

13. Kleberg, K., Jacobsen, F., Fatouros, D. G., Müllertz, A. (2010). Biorelevant media simulating fed state intestinal fluids: colloidal phase characterization and impact on solubility capacity, *J. Pharm. Sci.*, 99(8): 3522–3532.

14. Porter, C. J. H., Kaukonen, A. M., Taillardat-Bertschinger, A., Boyd, B. J., O'Connor, J. M., Edwards, G. A., Charman, W. N. (2004). Use of *in vitro* lipid digestion data to explain the *in vivo* performance of triglyceride-based oral lipid formulations of poorly water-soluble drugs: studies with halofantrine, *J. Pharm. Sci.*, 93(5):1110–1121.

15. O'Driscoll, C. M. (2002). Lipid-based formulations for intestinal lymphatic delivery, *Eur. J. Pharm. Sci.*, 15(5): 405–415.

16. Pouton, C. W. (2006). Formulation of poorly water-soluble drugs for oral administration: physicochemical and physiological issues and the lipid formulation classification system, *Eur. J. Pharm. Sci.*, 29(3–4):278–287.

17. Williams, H. D., et. al. (2012). Toward the establishment of standardized *in vitro* tests for lipid-based formulations. Part 1: Method parameterization and comparison of *in vitro* digestion profiles across a range of representative formulation, *J. Pharm. Sci.*, 101(9): 3360–3380.

18. USP Dissolution General Chapter <711> Dissolution, United States Pharmacopeia and National Formulary, USP 36-NF 31, United States Pharmacopoeial Convention, Inc., Rockville, MD, 2013, pp. 307–313.

19. USP General Chapter <1094> Liquid-Filled Capsules-Dissolution Testing and Related Quality Attributes, USP37-NF32 S1, United States Pharmacopoeial Convention, Inc., Rockville, MD, USP Liquid-Filled Capsules Expert Panel, 2014.

20. USP Dissolution Procedure: Development and Validation General Chapter <1092>, United States Pharmacopeia and National Formulary, USP 36-NF 31, United States Pharmacopeial Convention, Inc., Rockville, MD, 2013, pp. 735–741.

21. Rama Rao, K. V., Pakhale, S. P., Singh, S. (2003). A film approach for the stabilization of gelatin preparations against cross-linking, *Pharm. Technol.,* pp. 54–63.

22. Meyer, M. C. et al. (2000). The effect of gelatin cross-linking on the bioequivalence of hard and soft gelatin acetominophen capsules, *Pharm. Res.,* 17(8):962–966.

23. 40(6). In Process Revision: <2040> Disintegration and Dissolution of Dietary Supplements, First Supplement to United States Pharmacopeia and National Formulary USP 32–NF 32 and PF 39(3); United States Pharmacopeial Convention, Inc.: Rockville, MD, 2013; p. 6439.

24. Hu, J., Kyad, A., Ku, V., Zhou, P., Cauchon, N. (2005). A comparison of dissolution testing on lipid soft gelatin capsules using USP apparatus 2 and apparatus 4, *Dissolut. Technol.,* 12(2):6–9.

25. Guidance for Industry: Dissolution Testing of Immediate Release Solid Oral Dosage Forms, FDA Center for Drug Evaluation and Research, August 1997, BP1.

26. Price, J. C. (2003). Gelatin, in *Handbook of Pharmaceutical Excipients,* 4th ed., Rowe, R. C., Shesky, P. J., Weller, P. J. (eds.). London: Pharmaceutical Press and Washington, D. C.: American Pharmaceutical Association, pp. 252–254.

27. Pennings, F. H., Kwee, B. L., Vromans, H. (2006). Influence of enzymes and surfactants on the disintegration behavior of cross-linked hard gelatin capsules during dissolution, *Drug Dev. Ind. Pharm.,* 32(1):32–37.

28. ICH Q2(R1) – Validation of Analytical Procedures: Text and Methodology – ICH Expert Working Group – International Conference on Harmonization of Technical Requirements for Registration of Pharmaceuticals For Human Use – November 1996.

29. Chiwele, I., Jones, B. E., Podczeck, F., (2000). The shell dissolution of various empty hard capsules, *Chem. Pharm. Bull.,* 48(7):951–956.

30. Tarr, B. D., Yalkowsky, S. H. (1989). Enhanced intestinal absorption of cyclosporine in rats through the reduction of emulsion droplet size, *Pharm. Res.,* 6(1):40–43.

Chapter 12

Dissolution of Stabilized Amorphous Drug Formulations

Justin R. Hughey

Banner Life Sciences, 4125 Premier Drive, High Point, NC 27265, USA
justin.hughey@bannerls.com

12.1 Introduction

Current estimates indicate that about 90% of drugs in development pipelines exhibit a combination of poor water solubility and poor permeability.[1] About 70% of these exhibit good permeability but are solubility limited. These drugs, known by the Biopharmaceutical Classification System as class 2 compounds, are amenable to bioavailability enhancement through solubility enhancement pathways. By simply increasing drug concentration in the gastrointestinal tract (GIT), in vivo absorption can be enhanced. Hence, the application of solubility enhancement platforms to drugs in development has become increasingly common.

Strategies used to enhance solubility and, by default, dissolution rates of these compounds include particle size reduction, salt formation, lipid and co-solvent solubilization, self-emulsifying drug

Poorly Soluble Drugs: Dissolution and Drug Release
Edited by Gregory K. Webster, J. Derek Jackson, and Robert G. Bell
Copyright © 2017 Pan Stanford Publishing Pte. Ltd.
ISBN 978-981-4745-45-1 (Hardcover), 978-981-4745-46-8 (eBook)
www.panstanford.com

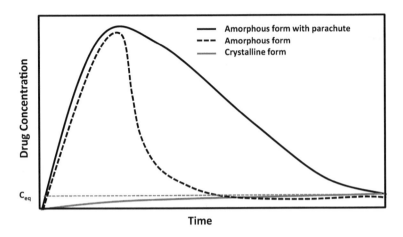

Figure 12.1 Drug concentration versus time profiles illustrating the spring and parachute effect imparted by stabilizing excipients. Adapted from Brouwers et al.[3]

delivery systems (SEDDS), cyclodextrin complexation, and stabilized amorphous drug formulations (SADFs).[2] SADFs, or amorphous solid dispersions, have garnered a substantial amount of attention in recent years due to their ability to provide a high degree of bioavailability enhancement. The mechanism by which bioavailability enhancement occurs is often referred to as the "spring and parachute" effect, which is illustrated in Fig. 12.1.[3,4] High free energy inherent to the amorphous form generally provides the "spring" effect into a supersaturated state where thermodynamic solubility can be exceeded by a significant amount. This kinetic solubility, however, is typically short-lived as there is oftentimes a strong driving force for precipitation. When precipitation inhibitors or concentration enhancing polymers are present, the rate at which precipitation occurs can be reduced such that a "parachute" effect is realized. This results in a longer exposure to elevated drug concentrations, which can translate into increased bioavailability when permeability isn't a limitation.

Precipitation of a drug from a supersaturated state generally occurs in two stages: nucleation and crystal growth. Precipitation inhibitors may interfere with one or both of these processes by interacting with the drug (i.e., hydrogen bonding), changing the properties of the medium or adsorbing onto potential nucleation points. Other factors such as polymer rigidity and distribution of functional groups in the polymer can play a significant role in polymer–drug interactions. In the case of hypromellose acetate succinate (HPMCAS)-based dispersions, Friesen et al. proposed that drug–polymer colloids form in aqueous media which are critical to improving bioavailability.[5] The strength and type of interactions that occur in both the solution and solid states are strongly dependent on the functional groups present on the drug substance and the polymer. Thus appropriate formulations are developed on a case-by-case basis. Estimations of drug-polymer compatibility are often predicted through the generation of phase diagrams based on the Flory–Huggins theory, but models such as these are typically only useful for selection of initial starting points. Polymers most commonly used in solubility enhancement applications include poly(vinylpyrrolidone) (PVP), poly(vinylpyrrolidone)-vinyl acetate copolymer (PVPVA), hypromellose (HPMC), enteric derivatives of hypromellose, including hypromellose acetate succinate (HPMCAS) and hypromellose phthalate (HPMCP), cellulose acetate phthalate (CAP), as well as methacrylic acid–based copolymers. It is also not uncommon for surfactants present in the composition to improve wettability and stabilization in the solution state. These are typically non-ionic materials such as polyethoxylated sorbitan esters, sorbitan esters, and tocopheryl derivatives, but may also include ionic surfactants such as sodium dodecyl sulfate.

The most common type of amorphous solid dispersion used in solubility enhancement applications is the glass solution. This is a single-phase system in which all molecules of the drug substance are intimately mixed with carrier molecules, typically a polymer. These systems are generally prepared by techniques such as spray drying, hot-melt extrusion, and anti-solvent precipitation, each having specific advantages. Specific examples of marketed products utilizing the SADF platform are shown in Table 12.1.

As bioavailability is often directly related to the extent and maintenance of supersaturation generated by a SADF, it stands to reason that dissolution performance can provide an indication of

Table 12.1 Marketed drug products utilizing the stabilized amorphous drug formulation approach

Product	Compound	Manufacturing method
Norvir®	Ritonavir	Hot melt extrusion
Kaletra™	Ritonavir/lopinavir	Hot-melt extrusion
Fenoglide™	Fenofibrate	Hot-melt extrusion
Noxafil®	Posaconazole	Hot-melt extrusion
Onmel™	Itraconazole	Hot-melt extrusion
Sporanox®	Itraconazole	Spray layering
Harvoni®	Ledipasvir/sofosbuvir	Spray drying
Intelence™	Etravirine	Spray drying
Incivek®	Telaprevir	Spray drying
Kalydeco™	Ivacaftor	Spray drying
Cesamet®	Nabilone	Spray drying
Prograf®	Tacrolimus	Spray drying
Nimotop®	Nimodipine	Spray drying
Zelboraf™	Vemurafenib	Antisolvent precipitation

in vivo performance. Hence, dissolution testing of these systems at all stages of development, from screening studies to evaluation of the final dosage form, can provide an indication of how formulation and process parameters can impact in vivo performance. The following sections outline various aspects of characterizing the in vitro dissolution performance of these systems.

12.2 Dissolution Considerations

Generally speaking, dissolution testing serves two main purposes: prediction of in vivo performance during formulation and process development and evaluation of batch-to-batch variability in a quality control (QC) environment once the dosage form is manufactured. Ideally the dissolution tests used for each purpose would be the same, but in the case of SADFs there tends to be substantial differences. This is primarily due to the evolving nature and complexity of the tests used during development to obtain in vitro-in vivo relationships and correlations (IVIVR/IVIVC).

12.2.1 Dissolution at Sink at Non-sink Conditions

The United States Pharmacopeia (USP) has traditionally been used as a guideline for dissolution method development for dosage forms containing new chemical entities (i.e., USP <711> and <1092>). Within the USP, there are four basic types of compendial dissolution apparatuses. USP apparatus 1 and 2 (basket and paddle) are robust, adequately standardized, and normally chosen due to their simple design and ease of use in the QC environment. However, these systems typically use large volumes of dissolution media (500 to 1000 mL) and their hydrodynamics are not necessarily physiologically relevant. These aspects can be improved upon somewhat with apparatus 3 (reciprocating cylinder) and apparatus 4 (flow-through cell). In these systems, a dosage form can be exposed to multiple types of dissolution media during one experiment to simulate the changing conditions during its passage through the GIT. Traditionally, methods utilizing these apparatus are carried out at sink conditions, wherein the volume of media is at least three times that required to form a saturated solution of drug substance, as described in USP <1092>. At these conditions, dissolution results reflect properties of the dosage form itself, removing aqueous solubility of the drug substance as a variable. When drug solubility is low, surfactants are commonly added to the dissolution media to generate sink conditions. Dissolution methodologies such as these have found utility in QC environments due to their ability to provide consistent results.

While useful in a QC environment, dissolution testing of SADFs at sink conditions is generally not predictive of in vivo performance due to poor water solubility of the drug substance. The need for an alternative dissolution methodology can be easily demonstrated by calculating the dose–solubility ratio of a compound with poor aqueous solubility.[6] For example, a drug substance with a solubility of 2.5 μg/mL in gastrointestinal (GI) fluid and a dose of 25 mg would require 10 L of GI fluid for complete solubilization and at least 30 L to be at sink conditions, as defined by the USP. Obviously, this condition will not be met in an in vivo environment, as GI volumes in the fasted state are normally less than 250 mL.[7] As previously discussed, in vivo absorption generally correlates well with the extent and maintenance of supersaturation achieved, either through the use of a high-energy form of the drug or driven by a pH transition, when permeability is not a limitation.[8] Clearly the extent and maintenance of supersaturation can only be achieved at non-sink conditions. Ultimately, different variations of both sink and non-sink methodologies should be carried out and combined with in vivo data to construct an IVIVR or IVIVC. Models such as these are particularly useful for poorly water-soluble drug substances as they can expedite the drug development process by reducing the number of animal studies required. The United States Pharmacopeia provides guidance on the application of these relationships and correlations in USP <1088>.

12.2.2 Media Selection

Selection of an appropriate dissolution media is critical for dissolution methodologies that are intended to be predictive of pharmacokinetic data. This requires that the media simulate the gastrointestinal (GI) environment in terms of pH as well as the type and concentrations of biorelevant species present such as bile salts and phospholipids. In general, the pH range should bracket the physiological range of about 1.8 to about 7.8. It is not necessary, however, to maintain the same pH during analysis. Dissolution methodologies may involve pH changes throughout the analysis to model conditions in the GI tract, as shown in Table 12.2. In addition to physiological factors, pH selection should also take into account the type of polymer utilized in the SADF. For example, carriers comprising enteric polymers such as HPMCAS or CAP have pH-

Table 12.2 pH ranges and residence times typical in the human GIT. Adapted from Newman et al.[9]

Region	pH	Residence time
Stomach	1.8–2.5	1–5 h
Duodenum	5.0–6.5	>5 min
Jejunum	6.9	1–2 h
Ileum	7.6	2–3 h
Colon	5.5–7.8	15–48 h

dependent solubility while water-soluble polymers such as PVPVA and HPMC have pH-independent solubility.

It is not uncommon for researchers to utilize simulated intestinal fluid (SIF) during dissolution testing, which is free of natural surfactants. However, bile salts and phospholipids in the GI tract act as natural surfactants, effectively providing solubilization capacity through the formation of mixed micelles. It is important that these biorelevant species be incorporated into dissolution media, as the increase in solubilization capacity may be substantial. For example, sodium taurocholate and lecithin are both incorporated into fed and fasted state simulated intestinal fluid (FeSSIF/FaSSIF), which are very common and have been used successfully by many researchers.[10] Similarly, the combination of sodium taurocholate and 1-palmitoyl-2-oleoyl-*sn*-glycero-3-phosphocholine (POPC) are used in model fasted duodenal fluid (MFDF).[5,11]

12.2.3 Speciation at Non-sink Conditions

During traditional sink dissolution testing of soluble drug substances, drug-containing species are generally limited to freely dissolved drug (free drug) and crystalline drug (undissolved drug). The process for measuring dissolved drug is as trivial as a filtration step to remove crystalline drug and other interfering components. Non-sink dissolution analyses, however, are substantially different in this respect as a number of physiologically relevant species responsible for supersaturation can form upon solvation.

The types of species that form under non-sink conditions may include free drug, drug in bile-salt micelles, drug-polymer complexes and drug in polymeric colloidal nanostructures and aggregates. These species are strongly dependent on the type of carrier used to stabilize the amorphous drug and the type of dissolution media used. Of these, free drug is the only species that is absorbed in vivo. However, other drug containing species are equally important as their ability to enhance in vivo bioavailability is well documented.[5,11,12] As free drug is continuously absorbed, these species can act as free drug "reservoirs", replenishing the supply of free drug during transit through the GIT. Hence, a critical aspect to non-sink dissolution testing is a comprehensive understanding of the drug containing species that can be formed and their relative sizes. While many techniques focus on the analysis of total dissolved drug (i.e., all drug containing species), analyses can be carried out in a way to measure the ability of a composition to generate free drug. Both types of analyses are discussed in the sections below.

12.3 Dissolution Case Studies

Application of non-sink dissolution testing extends from early stage carrier screening studies to formulation prototypes and final dosage forms (i.e., tablets or capsules). Utilization of non-sink dissolution testing on formulation prototypes (i.e., films and intermediate solid dispersion powder) allows the pharmaceutical scientist to remove variables such as tablet disintegration characteristics. The following sections describe techniques that are currently utilized in the industry to evaluate dissolution performance at all stages of drug development.

12.3.1 Carrier Screening Methodologies

With respect to carrier screening methodologies, pharmaceutical scientists are able to rapidly assess, by using miniaturized dissolution methods, the potential of excipients to provide and maintain supersaturation when limited amounts of the drug substance are available. Methodologies used for these small-scale studies are generally based on co-solvent precipitation and amorphous film dissolution approaches, both of which are discussed in the following sections.

12.3.1.1 Co-solvent precipitation

In the co-solvent approach, a drug substance is dissolved in a water-miscible organic solvent at a relatively high concentration. An aliquot of the drug-containing solution is then dispensed into an aqueous medium containing stabilizing excipients, creating a supersaturated state. Dissolved drug or turbidity can then be quantified as a function of time to evaluate the effectiveness of precipitation inhibitors. In reality, this methodology becomes a solubility study rather than a dissolution study. Nevertheless, it is a useful tool during early development. Vandecruys et al. described the application of the co-solvent methodology to 25 drug substances.[13] In this study, potential precipitation-inhibiting excipients were dissolved at 2.5% w/w in 10 mL of dissolution media (0.01 N HCl, pH 4.5 buffer, pH 6.8 buffer or water). Separately, drug substances were dissolved at 50 to 100 mg/mL in a water-miscible organic solvent. To initiate the study, a portion of the drug-containing solution was dispensed into the dissolution media until precipitate was observed. Samples were then collected as a function of time, filtered to remove precipitate, diluted to prevent recrystallization, and analyzed for drug content. Evaluation of the initial extent of supersaturation and maintenance of the same allowed the researchers to assign a rank order to carriers and surfactants, which correlated with canine pharmacokinetic data. Other researchers have reported using variations of this methodology for carrier screening studies.[14–16]

Miniaturization of the co-solvent approach has also been reported in order to reduce the amount of drug substance required and enable high-throughput data collection. For example, Yamashita et al. utilized a 96-well plate to evaluate the precipitation inhibition properties of 14 excipients in biorelevant media, measuring dissolved drug as a function of time.[17] Results of the miniaturized study were confirmed by USP apparatus 2 testing.

12.3.1.2 Amorphous film dissolution

While co-solvent precipitation methods are useful, the drug substance is instantaneously placed in a supersaturated state and thus kinetics of the dissolution process itself are not evaluated. Techniques that utilize an amorphous substrate, however, require that dissolution occur before precipitation inhibition can occur. Amorphous films containing a drug and stabilizing excipients can be prepared by techniques such as rotary evaporation or solvent

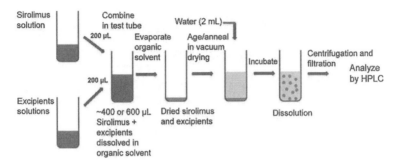

Figure 12.2 Illustration of a solvent-casting process followed by in situ dissolution analysis. Reprinted from Kim et al.[18] with permission from Elsevier.

casting and can be tested by a variety of dissolution methodologies. When drug substance is limited and the evaluation of multiple systems is desired, solvent casting may be more appropriate. Many scientists have reported casting films into the container in which the subsequent dissolution analysis is carried out. For example, Kim et al. utilized the methodology summarized in Fig. 12.2 to screen polymers and surfactants for their ability to stabilize a poorly water-soluble drug substance in solution.[18] In this study, 400 μg of drug substance was combined with stabilizing excipients and a dissolving solvent in a culture tube. After removal of the organic solvent to create dried films, 2 mL dissolution media was added directly to the culture tubes to initiate dissolution testing. After centrifugation/filtration to remove precipitate and dilution to prevent recrystallization, total dissolved drug was measured. The data generated in this study at the 1 and 24 hour timepoints are shown in Table 12.3. These data indicated that HPMC provided good solubilization of sirolimus. Furthermore, ternary compositions containing HPMC and TPGS or Suroester 15 provided the highest levels of solubilization. Lead formulations were manufactured by spray drying and dosed to male Sprague–Dawley rats. The data in Fig. 12.3 indicated that sirolimus blood levels followed the same rank ordering as the screening study.

A similar study was described by Barillaro et al. in which automated solvent casting was utilized to prepare films containing

Table 12.3 Summary of screening experiment results. Adapted from Kim et al.[18]

	None	**SLS**	**TPGS**	**G50/13**	**M52**	**P407**	**SE15**	**G44/14**
Concentration (μg/mL) after 1 h of dissolution of 1:8:1 ratio formulations								
HPC	16.75	25.25	*47.82*	26.43	28.56	<u>67.41</u>	*39.76*	27.09
HPMC	*35.97*	*42.45*	**91.33**	*41.68*	*42.79*	<u>63.83</u>	<u>67.97</u>	<u>63.21</u>
PVP VA64	16.20	15.03	*44.18*	25.96	27.68	*36.58*	*43.08*	21.65
PEG 8000	14.86	*42.53*	<u>54.40</u>	29.37	30.77	<u>55.60</u>	*39.77*	25.94
PVP K30	14.99	*32.32*	<u>51.99</u>	28.67	29.99	*47.12*	*38.75*	24.82
Concentration (μg/mL) after 24 h of dissolution of 1:8:1 ratio formulations								
HPC	9.92	12.27	*46.61*	21.80	21.04	27.13	*33.99*	0.96
HPMC	15.32	10.17	*47.31*	22.53	21.97	27.68	<u>54.87</u>	2.10
PVP VA64	11.01	11.37	*42.00*	1.18	1.11	24.87	*39.08*	0.89
PEG 8000	9.63	*39.32*	*46.73*	23.04	22.05	28.03	*42.79*	14.74
PVP K30	13.70	12.31	*46.06*	1.30	1.59	27.44	*48.17*	6.36

The font indicates whether the concentration was <30 μg/mL, between 30 and 50 μg/mL (*italic*), between 50 and 70 μg/mL (<u>underlined</u>), or >70 μg/mL (**bold**). SLS, sodium lauryl sulfate; G50/13, Gelucire 50/13; M52, Myrj 52; P 407, Poloxamer 407; SE15, Sucroester 15; Gelucire 44/14.

drug and stabilizing excipients in vials.[19] Dissolution analysis was carried out manually, in situ, after the addition of 4 mL of simulated gastric fluid to the vials. To confirm the utility of the methodology, films were prepared by rotary evaporation and analyzed separately. Dissolution performance of this material was analogous to the high-throughput approach, further indicating that the developed methodology was sound.

As with the co-solvent approach, miniaturization of the amorphous film dissolution methodology allows for minimal drug to be used while maximizing data. Shanbhag et al. successfully applied miniaturization to the evaluation of amorphous films in a two-tier dissolution testing approach.[20] Films containing only 60 μg of the drug substance were cast in 96-well plates and then analyzed for dissolution performance in situ with 300 μL of simulated intestinal fluid. After selection of lead solvent cast films from this study, melt compressed films were prepared to more closely resemble the hot-melt extrusion process. Dissolution performance of these materials was evaluated with a modified USP type 7 apparatus in 200 mL of simulated intestinal fluid.

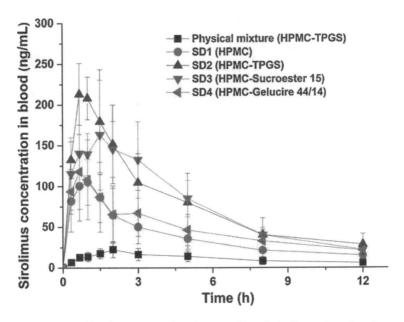

Figure 12.3 Blood concentration-time profile of sirolimus in rats after oral administration of the physical mixture and solid dispersion particles prepared by spray drying. Reprinted from Kim et al.[18] with permission from Elsevier.

12.3.2 Dissolution of Drug Product Intermediate and Solid Dosage Forms

It is not uncommon to forgo the carrier screening studies outlined in the previous sections and advance directly to the manufacture of prototype amorphous intermediates by methods such as spray-drying or hot-melt extrusion when sufficient drug is available. In this case, relative dissolution performance is used to assess compositions, including those that have differences in excipient type, particle size, and manufacturing processes. With a few exceptions, assessment of the intermediate powder, as opposed to the solid dosage form, provides a "best-case scenario" in terms of supersaturation performance as native particles are free of the effects normally encountered during tableting and encapsulation (i.e., agglomeration due to compression or muted release due to lubrication). Studies at this stage normally focus on assessment of total dissolved drug, although analysis of free drug is also possible, which is discussed in subsequent sections. Testing dissolution performance of drug product intermediates and solid dosage forms is a more straightforward process than the screening studies previously outlined. In these studies, materials are added directly to the dissolution media such that an excess amount is present to generate a supersaturated state. The amount of drug required for these studies is dependent on the specific methodology and can vary from milligram to gram quantities.

A non-sink dissolution study utilizing a USP apparatus 2 was described by DiNunzio et al. to analyze the relative performance of several stabilized amorphous compositions stabilized with concentration enhancing polymers.[21] In the study, the equivalent of 37.5 mg of itraconazole (ITZ), a weakly basic drug in the form of process intermediate powder, was added directly to dissolution vessels containing 0.1 N HCl at pH 1.2. To model GI transit, the pH was adjusted to 6.8 at 2 hours (1,000 mL total volume). Samples were removed, passed through a 0.2 µm filter to isolate total drug in solution, diluted with solvent and analyzed. Example non-sink dissolution profiles for cellulose acetate phthalate (CAP)-based compositions are shown in Fig. 12.4. Using area under the dissolution curve, the authors assigned a rank order to the prepared compositions and dosed the lead composition in a subsequent rat study.

Assessment of solid dispersion systems, prepared by various solvent and thermal-based techniques, using iterations of the dissolution methodology outlined above has been reported by a number of authors.[15,16,21–30] While effective, the methodology requires large dissolution volumes and is not conducive to rapid analysis of multiple compositions simultaneously due to the need for a USP-type dissolution apparatus. Friesen et al. presented a methodology in which dissolution of drug product intermediates were carried out in small centrifuge tubes to evaluate spray-dried dispersions containing HPMCAS.[5] In the described methodology, a small amount of the intermediate was placed in a centrifuge tube to which 1.8 mL of phosphate buffer or model fasted duodenal fluid (MFDF) was added. Samples were vortexed to disperse the intermediate, equilibrated, and left undisturbed. After a designated time, samples

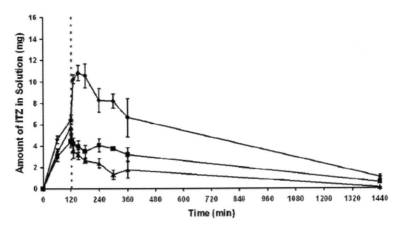

Figure 12.4 Non-sink dissolution profiles of itraconazole-based solid dispersion systems prepared by hot-melt extrusion utilizing the pH change method. Key: 1:2 ITZ:CAP (♦), 1:2 ITZ:CAP (■), 2:1 ITZ:CAP (▲). Each dissolution vessel contained 37.5 mg itraconazole, corresponding to 10 times the saturation solubility of ITZ in the acid phase. Reprinted with permission from from DiNunzio et al.[21] Copyright (2008) American Chemical Society.

were centrifuged to isolate total dissolved drug as an alternative to filtration. Subsequently, an aliquot of the supernatant was removed, diluted with an organic solvent to prevent recrystallization and analyzed. After removal of the aliquot, the centrifuge tubes were vortex mixed to disperse the solids and again allowed to equilibrate. An example concentration–time profile for this type of analysis is shown in Fig. 12.5. The use of centrifugation in this technique allows for very small dissolution media volumes relative to those that require filtration to isolate dissolved drug. Other researchers have reported using this methodology to successfully evaluate amorphous solid dispersion systems.[11,31]

12.3.3 Methodologies to Determine Free Drug

While total dissolved drug concentrations provide an indication of supersaturation, knowledge of the amount of free drug generated can paint a more complete picture of drug speciation, allowing a higher degree of in vivo performance predictability. Unfortunately,

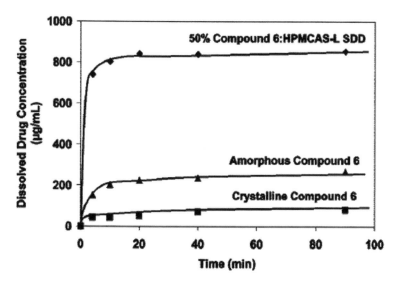

Figure 12.5 Microcentrifuge dissolution data comparing crystalline and amorphous drug with a spray-dried dispersion. The drug substance (compound 6) was present at 1000 µg/mL. Reprinted with permission from Friesen et al.[5] Copyright (2008) American Chemical Society.

measurement of free drug is not a trivial task due to the need to isolate it from colloidal species either analytically or physically. With respect to analytical isolation, various types of solution state nuclear magnetic resonance (NMR) spectroscopy techniques have been reported for the measurement of free drug.[5,32–34] However, this type of analysis is not ideal from a formulation-screening standpoint in which rapid analysis of many dissolution time points is preferred. A more practical approach to free drug measurement is to physically isolate free drug. In this type of analysis, free drug is isolated from colloidal species by partitioning, diffusion-partitioning or ultracentrifugation and subsequently analyzed by traditional techniques.

12.3.3.1 Free drug isolation by partitioning into organic solvent

Several researchers have applied biphasic dissolution analyses to amorphous solid dispersion systems in an attempt to generate a

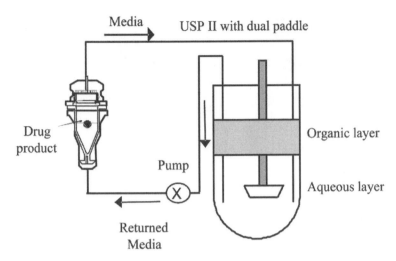

Figure 12.6 Example of biphasic dissolution apparatus. Reprinted with permission from Shi et al.[36] Copyright (2010) American Chemical Society.

meaningful IVIVR/IVIVC.[35–38] In these types of analyses, the organic phase acts as a sink to remove free drug from the aqueous phase as it is generated. In doing so, the organic phase acts as a model for in vivo drug absorption into systemic circulation. A modified USP 2 apparatus is typically used which incorporates stirring and sampling at each layer. Furthermore, the organic partitioning phase (i.e., octanol) should be completely immiscible with the aqueous phase to prevent interfering with dissolution kinetics. Vangani et al. demonstrated the utility of this technique for the analysis of an amorphous solid dispersion, among other dosage forms.[38] The authors utilized a hybrid USP 2/4 apparatus, similar to the apparatus shown in Fig 12.6. The USP 4 apparatus held the dosage form in place, while the USP 2 apparatus held biphasic media, which consisted of pH 6.8 buffer and nonanol/cyclohexane. As the only drug containing species in the organic phase was free drug, its concentration as a function of time was determined with fiber optic measurement. A rank order correlation was made between in vitro release and in vivo absorption, demonstrating the utility of the method.

12.3.3.2 Free drug isolation by membrane-partitioning

While partitioning-based analyses provide insight to free drug concentrations generated by a given composition, their ability to prevent transfer of crystalline or colloidal species to the organic phase is not fully known. This can be addressed, to a degree, by incorporating a microporous membrane at the interface to physically prevent species larger than a given size from traversing (e.g., 100 nm). A methodology similar to this was described by Alonzo et al. to determine free drug concentrations.[39] The authors utilized a dialysis membrane to create donor and receptor chambers, each containing pH 6.8 buffer (no organic solvent used). In addition to buffer, the receptor chamber also contained 500 µg/mL of pre-dissolved hypromellose to prevent recrystallization of the drug substance following diffusion. Each chamber was equipped with a UV fiber-optic probe to measure total drug in solution, as illustrated in Fig. 12.7.

Amorphous systems of felodipine prepared by rotary evaporation as well as felodipine spiked organic solutions were loaded into the donor chamber. Figure 12.8 illustrates the concentration-time data for the donor chamber. These data indicated that the

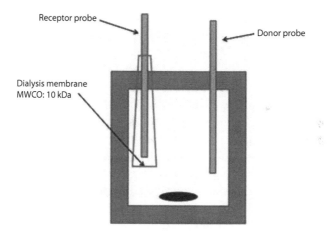

Figure 12.7 Experimental schematic for diffusion experiments to determine free drug. Reprinted with permission from Alonzo et al.[39] Copyright (2011) Wiley-Liss, Inc.

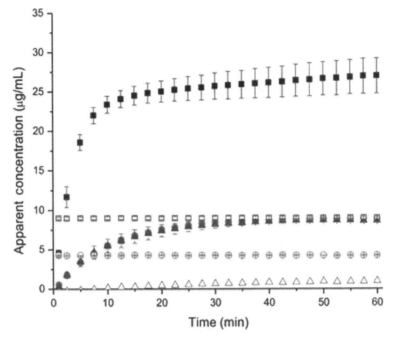

Figure 12.8 Apparent concentration-time profiles for the dissolution of 90:10 (■) and 50:50 (●) HPMC-felodipine solid dispersion, pure amorphous felodipine (▲), and crystalline felodipine (△). Concentration–time profiles of artificial supersaturations of felodipine at 5 (○) and 10 (□) μg/mL. Reprinted with permission from Alonzo et al.[39] Copyright (2011) Wiley-Liss, Inc.

composition containing 90:10 hypromellose-felodipine generated a substantially higher total felodipine concentration than the 50:50 hypromellose-felodipine composition as well as amorphous and crystalline felodipine. However, free drug concentrations in the receptor chamber indicated that higher total drug concentrations did not necessarily translate into higher free drug concentrations, as shown in Fig. 12.9. Despite substantial differences in total felodipine in solution, the three amorphous solid systems studied provided similar levels of free drug in solution.

In a similar approach, Babcock et al. developed a small volume methodology in which donor and receptor chambers were separated by a semipermeable membrane, as shown in Fig. 12.10.[40] In an

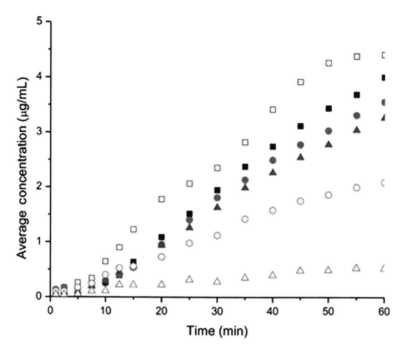

Figure 12.9 Diffusion profiles (receptor chamber) for the dissolution of 90:10 (■) and 50:50 (●) HPMC-felodipine solid dispersion, pure amorphous felodipine (▲), and crystalline felodipine (△). Concentration-time profiles of artificial supersaturations of felodipine at 5 (○) and 10 (□) μg/mL. Reprinted with permission from Alonzo et al.[39] Copyright (2011) Wiley-Liss, Inc.

example provided by the inventors, 5 mL of model fasted duodenal fluid was charged into the donor chamber while a 20% (w/w) decanol in decane solution was charged into the receptor chamber. A polypropylene membrane (100 nm nominal pore size) with a hydrophilic surface treatment on the donor side was used to isolate the two chambers. In doing so, an unstirred water layer was created on the donor side while the membrane itself was allowed to fill with the decane/decanol solvent system. The ability of a spray-dried dispersion to generate free drug was evaluated by adding 2.4 mg to the donor chamber and measuring drug concentrations in the receptor chamber as a function of time. A number of additional examples were provided that demonstrated

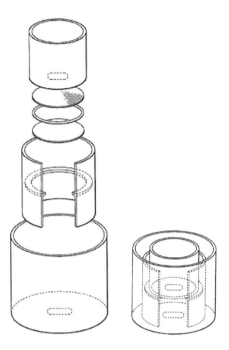

Figure 12.10 Membrane-permeation apparatus used for the determination of free drug: expanded and collapsed. Adapted from United States Patent 7,611,630.

the utility of this methodology. Pharmacokinetic studies on beagle dogs demonstrated good agreement with in vitro studies. As further confirmation of its utility, Thombre et al. applied the methodology to evaluate several ziprasidone-based systems with success.[29]

12.3.3.3 Free drug isolation by ultracentrifuge

While total dissolved drug can be separated from precipitated or crystalline species by microcentrifugation, further isolation of free drug from colloidal species is less trivial. Ultracentrifugation, however, is capable of separating these species entirely; leaving only truly dissolved species (free drug) in the supernatant. This methodology was applied in a small volume pH dilution dissolution methodology by Gao et al.[41] The specific technique utilized by the authors modeled the pH profile of rat GI tracts. Over the

Section 1 - Stomach

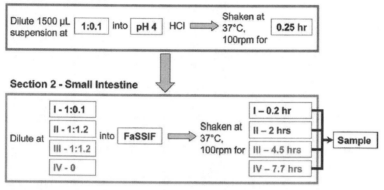

I - duodenum, II – jejunum / ileum, III – cecum **and** IV - colon

Figure 12.11 In vitro pH-dilution methodology to model rat gastrointestinal conditions. Reprinted with permission from Gao et al.[41] Copyright (2010) American Chemical Society.

course of the analysis, dissolution volume and pH was progressively increased, as shown in Fig. 12.11. At specific time points, the authors sampled the media, ultracentrifuged at $\sim$150,000 g to isolate free drug and quantified drug content. Dissolution data was input into a physiologically based pharmacokinetic modeling tool, which resulted in excellent predictions of in vivo absorption. While this technique was specific to rat physiology, similar methods could be developed for other species.

12.4 Conclusion

Stabilized amorphous drug formulations have been demonstrated to be an effective platform to enhance the bioavailability of poorly water-soluble drug substances. Therefore, the need to utilize dissolution testing methodologies throughout the development process that provide insight to in vivo performance and provide an IVIVR/IVIVC exists. As discussed, these methodologies are generally more complicated than traditional dissolution methodologies due to the need for non-sink conditions and the presence of multiple types physiologically relevant species in solution that vary in size. Despite their relative complexity, a multitude of methodologies exists for evaluating stabilizing excipients and compositions at all stages of the development process. By performing these analyses during development, it is possible to optimize formulation and process parameters to achieve desired supersaturation profiles.

References

1. Hauss, D. J. (2013). Are we fully leveraging existing technologies to overcome poor drug solubility? in *7th Improving Solubility Conference*. Philadelphia, Pennsylvania.

2. Hughey, J. R., Williams III, R. O. (2012). Solid-state techniques for improving solubility, in *Formulating Poorly Water Soluble Drugs*, ed. Williams III, R. O., Watts, A. B., Miller, D. A. (Springer, New York), pp. 95–131.

3. Brouwers, J., Brewster, M. E., Augustijns, P. (2009). Supersaturating drug delivery systems: the answer to solubility-limited oral bioavailability? *J. Pharm. Sci.*, **98**, 2549–2572.

4. Guzmán, H. R., Tawa, M., Zhang, Z., Ratanabanangkoon, P., Shaw, P., Gardner, C. R., et al. (2007). Combined use of crystalline salt forms and precipitation inhibitors to improve oral absorption of celecoxib from solid oral formulations, *J. Pharm. Sci.*, **96**, 2686–2702.

5. Friesen, D. T., Shanker, R., Crew, M., Smithey, D. T., Curatolo, W. J., Nightingale, J. A. (2008). Hydroxypropyl methylcellulose acetate succinate-based spray-dried dispersions: an overview, *Mol. Pharm.*, **5**, 1003–1019.

6. Xia, D., Cui, F., Piao, H., Cun, D., Piao, H., Jiang, Y., et al. (2010). Effect of crystal size on the in vitro dissolution and oral absorption of nitrendipine in rats, *Pharm. Res.*, **27**, 1965–1976.

7. CDER (2000). Waiver of in vivo bioavailability and bioequivalence studies for immediate-release solid oral dosage forms based on a biopharmaceutics classification system, pp. 1–13.

8. Amidon, G. L., Lennernäs, H., Shah, V. P., Crison, J. R. (1995). A theoretical basis for a biopharmaceutic drug classification: the correlation of in vitro drug product dissolution and in vivo bioavailability, *Pharm. Res.*, **12**, 413–420.

9. Newman, A., Knipp, G., Zografi, G. (2012). Assessing the performance of amorphous solid dispersions, *J. Pharm. Sci.*, **101**, 1355–1377.

10. Dressman, J., Amidon, G., Reppas, C., Shah, V. (1998). Dissolution testing as a prognostic tool for oral drug absorption: immediate release dosage forms, *Pharm. Res.,* **15**, 11–22.

11. Curatolo, W., Nightingale, J. A., Herbig, S. M. (2009). Utility of hydroxypropylmethylcellulose acetate succinate (HPMCAS) for initiation and maintenance of drug supersaturation in the GI milieu, *Pharm. Res.,* **26**, 1419–1431.

12. Kennedy, M., Hu, J., Gao, P., Li, L., Ali-Reynolds, A., Chal, B., et al. (2008). Enhanced bioavailability of a poorly soluble VR1 antagonist using an amorphous solid dispersion approach: a case study, *Mol. Pharm.,* **5**, 981–993.

13. Vandecruys, R., Peeters, J., Verreck, G., Brewster, M. E. (2007). Use of a screening method to determine excipients which optimize the extent and stability of supersaturated drug solutions and application of this system to solid formulation design, *Int. J. Pharm.,* **342**, 168–175.

14. Janssens, S., de Armas, H. N., Roberts, C. J., Van den Mooter, G. (2008). Characterization of ternary solid dispersions of itraconazole, PEG 6000, and HPMC 2910 E5, *J. Pharm. Sci.,* **97**, 2110–2120.

15. DiNunzio, J. C., Hughey, J. R., Brough, C., Miller, D. A., Williams III, R. O., McGinity, J. W. (2010). Production of advanced solid dispersions for enhanced bioavailability of itraconazole using KinetiSol® dispersing, *Drug Dev. Ind. Pharm.,* **36**, 1064–1078.

16. Hughey, J. R., Keen, J. M., Miller, D. A., Brough, C., McGinity, J. W. (2012). Preparation and characterization of fusion processed solid dispersions containing a viscous thermally labile polymeric carrier, *Int. J. Pharm.,* **438**, 11–19.

17. Yamashita, T., Kokubo, T., Zhao, C., Ohki, Y. (2010). Antiprecipitant screening system for basic model compounds using bio-relevant media, *J. Assoc. Lab. Autom.,* **15**, 306–312.

18. Kim, M.-S., Kim, J.-S., Cho, W., Cha, K.-H., Park, H.-J., Park, J., et al. (2013). Supersaturatable formulations for the enhanced oral absorption of sirolimus, *Int. J. Pharm.,* **445**, 108–116.

19. Barillaro, V., Pescarmona, P. P., Van Speybroeck, M., Thi, T. D., Van Humbeeck, J., Vermant, J., et al. (2008). High-throughput study of phenytoin solid dispersions: formulation using an automated solvent casting method, dissolution testing, and scaling-up, *J. Comb. Chem.,* **10**, 637–643.

20. Shanbhag, A., Rabel, S., Nauka, E., Casadevall, G., Shivanand, P., Eichenbaum, G., et al. (2008). Method for screening of solid disper-

sion formulations of low-solubility compounds: miniaturization and automation of solvent casting and dissolution testing, *Int. J. Pharm.,* **351**, 209–218.

21. DiNunzio, J. C., Miller, D. A., Yang, W., McGinity, J. W., Williams III, R. O. (2008). Amorphous compositions using concentration enhancing polymers for improved bioavailability of itraconazole, *Mol. Pharm.,* **5**, 968–980.

22. Hughey, J. R., DiNunzio, J. C., Bennett, R. C., Brough, C., Miller, D. A., Ma, H., et al. (2010). Dissolution enhancement of a drug exhibiting thermal and acidic decomposition characteristics by fusion processing: a comparative study of hot melt extrusion and KinetiSol® dispersing, *AAPS PharmSciTech,* **11**, 760–774.

23. Hughey, J. R., Keen, J. M., Brough, C., Saeger, S., McGinity, J. W. (2011). Thermal processing of a poorly water-soluble drug substance exhibiting a high melting point: the utility of KinetiSol® Dispersing, *Int. J. Pharm.,* **419**, 222–230.

24. DiNunzio, J. C., Brough, C., Miller, D. A., Williams III, R. O., McGinity, J. W. (2010). Fusion processing of itraconazole solid dispersions by KinetiSol® dispersing: a comparative study to hot melt extrusion, *J. Pharm. Sci.,* **99**, 1239–1253.

25. DiNunzio, J. C., Brough, C., Miller, D. A., Williams III, R. O., McGinity, J. W. (2010). Applications of KinetiSol® dispersing for the production of plasticizer free amorphous solid dispersions, *Eur. J. Pharm. Sci.,* **40**, 179–187.

26. Miller, D. A., DiNunzio, J. C., Yang, W., McGinity, J. W., Williams III, R. O. (2008). Targeted intestinal delivery of supersaturated itraconazole for improved oral absorption, *Pharm. Res.,* **25**, 1450–1459.

27. Bennett, R. C., Brough, C., Miller, D. A., O'Donnell, K. P., Keen, J. M., Hughey, J. R., et al. Preparation of amorphous solid dispersions by rotary evaporation and KinetiSol dispersing: approaches to enhance solubility of a poorly water-soluble gum extract, *Drug Dev. Ind. Pharm.,* **0**, 1–16.

28. Overhoff, K., McConville, J., Yang, W., Johnston, K., Peters, J., Williams, R., III (2008). Effect of stabilizer on the maximum degree and extent of supersaturation and oral absorption of tacrolimus made by ultra-rapid freezing, *Pharm. Res.,* **25**, 167–175.

29. Thombre, A. G., Shah, J. C., Sagawa, K., Caldwell, W. B. (2012). In vitro and in vivo characterization of amorphous, nanocrystalline, and crystalline ziprasidone formulations, *Int. J. Pharm.,* **428**, 8–17.

30. Miller, D. A., DiNunzio, J. C., Yang, W., McGinity, J. W., Williams III, R. O. (2008). Enhanced in vivo absorption of itraconazole via stabilization of supersaturation following acidic-to-neutral pH transition, *Drug Dev. Ind. Pharm.*, **34**, 890–902.

31. Qian, F., Wang, J., Hartley, R., Tao, J., Haddadin, R., Mathias, N., et al. (2012). Solution behavior of PVP-VA and HPMC-AS-based amorphous solid dispersions and their bioavailability implications, *Pharm. Res.*, **29**, 2766–2776.

32. Kojima, T., Higashi, K., Suzuki, T., Tomono, K., Moribe, K., Yamamoto, K. (2012). Stabilization of a supersaturated solution of mefenamic acid from a solid dispersion with EUDRAGIT® EPO, *Pharm. Res.*, **29**, 2777–2791.

33. Banerjee, A., Chandrakumar, N. (2011). Real-time in vitro drug dissolution studies of tablets using volume-localized NMR (MRS), *Appl. Magn. Reson.*, **40**, 251–259.

34. Zhang, Q., Gladden, L., Avalle, P., Mantle, M. (2011). In vitro quantitative 1H and 19F nuclear magnetic resonance spectroscopy and imaging studies of fluvastatin™ in Lescol® XL tablets in a USP-IV dissolution cell, *J. Control. Release*, **156**, 345–354.

35. Grundy, J. S., Anderson, K. E., Rogers, J. A., Foster, R. T. (1997). Studies on dissolution testing of the nifedipine gastrointestinal therapeutic system. I. Description of a two-phase in vitro dissolution test, *J. Control. Release*, **48**, 1–8.

36. Shi, Y., Gao, P., Gong, Y., Ping, H. (2010). Application of a biphasic test for characterization of in vitro drug release of immediate release formulations of celecoxib and its relevance to in vivo absorption, *Mol. Pharm.*, **7**, 1458–1465.

37. Grundy, J. S., Anderson, K. E., Rogers, J. A., Foster, R. T. (1997). Studies on dissolution testing of the nifedipine gastrointestinal therapeutic system. II. Improved in vitro-in vivo correlation using a two-phase dissolution test, *J. Control. Release*, **48**, 9–17.

38. Vangani, S., Li, X., Zhou, P., Del-Barrio, M.-A., Chiu, R., Cauchon, N., et al. (2009). Dissolution of poorly water-soluble drugs in biphasic media using USP 4 and fiber optic system, *Clinl. Res. Regul. Aff.*, **26**, 8–19.

39. Alonzo, D. E., Gao, Y., Zhou, D., Mo, H., Zhang, G. G. Z., Taylor, L. S. (2011). Dissolution and precipitation behavior of amorphous solid dispersions, *J. Pharm. Sci.*, **100**, 3316–3331.

40. Babcock, W. C., Friesen, D. T., McCray, S. B. (2009) Bend Research, Inc., *Method and device for evaluation of pharmaceutical compositions*, U.S. Patent 7611630 B2.

41. Gao, Y., Carr, R. A., Spence, J. K., Wang, W. W., Turner, T. M., Lipari, J. M., et al. (2010). A pH-dilution method for estimation of biorelevant drug solubility along the gastrointestinal tract: application to physiologically based pharmacokinetic modeling, *Mol. Pharm.,* **7**, 1516–1526.

Chapter 13

Dissolution of Pharmaceutical Suspensions

Beverly Nickerson, Michele Xuemei Guo, Kenneth J. Norris, and Ling Zhang

Pfizer Inc., MS 8220-3467, Eastern Point Road, Groton, CT 06340, USA

beverly.nickerson@pfizer.com, michele.guo@pfizer.com, kenneth.j.norris@pfizer.com, ling.zhang@pfizer.com

13.1 Introduction

A suspension consists of insoluble solid particles dispersed in a liquid medium. The suspension typically consists of uniform particles that are readily suspended and easily dispersed. Suspension dosage forms include oral suspensions, topical suspensions, and suspensions for aerosols. Sterile suspension dosage forms include suspensions for injection (subcutaneous, intramuscular, or intra-articular) and suspensions for ophthalmic or otic administration. Some suspensions are ready for use, while others consist of solid powder mixtures which need to be reconstituted with a vehicle prior to use. In addition, some suspensions are prepared from tablets.[1]

Poorly Soluble Drugs: Dissolution and Drug Release
Edited by Gregory K. Webster, J. Derek Jackson, and Robert G. Bell
Copyright © 2017 Pan Stanford Publishing Pte. Ltd.
ISBN 978-981-4745-45-1 (Hardcover), 978-981-4745-46-8 (eBook)
www.panstanford.com

The most common reason to develop a suspension dosage form is limited aqueous solubility of the active pharmaceutical ingredient (API) at the dosage required. Another common reason to use suspensions is that they typically offer improved chemical stability compared to solutions. Suspensions are also used to achieve accurate weight-based dosing, which is limited for unit dosage forms such as tablets, by varying the volume of suspension delivered. In addition, suspensions offer advantages in taste masking compared to solutions and a more convenient dosage form compared to tablets for certain patients (e.g., pediatrics). Injectable suspensions offer prolonged duration of action and avoidance of first-pass metabolism.

This chapter focuses on dissolution and drug release from suspension dosage forms that are dosed by the oral route of administration and include oral suspensions, suspensions for reconstitution, and suspensions in capsules. Dissolution or drug release testing is required for these dosage forms. As specified in USP general chapter <1088> "In vitro and in vivo Evaluation of Dosage Forms," "Dissolution testing is required for all non-solution oral, including sublingual, Pharmacopeial dosage forms in which absorption of the drug is necessary for the product to exert the desired therapeutic effect. Exceptions include tablets that meet a requirement for completeness of solution, products that contain radio-labeled drugs, or products that contain a soluble drug and demonstrate rapid (10–15 min) disintegration".[2]

13.2 Types of Oral Suspension Formulations

13.2.1 Oral Suspensions

Oral suspensions can be available as "ready to use" or as a powder for constitution, commonly referred to as a powder for oral suspension (POS). The typical oral suspension formulation generally contains a sweetener, flavoring agent, buffers, preservatives, and a suspending agent along with the API. The suspending agent, often xanthan gum, provides a suitable fluid viscosity such that the

API is suspended in a homogeneous fluid. This enables adequate dose delivery throughout the use period for the product. This type of formulation is developed for the patient population that has difficulties swallowing tablets or capsules (e.g., pediatrics and geriatrics). It is also used to prepare veterinary formulations and formulations for dosing animals in pre-clinical studies.[3] An added benefit to this type of formulation is that the dosage volume can be easily manipulated allowing for individualized dose flexibility.

The term "POS" is used to describe both dry powder mixes and formulations prepared by granulation techniques. POS formulations are solid oral dosage forms that yield suspensions following constitution with suitable diluents. The POS formulation approach is typically used where chemical or physical instability of the drug precludes preparations of a "ready to use" suspension. Physical instability can include changes in the particle size of the API or a change in the polymorphic form of the API. Chemical or physical changes can affect the performance of the drug product.

13.2.2 Extemporaneous Preparations

Another type of suspension prepared from powders is extemporaneous preparations prepared from tablets or capsules. The tablets or the contents of capsules are thoroughly and uniformity pulverized by trituration. Trituration is a process in which the solid substance is reduced to fine particles in a mortar with a pestle. The resulting powder is then suspended in commercially available suspending agents and sweeteners such as Ora-Sweet® and Ora-Plus®.[4,5] Both of these commercially available solutions are mixtures containing coloring, flavoring, suspending agents, as well as preservatives. Previous reports have demonstrated that many drugs are compatible with these suspending agents.[4,5] In the book *Extemporaneous Formulations for Pediatric, Geriatric, and Special Needs Patients*, published by the American Society of Health-System Pharmacists, the majority of the oral preparations described are suspensions.[6] This formulation method is also used in early stages of drug development where the API is mixed with sweetener and suspending agents prior to dosing API.[7] Because extemporaneous preparations are typically prepared in a pharmacy or clinic with commercial tablets, there is limited information regarding the dissolution behavior of this type of formulation both initially and after aging. During the preparation of a suspension from tablets, the procedure of crushing and suspending the material in a vehicle prior to administration may alter the dissolution profile compared to the original dosage form. There are reports that grinding can cause polymorph changes of the drug substance or lead to reduced crystallinity or the formation of amorphous material.[8, 9] It is therefore recommended that dissolution studies be performed for all proposed extemporaneous preparations made from commercially approved products. However, this is not required in phase 1 and 2a studies, because the dissolution test is not a common attribute at that stage.

13.2.3 Nanosuspensions

Nanosuspensions are colloidal dispersions of nanometer sized (between a few nanometers and 1000 nm in size) drug particles which are stabilized by a suitable stabilizer (such as surfactants or polymeric stabilizers). The dispersion media can be water, aqueous solutions, or nonaqueous media. Reduction of drug particle size to the nanometer range leads to enhanced dissolution rates because of the increased surface area based on the Noyes Whitney equation[10] and because of the increased saturation solubility as described by the Freundlich–Ostwald equation.[11, 12]

There are several techniques available to manufacture nanosuspensions. Bottom-up techniques to prepare nanosuspenions start with the drug dissolved in a solvent and the drug is then precipitated in different ways. The top-down techniques reduce the drug particles to submicrometer units using techniques such as high-pressure homogenization or media/wet milling.[11, 13, 14] The nanosuspension can then be dosed as a suspension or the suspension can be converted to a powder (through spray drying, lyophilization, etc.) and then formulated as a tablet, capsule, or other solid oral dosage form.

13.2.4 Suspension in Gelatin Capsules

Gelatin capsules have become popular in pharmaceutical development. There are soft gelatin, often refered to as softgel, and hard gelatin capsules. The gelatin shell encapsulates solubilized, suspended, or semisolid API in nonaqueous vehicles. Capsules with suspension fill can overcome several pharmaceutical development limitations and offers advantages for development of low-solubility compounds. The low-solubility API can be suspended in nonaqueous excipient vehicles at the dosage required for therapeutic strength. The suspension formulation may provide better stability compared to other approaches for low solubility compounds, such as dissolving in co-solvent, surfactant, etc. The gelatin shell can mask odors and tastes associated with the drug, hold dye color and provide convenience for processing. It also protects the drug substance from light and oxygen.

The gelatin powder is made from hydrolyzed collagen from animal bones or hides using either an acidic or basic process. As with proteins, gelatin contains both amine and carboxylic groups. Type A gelatin, made from the acidic hydrolysis process, has isoelectric points (IEPs) in the range of 7–9 and type B gelatin, made from the basic hydrolysis process, has IEP values in the range of 4.7–5.3.[15] Gelatin is readily dissolved in gastric fluid in the GI tract.

Different from hard gelatin shells which are premade, soft gelatin shells are prepared in the encapsulation station. Prior to encapsulation, the gelatin powder is prepared into a molten mass with water, plasticizer, and other minor additives added. During encapsulation, the gelatin mass is pressed into two ribbons, a semimolten-state gelatin. The ribbons are lubricated and drawn into the encapsulation station. The fill material (drug formulation) is fed from the fill reservoir to the encapsulation station and accurate volumes of the fill are injected into the space between the two ribbons and encapsulated. With newer equipment, inline printing on the capsules can be performed. During encapsulation, fill weight, ribbon thickness, and seam thickness are the important measurements to ensure product quality, such as content uniformity and capsule integrity. When the softgel capsules are just made, they are very soft with strong elasticity. The capsules are then dried in tumbler dryers with low heat and low humidity air for a short period of time, followed by tray drying at ambient temperature but low humidity typically for 3–10 days depending on the product. At the end of drying process, equilibrium of the moisture in the capsule shell and fill material is reached.

13.3 Dissolution Mechanisms for Suspensions

Mechanisms proposed for the dissolution of suspensions, based on a quasi-steady-state diffusion model, have been summarized by Abdou and are based on the assumption of sink conditions, monodisperse particles with a spherical particle shape, and a diffusion coefficient independent of concentration.[16] These models are summarized in Table 13.1, where h is the diffusion layer thickness, and show that the rate of change (da/dt) of the particle diameter (a) is dependent upon the drug solubility (C_s), the diffusion coefficient (D), and the particle density (ρ). k is a constant.

As shown in Table 13.1, drug solubility will affect the rate of change of the particle diameter, or the dissolution rate of the suspensions, as it does for other formulations types. The dissolution rate of suspensions is also affected by API particle diameter; hence particle size and particle size distribution of the API can impact dissolution of the suspension. As with tablets and capsules, a reduction in API particle size, in general, will increase the dissolution rate. This occurs because of the increased API surface area in contact with the dissolution medium. The rate of dissolution

Table 13.1 Dissolution rate models for suspensions[16]

Model assumptions	Rate of change of particle diameter
Diffusion layer thickness remains constant as the particle dissolves	$\dfrac{da}{dt} = -\dfrac{2DC_s}{h\rho}$
Diffusion layer thickness is proportional to the square root of the particle diameter	$\dfrac{da}{dt} = -\dfrac{2DC_s}{ka^{1/2}\rho}$
Diffusion layer thickness is proportional to the particle diameter	$\dfrac{da}{dt} = -\dfrac{2DC_s}{a\rho}$

of a suspension is dependent on the diffusion coefficient of the API which in turn has an inverse dependency on the viscosity of the fluid. An increase in the suspension viscosity in general results in a decrease in the rate of dissolution of the suspension. The type and concentration of the suspending agent can affect the dissolution rate of suspensions as this will impact the viscosity of the suspension. These factors and examples are discussed in more detail later in the chapter.

13.4 Properties That Affect Dissolution of Suspensions

Several important API properties influence the dissolution of the API from suspensions, including solubility, particle size distribution, and solid form. An increase in the effective particle size upon settling and aggregation may decrease the dissolution rate. Likewise, polymorphs, solubility, particle size distribution, and changes in polymorphic form have been shown to influence dissolution and subsequent drug adsorption.[17] In addition, a critical formulation factor that influences the dissolution of the API from a suspension formulation is the viscosity of the formulation.

13.4.1 API Particle Size

Particle size of the drug substance has been found to significantly influence suspension dissolution.[18,19] As an example, Fig. 13.1 shows impact of particle size on dissolution rate of a low-solubility compound in suspension where the smaller particles have the faster dissolution rate.[18] In another example, Mauger et al. studied dissolution profiles of four finely divided particle suspensions obtained by the technique of centrifugal elutriation.[19] The dissolution profiles are substantially different among the four suspension samples which have mean particle sizes of 4.9, 7.1, 10.6, and 13.5 μm, with the fastest dissolution profile observed from the smallest particle suspension.

For poorly soluble compounds, one way to increase the dissolution rate is to prepare a nanosuspension which reduces API

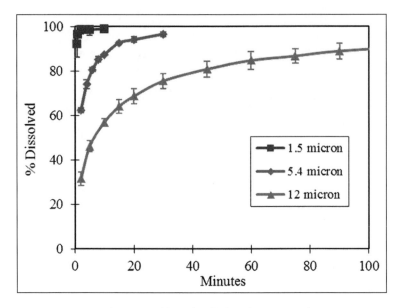

Figure 13.1   Dissolution profiles of milled and micronized API in suspensions showing particle size effect for a compound with 77 μg/mL solubility. Reprinted with permission from Rohrs.[18]

particle size below 1 μm and significantly increases API surface area. Various studies report significantly increased dissolution rates from nanosuspension dosage forms.[20,21] As an example, Lai and coworkers evaluated the dissolution of diclofenac (DFC), a potent nonsteroidal anti-inflammatory drug with very low aqueous solubility which increases with pH.[20] They prepared a coarse suspension of DCF and a nanosuspension of DCF, then lyophilized the suspensions. Samples of the lyophilized coarse suspension, lyophilized nanosuspension, physical mixture of DCF and excipients, and DCF equivalent to 25 mg DCF were each placed into gelatin capsules. These filled capsules were then studied by dissolution testing using USP apparatus 1 (rotating baskets) at 100 rpm with 500 mL of media at 37°C. As shown in Fig. 13.2, the drug dissolution rate increased with pH of the dissolution medium and the dissolution profile for the nanosuspension sample was faster and reached a higher percentage dissolved level than for the other samples.[20]

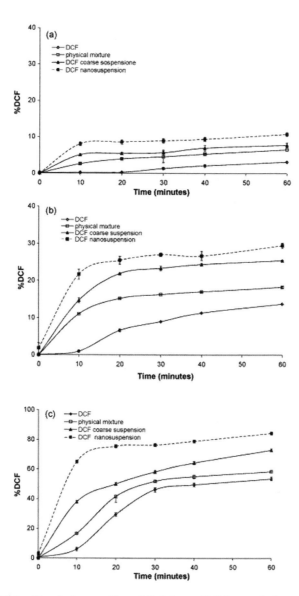

Figure 13.2 Dissolution profiles of diclofenac (DCF) capsule formulations (bulk DCF, physical mixture, lyophilized coarse suspension, lyophilized nanosuspension): (a) in SGF (pH 1.2); (b) in water (pH 5.5); (c) in SIF (pH 7.5). All dissolution experiments were carried out at 37°C. Error bars represent standard deviation of three independent experiments. Reprinted with permission from Lai et al.[20] Copyright (2009) Elsevier.

A potential issue for suspension formulations not typically observed with some other formulation types, such as tablets and solid filled capsules, is that the API particle size in the suspension could change after the suspension is prepared. An increase in the effective particle size upon settling and aggregation may decrease the dissolution rate. Crystallization of a solubilized compound from the solution fill in a capsule can occur. It is known that the smaller particles in a suspension will dissolve preferentially in the surrounding vehicle compared to larger particles. Driven by thermodynamic equilibrium, the dissolved API may crystallize on the surface of the larger crystals, which results in larger particles or agglomerates (e.g., Ostwald ripening). This process may be accelerated by temperature fluctuations.

It should be noted that processing conditions can alter the solution dynamics of the filled softgel in a manner that can impact the dissolved fraction and saturation conditions resulting in physical changes in the formulation/API. During the softgel encapsulation process, a significant amount of water is introduced into the suspension fill and then is pulled out during the drying process, leaving the water content of the drug product at a level where the fill and shell reach equilibrium. With the significant amount of water going in and out of the fill formulation, special attention should be paid to the API solubility. The drug could dissolve and precipitate, forming large particles. Young and Buckton reported the increase in particle size for a suspension with high interfacial surface energy in saturated aqueous solution which occurred in the course of hours and days.[22] An increase in particle size may reduce the dissolution rate. Dissolution testing for suspensions and suspension fills in softgel capsules is a critical test to monitor product performance throughout the shelf life.

13.4.2 API Crystalline Form

Haleblian and McCrone reviewed pharmaceutical applications of drug polymorphism.[23] Different API polymorphic forms may have varied solubilities, melting points, densities, hardnesses, crystal shapes, optical and electrical properties, vapor pressures, etc.[23] The crystalline form, or polymorphism, of an API influences its

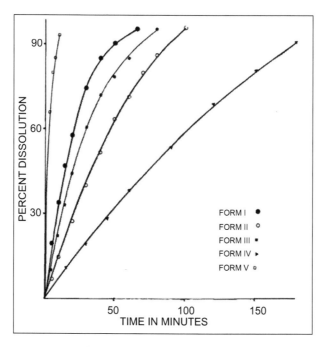

Figure 13.3 Dissolution rate curves of the five polymorphic forms of chlorpropamide. Reprinted with permission from Al-Saier and Riley.[27] Copyright (1982) Elsevier.

dissolution rate.[24–28] Al-Saier and Riley studied dissolution profiles of chlorpropamide (CPM) polymorphs using a rotating basket method and reported that the four metastable forms showed faster dissolution rates than that of the stable form.[27] The dissolution rates of the five polymorphic forms are shown in Fig. 13.3. Ueda and coworkers extended the study of CPM.[26] Six polymorphic forms of CPM were subjected to stationary disk dissolution. The study reported that meta stable form II had a much faster dissolution rate than the other forms initially, but then the concentration gradually decreased due to conversion to the stable form during dissolution.

During storage, crystal changes as a result of hydration, solvation, and polymorph transformation may occur. When working with suspensions, companies strive to develop the thermodynamically stable form of the API. It is possible, however, that the most thermodynamically stable form of the API is not identified during development activities. In some cases the amorphous form is selected for formulation development due to its higher solubility compared to the crystal form. The solubility difference in forms may result in in vivo absorption differences. Mullins and Macek studied novobiocin polymorphism and its oral absorption in dogs.[29] The study showed that the amorphous form has significantly higher solubility than the crystaline form. Following oral administration in dogs, the amorphous form is readily absorbed while the plasma concentration for the crystal form is not detected.

In cases where the metastable form is used in the formulation, the change of a metastable form to a more stable form during suspension manufacture and/or during storage may occur. For poorly soluble compounds for which the solubilty is the dissolution rate–limiting factor, when the solid form changes product dissolution can be reduced as the more stable form has lower solubility. Dissolution may be the rate-limiting step for absorption of these compounds. It is critical to monitor dissolution during product stability to ensure that product performance does not change during storage. Dissolution as product performance testing may or may not be biorelevent. An example of a change in API crystal form in a suspension is amorphous metronidazole, which was formulated as an aqueous suspension that converted to the monohydrate form, leading to crystal growth.[30] Carbamazepine anhydrate formulated as a suspension converted to the dihydrate form, which has a lower solubility and slower dissolution profile, which affected the overall exposure.[3]

13.4.3 Viscosity

A critical formulation factor that influences the dissolution of the API from a suspension formulation is the viscosity of the formulation. The viscosity of the suspension is dependent on the suspending agent used and its concentration. The suspending agent helps reduce the sedimentation rate of particles in the suspension by increasing the viscosity of the vehicle. Typical suspending agents that are used in suspensions include xanthan gum, sodium carboxymethylcellulose, cyclodextran, methylcellulose,

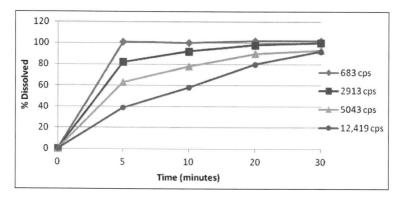

Figure 13.4 Dissolution profiles of a powder for oral suspension formulated at four different viscosity levels. Dissolution performed using USP apparatus 2 (paddles) at 50 rpm with 900 mL 0.01 N HCl.[33]

and high molecular weight povidone. The viscosity of suspensions can affect the in vitro dissolution of the drug as the diffusion coefficient decreases with the increasing viscosity. Other properties of the suspending agent, such as anionic charge, can influence the dissolution properties of the API.[31] Because of the patient population using suspension formulations, taste masking may be needed to improve patient compliance. Taste masking techniques such as encapsulation, complexation to cyclodextrin, or improved mouth feel via increased viscosity can retard the dissolution of the API.[32]

In the results presented in Fig. 13.4 four suspension formulations of a BCS class II API (based on solubility at neutral pH) were prepared at different viscosities.[33] The viscosity of the formulation was governed by the amount of xanthan gum present in the formulation. The range of viscosities that were evaluated were between 683 cps to 12,419 cps. Greater than 80% of the API was dissolved in less than 5 minutes for the lowest viscosity sample, while the highest viscosity sample took over 20 minutes to reach this level.

The relationship between higher suspension viscosity and slower rate of dissolution has been reported previously.[34,35] As discussed in Section 3.3, dissolution rate is dependent on the diffusion coefficient which has an inverse relationship with the fulid viscosity. It has been postulated that this retardation of dissolution in viscous suspensions is due to the hydrated polymer chains of the suspending agent causing resistance in the diffusion process of the API.[36] The difference could also be explained by how quickly the sample disperses. The lower the viscosity the quicker the sample is dispersed in the dissolution media.[34]

13.5 Dissolution Method Development for Suspensions for Oral Administration

In 2003 the Federation International Pharmaceutique (FIP) and the American Association of Pharmaceutical Scientists (AAPS) published a position paper which discussed recommendations for dissolution and in vitro release testing for novel and special dosage forms, including suspensions.[37,38] In this paper the rotating paddle method using an aqueous medium is recommended for oral suspensions intended for systemic use. The selection of dissolution medium and agitation rate should allow discrimination between batches with different release properties, and be able to identify manufacturing process changes, and product changes during storage over the shelf life.

13.5.1 Selection of Apparatus and Agitation Speed

Oral suspensions share many physiochemical characteristics of tablets and capsules with respect to the process of dissolution.[17] Although USP Apparatus 2 was mainly designed for tablets and capsules, it has also been the major technique used to evaluate the dissolution performance of suspensions.[7,32,36–41] Because suspensions are similar to a tablet that has disintegrated, a slower paddle speed is used for the dissolution testing of suspensions, typically ranging from 25–75 rpm.[37,38] Typically an agitation rate of 25 rpm is recommended for low viscosity suspensions,[42] while higher rates such as 50 or 75 rpm may be required for high viscosity suspensions. For an immediate release tablet, the disintegration of the tablet may be rate limiting for dissolution. Since there is not a disintegration step needed for the dissolution of the API to begin for a suspension,

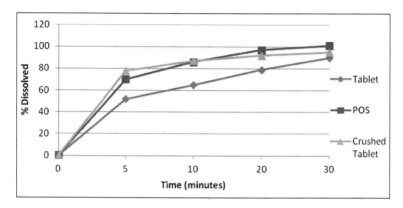

Figure 13.5 Comparison of dissolution profiles for a BCS class II compound formulated as a tablet, crushed tablet and a powder for oral suspension. Dissolution testing was performed using USP apparatus 2 (paddles) at 50 rpm with 900 mL 0.01 N HCl.[33]

it can be difficult to obtain a reasonable dissolution profile for suspension formulations under sink conditions at higher agitation speeds.

When developing a dissolution method for a suspension formulation, it is not uncommon to start with the conditions used for the tablet or capsule formulation if it exists. Typically these conditions are sink conditions and are based on substantial compound understanding. Based on the authors unpublished work, the dissolution profiles for a BCS class II API from a immediate release tablet, crushed tablet, and a constituted POS at equal strength are shown in Fig. 13.5.[33] The POS formulation was prepared at commercial scale by blending the API with the excipients and bottling the powder mixture. Once constituted with water the POS formulation had an API concentration of 10 mg/mL. As can be seen in Fig. 13.4, the tablet formulation is the slowest to release of either of the other two formulations. This statistical difference is not unexpected as suspensions correspond to post-disintegration stage of the tablet so only the dissolution rate is being observed.

The dissolution profiles obtained for the same suspension formulation of a BCS class II compound under the agitation conditions of 25, 35, and 50 rpm are shown in Fig. 13.6.[33] As can

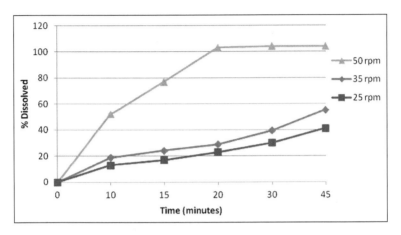

Figure 13.6   Effect of paddle speed on the dissolution profile of a BCS class II compound formulated as a suspension (viscosity 2900 cps). Dissolution performed using USP apparatus 2 (paddles) in 500 mL McIlvaine buffer, pH 5.[33]

be seen for this formulation having a viscosity if 2900 cps, a paddle speed of 50 rpm was required to obtain complete dissolution. The viscosity of this formulation requires additional energy to break up the suspension so the dissolution of the API can take place.

As with other types of oral suspensions, USP apparatus 2 (paddles) is the most commonly used method for dissolution testing of nanosuspensions. A few examples are discussed below.

Liu and colleagues studied the effect of sink and non-sink conditions on dissolution profiles for indomethacin nanosuspensions.[43] In this work several indomethacin nanosuspension formulations were prepared with the same chemical components but with different particle sizes. Dissolution of the suspensions were tested using USP apparatus 2 (paddles) at sink conditions and close to the solubility limit of the drug. The dissolution media studied were 600 mL hydrochloric acid pH 1.2 and phthalate buffer pH 5 with agitation rates of 50 and 120 rpm. Lower dissolution rates and more discriminating dissolution profiles were observed when the drug amount used for dissolution testing was close to the solubility of the drug in the dissolution medium compared to those profiles obtained under sink conditions.

Megace ES® (Enhanced Solubility) is an example of a commercially available nanosuspension product using NanoCrystal™ technology to enhance the performance of the poorly water-soluble megestrol acetate. Megace ES® (125 mg megestrol acetate/mL) is an oral suspension used for the treatment of anorexia (loss of appetite), cachexia (severe malnutrituion), or an unexplained, significant weight loss in patients diagnosed with AIDS. Megace ES® is a reformulation of Megace® oral suspension to obtain a formulation with a higher dissolution rate and improved bioavailability in the unfed state and to improve dose administration by reducing the dosing volume and suspension viscosity.[12,44]

Since megestrol acetate is generally insoluble in aqueous media, a surfactant in an aqueous medium is needed to aid in solubilization of the drug during dissolution testing. The USP monograph for megestrol acetate oral suspension (regular suspension, not nanosuspension)[45] contains three different dissolution test options using USP <711>.[46] Tests 1 and 2 both specify USP apparatus 2 (paddles) at 25 rpm with 900 mL of 0.5% sodium lauryl sulfate in water at 37°C. The sampling time is 30 minutes with a Q of NLT 80% dissolved. Withdrawn dissolution samples are filtered through a 0.45 µm pore size filter. In test 1, sample is introduced to the dissolution vessel by transferring an accurately measured volume of oral suspension (freshly mixed and free of air bubbles), equivalent to 160 mg of megestrol acetate, to the surface of the medium in the dissolution vessel. In test 2, 10 mL of oral suspension is withdrawn using a 10 mL syringe, bubbles are removed from the syringe, the needle is removed and a gross weight is obtained. The oral suspension is then rapidly dispensed to the side of the vessel about halfway from the bottom of the vessel. The syringe is weighed to obtain the tare weight and the sample weight is determined. Test 3 uses the same apparatus, conditions, time point, and Q value as tests 1 and 2, but agitation is performed at 50 rpm. Sample is introduced to the dissolution vessel as in test 2 with sample being introduced into the vessel over a 10- to 15-second time period.

For Megace ES® suspension (the nanosuspension), the CDER recommended dissolution method is USP apparatus 2 (paddles) at 25 rpm with 900 mL of 1.0% sodium lauryl sulfate in water at 37°C.

The sample amount recommended is 5 mL (one unit dose) with a Q at 10 minutes.[47]

13.5.2 Media Selection (pH, Surfactant)

The choice of medium has a significant impact on the dissolution of poorly soluble compounds. For a quality control dissolution test an aqueous media with a pH within the physiological range from 1.2 to 7.4 is typically used if at all possible. The final choice of media pH should be based on the drug's solubility pH profile. Sink conditions are preferred to mimic in vivo situation and for improved method robustness.

For poorly soluble compounds, a suitable surfactant can be used to aid solubility. The use of a surfactant is justified as there are natural surfactants present in the gastrointestinal tract. There are anionic (e.g., sodium lauryl sulfate, sodium deoxycholate), cationic (e.g., cetyl trimethylammonium bromide, hexadecyltrimethyl ammonium bromide) and neutral (e.g., Tween 20, Tween 80, Triton X-100, Brij-35) surfactants commercially available. The type of the surfactant used in the method should be based on the compatibility of the surfactant with the drug and excipients in the formulation. The minimal amount of surfactant which is effective in improving the compound solubility should be used. As examples, methods using 0.1–2% sodium lauryl sulfate or sodium dodecyl sulfate have been reported.[48]

Since dissolution for poorly soluble compounds can be rate limiting to absorption and can be dependent on factors such as surfactants, pH, and ionic strength, biorelevant media may be needed to mimic product in vivo performance and establish an in vitro–in vivo correlation (IVIVC). Several biorelevent media are reported:[49–53] fasted state simulated gastric fluid (FaSSGF), fed state simulated gastric fluid (FeSSGF), fasted state simulated intestinal fluid (FaSSIF), fasted state simulated intestinal fluid – V2 (FaSSIF-V2), fed state simulated intestinal fluid (FeSSIF), and fed state simulated intestinal fluid – V2 (FeSSIF-V2). While dissolution in biorelevant media may have a good chance of predicting in vivo performance, the preparation of biorelevant media is costly and tedious. Biorelevant media, therefore, are usually not practical for a quality control method. More detailed information on biorelevant dissolution is provided in Chapters 14 and 15.

A quality control dissolution method could have biorelevance as well if proved to be so. The typical way to develop an IVIVC is to develop two or three significantly different release profile products. The products are subjected to various dissolution conditions (apparatus, rotation speed, media with different pHs, surfactants if needed) to understand the effects of various dissolution parameters on the product release. The products with different release profiles are then used in a pharmacokinetic study. The plasma concentrations are deconvoluted to provide the absorption profile. The product absorption profile and dissolution profiles are compared and a possible IVIVC may be established as outlined in an FDA guideline.[54] The details of developing an IVIVC is not in the scope of this chapter. More discussion on IVIVCs can be found in Chapter 15 this book.

13.5.3 Sample Preparation and Introduction into the Dissolution Vessel

As noted in the FIP/AAPS position paper, sample preparation for oral suspensions is important and the sample should be shaken or mixed appropriately (e.g., as specified on the product label direction for use) to ensure a representative sample can be obtained for the dissolution test. Typically a sample weight or volume equivalent to a single dose of the product is used; however, a partial dose can be used to avoid the use of a surfactant to obtain sink conditions.[37,38] Sample introduction to the dissolution vessel should be evaluated and should ensure accurate, precise, and reproducible results based on the viscosity and composition of the suspension matrix. The location within the dissolution vessel where the sample should be introduced needs to be evaluated and should be specified in the method. In some cases the rate of sample introduction may be important.

Introduction of the sample to the bottom of the dissolution vessel is recommended for low-viscosity suspensions.[37,38] Dissolution methods exist that introduce the suspension sample to the bottom of the vessel,[55,56] midway between the surface of the medium and the top of the rotating blade[45,57,59,60] or at the surface of the dissolution medium.[45,58,61]

High-viscosity suspensions may require that the dose used in the test be determined by weight (i.e., weigh by difference)[45,56−58,62,63] or by using a quantitative transfer to the dissolution vessel.[55,60] Alternatively, a positive displacement pipet may be used.[59] The sample weight or volume used in the dissolution test should represent a typical single dose of the product; however, using a partial dose of the product (10–20%) is preferred over using a surfactant to obtain sink conditions.[37,38]

13.6 Additional Dissolution Method Development Considerations for Suspension Filled Gelatin Capsules

The therapeutic drug should be released from the formulated drug product and thus be available for absorption upon oral administration. Dissolution testing monitors the drug release process from the formulation. For hard gelatin shell or softgel products with a suspension fill, the dissolution process includes shell rupture and subsequent release and dissolution of the fill contents. Dissolution of the suspension fill in capsules should be conducted for both release and stability to monitor product quality over its shelf life.

The softgel shell typically ruptures and dissolves within 15 minutes in the dissolution media during testing.[63] The rupture test is listed in USP general chapter <2040> "Disintegration and Dissolution of Dietary Suppliments".[64] The rupture tolerance is set as NMT 15 minutes. Fresh gelatin shells dissolve within a few minutes, while aged gelatin shells may take longer due to shell crosslinking. The dissolution rate of the gelatin shell for both hard shell and softgel is influenced by a number of factors, such as gelatin type, bloom strength, dissolution media pH and temperature, and agitation rate. Hom's study showed that type B gelatin dissolves faster than type A gelatin under all conditions.[65] Both types of gelatin showed a much faster dissolution rate in media with pH below 3, which may be attributed to the protonation of the amine groups in the gelatin. While in media with pH 3–10, no significant changes in dissolution rate were seen except that the type B gelatin dissolution rate is at a minimum at pH 5, which might be attributed to the isoelectric point of type B gelatin being in the range of pH 4.7–5.1. The gelatin bloom strength also affects the dissolution rate. The higher the bloom strength of the gelatin, the slower it dissolves. Hom's study also showed that the gelatin dissolution is a diffusion-controlled process. Thus, it is not surprising that the gelatin dissolution rate increases as media temperature or agitation rate increases.[65]

The gelatin shell may undergo crosslinking when stored at accelerated conditions for long periods of time, which is also accelerated under higher temperature and/or humidity. Intra- and interpeptide crosslinking can be formed. Impurities present in the formulation components, such as aldehydes and peroxides, are reportedly involved in the chemical induced crosslinking. Digenis et al. published a review on gelatin crosslinking with proposed chemistry schemes.[66] Gelatin shell crosslinking can also form when the capsules are exposed to elevated temperature and occurs via condensation reaction between carboxylic and amino groups in the peptide chain.[67] When the three dimensional crosslinking forms, the shell solubility is significantly reduced. During dissolution, a thin rubbery pellicle layer can be observed inside the capsule shell. Although it is fragile, the pellicle is not readily dissolved in dissolution media under normal operation conditions. The pellicle impedes the release and dissolution of the encapsulated drug during dissolution testing.

A gelatin capsule working group from USP, FDA, and industry was formed to address the gelatin crosslinking. Studies have shown that capsules with a moderate amount of crosslinking of the gelatin shell are bioequivalent to capsules without crosslinking.[68–72] Studies on the bioavailability of capsules with crosslinked gelatin shells led to the use of a digestive enzyme in the dissolution media to dissolve the gelatin shell. If the dissolution of a formulation with gelatin shell fails to meet the specification due to crosslinking, USP <711>[46] and USP <2040>[64] provide instructions for tier 2 dissolution test using digestive enzymes in the dissolution media. The enzymes cleave the peptide bonds so to have shell ruptured and have the content

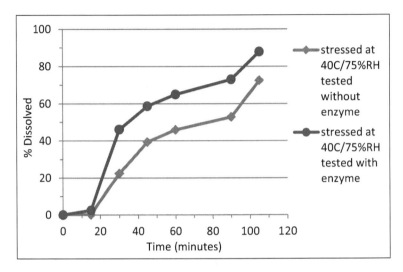

Figure 13.7 Dissolution profiles of a soft gelatin capsule formulation which show dissolution rate slowdown as a result of crosslinking formed in the gelatin shell.[33]

released. Up to 750,000 activity/L pepsin and 1725 activity/L of pancreatin are suggested to be used in dissolution media with pH less than 6.8 and greater than or equal to 6.8, respectively. Figure 13.7 shows dissolution profiles from tier 1 and 2 testing of a softgel product stored at 40°C/75%RH for 6 months.[33] The tier 2 dissolution rate is significantly faster than that of tier 1, which shows crosslinking was formed during the extended period storage at accelerated storage condition. The pancreatin added in the tier 2 media cleaved the peptide bond on gelatin shell thus enabling additional release of the drug.

When the suspension content of a soft gelatin capsule is exposed to the dissolution media, the dissolution is a process of drug substance dispersion, diffusion, and dissolving in the media. As shown in the equations in Table 13.1, the dissolution rate is influenced by a number of factors: saturated solubility, diffusion layer thickness, surface area, dose (concentration in the bulk media), and media viscosity (dissolution constant). For low-solubility compounds, the difference between the saturated solubility and bulk concentration might be small (non-sink condition). In these cases, it is important to use other means, such as a surfactant, to increase solubility. The physical properties of the API, such as particle size and polymorphism, significantly influence the API solubility, and thus dissolution rate, and possibly in vivo performance. In the suspension formulation, the particle size and polymorphism and their stability during shelf life are important parameters to consider in developing a stable product. Capsule dissolution should be monitored during stability to ensure product performance.

13.7 Compendial Dissolution Methods for Suspensions

While several USP monographs exist for oral suspensions and suspensions for reconstitution (e.g., POS), not all of these monographs contain dissolution as a performance test. Eight monographs for oral suspension[45,55–57,59–62] and five monographs for powders for oral suspensions[58,72–75] require dissolution testing and dissolution conditions for these monographs are summarized in Tables 13.2–13.3. In addition four monographs for tablets for oral suspension[76–79] require dissolution. As shown in Tables 13.2 and 13.3, the majority of the monographs for oral suspensions and for powders for oral suspension use USP apparatus 2 at 25 or 50 rpm. Samples for oral suspensions and powders for oral suspension are introduced to the surface of the medium, to the middle of the medium or to the bottom of the medium. Both syringes and positive displacement pipets are used for sample introduction in the monographs.

The FDA Dissolution Methods Database provides recommended dissolution methods for 56 oral suspension products.[48] The recommended methods are consistent with the FPI/AAPS recommendations.[37,38] All of these methods use USP apparatus 2 with 900 mL of dissolution medium. The majority of the methods use a paddle speed of 25 to 75 rpm. The dissolution media vary and include water, 0.1 N HCl, simulated gastric fluid without enzyme, simulated intestinal fluid, and buffers with pH values ranging from 3.0 to 7.5. Thirteen methods include surfactant (e.g.,

Table 13.2 USP monographs for oral suspensions[45,55-57,59-62]

USP monograph	USP apparatus (speed)	Dissolution medium (volume)	Tolerance	Sample preparation and introduction
Felbamate oral suspension	Apparatus 2 (50 rpm)	Water (900 mL)	NLT 80% (Q) at 15 min	Using a syringe, accurately weigh by difference approximately 5 mL of well mixed suspension. Introduce the sample into the dissolution vessel with the paddles rotating, and avoid getting the sample on the paddle or shaft.
Ibuprofen oral suspension	Apparatus 2 (50 rpm)	Phosphate buffer, pH 7.2 (900 mL)	NLT 80% (Q) at 60 min	Withdraw about 10 mL of well mixed suspension using a syringe and accurately weigh by difference to determine the sample weight. Transfer the sample to the vessel between the surface of the medium and the top of the rotating blade. Reweigh the syringe to determine the sample weight.
Indomethaicin oral suspension	Apparatus 2 (50 rpm)	0.01M Phosphate buffer, pH 7.2 (900 mL)	NLT 80% (Q) at 20 min	Transfer to the surface of medium in the vessel a volume of suspension, freshly mixed and free of air bubbles, equivalent to about 25 mg indomethacin.

Megestrol acetate oral suspension				
Test 1	Apparatus 2 (25 rpm)	0.5% SLS in water (900 mL)	NLT 80% (*Q*) at 30 min	Sample should be freshly mixed and free of air bubbles. Transfer sample to surface of medium.
Test 2	Apparatus 2 (25 rpm)	0.5% SLS in water (900 mL)	NLT 80% (*Q*) at 30 min	Withdraw sample using syringe and weigh. Rapidly dispense to the side of the vessel about halfway from the bottom. Reweigh the syringe to obtain the sample weight.
Test 3	Apparatus 2 (50 rpm)	0.5% SLS in degassed water (900 mL)	NLT 80% (*Q*) at 30 min	Same as test 2.
Meloxicam oral suspension	Apparatus 2 (25 rpm)	Phosphate buffer, pH 7.5 (900 mL)	NLT 75% (*Q*) at 15 min	Weigh sample in tared beaker. Introduce sample to the middle of the vessel. Rinse beaker with medium withdrawn from vessel and return to vessel. Lower paddle and begin rotation.
Nevirapine oral suspension	Apparatus 2 (25 rpm)	0.1N HCl (900 mL)	NLT 80% (*Q*) at 45 min	Gently shake by inversion and rotation. Ensure free of air bubbles. Use a positive displacement pipet to withdraw sample. Remove excess by wiping the outside of the tip without touching the opening of the tip. Introduce the sample to the vessel over 1-2 s by immersing the tip of the pipet into the medium midway between the shaft and vessel wall about 1 cm below the meniscus.

(Contd.)

Table 13.2 (*Contd.*)

USP monograph	USP apparatus (speed)	Dissolution medium (volume)	Tolerance	Sample preparation and introduction
Oxcarbazepine oral suspension	Apparatus 2 (75 rpm)	1% SDS in water (890 ml)	NLT 80% (Q) at 30 min	Shake bottle. Withdraw sample using a syringe. Using a long needle on the syringe, deliver the sample to the bottom of the vessel. Use media from the vessel to clean the syringe and place back into the vessel. Start rotation of the paddle.
Phenytoin oral suspension	Apparatus 2 (35 rpm)	0.05M Tris buffer, pH 7.5, degassed (900 mL)	NLT 80% (Q) at 60 min	Shake well. Withdraw sample in syringe and weigh. With paddles lowered, gently empty syringe contents into the bottom of each vessel containing medium. Start rotation. Reweigh syringe to obtain sample weight.

Table 13.3 USP monographs for powders for oral suspension[58,72–75]

USP monograph	USP apparatus (speed)	Dissolution medium (volume)	Tolerance	Sample preparation and introduction
Cefadroxil for oral suspension	Apparatus 2 (25 rpm)	Water (900 mL)	NLT 75% (Q) at 30 min	Transfer 5.0 mL of the constituted suspension (weighed) to the dissolution vessel.
Cefdinir for oral suspension	Apparatus 2 (50 rpm)	0.05M Phosphate buffer, pH 6.8 (900 mL)	NLT 80% (Q) at 30 min	Transfer 5 mL, by weight, of the constituted suspension into the vessel.
Cefuroxime axetil for oral suspension	Apparatus 2 (50 rpm)	0.07M Phosphate buffer, pH 7.0 (900 mL)	NLT 60% (Q) at 30 min	Test 5.0 mL of constituted suspension equivalent to 125 or 250 mg of cefuroxime.
Flucanozole for oral suspension	Apparatus 2 (50 rpm)	Water (900 mL or 500 mL depending on dose)	NLT 85% (Q) at 30 min	Reconstitute the suspension according to the label instructions. Weigh, and transfer an amount of the reconstituted suspension equivalent to one dose to the vessel.
Mycophenolate mofetil for oral suspension	Apparatus 2 (40 rpm)	0.1N HCl, deaerated (900 mL)	NLT 80% (Q) at 20 min	Reconstitute the suspension according to the labeling instruction and shake well. Using a separate syringe for each vessel, withdraw 2 mL of suspension and remove air bubbles. Adjust the volume to 1.2 mL and determine the sample weight by difference. Operate the apparatus, hold the syringe above the surface of the medium halfway between the paddle shaft and the vessel wall and introduce the sample to the vessel over a 5–10 s period.

Brij-35, hexadecyltrimethyl ammonium bromide, polysorbate 80, polysorbate 20, SLS) in the dissolution medium.

13.8 Non-compendial Dissolution Methods and Alternative Methods for Suspensions

As previously discussed, USP apparatus 2 is the most commonly reported method for suspensions and the only method reported in USP monographs and the FDA Dissolution Method Development Database. Literature is available describing non-compendial dissolution methods and alternative methods to dissolution to characterize suspensions and some examples are discussed below.

A dialysis adapter which uses dialysis tubing to sequester a sample, is available for use with USP apparatus 4. The intended applications of the dialysis adapter are for in vitro release testing of dispersed dosage forms. Bhardwaj and Burgess evaluated the dialysis adapter with USP apparatus 4 and showed that the method could discriminate the release of dexamethasone from solutions, suspension and liposome formulations and the method also showed discrimination of dexamethasone release between extruded and non-extruded liposomes.[80]

Gao and Westenberger evaluated the use of a dialysis adapter with USP apparatus 4 to monitor the dissolution of acetaminophen solutions and suspensions.[81] Sample in the dialysis adapter must dissolve and migrate to the dialysis membrane and then permeate through the dialysis membrane to the bulk of the dissolution medium in order to be sampled and detected during the analysis. Dissolution results for tylenol suspension showed that drug dissolution and migration to the membrane surface were slow and were rate-limiting steps. Medium pH and ionic strength as well the molecular weight cut-off of the membrane showed some effects on the dissolution profile of the suspension formulation.

Kayaert and colleagues studied the use of solution calorimetry as an alternative to dissolution testing for nanosuspensions.[82] Solution calorimetry was used to measure the dissolution rate of nanosuspensions by recording the temperature change in the dissolution vessel for three drugs formulated as nanosuspensions (naproxen, cinnarizine, and compound A). Although the dissolution process was completed in less than 1 minute for naproxen and cinnarizine, the data sampling (one data point per second) was high enough to obtain a dissolution curve. The limitations of this method are the long time needed to conduct the experiment and the fact that this technique measures total heat produced or consumed by the processes that occur during dissolution so all phenomena other than dissolution must be taken into account.

There are several reports of using a transmittance based method to obtain dissolution profiles.[21,83,84] As an example, Chaubal and Popescu reported using this method for samples of spray dried nanosuspensions, spray dried drug and micronized drug.[21] Powders were first wetted in a buffer for a fixed time period at 37°C, then the samples were introduced into the dissolution cell. The % transmittance through the solution was monitored using a UV–vis spectrophotometer at 400 nm. With the addition of the sample to the dissolution cell, the percentage transmittance goes down, then increases as the drug particles dissolve. Dissolution profiles are generated by deconvoluting the transmittance data to obtain percentage dissolved vs. time. The authors considered this approach more convenient than traditional dissolution with HPLC end analysis as the dissolution of the samples was fast, with about 80% of the drug dissolved by 10 minutes for the nanosuspension sample.[21]

13.9 Summary

API properties that impact dissolution of suspension formulations include API solubility, particle size distribution, and polymorphic form. Properties of the formulation which can influence dissolution of the dosage form include viscosity, type and concentration of suspending agents, potential for settling and aggregation, and taste-masking techniques used for the formulation. In addition, if the suspension is capsulated in gelatin shells, shell crosslinking during storage may slow the drug dissolution.

A dissolution or drug release test is generally expected as a part of release and stability testing for suspensions. An ideal dissolution method should be sensitive enough to detect relevant product changes to ensure the quality and consistent performance of the product as well as predictive of in vivo performance of the drug product. An IVIVC dissolution method may reduce unnecessary human studies, accelerate drug development, and speed up evaluation of post-approval changes.

Most common queries from regulatory agencies are related to acceptance criteria and justification of method conditions. It is important to take a systematic approach in developing dissolution/drug release test methods. The critical quality attributes that affect drug release should be well understood. Dissolution test media ideally should be physiologically meaningful. Dissolution/drug release test conditions (e.g., agitation), duration, and sample collection times should be consistent with its intended use (as in Quality Target Product Profile). Additionally, factors that affect the performance of the drug release method, such as variability and the impact of changes, should be fully understood and well characterized.

References

1. USP General Chapter <1151> Pharmaceutical Dosage Forms, USP37, 1010–1033. Accessed December 4, 2013. (www.uspnf.com)

2. USP General Chapter <1088> In vitro and in vivo Evaluation of Dosage Forms, USP37, 834–843. Accessed January 19, 2014. (www.uspnf.com)

3. Garad, S., Wang, J., Joshi, Y., Panicucci, R. (2010). Preclinical development for suspensions. In *Pharmaceutical Suspensions: From Formulation Development to Manufacturing.* Kulshreshtha, A. K., Singh, O. N., Wall, G. M. (eds.). Springer, New York.

4. Allen, Jr., L. V., Erickson, M. A. (1996). Stability of baclofen, captopril, diltiazem hydrochloride, dipyridamole, and flecainide acetate in extemporaneously compounded oral liquids. *Am. J. Health Syst. Pharm.*, **53**, 2179–2184.

5. Allen, Jr., L. V., Erickson, M. A. (1998). Stability of bethanechol chloride, pyrazinamide, quinidine sulfate, rifampin, and tetracycline hydrochloride in extemporaneously compounded oral liquids. *Am. J. Health Syst. Pharm.*, **55**, 1804–1809.

6. Jew, R., Soo-Hoo, W., Erush, S. (2010). *Extemporaneous Formulations for Pediatric, Geriatric, and Special Needs Patients*, Second edition, American Society of Health-System Pharmacists, Inc.

7. Aubury, A.-F., Sebastian, D., Hobson, T., Xu, J. Q., Rabel, S., Xie, M., Gray V. (2000). In-use testing of extemporaneously prepared suspensions of second generation non-nucleoside reversed transcriptase inhibitors in support of Phase I clinical studies. *J. Pharm. Biomed. Anal.*, **23**, 535–542.

8. Takahashi, Y., Nakashina, K., Nakgawa, H., Sugimoto, I. (1984). Effects of grinding and drying on the solid-state stability of ampicillin trihydrate. *Chem. Pharm. Bull.*, **32**(12), 963–4970.

9. Lefebvre, C., Guyot-Herman, A. M., Draguet-Brughmans, M., Bouche, R., Guyot, J. C. (1986). Polymorphic transitions of carbamazepine during grinding and compression. *Drug Dev. Ind. Pharm.*, **12**(13), 1913–1927.

10. Noyes, A., Whitney, W. (1897). The rate of solution of solid substances in their own solutions. *J. Am. Chem. Soc.*, **19**, 930–934.

11. Rabinow, B. E. (2004). Nanosuspensions in drug delivery. *Nat. Rev. Drug Discovery*, **3**, 785–796.

12. Junghanns, J.-U. A. H., Muller, R. H. (2008). Nanocrystal technology, drug delivery and clinical applications. *Int. J. Nanomed.*, **3**, 295–309.

13. Wagh, K. S., Patil, S. K., Akarte, A. K., Baviskar, D. T. (2011). Nanosuspensions – a new approach of bioavailability enhancement. *Int. J. Pharm. Sci. Rev. Res.*, **8**(2), 61–65.

14. Sutradhar, K. B., Khatun, S., Luna, I. P. (2013). Increasing possibilities of nanosuspensions. *J. Nanotechnol.*, Article 346581, 12 pages.

15. Gelatin Handbook, Gelatin Manufacturers Institute of America (2012). (http://www.gelatin-gmia.com/technical.html)

16. Abdou, H. M. (1989). Dissolution of Suspensions in Dissolution, Bioavailability. Bioequivalence. Mack Printing Company, Easton, PA.

17. Umesh V. Banakar. Pharmaceutical Dissolution Testing. 1992. Marcel Dekker, Inc., New York, New York.

18. Rohrs, B. (2001). Dissolution method development for poorly soluble compounds. *Dissolut. Technol.*, **8**(3), 1–5.

19. Mauger, J. W., Howard, S. A., Amin, K. (1983). Dissolution profiles for finely divided drug suspensions. *J. Pharm. Sci.*, **72**(2), 190–193.

20. Lai, F., Sinico, C., Ennas, G., Marongiu, F., Marongiu, Fadd, A. M. (2009). Diclofenac nanospenions: influence of preparation procedure and crystal form on drug dissolution behavior. *Int. J. Pharm.*, **373**(1-2), 124–132.

21. Chaubal, M. V., Popescu, C. (2008). Conversion of nanosuspensions into dry powders by spray drying: a case study. *Pharm. Res.*, **25**, 2302–238.

22. Young, S. A., Buckton, G. (1990). particle growth in aqueous suspensions: the influence of surface energy and polarity. *Int. J. Pharm.*, **60**, 235–241.

23. Haleblian, J., McCrone, W. (1969). Pharmaceutical applications of polymorphism. *J. Pharm. Sci.*, **58**(8), 911–929.

24. Simmons, D. L., Ranz, R. J, Gyanchandani, D. N., Picotte, P. (1972). Polymorphism in pharmaceuticals II Tolbutamide. *Can. J. Pharm. Sci.*, **7**(4), 121–123.

25. Ueda, H., Nambu, N., Nagai, T. (1982). Thermodynamic properties of polymorphs of tolbutamide. *Chem. Pharm. Bull.* **30**(7), 2618–2620.

26. Ueda, H., Nambu, N., Nagai, T. (1984). Dissolution behavior of chlorpropamide polymorphs. *Chem. Pharm. Bull.*, **32**(1), 244–250.

27. Al-Saier, S. S., Riley, G. S. (1982). Polymorphism in sulphonylurea hypoglycaemic agents: II. Chlorpropamide. *Pharm. Acta Helv.*, **57**, 8–11.

28. Al-Saier, S. S., Riley, G. S. (1981). Polymorphism in sulphonylurea hypoglycaemic agents: I. Tolbutamide. *Pharm. Acta Helv.*, **56**, 125–129.

29. Mullins, J., Macek, T. J. (1960). Some pharmaceutical properties of novobiocin. *J. Pharm. Sci.*, **49**, 245–248

30. Zietsman, S., Kilian G., Worthington, M, Stubbs, C. (2007). Formulation development and stability studies of aqueous metronidazole benzoate suspensions containing various suspending agents. *Drug Dev. Ind. Pharm.*, **33**(2), 191–197.

31. Hashem, F., El-Said, Y. (1987). Effect of suspending agents on the characteristics of some anti- inflammatory suspensions. *Pharmazie*, **42**, 732–735.

32. Andersen, F., Bundgaard, H., Mengel, H. (1984). Formation, bioavailability and organoleptic properties of an inclusion complex of Femoxetine with B-cyclodextrin. *Int. J. Pharm.*, **21**, 51–60.

33. Unpublished data from the authors.

34. Azam, M. G, Haider, S. S. (2008). Evaluation of dissolution behavior of paracetamol suspension. *J. Pharm. Sci.*, **7**, 53–58.

35. Barzegar-Jalali, M., Richards, J. H. (1979). The effect of suspending agents on the release of aspirin from aqueous suspensions in vitro. *Int. J. Pharm.*, **2**, 195–201.

36. Shah, N. B, Sheth, B. B. (1976). Effect of polymers on dissolution from drug suspensions. *J. Pharm Sci.*, **65**, 1618–1623.

37. Siewert, M., Dressman, J., Brown, C., Shah, V. (2003). FIP/AAPS Guidelines for dissolution/in vitro release testing of novel/special dosage forms. *Dissolut. Technol.,* **10**(1), 6–15.

38. Siewert, M., Dressman, J., Brown, C., Shah, V., Aiache, J.-M., Aoyagi, N., Bashaw, D., Brown, W., Durgess, D., Crison, J., DeLuca, P., Djerki, R., Foster, T., Gjellan, K., Gray, V., Hussain, A., Ingallinera, T., Klance, J., Kraemer, J., Kristense, H., Kuni, K., Leuner, C., Limber, J., Loos, P., Marulis, L., Marroum, P., Mueller, B., Okafo, N., Ouderkirk, L., Parsi, S., Qureshi, S., Robinson, J., Uppoor, R., Williams, R. (2003). FIP/AAPS guidelines for dissolution/in vitro release testing of novel/special dosage forms. *AAPS PharmSciTech,* **4**(1), 43–52.

39. The United States Pharmacopeia (2007). 31st ed. United States Pharmacopeia Convention, Inc. Rockville, MD.

40. Gonzalez Vidal, N. L., Zubata, P. D., Siionato, L. D., Pizzorno, M. T. (2008). Dissolution stability study of cefadroxil extemporaneous suspension. *Dissolut. Technol.,* **15**(3), 29–36.

41. Noelia L. Gonzalez Vidal, Starkloff, W. J., Benancor, S., Castro, S., Suarez, G., Palma, S. D. (2013). Comparative dissolution studies of albendazole oral suspensions for veterinary use. *Dissolut. Technol.,* **20**(4), 27–30.

42. USP <1088> In vitro and in vivo evaluation of dosage forms. USP 24/NF19 (2000) p. 2051.

43. Liu, P., De Wulf, O., Laru, J., Heikkilä, T., van Veen, B., Kiesvaara, J., Hirvonen, J., Peltonen, L., Laaksonen, T. (2013). Dissolution studies of poorly soluble drug nanosuspensions in non-sink conditions. *AAPS PharmSciTech,* **14**, 748–756.

44. Van Eerdenbrugh, B., Van den Mooter, G., Augustijns, P. (2008). Top-down production of drug nanocrystals: nanosuspension stabilization, miniaturization and transformation into solid products. *Int. J. Pharm.,* **364**, 64–75.

45. Monograph for Megestrol Acetate Oral Suspension. USP36, The United States Pharmacopeial Convention, 2013, pp. 4423–4224. Accessed December 22, 2013. (www.uspnf.com)

46. USP General Chapter <711> Dissolution, USP 37, 344–351. Accessed January 19, 2014. (www.uspnf.com)

47. Center for Drug Evaluation and Research, Office of Clinical Pharmacology and Biopharmaceutics Review, Application Number 21–778, Megace ES™. Accessed December 22, 2013. (www.accessdata.fda.gov/drugsatfda_docs/nda/2005/021778s000_ClinPharmR.pdf)

48. FDA Dissolution Methods Database. Accessed January 19, 2014. (http://www.accessdata.fda.gov/scripts/cder/dissolution/dsp_Search Results_Dissolutions.cfm?PrintAll=1)

49. Jantratid, E., Janssen, N., Reppas, C., Dressman, J. B. (2008). Dissolution media simulating conditions in the proximal human gastrointestinal tract: an update. *Pharm. Res.*, **25**, 1663–1676.

50. Nikoletta, F., Vertzoni, M. (2010). Biorelevant dissolution methods and their applications in *in vitro–in vivo* correlations for oral formulations. *Open Drug Delivery J.*, **4**, 2–13.

51. Biorelevant media powder from Biorelevant.Com Ltd (UK). Accessed September 4, 2013. (www.biorelevant.com).

52. Galia, E., Nicolaides, E., Hörter, D., Löbenberg, R., Reppas, C., Dressman, J. B. (1998). Evaluation of various dissolution media for predicting *in vivo* performance of class I and II drugs. *Pharm. Res.*, **15**(5), 698–705.

53. Klein, S., Stippler, E., Wunderlich, M., Dressman, J. Development of dissolution tests on the basis of gastrointestinal physiology. (2005) In *Pharmaceutical Dissolution Testing*, Dressman, J., Krämer, J. (eds.). Taylor. Francis: New York, 206.

54. Guidance for Industry, Extended release oral dosage forms, development, evaluation and application of in vitro/in vivo correlations. September 1997.

55. Monograph for Oxcarbazepine Oral Suspension. USP36, The United States Pharmacopeial Convention, 2013, pp. 4625–2646. Accessed December 22, 2013. (www.uspnf.com)

56. Monograph for Phenytoin Oral Suspension. USP36, The United States Pharmacopeial Convention, 2013, pp. 4784–4785. Accessed December 22, 2013. (www.uspnf.com)

57. Monograph for Ibuprofen Oral Suspension. USP36, The United States Pharmacopeial Convention, 2013, pp. 3876–3977. Accessed January 19, 2014. (www.uspnf.com)

58. Monograph for Mycophenolate Mofetile for Oral Suspension. USP36, The United States Pharmacopeial Convention, 2013, pp. 4425–4426. Accessed December 22, 2013. (www.uspnf.com)

59. Monograph for Nevirapine Oral Suspension. USP36, The United States Pharmacopeial Convention, 2013, pp. 4493–4495. Accessed December 22, 2013. (www.uspnf.com)

60. Monograph for Meloxicam Oral Suspension. USP36, The United States Pharmacopeial Convention, 2013, pp. 4228–4230. Accessed December 22, 2013. (www.uspnf.com)

61. Monograph for Indomethaicin Oral Suspension. USP36, The United States Pharmacopeial Convention, 2013, pp. 3907–3908. Accessed December 22, 2013. (www.uspnf.com)

62. Monograph for Felbamate Oral Suspension. USP36, The United States Pharmacopeial Convention, 2013, pp. 3535–3537. Accessed December 22, 2013. (www.uspnf.com)

63. Almukainzi, M., Salehi, M., Chacra, N. A. B., Löbenberg, R. Dissolution Technologies (2001). Comparison of the rupture and disintegration tests for solf-shell capsules. *Dissolut. Technol.,* **18**(1), 21–25.

64. USP General Chapter <2040> Disintegration and Dissolution of Dietary Supplements, USP37, pp. 1338–1344. Accessed February 11, 2014. (www.uspnf.com)

65. Hom, F. S., Veresh, S. A., Miskel, J. J. (1973). Soft gelatin capsules. I. Factors affecting capsule shell dissolution rate. *J. PharmSci.,* **62**(6), 1001–1006.

66. Digenis, G. A., Gold, T. B., Shah, V. P. (1994). Cross-linking of gelatin capsules and its relevance to their in vitro–in vivo performance. *Pharm. Sci.,* **83**(7), 915–921.

67. Welz, M. M., Ofner III, C. M. (1992). Examination of self-crosslinked gelatin as a hydrogel for controlled release. *J. Pharm Sci.,* **81**(1), 81–90.

68. Gelatin Capsule Working Group. (1998). Collaborative development of two-tier dissolution testing for gelatin capsules and gelatin-coated tablets using enzyme-containing media. *Pharmacop. Forum,* **24**(5), 7046–7050.

69. Meyer, M. C., straughn, A. B., Mhatre, R. M., Hussain, A., Shah, V. P., Bottom, C. B., Cole, E. T., Lesko, L. L, Mallinowski, H., Williams, R. L. (2000). The effect of gelatin crosslinking on the bioequivalence of hard and soft gelatin acetaminophen capsules. *Pharm. Res.,* **17**(8), 962–966.

70. Dey, M., Enever, R., Kraml, M., Prue, D. G., Smith, D., Weierstall, R. (1993). The dissolution and bioavailability of etodolac from capsules exposed to conditions of high relative humidity and temperatures. *Pharm. Res.,* **10**(9), 1295–1300.

71. Brown, J., Madit, N., Cole, E. T., Wilding, I. R., Cadé, D. (1998). The effect of crosslinking on the in vivo disintegration of hard gelatin capsules. *Pharm. Res.,* **15**(7), 1026–1030.

72. Monograph for Cefadroxil for Oral Suspension. USP36, The United States Pharmacopeial Convention, 2013, pp. 2840–2841. Accessed December 22, 2013. (www.uspnf.com)

73. Monograph for Cefdinir for Oral Suspension. USP36, The United States Pharmacopeial Convention, 2013, pp. 2852–2855. Accessed December 22, 2013. (www.uspnf.com)

74. Monograph for Cefuroxime Axetil for Oral Suspension. USP36, The United States Pharmacopeial Convention, 2013, pp. 2893–2894. Accessed December 22, 2013. (www.uspnf.com)

75. Monograph for Flucanozole for Oral Suspension. USP36, The United States Pharmacopeial Convention, 2013, pp. 5984–5986. Accessed December 22, 2013. (www.uspnf.com)

76. Monograph for Amoxicillin Tablets for Oral Suspension. USP36, The United States Pharmacopeial Convention, 2013, pp. 2482–2485. Accessed December 22, 2013. (www.uspnf.com)

77. Monograph for Cephalexine Tablets for Oral Suspension. USP36, The United States Pharmacopeial Convention, 2013, pp. 2905. Accessed December 22, 2013. (www.uspnf.com)

78. Monograph for Didanosine Tablets for Oral Suspension. USP36, The United States Pharmacopeial Convention, 2013, pp. 3232–3233. Accessed December 22, 2013. (www.uspnf.com)

79. Monograph for Lamotrigine Tablets for Oral Suspension. USP36, The United States Pharmacopeial Convention, 2013, pp. 4058–4060. Accessed December 22, 2013. (www.uspnf.com)

80. Bhardwaj U., Burgess D. (2010). A novel USP apparatus 4 based release testing method for dispersed systems. *Int. J. Pharm.*, **388**, 287–294.

81. Gao, Z., Westenberger, B. (2012). Dissolution testing of acetoaminophen suspension using dialysis adapter in flow-through apparatus: a technical note. *AAPS PharmSciTech*, **13**, 944–948.

82. Kayaert, P., Li, B., Jimidar, I., Rombaut, P., Ahssini, F., Van den Mooter, G. (2010). Solution calorimetry as an alternative approach for dissolution testing of nanosuspensions. *Eur. J. Pharm. Biopharm.*, **76**, 507–513.

83. Rabino B. (2004). Nanoedge drug delivery solves the problems of insoluble injectable drugs. Supplement to Scrip World Pharmaceutical News, October, 13–16.

84. Crisp, M. T., Tuckers, C. J., Rogers, T. L., Williams III, R. O., Johnston, K. P. (2007). Turbidimetric measurement and prediction of dissolution rates of poorly soluble drug nanocrystals. *J. Control. Release*, **117**, 351–359.

Chapter 14

Dissolution Testing of Poorly Soluble Drugs: "Biorelevant Dissolution"

Mark McAllister and Irena Tomaszewska

Pharmaceutical Sciences, Pfizer Ltd, Sandwich, Kent, CT13 9NJ, UK

mark.mcallister@pfizer.com

14.1 Introduction

Since the introduction of USP compendial dissolution equipment in the 1970s, dissolution studies have become the main tool used by pharmaceutical scientists when attempting to understand the relationship between formulation performance and the complex gastrointestinal physiological factors which influence oral absorption. Since the first description in 1951 by Edwards of how dissolution can impact the absorption of a drug substance in the gastrointestinal tract,[1,2] scientists have continued to adapt both dissolution equipment and methodology to more closely mimic the luminal environment particularly with regards to media composition, and hydrodynamics/physical stress. The basket (USP apparatus 1) and paddle (USP apparatus 2) compendial apparatus have been at the forefront of this effort and despite their limitations,[3] have proved

Poorly Soluble Drugs: Dissolution and Drug Release
Edited by Gregory K. Webster, J. Derek Jackson, and Robert G. Bell
Copyright © 2017 Pan Stanford Publishing Pte. Ltd.
ISBN 978-981-4745-45-1 (Hardcover), 978-981-4745-46-8 (eBook)
www.panstanford.com

very useful in understanding how formulation type, composition, and process variation can impact drug release. Over the last two decades of pharmaceutical research,[4,5] the emergence in industrial drug development pipelines of a dominant class of poorly soluble compounds has created a multifaceted challenge to the continued use of compendial apparatus to accurately predict oral formulation performance and has driven the need to improve the biorelevance of media and equipment. The concurrent development of increasingly sophisticated oral drug delivery systems[6] to improve absorption of poorly soluble compounds has added another layer of complexity which needs to be adequately addressed by dissolution approaches seeking to provide robust predictions of in vivo performance.

In general, the approaches adopted to improve the physiological relevance of a dissolution test can be broadly categorized into one of two groups. The first group encompasses mechanistic approaches in which a single aspect of the dissolution process is controlled to study the impact of a physiological variable such as media or hydrodynamics. The second group includes methods and equipment designed to simulate multiple aspects of the gastrointestinal tract and deliver a holistic simulation of luminal conditions encompassing fluids, digestion, transit, and absorption. Examples of this type of approach range from simple two-compartment transit models such as the artificial stomach duodenal model[7,8] to the more complex gastrointestinal simulators such as the human gastric simulator,[9,10] TNO TIM-1/TIM-2[11,12] and the dynamic gut model.[13]

It is recognized that for particular aspects of dosage form or API dissolution that a mechanistic (reductionist) approach may allow a detailed assessment of individual phenomena such as supersaturation, precipitation and re-dissolution. In contrast, a holistic simulation approach considers the summation or net effects of multiple processes such as digestion, transit, and absorption on dosage form performance, for example when assessing complex food effects. This chapter describes the development of biorelevant dissolution testing for both mechanistic and holistic approaches and assesses the biological relevance of modifications made to media (composition and volume), hydrodynamics and integration of an absorptive component within the dissolution test.

14.2 Evolution of Biorelevant Media

14.2.1 Simulating the Gastric Environment

There are many factors that can affect the characteristics and composition of the luminal fluids in the GI tract. Generally, they can be classified into two categories: endogenous and exogenous factors. Endogenous factors include the various fluids that are secreted and reabsorbed by the GI tract, GI transit, and motility. Exogenous factors are those parameters that can be controlled by the individual such as the time of administration of a dosage form with meals, the type of meal, and the volume of the beverages ingested. In both cases, inter- and intra-subject variability are important to consider.

Over the decades gastric fluids have been characterized in terms of their composition, pH, and surface tension. Stomach content is composed of saliva, gastric secretions (hydrochloric acid, digestive enzymes, including pepsin and gastric lipase which aids release from lipid-based dosage forms,[14,15]) dietary food and refluxed fluids from the duodenum.[16] The pH of gastric fluids in fasted state can be very unstable. It has been reported that the median pH of gastric fluids in the fasted state is in the range of 1.5–1.9,[17–19] although in some cases the extremes of pH reported can vary between pH 1 to 6; however, the exact pH can be hard to define as its values fluctuate on a minute-to-minute basis.[20,21] The pH of the gastric fluids in healthy Caucasians stays below pH 3. However, certain populations, i.e., Japanese or patients with specific health-related conditions such as achlorhydria, can have gastric fluid pH values above 6.[20,22] Surface tension in gastric fluids varies between 35–46 mN/m,[18,23] which indicates presence of surface active agents such as lecithin and lysolecithin.[24] According to Solvang and Finholt, the levels of these surfactants in vivo are below critical micelle concentration (CMC).[25] In a review of a number studies which measured bile salts levels in the stomach, Vertzoni et al. concluded that bile salt reflux into the stomach occurs sporadically and that concentrations in the stomach are very low compared with those in the small intestine.[21]

Table 14.1 Composition of artificial gastric fluids[27]

AGF	
Sodium citrate (mM)	2.34
Sodium malate (mM)	2.81
Lactic acid (μL)	420
Acetic acid (μL)	500
Pepsin (g)	1.25
Hydrochloric acid	qs
pH	1.2

14.2.1.1 Fasted state media

One of the earliest attempts to develop biorelevant media that would simulate gastric fasted state conditions was made by Ruby et al., who proposed artificial gastric fluids (AGF) in 1996 (Table 14.1).[26] This medium was used to estimate lead and arsenic bioavailability using a physiologically based extraction test.

Over a decade later, this medium (AGF pH 1.2) and its variants (pH 2 and pH 3.0), compendial (SGF, SGF$_{sp}$), and modified media (SGF+0.1%$_{Triton X 100}$; SGF$_{SLS}$) (Table 14.2) were used to investigate the effect of biorelevant media on the disintegration process.[27] These studies were performed to examine the influence of physiologically relevant media on the disintegration stage and its contribution in the overall dissolution process.[27] Other studies have been performed to assess the replacement of physiologically relevant surfactants with synthetic alternatives to reduce the cost of media production, ease preparation, or improve shelf-life. Galia

Table 14.2 Modified dissolution medium to simulate gastric conditions[27]

	SGF$_{SLS}$	SGF$_{Triton X}$
Sodium chloride (mM)	34.22	34.22
Sodium lauryl sulphate (mM)	8.57	–
Triton X 100	–	1.55
Hydrochloric acid (mM)	qs	qs
pH	1.2	1.2
Surface tension (mN/m)	33.30	30.83

et al. added small amounts (0.1%) of Triton X 100 in order to closely mimic surface tension of human gastric fluids. The addition of surfactant resulted in dissolution enhancement of albendazole but compromised the discriminatory power of the method.[28]

It was found that most of the media had no effect on the disintegration time of fast disintegrating tablets. The shortest disintegration time was recorded for slowly disintegrating tablets when tests were performed in SGF_{SLS} (7.9 ± 1.8 min) and water (8.1 ± 1.5 min). It would appear the type of surfactant used in the medium plays an important role in disintegration stage, as slowly disintegrating tablets tested in $SGF_{TritonX}$ had prolonged disintegration times relative to those tested in SGF_{SLS} (13.0 ± 2.5 min vs. 7.9 ± 1.8 min). The type of medium used in disintegration test plays an important role in determining dissolution rates of slowly disintegrating tablets.[27]

However, studies reported elsewhere indicated the risk associated with the use of surfactants such as SLS in dissolution medium.[29] Zhao et al. claimed that SLS present in the medium below pH 5 interacts with the hard gelatin capsules causing gelatin precipitation, which results in much slower dissolution profile.[29] Therefore, it is important to be aware of the impact that different types of surfactant can have on formulation performance and that often their use in dissolution media can lead to erroneous predictions for drug dissolution.[30,31]

However, at this stage of media development, the compendial gastric media that were intended to mimic fasting gastric contents still did not fully simulate the range of conditions reported from in vivo investigations. As a result, more physiologically relevant media, based on newly published in vivo data were further developed (Table 14.3).[30] It has been reported that median pH values in fasted stomach of healthy volunteers vary between pH 1.5–1.9.[17,18,30,32] However, in patients with certain health conditions, such as achlorhydria that often occurs in Japanese population, elevated pH (pH 5-6) in the fasted state has been observed.[22] The addition of bile salts and lecithin in 4:1 ratio resulted in a lowering of surface tension to a more physiologically relevant range.[18,30,33] The proposed media which simulated the gastric environment more closely resulted in an improved prediction of the cumulative amount dissolved in vivo.[30]

Table 14.3 Composition of media-simulating fasted state gastric conditions[30]

FaSSGF	
Sodium chloride (mM)	34.2
Pepsin (mg/mL)	0.1
Sodium taurocholate (μM)	80
Lecithin (μM)	20
Surface tension (mN/m)	42.6
Osmolarity (mOsm/kg)	120.7 ± 2.5
pH	1.6

The ability of FaSSGF to mimic the human intragastric environment was evaluated by measuring the solubility of one non-ionizable drug (felodipine) and three weak bases (ketoconazole, dipyridamole, and miconazole)[31] and comparing the values with those measured in human gastric fluids (HGF), canine gastric aspirates (CGF), and other compendial media such as SGF_{SLS}, $SGF_{TritonX}$, or $FaSSGF_{NaCl}$. The composition of FaSSGF was modified by doubling the molar concentration of chloride salt in order to reflect in vivo chloride levels ($FaSSGF_{NaCl}$).[31,34] It was found that for all solubility measurements, FaSSGF correlated closer to HGF than canine aspirates and other compendial media. Interestingly, the solubility values in FaSSGF pH 1.6 were similar to those tested in HCl pH 1.6. Therefore, it would appear that FaSSGF is advantageous when wettability is a critical parameter for formulation performance.[31]

14.2.1.2 Fed state gastric media

The pH value of gastric fluids after food ingestion varies between pH 3 and pH 7. The pH value and rate of change with time is strongly related to the type and size of the meal.[20] The pH of gastric fluids returns to the fasted state range within two to three hours. In many cases, compendial media were not able to distinguish between formulation performance in fasted and fed states and consequently a number of alternative media compositions have been investigated.

Early studies in this area used milk as a medium that can simulate gastric components in the fed state.[28,35-39] It contains similar ratios of fat, protein and carbohydrates that are present in a typical western diet (according to FDA guidance on food effect bioavailability and fed bioequivalence studies, a high fat meal of 1000 kcal of which 50% comes from fat should be ingested to provide a standarized fed state.[39,40])

A practical consideration for performing dissolution in milk media is that an additional extraction step may be required as highly lipophilic compounds will tend to bind to the lipid components in milk.[36] Macheras et al. reported that the higher the fat content of milk the greater binding of the drug to milk component.[41] Batch-to-batch milk composition variability can contribute to variable dissolution data.[39] Alternative approaches to using milk have assessed homogenized meals, nutrient drinks and emulsions with proportions of carbohydrates, proteins and fats matching those in typical Western diets.[42,43] The biggest challenge in developing gastric media that would simulate the fed state is to capture the compositional changes associated with dynamic digestion and gastric emptying processes. Fotaki et al. attempted to simulate these fluctuating conditions by periodical addition of pepsin to milk.[44] Three years later, the "snapshot" approach was introduced by Jantratid et al.[37] It has been reported that gastric pH after meal ingestion declines from pH 6.4 to 2.7.[23] On the basis of these findings, Jantratid proposed three "snapshot" media which were designed to reflect the composition of gastric fluids in early, middle and late stages of the fed state[37] (Table 14.4). In order to mimic ongoing secretion of gastric juices and gastric emptying of food, dilution of milk with buffer in the ratios 1:1 and 1:3 was introduced in middle and late snapshot media, respectively.[37] The composition of early, middle, and late FeSSGF media reflected composition of the specific time-framed composition of human gastric fluids. Early stage media corresponded to the composition of aspirates within the first 75 min following meal ingestion. FeSSGF$_{middle}$ (also referred as FeSSGF) mimics the composition of HGF between 75 and 165 min, whereas late stage medium reflected the composition after 165 min post food ingestion.[37]

Table 14.4 Composition of the snapshot media simulating fed state gastric environment[37]

	FeSSGF$_{early}$	FeSSGF$_{middle}$	FeSSGF$_{late}$
Sodium chloride (mM)	148	237.02	122.6
Acetic acid (mM)	–	17.12	–
Sodium acetate (mM)	–	29.75	–
Orthophosphoric acid (mM)	–	–	5.5
Sodium dihydrogen phosphate (mM)	–	–	32
Milk/Buffer	1:0	1:1	1:3
Hydrochloric acid/sodium hydroxide	qs	qs	qs
pH	6.4	5.0	3.0
Osmolality (mOsm/kg)	559	400	300
Buffer capacity (mmoL/L/pH)	21.33	25	25
Surface tension (mN/m)	49.7 ± 0.3 [45]	52.3 ± 0.3 [45]	58.1 ± 0.1 [45]

14.2.2 Simulating the Small Intestinal Environment

The pH in the intestinal regions of the GI tract are higher than in the gastric environment due to the neutralization of the gastric fluids with bicarbonate ions secreted mainly by the pancreas. The pH values in the intestinal segments gradually increase throughout duodenum (pH 6.5), jejunum (pH 6.8), and ileum (pH 7.5).[20] In the upper small intestine surface tension can vary between 28 and 46 mN/m.[46,47] In the fasted state, bile salts secreted by the liver form mixed micelles with lecithin that contribute to the much lower surface tension values reported for small intestinal aspirates relative to those reported for gastric fluids.[20] The value in the fed state is even lower than in the fasted state (28 vs. 33.6 mN/m), corresponding to the higher levels of bile salts present in the fed state.[48] The concentration of bile components in the fed state has been reported to peak 30 min after meal ingestion.[49] Thereafter, the levels decline as a result of dilution with chyme.[20]

Bile salts and lecithin mixed micelles enhance the wetting of solids and the solubilization of lipophilic drugs.[50] Cholic acid is the dominant bile salt in human bile.[20] In most simulated media, the sodium salt of taurocholate acid, which is a conjugate of cholic acid with taurine, was selected to be the most representative bile salt. Sodium taurocholate has a low pK_a value of 1.4 and good solubility in the physiological pH range.[37] In the fasted state, the concentration of bile salts varies between 3 and 5 mM.[20] In vivo, the ratio of bile salt to lecithin varies between 2:1 and 5:1. Therefore, a 3:1 ratio was chosen for fasted state simulated intestinal fluids.[20,38] In the lumen of the small intestine, the dominant buffer species is bicarbonate but due to the practical issues with using bicarbonates in vitro (pH instability caused by reaction with oxygen), phosphate buffers are typically used to mimic the physiological pH of intestinal fluids.[20]

In 1998, Dressman and co-workers developed fasted state simulated intestinal fluid (FaSSIF) and fed state simulated intestinal fluid (FeSSIF) to simulate fasting and fed conditions in the small intestine (Table 14.5).[20] This medium has been shown to provide a more accurate simulation of pharmacokinetic profiles than compendial SI.[51,52]

In contrast to FaSSIF, FeSSIF uses acetate buffer in order to adequately adjust pH to the value of 5.0. In practice, equivalent molar concentrations of sodium salts can be used instead of potassium salts.[53]

Table 14.5 Composition of medium simulating fasted and fed conditions of small intestine[38]

	FaSSIF	FeSSIF
Sodium taurocholate (mM)	3	15
Lecithin (mM)	0.75	3.75
Potassium dihydrogen orthophosphate (mM)	28.66	–
Potassium chloride (mM)	103.29	203.89
Sodium hydroxide (mM)	qs	qs
Acetic acid (mM)	–	144.05
pH	6.5	5.0
Osmolality (mOsmol)	270 ± 10	635 ± 10 – 670 [55]
Buffer capacity (mmol/L/pH)	10 [53]	75 [53]
Ionic strength	0.15 [56]	0.3 [56]
Surface tension (mN/m)	48[55] – 49[56]	–

Fagerberg et al. claimed that performing solubility tests in biorelevant media provided a more accurate estimation of the in vivo solubility of poorly soluble compounds.[54] Solubility and dissolution rate studies of ten different compounds revealed that the accuracy of prediction in biorelevant media was particularly improved for bases and neutral compounds. These compounds typically display higher solubility in FeSSIF than FaSSIF. The opposite trend was observed for weakly acidic drug.[54] These observations may be particularly useful for anticipating the directions of a food effect.

Selecting a medium with mixed micelles will have an important role on the dissolution of the BCS class II drugs.[38] Dissolution of this class of compounds can be the rate limiting step to absorption and thus various factors such as presence of surfactants, pH, ionic strength, buffer capacity, and volume of media are crucial to accurately replicate the physiological environment.[38] Galia et al. tested the effect of media composition on the dissolution of immediate release formulations of BCS class I (acetaminophen, metoprolol) and BCS class II compounds (danazol, mefenamic acid, ketoconazole).[38] It was found that the presence of mixed micelles has a predominant effect on compounds in the BCS class II category. For example, a 30-fold increase in % dissolved within 90 min was observed for a danazol formulation tested in FaSSIF, in contrast to aqueous buffer.[38]

In some cases, when attempting to predict in vivo performance of ionized compounds, pH of the buffer may have more of an effect than the presence of the bile salts. This has been reported for the weak base, loratadine.[57] Dissolution in compendial acidic buffers (pH 1.2 and 2.0) resulted in better in vivo prediction than in biorelevant FaSSIF medium.[57] In contrast, Löbenberg et al. successfully developed a Level C IVIVC for two immediate-release formulations of the poorly soluble weak acid, glibenclamide (formulated as 3.5 mg Euglucon N® tablets and 3.5 mg Glukovital® tablets) using FaSSIF as a medium (USP 2, 75 rpm 500 mL).[56] These studies were also performed in SIF but dissolution profiles did not achieve an IVIVC.

Persson et al. performed solubility studies and measured intrinsic dissolution rates with four poorly soluble compounds (cyclosporine, danazol, griseofulvin and felodipine) in fasted and fed human intestinal fluids (HIF) and compared values with those tested in dog intestinal fluids (DIF) and FeSSIF.[47] Detailed characterization of the HIF and DIF was also conducted. According to the studies, buffer capacities of fasted HIF are four to six times lower than reported for fed HIF. Both fluids have similar surface tensions (ranging from 26 to 29 mN/m). Total protein and bile salts concentrations are four to five fold greater in the fed than in the fasted state, with a 15-fold greater total phospholipid concentration reported in the fed state. Interestingly, it has been found that fed DIF has a very similar composition to fed HIF. On this basis, the dog would seem to be a good model for man in terms of dissolution in the small intestine during the fed state.[47] However, it is important to note that the proportions of different bile salts (taurocholic acids, cholic acids, glycocholic acids, etc.), phospholipids (lyso-phosphatidylcholine and phosphatidylcholine), and neutral lipids (free fatty acids, cholesterol, triglycerides, etc.) are different.[47] Moreover, solubility values of all the compounds in the fed HIF were two to fivefold higher than in FeSSIF. This might be attributed to the lack of neutral lipids in the FeSSIF composition.

In the same year, Sunesen et al. published dissolution data for a danazol capsule formulation, using different biorelevant media in a flow-through cell dissolution apparatus (USP apparatus 4).[58] In these studies, biorelevant media were modified by altering concentrations of bile salts and phospholipids in addition to lipolytic products such as fatty acids and monoglycerides (Table 14.6). Dissolution of danazol in the single medium of Fasted (high) and Fed (30:4) at flow rates of 8 mL/min resulted in a successful Level

Table 14.6 Biorelevant media proposed by Sunesen et al.[58]

	Fasted (high)	Fed (30:4)
Potassium dihydrogen orthophosphate (mM)	29	–
TRIZMA® maleate (mM)	–	12
Bile salts (mM)	6.3	18.8
Phospholipids (mM)	1.25	3.75
Fatty acids (mM)	–	30
Monoglycerides (mM)	4	4
pH	6.8	5.5 ± 0.4

A IVIVC for both fasted and fed state.[58] It is worth noting that the prediction of in vivo performance was achievable using only a single medium and flow rate throughout the experiment. Given that danazol is unionized across the pH range and its dissolution is therefore not impacted by changes in media pH, these conditions may be more generally applicable to neutral compounds than those from other physicochemical classes.

The Fasted (high) medium proposed by Sunesen et al. contains elevated concentration of bile salts and phospholipids relative to the previously described FaSSIF medium. The main alterations to the Fed state media proposed by Sunesen et al. are an increased concentration of bile salts (an increase of 3.8 mM over FeSSIF media), the replacement of phosphate with maleate buffer components and the additions of monoglycerides and fatty acids. The authors also used inexpensive crude bile components such as porcine bile extract and soybean phospholipids. This approach to substitute expensive components of biorelevant media was also advocated by Vertzoni et al.[59]

The purity of bile salts used in FaSSIF media has been shown to impact the solubility of poorly soluble drugs. The solubility of glyburide in biorelevant media containing crude bile salts was over twofold higher than when pure bile salts were used.[60] Dissolution data in FaSSIF containing crude bile salts was able to differentiate between two formulations of glyburide.[60] The opposite trend was found when the same formulations were tested in a previous study.[56] This can be attributed to the different volumes of media and batches of bile salts and lecithin used. It was also found by the same group of researchers that the use of crude bile salts in sequential dissolution testing prevented precipitation of glyburide caused by the media and pH change.[60]

The high cost of key components of FaSSIF and FeSSIF such as sodium taurocholate and lecithin (it has been estimated that the cost of preparing 1L of FeSSIF is approximately US$700) has prompted several groups to evaluate alternative surfactant combinations.[55] It has been found that release from Nizoral® tablets (ketoconazole) tested in blank FeSSIF in combination of 0.25% Tween® 80 and 0.25% triethanolamine resulted in a similar dissolution profile to

Table 14.7 Composition of the revised medium simulating fasted conditions of small intestine[37]

	FaSSIF-V2
Sodium taurocholate (mM)	3
Lecithin (mM)	0.2
Maleic acid (mM)	19.12
Sodium hydroxide (mM)	34.8
Sodium chloride (mM)	68.62
pH	6.5
Osmolality (mOsm/kg)	180 ± 10
Buffer capacity (mmoL/L/pH)	10
Surface tension (mN/m)	54.3 [45]

that obtained in FeSSIF.[55] The cost of this alternative medium is as little as US$1 but equivalency to standard media needs to be assessed on a case-by-case basis. In 2008, Jantratid et al. revised the composition of FaSSIF based on data from in vivo studies[23] which reported lower bile salt concentrations than those present in FaSSIF[37] (Table 14.7). The revised version (FaSSIF-V2) of FaSSIF is also characterized by a lower osmolality, with the use of a maleate buffer in place of phosphate buffer. It has been reported that maleic acid is able to delay the rancidity of fats and oils,[37] thereby improving the shelf-life of the media. The concentration of lecithin in fasted state simulating intestinal fluids was also changed with a reduction in concentration to 0.2 mM.

Jantratid applied the same snapshot approach adopted for development of gastric media simulating the fed state. This approach allowed the presence of lipolysis products, changes in bile salts concentration, osmolality, buffer capacity, and fluid pH after food intake to be incorporated within media design and proposed three media that simulated early, medium and late stage intestinal fluids in the fed state (Table 14.8).[37] These media reflect the decrease of the pH of the fluids in the distal duodenum from pH 6.6 to 5.4.[23]

The three FeSSIF snapshot media provide a platform to assess the performance of formulations in a methodical, step-wise fashion, and it is possible to configure USP 3 or USP 4 systems to provide sequential exposure and mimic the changing luminal environment

Table 14.8 Composition of the snapshot media simulating fed conditions of small intestine[37]

	FeSSIF$_{early}$	FeSSIF$_{middle}$	FeSSIF$_{late}$
Sodium taurocholate (mM)	10	7.5	4.5
Lecithin (mM)	3	2	0.5
Glyceryl monooleate (mM)	6.5	5	1
Sodium oleate (mM)	40	30	0.8
Maleic acid (mM)	28.6	44	58.09
Sodium hydroxide (mM)	52.5	65.3	72
Sodium chloride (mM)	145.2	122.8	51
pH	6.5	5.8	5.4
Osmolality (mOsm/kg)	400 ± 10	390 ± 10	240 ± 10
Buffer capacity (mmoL/L/pH)	25	25	15

as a drug (or API) transits the length of the small intestine. However, for routine use, the three media may be time-consuming to prepare and often an analyst will prefer to use a single medium to represent the average of the three snapshot conditions. To meet this requirement, Jantratid revised the FeSSIF medium composition to FeSSIF-V2 (Table 14.9).[37] In this revised composition, three changes were implemented: (i) concentrations of bile salts and lecithin were decreased based on the findings from the studies by Kalantzi et al.,[23] (ii) maleate buffer replaced phosphate buffer resulting in lower osmolality and buffer capacity values, and (iii) glyceryl

Table 14.9 Composition of the revised medium simulating fed conditions of small intestine[37]

	FeSSIF-V2
Sodium taurocholate (mM)	10
Lecithin (mM)	2
Glyceryl monooleate (mM)	5
Sodium oleate (mM)	0.8
Maleic acid (mM)	55.02
Sodium hydroxide (mM)	81.65
Sodium chloride (mM)	125.5
pH	5.8
Osmolality (mOsm/kg)	390 ± 10
Buffer capacity (mmoL/L/pH)	25

monooleate and sodium oleate were added to reflect presence of lipolysis products. All fed state media were reported to be stable for at least 72 hours at ambient temperature and at least eight hours of experiment at $37°C$.[37]

Kleberg et al. further emphasized the importance of the addition of digestion products to media designed to simulate the fed state.[61] Dietary triglycerides which hydrolyze to monoglycerides and free fatty acids, form mixed micelles with bile salts and lecithin. Depending on the total concentration of the surface active agents and ratio of lipolysis products, different shapes and sizes of colloidal structures were formed including mixed micelles, vesicles, or other colloidal species. Cryo-TEM imaging was a useful technique to study the impact of different lipolysis products on the types of colloidal structure formed. Oleic acid (free fatty acid) and monoolein (monoglyceride) were found to be responsible for formation of colloidal structures in biorelevant media simulating the fed state.[61] These findings have a particular importance for lipophilic compounds. Söderlind et al. investigated the role of bile salts and lecithin in FaSSIF-V2 and its ability to mimic in vivo solubility.[52] Five different bile acids, including glycocholic acid, taurocholic acid, glycodeoxycholic acid, taurodeoxycholic acid, and glycochenodeoxycholic acid, were subject of this research. Each bile salt formed mixed micelles with lecithin in ratio 4:1 and were diluted to various concentrations ranging between 1 and 15 mM. As expected, the solubility of neutral compounds increases with increased concentration of bile acids. According to this study, the solubilization capacity of mixed micelles is affected by the degree of hydroxylation of the steroid ring system but not by the type of conjugating amino acid (glycine or taurine). Furthermore, solubility studies of 24 model compounds in FaSSIF, FaSSIF-V2 and HIF demonstrated that FaSSIF-V2 solubilities correlated more closely with solubilities in HIF.[52] This was particularly evident for neutral compounds, whereas solubility values of acidic and basic compounds in FaSSIF-V2 are similar to FaSSIF.

Recently, biorelevant transport media were proposed for testing intestinal drug permeability using the Caco-2 cell model that could distinguish between fasted and fed conditions.[62] Those media were prepared based on the compositions of FaSSIF-V2 and FeSSIF-V2.

It has been reported that the use of these media is particularly useful when testing lipophilic compounds, as greater discrepancies between the permeability values determined using aqueous media and biorelevant transport media were observed for compounds characterized by high lipophilicity.[62]

Over the last two decades, the widespread adoption of biorelevant media such as FaSSIF and FeSSIF is clearly reflected in the numerous publications that are subject of this chapter. However, the preparation of biorelevant media can be time-consuming given the need in some cases to evaporate lecithin from organic solvent and the requirement for frequent preparation as a result of the short-shelf life associated with most compositions (typically up to 72 h). It has been recognized for some time that more practical and simplistic methods of preparation are needed.[63] Different methods to prepare instant FaSSIF and FeSSIF powders were investigated and were compared with conventional biorelevant media. Biorelevant media prepared using instant powders were found to have similar performance to conventional media. Further work by Kloeffer et al. revealed that instant powders are stable for at least 12 months when stored in original container at 2–8°C.[64] Because colloidal structures such as micelles in FaSSIF are influenced by the temperature changes, FaSSIF should be equilibrated for two hours at room temperature after preparation. However, in case of FeSSIF the equilibration is not needed as its particle sizes are independent of the temperature.[64]

14.2.3 Simulating the Colonic Environment

The pH in the colon is influenced by products of anaerobic bacterial reactions and is reported to be around pH 6.0.[20] Conversion of undigested carbohydrates to short chain fatty acids can be a significant contributor to the overall colonic pH.[65] For MR formulations, the colon is a significant site for absorption and dissolution conditions that would simulate the colonic environment are needed. In 2005 Fotaki et al. proposed simulated colonic fluid (SCoF), which mimics the pH and buffer capacity of the colonic environment[66] (Table 14.10). A biorelevant dissolution method for testing extended-release (ER) tablets of isosorbide-5-mononitrate using USP apparatus 4 was developed.[66] This method specified sequential media and flow rate changes after a specified time of exposure. Initially, the tablet was exposed to SGF at flow rate of 8 mL/min for 60 min, and then the flow rate was decreased to 4 mL/min for 210 min, during which time the formulation was exposed to FaSSIF.[66] The final stage of this dissolution protocol used SCoF at 4 mL/min for the last 150 min of the experiment. This method not only utilized physiologically relevant volumes and fluid compositions but also considered hydrodynamics and physiologically relevant residence times (for fasted state conditions) in each of three segments (stomach, small intestine, and ascending colon) of the GI tract.[66]

A later study by Vertzoni et al. proposed a composition for fasted state simulated colonic fluid (FaSSCoF) that was designed to reflect the composition of fluids collected from the ascending colon in healthy adults (Table 14.10).[67] It has been reported that FaSSCoF more closely predicts the solubility of poorly soluble compounds in human colonic fluids (HCF) than the use of simple buffers.[67] However, it should be noted that none of the media proposed to simulate the colonic environment contain colonic bacteria due to

Table 14.10 Composition of the simulated colonic fluid and fasted state simulated colonic fluid[66,67]

	SCoF	FaSSCoF
TRIS (mM)	–	45.40
Maleic acid (mM)	–	75.82
0.5N sodium hydroxide (mL)	–	240
Sodium hydroxide (mM)	157	–
Acetic acid (mM)	170	–
Ox bile salt extract (g)	–	0.113
Phosphatidylcholine (g)	–	0.222
Palmitic acid (μM)	–	101.40
Dichloromethane (mL)	–	6
Bovine serum albumin (g)	–	3
pH	5.8	7.8
Osmolality (mOsmol/kg)	295	–
Buffer capacity (mmoL/L/pH)	29.1	–
Ionic strength	0.16	–

the significant practical challenges associated with simulating the anaerobic environment required for microbial growth.[68]

14.2.4 Biorelevant Volumes

Dissolution is a dynamic process that depends not only on the composition of the medium but also on volume of the media and hydrodynamics.[39,69] In vivo drug dissolution is also dependent on the fluid volume present in the gastrointestinal tract. The fluid volume present is the net result of the oral intake of fluid and food, secretion into the tract and water flux across the gastrointestinal epithelium. On average 2000 mL of liquid is ingested every day and 6000 mL is secreted by para- and gastrointestinal organs such as salivary glands, liver, pancreas and stomach.[20] Approximately 600 mL of bile salts and 1000–2000 mL of pancreatic juices are secreted every day. The volume of the stomach fluids in the fasted state varies between 20 and 30 mL and is present as mucus rather than a fluid pool.[20] Depending on the fluid intake, the volume of fluids in the jejunum and ileum varies between 120 to 350 mL.[20]

The simple basket or paddle stirred USP 1/2 systems provide a well-stirred, media-rich environment in which dosage form disintegration and dissolution can be evaluated. The resulting static, closed environment is limited by the absence of an absorptive sink and the relevance of the resulting hydrodynamics is also questionable given the continuous stirring and large media volumes often deployed.[2,70,71] With the introduction of USP apparatus 1 as a compendial test in 1970 and the latter introduction of USP apparatus 2 equipment in 1978, it is interesting to reflect that the 900 mL volume widely adopted for both methods is based not on physiological considerations but on an assessment of what was required to achieve sink conditions for most active pharmaceutical ingredients in development at that time.[71]

The volumes used by compendial systems are in stark contrast to images of the gastrointestinal tract obtained by water-sensitive MRI which clearly show that the in vivo gut lumen is a rapidly changing dynamic environment with fluctuating fluid volumes much less than the 500 mL or 900 mL typically used in compendial dissolution testing.[72] This study, which evaluated the gastrointestinal transit of

Table 14.11 Gastrointestinal fluid volumes in healthy subjects ($n = 12$) as determined by magnetic resonance imaging (MRI) under fasting conditions and 1 hour after a meal. Means stated with standard deviations in parentheses. Reproduced with permission from Schiller et al.[72]

	Stomach volumes (mL)	Small intestinal volumes (mL)	Large intestinal volumes (mL)
Fasted state			
Min	13	45	1
Max	72	319	44
Median	47	83	8
Mean (s.d.)	45 (18)	105 (72)	13 (12)
Fed state			
Min	534	20	2
Max	859	156	97
Median	701	39	18
Mean (s.d.)	686 (93)	54 (41)	11 (26)

sequentially administered capsules to healthy subjects in both fasted and fed states, showed that fluid was distributed along the length of the intestine as separated pockets. Moreover, it was observed that both gastric and intestinal fluid volumes in the fasted and fed states were highly variable (Table 14.11). In the fasted state, gastric volumes ranged from 13 mL to 72 mL, which is broadly consistent with earlier studies which used aspiration of gastric contents to determine resting gastric fluid volumes.[73,74] Only in the fed state do gastric volumes (534–859 mL) begin to approach the volumes used in compendial dissolution testing (although it should also be noted that the MRI measurement of gastric volume represents the total fill volume of which a proportion will be attributable to the ingested bulk food material). The small intestinal volumes ranged from 45 to 319 mL in the fasted state, with fluid clearly distributed along the length of the tract. In the fed state, small intestinal fluid volumes decreased significantly and varied between 20 mL and 156 mL. In the large intestine, segmented distribution of fluid pockets was also observed with volumes less than those seen in the small intestine, consistent with fluid absorption rather than secretion being the dominant physiological process in this region of the gastrointestinal tract. It was concluded that conventional in vitro dissolution tests are not suitable to predict in vivo release because dosage forms are not in contact with fluid pockets for significant periods of transit through both the small and large intestine. Although the imaging technique used in this study has limitations, which may compromise the absolute accuracy of the volume determinations, it demonstrates that the free fluid is not homogeneously distributed along the gut and that the fluid volumes are considerably less than those typically used in compendial dissolution testing. Both findings have significant implication for biorelevant dissolution test design, particularly for extended release dosage forms, for which it is clear that periodic contact with luminal fluid is a more realistic proposition than the constant immersion seen with compendial methodology.

The volume of the media is particularly important for poorly soluble drugs that require large amounts of fluid in order to maintain sink conditions for dissolving material.[3] For BCS II compounds, dissolved compound in the gastrointestinal lumen rapidly permeates across the absorptive epithelial layer providing an in vivo absorptive sink which facilitates on-going dissolution. Replicating this scenario in an in vitro dissolution test can be challenging for this class of compound as it is not always simply possible to address the sink limitation through the use of increased aqueous volumes. Sink conditions are typically defined as the conditions in which the volume of the medium is 3 to 10 times the saturation solubility of the tested drug. Calculation of dose/solubility (D/S) ratio can be a useful tool to assess if sink condition is an issue.[75] It provides an estimate of the medium required to dissolve the dose. When D/S is greater than 1 L dissolution may be problematic[32] as compendial dissolution systems can typically only accommodate up to 1L of medium providing finite sink conditions. For instance, Kostewicz et al. calculated D/S ratios for three weakly basic compounds (dipyridamole, BIMT 17BS, and BIBU 104XX) in compendial (pH 2 and pH 5 buffers) and biorelevant (FaSSIF pH 6.5, FeSSIF pH 5.0) media.[76] Calculated D/S ratios clearly illustrated that all formulations were highly soluble at low pH resulting in low D/S ratios of 11, 5, and 14 mL, respectively. Despite of the role of the pH on the weakly basic drugs, the effect of the elevated concentrations of bile salts and lecithin present in the media simulating fed state (FeSSIF pH 5.0) was clearly seen. The presence of bile salts and lecithin had more predominant effect than the pH. Sink conditions are no longer problematic when formulations are tested using USP Apparatus 4, which is able to provide infinite sink conditions by supplying continuous flow of fresh media when operates in open loop mode.[77,78]

14.2.5 Biorelevant Hydrodynamics

The physiological processes and peristaltic pressures which ultimately drive the segmented distribution of fluid along the GI tract will also impact dosage form performance, particularly for delayed or extended-release dosage forms. Sunesen et al. reported that small intestinal transit time in fasted and fed states in vivo were 241 ± 58 min and 268 ± 106 min, respectively.[58,76,79] Assuming that the total length of the small intestine is 350 cm,[58] the estimated average axial velocity was 1.5 cm/min in fasted state and 1.3 cm/min under fed state conditions.

The transport of dosage forms is characterized by a combination of relatively long static phases with short dynamic transport events during which dosage forms move with very high velocities.[80] High transport velocities are seen particularly during gastric emptying and transit from the terminal ileum to the ascending colon via the ileocaecal junction.[81,82] As a result, the shear rate and shear forces in the GI luminal environment need to be considered when developing a biorelevant dissolution method for delayed or extended-release systems. Conventional apparatus are not capable of mimicking such stresses and while adaptations such as the pressure cell described by Burke et al.[83] allow some aspects of peristaltic stress to be reproduced, a fundamental equipment re-design is required so that pressure events and rapid dosage form transit can be appropriately assessed. The leading example of equipment developed to meet such challenges is the biorelevant stress test device described by Garbacz et al.[84] This system, defined as the "Physiostress" apparatus in a recent publication,[85] simulates the physiological mechanical stresses described above by encasing the dosage form in a spherical

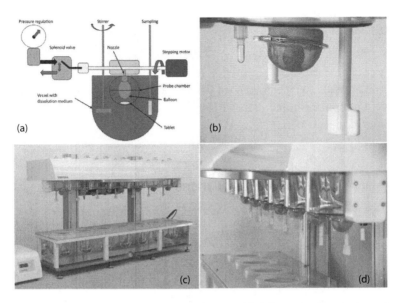

Figure 14.1 Dissolution stress test device. (a) Schematic description of the apparatus, (b) photograph of one opened chamber with the paddle stirrer and inflation device, (c, d) complete instrumentation. Figure kindly provided by Prof. Werner Weitshies, Greifswald University, Germany.

steel wire mesh holder which is rotated around a central apparatus axis (Fig. 14.1).

Pressure fluctuations are generated by periodic inflation and deflation of an air-filled balloon inside each holder. Discontinuous contact with media (intended to replicate the discontinuous contact envisaged in vivo with multiple pockets of fluid distributed along the tract) is also possible by controlling the positioning of the dosage form holder within each dissolution media bath. Computer control of the rotational movement of the holder and the pressure profiles generated by the balloon devices allow a range of profiles to be tested which simulate the media flow rates and peristaltic pressures experienced during quiescent and rapid periods of dosage form transit. This system has been studied with a range of dosage forms, including extended-release diclofenac and nifedipine tablets,[86–88] hard shell HPMC and gelatin capsules,[89] pressure-sensitive capsules,[90] and extended-release quetiapine HPMC matrix tablets.[85] For both the diclofenac and nifedipine products, it was seen that the dissolution profiles could be influenced by the application of mechanical stress. The authors attributed the sensitivity to stress as a key cause of variable in vivo release profiles and a significant factor in dose-dumping events. The study with quetiapine HPMC matrix extended-release tablets assessed the performance in both the Physiostress equipment and USP apparatus 2 under conditions intended to simulate the fasted state. Two gastric residence times (30 min and 60 min) were evaluated with the process of gastric emptying simulated as a stress phase of high-intensity pressure and agitation. The pressure component of the gastric emptying event was simulated by three consecutive inflations of the balloon and accelerated tablet transport mimicked by short periods of rapid rotation of the dosage form holder around the apparatus's central axle. A similar approach was adopted to simulate the small intestinal conditions and transit via the ileocaecal valve. As quetiapine is a weak base with two basic nitrogens providing high solubility under acidic conditions, the authors unsurprisingly found pH-dependent drug-release using USP apparatus 2. At pH 1, complete dissolution was observed after 8 hours with significantly slower dissolution in pH 6.8 phosphate buffer. Release rates for the 50 mg and 400 mg strengths were noted to be significantly different at pH 6.8, with the lower strength 50 mg tablet achieving complete dissolution after 16 hrs whilst the higher strength 400 mg tablet did not achieve complete release even after 24 hours. Differences between the tablet strengths were amplified when profiles generated using the Physiostress equipment were compared. When tested using pH 6.8 phosphate buffer, with a gastric emptying event simulated after 60 min by balloon-induced pressure events and a short period of intense rotation of the dosage form holder, it was observed that the two strengths displayed differing degrees of robustness. The gastric emptying event had only a minor impact on the release from the lower strength 50 mg tablet but a significant burst release of approximately 7.5% of the higher strength tablet was observed (Fig. 14.2). A second stress event after 5 hours, intended to simulate the passage of the dosage form from the terminal ileum to the ascending colon via the ileocaecal valve, resulted in a significant burst release of approximately 40% for the

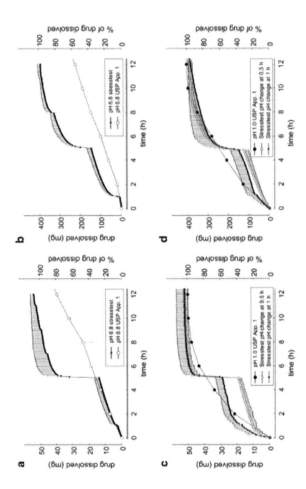

Figure 14.2 Dissolution profiles of quetiapine 50 mg ER tablets (a) and quetiapine 400 mg ER tablets (b) in the dissolution stress test device under test program 2 and the USP apparatus 2 at 100 rpm, 37°C, 1000 mL fill volume, using USP phosphate buffer pH 6.8 as dissolution media quetiapine 50 mg ER tablets (c) and quetiapine 400 mg ER tablets (d) tablets in the dissolution stress test device under test program 1 and 2 and the USP apparatus 2 at 100 rpm, 37°C, 1000 mL fill volume, using HCl pH 1.0 as dissolution media. Given are means of $n = 6$; the standard deviation is indicated by the error bars. Reproduced with permission from Garbacz et al.[85]

50 mg strength and approximately 30% for the higher strength. Profiles were noted to be significantly different from those obtained in USP apparatus 2 using the same media with much higher levels of drug release observed in the Physiostress system after a six hours period. When a media change was used with the same pressure and agitation parameters in the Physiostress, both dosage forms were clearly impacted by simulated stresses of physiological intensity. While differences in sensitivity to simulated gastric emptying between dosage strengths were reported, the impact of a simulated ileocaecal stress at 5 hours is clearly evident in both profiles (Fig. 14.2). The greater propensity of both tablet strengths to exhibit burst release after prolonged media exposure was suggested to be related to the mechanical strength of the HMPC rate-controlling matrix after five hours of media hydration. The almost pulsatile-shaped release profiles generated by the Physiostress equipment are clearly differentiated in terms of overall shape from those derived from the USP apparatus 2 testing protocol. Such data may provide additional support during formulation development of extended-release to understand sources of in vivo variability by providing a measure of dosage form physical robustness in addition to rate-controlling properties. Further improvements to the system proposed by the study authors include the use of smaller volumes and simulation of both pre- and postprandial states to achieve a more accurate simulation of the luminal environment.[85]

14.3 Prediction of Food Effects

Because of the simplistic composition of compendial media, it is not possible to use these media to distinguish between fasted and fed states. Currently, there are no official in vitro methods in pharmacopoeias that describe the requirements for testing for food effects. However, USP general chapter 1092 recommends that the compositions of media which simulate fed states can be found in literature.[91] This section of the review will focus on the role of previously described aspects of biorelevant dissolution (media selection, volume, hydrodynamics, system selection) in predicting food effects.

The impact which food has on drug absorption is affected by various physiochemical and physiological factors.[92] When increased drug absorption is observed it is often referred to as a positive food effect. Enhancement of absorption after food intake is particularly important for formulations that have dissolution rate or solubility limited absorption.[20] In contrast, a negative food effect is characterized by decreased drug release. Pharmacokinetics of drugs in the fed state can be impacted by several mechanisms such as changes in the gastrointestinal pH, retarded gastric emptying, stimulated lymphatic transport, changes in permeability across the luminal wall, bile salt secretion, as well as food-drug interactions.[46,93,94]

14.3.1 Impact of Media Selection

Various attempts to predict food effects have been reported.[95,96] For instance, AUC values for fasted and fed states were correlated with selected physicochemical properties of 100 immediate release formulations. It was concluded that aqueous solubility, lipophilicity, and dose/solubility ratio can help to predict a food effect.[95,97] BCS classification can also be a useful tool for predicting food effect.[98,99] Compounds that belong to BCS class I are unlikely to be affected by meal intake. Gu et al. reviewed 92 sets of clinical data and reported that 62% of BCS class I compounds did not show a food effect.[100] In contrast, compounds that are categorized in BCS class II are the most likely to exhibit a positive food effect (71%) as their in vivo solubility is increased due to higher concentration of the bile salts.[100] In this group three types of drugs should be identified based on their physicochemical properties including weak acids, weak bases and lipophilic compounds.[96] A positive food effect is expected for weak acids and for weak bases with high pKa values, whereas negative food effect can be observed for compounds with low pKa due to potential precipitation either in the gastric or intestinal conditions. A negative food effect is foreseen for a majority (61%) of BCS class III compounds as a result of food-drug interactions. In the case of BCS IV compounds, the impact of the food effect can be unpredictable.[96]

The direct comparison of solubilities in HIF and biorelevant media can be a useful tool. Clarysse et al. attempted to determine the solubilizing capacities of HIF in the fasted and fed state for five compounds (danazol, diazepam, nifedipine, ketoconazole and indomethacin) and correlate those values with solubility studies performed in biorelevant media.[101] In the case of non-ionized compounds, solubility in simulated intestinal fluids did not accurately predict the solubilizing capacity of the early postprandial stage. In the case of weakly acidic drugs like indomethacin, solubility strongly depends on the pH of the medium.[101] This study indicated that under- or over-estimations of solubility values are due to other intraluminal parameters, such as the presence of free fatty acids. Thus, a better understanding of the role of gastrointestinal components is needed. Ghazal et al. addressed this issue by investigating the effect of various types and concentrations of carbohydrates, amino acids and proteins on intrinsic dissolution rate of itraconazole.[102] Itraconazole is a very lipophilic weak base and therefore its dissolution is strongly affected by the pH of the medium. A significantly higher intrinsic dissolution rate was observed in SGF media at pH 1.2 than in pH 3.0. It would be expected that when itraconazole was administered with food, an elevated pH would result in poorer bioavailability. A comparison of four intrinsic dissolution profiles in a 1:1 mixture of SGF media (pH 3.0) and skimmed, semi-skimmed and whole milk revealed that intrinsic dissolution is significantly enhanced in media containing milk when compared to control medium (SGF pH 3.0). This can be explained by the high lipophilic nature of itraconazole and its affinity to fat content.[102] The solubilization effect of co-administered food or beverages was also demonstrated for the extended-spectrum tri-azole antifungal agent, posaconazole. In addition to plasma samples, gastrointestinal fluids were collected and characterized for pH and drug concentration. It was found that compared to administration with water, co-administration of Coca-Cola™ did not alter the pH of the intraluminal environment but did significantly increase posaconazole gastric concentrations and systemic exposure. This enhancement was attributed to improved posaconazole solubility in Coca-Cola™ and prolonged gastric residence.[103]

Furthermore, elevated levels of bile salts and lecithin found in the media that simulate the fed state (FeSSIF) were found to be very effective in predicting danazol food effect. In vitro dissolution results were in agreement with in vivo data published elsewhere.[104] Nicolaides et al. performed similar studies using four poorly soluble compounds, including troglitazone, atovaquone, sanfetrinem cilexetil, and GV150013X.[39] Media used in the studies (FaSSIF, FeSSIF) confirmed the utility of biorelevant media to predict food effects. Furthermore, Kostewicz et al. confirmed the effectiveness of using biorelevant media to predict a positive food effect for the weak base, dipyridamole.[76]

14.3.2 Impact of Methodology and Apparatus

Biorelevant media have evolved to be a powerful tool for in vitro biorelevant tests.[20,38,39,59] However, use of instruments that can also mimic more complex GI physiology can be advantageous in predicting complete in vivo performance. Transfer of the orally administered formulation to the site of absorption is a dynamic process. Therefore, a dissolution method that would mimic this process should incorporate the key elements of transit and a changing, dynamic, luminal environment. There are a number of instruments which can closely simulate in vivo environments in terms of sequential media change, low media volume, physiological hydrodynamics or even enzyme secretions. Some of them are officially approved by USP systems such as Bio-Dis (reciprocating cylinder, USP apparatus 3) and flow-through cell dissolution system (USP apparatus 4). In 2005, Klein et al. published dissolution data for poorly soluble mesalazine and budesonide formulations that were tested using physiologically relevant pH gradient dissolution.[105] The combination of biorelevant media with a dissolution apparatus (USP apparatus 3) which allowed the hydrodynamics and GI transit times to be closely replicated.[105] The experimental protocol allowed the simulation of transit through the stomach, small intestine and proximal colon. Interestingly, it has been found that in the case of pH-dependent release formulations, the pH and the residence time in the different segments of the GI tracts plays a more important role than the composition of the media used. Moreover, Jantratid et al. evaluated the ability of both USP apparatus 3 and 4 to predict the in vivo performance of MR Diclofenac sodium pellets.[106] It was found that both dissolution systems were able to successfully predict a food effect. However, the release of diclofenac tested using the flow-through cell apparatus was closer to in vivo absorption in both the fasted and fed states.[106] It has also been reported that in some cases such as the studies performed by Samaha et al. that there was no difference observed between release profiles tested in fasted media obtained using USP apparatus 2 and USP apparatus 3.[107] However, USP apparatus 3 provided a more discriminating and comprehensive evaluation of the release performance of Diltiazem ER products.[107]

14.3.3 Impact of Biomodeling/in silico Tools

Physiologically based pharmacokinetic modeling plays an important role in predicting food effects by integrating physicochemical drug properties, physiological parameters, and formulation characteristics.[108] It has been reported that in vitro biorelevant testing using conventional dissolution systems coupled with absorption modeling tools can result in successful prediction of food effect.[109] Shono et al. predicted the in vivo performance of Celecoxib capsules using solubility and dissolution data obtained in FaSSGF/FaSSIF-V2 and FeSSGF/FeSSIF-V2 media as inputs for a biomodel developed using the STELLA software platform.[109] Moreover, solubility values obtained in biorelevant media aided predictions of food effects of six Roche compounds using in silico modeling (GastroPlus),[92] where the models were able to accurately differentiate between limited (observed with four of the test compounds) and significant (observed for two of the test compounds) food effects.

Wagner et al. demonstrated that supporting biorelevant dissolution data with precipitation testing can aid the development of successful in silico models for predicting food effects with BCS class IV weakly basic compounds.[110] Furthermore, dynamic dissolution testing using biorelevant media proved to be a very promising tool, which in combination with in silico modeling allowed for the successful prediction of complete in vivo profiles.[58, 109, 111–113] For instance, Okumu et al. performed dissolution of a montelukast sodium formulation using the flow-through cell apparatus.[111] The formulation was exposed to SGF for 15 min at a flow rate of 3.3 mL/min, followed by three phases of FaSSIF media exposure (pH 6.8 for 45 min at a flow rate of 5.8 mL/min, pH 7.5 for 150 mins and then pH 5.0 to a total experiment time of 240 min).[111] The resulting dissolution data with physiochemical and physiological parameters were used as inputs to GastroPlus software and resulted in improved predictions of plasma concentrations of montelukast sodium.

Non-compendial Dissolution Apparatus

Drug dissolution in the GI tract occurs in parallel to permeation and if the dissolution rate is slower than the permeation rate through the intestinal membrane, oral absorption may be dissolution rate limited.[114] Refinement of media composition, volumes, and system hydrodynamics are important components of a biorelevant method. However, for poorly soluble molecules, the impact of drug permeation on concentration gradients within the dissolution medium is often overlooked. Simulating the drug absorption step is often confused with achieving sink conditions and while it is true that the strategy for providing an absorptive and/or dissolution sink are closely aligned, the approaches to achieve either can be very different. For example, common approaches to achieve sink conditions include manipulation of pH, use of surfactants, organic media addition and large volume dissolution (e.g., 2 or 4 L vessels).[71,115] By contrast, the approaches for implementing a concurrent absorption step in dissolution models rely upon transport of dissolved active from the bulk media into another phase or compartment. A number of systems have been described in the literature and range from the use of a lipid-like layer to replicate drug partitioning, dialysis membrane systems using molecular weight cut-offs and counterflow buffer exchange to generate diffusion gradients for separation, and combined dissolution-permeation tests which utilise an artificial or cell-based membrane.

Simple systems which provide an absorptive sink using a partitioning approach with octanol have shown some promise in improving the correlation to in vivo performance.[116,117] A novel extension of this concept has been reported by Vangani et al., who described a combined USP apparatus 4 and USP apparatus 2, which incorporated an organic phase to mimic the absorption process.[118] This system utilized flow-through USP apparatus 4 equipment as the primary dissolution apparatus from which dissolved eluent was transferred to a connected secondary USP apparatus 2 chamber containing a biphasic dissolution medium comprising of an aqueous buffer (pH 6.8 phosphate buffer) and organic solvent (either nonanol or a 1:1 mixture of cyclohexhane and nonanol). The USP apparatus 2 paddle was modified with an additional small paddle to provide stirring in both aqueous and organic phases and the stirring rate was optimised to facilitate transfer of compound into the organic phase. The return line to the USP apparatus 4 equipment was connected to the aqueous phase of the biphasic medium, completing the flow-through loop. This approach was used to characterise the dissolution profile of a poorly soluble experimental compound, AMG 517. It was reported that it had not been possible to generate meaningful data using USP apparatus 4 for this compound due to the very low solubility across the physiological pH range. By using the combined system, it was found that in vitro release profiles for AMG 517 capsule formulations could be superimposed with absorption–time profiles produced by deconvolution and allowed a rank order of formulations to be established.

A modification of the multi-compartment/transfer model which incorporated a third module intended to simulate the absorption compartment was reported by Gu et al.[119] This system, based on a modified USP apparatus, comprised gastric and small intestinal compartments supplemented with a third compartment to simulate absorption. Hydrodynamics were controlled by paddles at 100 rpm and a pH stat was used to maintain pH in the small intestinal chamber (pH 5.5 +/− 0.2), which increased to pH 6.5 over a 1 hour period. This was intended to simulate firstly the lower pH in the duodenum before exposing the compound to the higher pH in the jejunum.[120,121] The flow rate from the intestinal chamber to the absorption compartment could be adjusted based on the permeability of the compound. The amount of drug transferred to the absorption compartment after three hours was determined as the in vitro absorbed amount which was correlated to in vivo bioavailability data for two weak bases with low intrinsic solubility (cinnarizine and dipyridamole). For these model compounds, it was possible to estimate the precipitation potential to diagnose whether precipitation is a contributory factor in the cause of poor oral bioavailability. This was clearly demonstrated with cinnarizine where the difference in the amount transferred to the in vitro absorption compartment was attributed to precipitation in the intestinal compartment despite complete prior dissolution. However, although dipyridamole and cinnarizine were shown to exhibit significant differences in precipitation in vitro, the correlation to in vivo data was hampered by the lack of information on the absolute oral bioavailability and first-pass metabolism of these compounds.

A miniscale biphasic dissolution system which incorporated a pH shift in the aqueous compartment was described by Frank et al.[122] The ability to induce a pH shift to mimic the transit from the gastric compartment to the small intestine was implemented to allow the supersaturation and precipitation behavior of poorly soluble weak bases to be studied under sink conditions provided by the octanol absorptive layer. In the case of dipyridamole, precipitation after the pH shift from acidic conditions to pH 5.5 and then pH 6.8 was much less than that observed in a single phase dissolution model with pH change (35% vs. 90%, respectively). The authors suggested that the miniscale biphasic results for precipitation were much closer to in vivo results published by Psachoulias et al. (7% of the administered dose of dipyridamole was observed to precipitate in human in vivo studies) as a result of transport of dipyridamole into the octanol layer and a reduction in the time available for dipyridamole molecules in solution to form a seed for precipitation. Four formulations of a weak base were also studied.[122] This molecule, named only as BIXX in the publication, had a solubility of 4.7 mg/mL at pH 1 and a substantially lower solubility (0.008 mg/mL) at pH 5.5. Consequently, supersaturation was expected to be a critical component in determining the overall extent of absorption. Using a single phase USP apparatus 2 dissolution test (pH 5.5 using McIlvaine's buffer, 50 rpm, 900 mL), some differences between formulations were noted, with those pH-modified formulations containing an organic acid found to be significantly faster dissolving than a control formulation which contained only microcrystalline cellulose and lactose fillers. However, in vivo dog studies showed that neither the ranking of formulations nor the dissolution rate observed in the USP apparatus 2 single-phase dissolution were predictive for in vivo performance. By contrast, the miniscale biphasic test with pH shift method produced a level A correlation

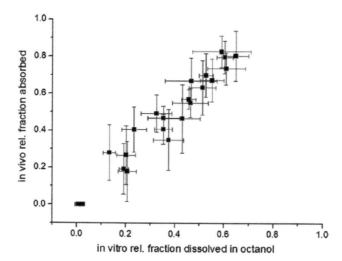

Figure 14.3 Level A IVIVC. Correlation of relative fraction absorbed in vivo and relative fraction dissolved in octanol in the miBIdi-ph between 0.00 h and 3.00 h. Mean values of the four formulations $\pm$ SD; $R^2 = 0.95$. Reprinted with permission from Frank et al.[122]

(Fig. 14.3) between the relative fraction dissolved in the octanol layer and the relative fraction absorbed in vivo. The authors concluded that the miniscale biphasic test with pH shift method was able to capture the kinetics of both supersaturation and dissolution of precipitated API and was much more accurate for predicting in vivo exposure than the single-phase dissolution method.

In general, it would appear that biphasic systems using an organic layer to simulate drug absorption via partitioning have a role to play in the search for a truly biorelevant dissolution method. Practical disadvantages, however, do need to be overcome to fully realize their potential. For example, saturation of the aqueous phase with octanol can influence formulation dissolution, partitioning of components of biorelevant media (taurocholate and lecithin in particular) can be expected to orientate at the phase boundary, and the relevance of mass transport of dissolved active into the organic layer needs to be carefully evaluated in the context of apparent permeability values.

Lamberti et al. proposed a modification to a conventional USP 2 apparatus to incorporate a hollow-fibre dialysis cartridge in order to simulate the absorption process.[123] Using theophylline as a model drug, it was found that the concentrations of drug in the donor compartment were lower than those found in a conventional reference test, reflecting significant mass-transport of dissolved theophylline across the dialysis membrane into the acceptor compartment. It was suggested that the lower concentrations seen in the aqueous phase of the dissolution medium when a dialysis cartridge was used, may provide a more accurate input for PBPK modeling than a conventional USP apparatus 2 dissolution profile. An extension of the dialysis approach to the study of nanocrystalline suspensions (a commonly used approached for bioenhancement of BCS II/IV molecules) was described in a paper by Kumar et al.[124] In this case, a dialysis sac was used to retain spray-dried nanosuspension formulations and act as an in situ membrane filter for dissolved and suspended API. This approach allows the bulk volume in the USP 2 vessel to act as a sink for dissolved material and highlights the ability of a dialysis membrane to efficiently separate dissolved and suspended material. A schematic representation of the assembly of the dialysis sac within a USP apparatus 2 is shown in Fig. 14.4. Dialysis is also used to separate products of digestion and dissolved API from luminal contents in the TNO TIM-1 system.

However, the use of a dialysis membrane to mimic the absorptive surface presented by the epithelial layer of the small intestine has obvious limitations in terms of surface composition, lipophilicity, and the use of molecular weight cut-off as the sole determinant of partitioning. Alternative membranes have been evaluated in the small-volume dissolution model known as the dissolution-permeation (D-P) system.[125–127] This two-compartment model consists of a dissolution and receiver compartment separated by a Caco-2 monolayer which can be used to perform simultaneous analysis of drug dissolution and permeation. Several publications have looked at the utility of this system for the evaluation of formulation effects on the oral absorption of poorly soluble drugs. Katoka et al. used the D-P system to study the impact of formulation (including suspension, SEDDS, and supersaturable SEDDS) on the dissolution and permeation behavior of two poorly soluble

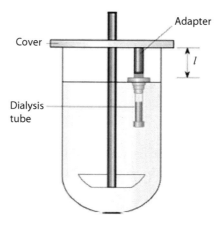

Figure 14.4 Schematic representation of the assembly of dialysis sac and its adapter with USP apparatus 2. Reprinted with permission from Kumar et al.[124]

compounds, danazol and pranlukast.[127] The permeated amount of compound was used to calculate fraction absorbed in humans. The equation used to calculate this value utilised data from previous studies to underwrite a correlation between oral absorption in humans with the percentage of applied dose permeated after 2 hours in the D-P assay. For danazol, it was found that the estimated absorption (%) correlated reasonably well ($R^2 = 0.83$) with AUC values obtained from in vivo rat exposure studies. In the case of pranlukast, significant formulation effects were not observed and a correlation could not be determined. A similar study with the three poorly soluble compounds, celecoxib, montelukast, and zafirlukast, has also been reported.[128] The relationship between the clinical dose and the percentage oral absorption predicted by the D-P system corresponded well to available clinical data. The ability to study dissolution in tandem with permeation allowed further insight to be gained on the rate-limiting processes for the absorption of each compound with celecoxib and montelukast shown to be solubility limited and zafirlukast limited by dissolution rate. The D-P system has also been used to correlate the in vitro dissolution and permeation of fenofibrate, a BCS class II compound with in vivo data in rats.[129] This study screened a number of bio-enhanced formulations including solid dispersion, nanoparticulate, and micronized approaches. All showed that the D-P system, when used with biorelevant media, could predict formulation performance in rats. However, the majority of D-P models reported in the literature are limited in their ability to reproduce in vivo relevant hydrodynamics and cannot deal with complex food materials or digestive processes. Additionally, despite their use of a biological membrane, the D-P model configuration provides a single compartment for dissolution to occur and does not mimic the dynamic range of conditions seen during gastrointestinal transit.

It is clear that in order to address the limitations of simple mechanistic systems such as the dissolution-permeation or biphasic methods a step-change in terms of equipment complexity is required. A number of more complex GI-like simulators, offering full control of physiological parameters such as temperature, pH, peristaltic mixing and transit, gastric secretion (lipase, pepsin, HCl), and small intestinal secretion (pancreatic juice, bile, and sodium bicarbonate) have been developed. These include the TNO gastro-intestinal model (TIM),[12] the ModelGut/dynamic gastric model (DGM)[130] and, most recently, the human gastric simulator [HGM].[9, 10] It is interesting to consider that all of these systems have originated from the nutritional/food science sector and this should be an area for dissolution scientists to actively monitor for relevant research given that the common goal of both scientific communities is to accurately model the gastrointestinal environment. The TIM-1 and DGM have been available to pharmaceutical scientists for a number of years and their ability to study the impact of multiple physiological parameters in a parallel and holistic fashion has prompted their use to assess the dissolution of a range of pharmaceutical dosage forms.

The Modelgut™ (Institute for Food Research, Norwich, UK) dynamic gastric model (DGM) was developed from insights gained from echo-planar magnetic resonance imaging studies on the gastric processing of complex meals.[131−134] These studies showed that mixing in the stomach is inhomogeneous and hydrated regions of a meal bolus are selectively processed to the antral region of the stomach. The DGM is claimed to provide an accurate in vitro

Figure 14.5 IFR dynamic gut model apparatus showing main body/fundal area (1) and antrum (2). Picture courtesy of ModelGut ©, IFR, Norwich, UK. Reproduced with permission.

simulation of gastric mixing (including digestive addition around the gastric bolus), shear rates and forces, peristalsis, and gastric emptying.[135] The equipment is composed of two main stages as shown in Fig. 14.5. The first stage mimics the gastric mixing and dynamic secretory profiles of the fundal regions of the stomach with the second-stage replicating the shear forces and mixing observed in the antral region. It is possible to couple the output from the first two stages to a biorelevant simulation of the small intestine to generate a complete profile of digestion and dissolution. However, to date, application to pharmaceutical dosage forms and in particular to poorly soluble compounds has been limited. One study which explored the ability of the DGM to replicate the dynamic digestion of a SEDDS formulation suggested that the DGM provided a more accurate simulation of SEDDS digestion (at least in terms of droplet size) than conventional USP apparatus 2.[136] A second study assessed the relative performance of gelatin and HPMC capsules in the fed and fasted states and concluded that the capture rupture times obtained from the DGM were similar to those observed by gamma scintigraphy in vivo studies in the fasted state and were delayed in the fed state, although the comparison to in vivo scintigraphy results in this case was affected by the impact of food on the dispersion of contents and subsequent sampling in the DGM.[137] The DGM has also been used to assess the release of a complex dosage form containing multiple APIs in immediate-release and controlled-release layers with some advantages observed for prediction of performance over conventional USP apparatus 2.[138] White still at an early-stage of evaluation for pharmaceutical applications, the ability of the DGM to simulate gastric forces and meal processing should have value in accurately assessing the potential for food effects, alcohol interactions, and the mechanical robustness of modified-release formulations.

The TNO TIM-1 system is a multi-compartmental, dynamic, computer-controlled model of the human upper gastrointestinal tract.[12] This system simulates the in vivo dynamic digestive and physiological processes which occur within the human stomach and small intestine and allows control of all the main parameters of digestion, including temperature, pH, peristaltic mixing and transit, gastric secretion (lipase, pepsin, HCl), and small intestinal secretion (pancreatic juice, bile, and sodium bicarbonate).[139] The absorption phase is simulated by dialysis or membrane filtration systems which remove dissolved (low molecular weight) water-soluble compounds or lipid-soluble compounds from the jejunal and ileal compartments of the system. The amount of dissolved active recovered in the dialysate or filtration fluid is used to calculate the bioaccessibility provided by a formulation (i.e., the amount of digested product or drug substance in solution and therefore available for absorption).[139-142] Hydrodynamics are controlled by changes in water pressure on flexible membranes which contain

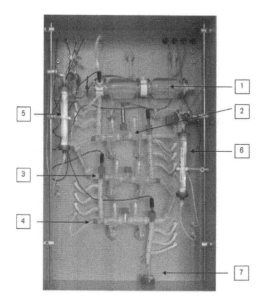

Figure 14.6   TIM-1 apparatus (1 – gastric compartment; 2 – duodenum; 3 – jejunum; 4 – ileum; 5 – jejunal dialysis cartridge; 6 – ileal dialysis cartridge; 7 – ileal eluent). Picture courtesy of TNO Triskelion, Zeist.

the luminal contents and enable mixing by alternate cycles of compression and relaxation, simulating in vivo muscular peristaltic contractions. Additionally, transit is regulated by opening or closing peristaltic valves that connect each compartment, allowing the controlled passage of liquids and food/drug particle. The system components are shown in Fig. 14.6 and a schematic of the system is provided in Fig. 14.7.

While there are multiple examples of where TIM-1 has been used to study the absorption of nutritional materials over the last few years,[143–147] there are only a limited number of examples in the literature describing its use for the evaluation of pharmaceutical dosage forms. Blanquet et al. and Souliman et al. used TIM-1 to evaluate the impact of transit time and food on the absorption of paracetamol and theophylline following administration as either the free powder form or as a sustained release tablets.[139,141,142] These studies demonstrated that the profiles of jejunal absorption found in vitro were consistent with in vivo data and a good correlation

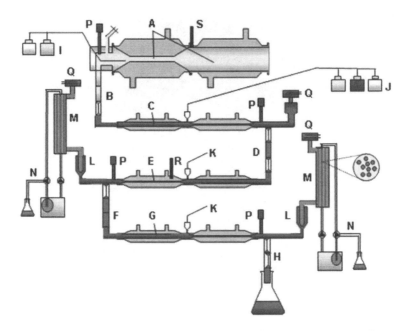

Figure 14.7 TIM-1 system schematic. A. stomach compartment; B. pyloric sphincter; C. duodenum compartment; D. peristaltic valve; E. jejunum compartment; F. peristaltic valve; G. ileum compartment; H. ileocecal sphincter; I. stomach secretion (HCL, electrolytes and enzymes); J. duodenum secretion (bicarbonate, bile and pancreatin); K. jejunum/ileum secretion (bicarbonate); L. pre-filter; M. semipermeable membrane; N. filtrate; P. pH electrodes; Q. level sensors; R. temperature sensor; S. pressure sensor. Picture courtesy of TNO Triskelion, Zeist.

was seen with T_{max} values for the immediate-release form. It was also shown that food intake (in the form of a standard breakfast) reduced the amount of paracetamol available for absorption. This was judged to be similar to clinical studies which showed a lower C_{max} and delayed T_{max} in the fed state compared to intake with water in the fasted state.[148–151] A further study evaluated the use of this dynamic model to improve the predictability of physiologically based pharmacokinetic (PBPK) simulation and modeling software for a paroxetine hydrochloride immediate-release tablet.[152] Using the bioaccessibility profile from TIM-1 instead of USP apparatus 2 dissolution data as the input rate for the in silico absorption model improved the predicted plasma profile. In a study with an immediate-release fosamprenavir tablet,[153] tested under fed state and fasted state conditions in the TIM-1, it was demonstrated that disintegration and fosamprenavir dissolution was significantly postponed in the fed state compared to the fasted state. This resulted in a lag in the appearance of bioaccessible fosamprenavir but no effect on the cumulative bioaccessibility. These results were in agreement with the data observed in a study with healthy volunteers.

The use of the TIM-1 to screen and select optimal formulations for a poorly soluble BCS class II GSK development compound was described by David et al.[140] It was found that sampling of material from the dialysate fluids was not possible due to very low recoveries, likely to be a result of hydrophobic interactions between the API and the dialysis membrane. In this case, direct sampling from the jejunal and ileal compartments was used to overcome this issue. Binding to dialysis membranes or the plastic components of the system impacts the ability to achieve full mass-balance for the analyte and can be a more general issue for the testing of hydrophobic poorly soluble compounds in this system. Dickinson et al. described limitations of the TIM-1 system in the standard dialysis configuration for the testing of poorly soluble drugs.[154] In this mode, with the use of dialysis to remove dissolved drug and thereby create a sink for dissolution, it is conceivable that for some poorly soluble drugs, the rate of drug dialysis could be limited by drug solubility in the dialysis buffer. This would not be the case in vivo due to a number of additional factors such as plasma protein binding, drug distribution, and clearance. To overcome this limitation, TNO has developed a novel lipid membrane configuration in which lipid digestion and the bioaccessibility of lipid-soluble compounds can be studied by applying filtration through a 50 nm pore filter at a predetermined filtration rate (the standard dialysis membrane configuration uses a membrane with a molecular weight cut-off of 10 kDa). The use of this filter allows removal of mixed micelles containing lipophilic compounds such as products of fat digestion and APIs, while undigested fat and undissolved compounds are retained in the luminal compartment.[155,156] Dickinson et al. demonstrated that it is possible to use the lipid membrane filter in place of dialysis to simulate absorption a BCS class II compound in both fasting and achlorhydric conditions.[154] The authors commented that at AstraZeneca the TIM-1 system had been used to successfully to support formulation design and development for numerous poorly soluble drug candidates. They also noted that for some candidate molecules, binding to plastic components was problematic and improvement to the TIM-1 stomach chamber design was required to improve hydrodynamics and reproduction of antral grinding processes. However, overall, the examples of formulation studies reported in the literature suggest that the TNO TIM-1 system, which provides an advanced level of control over a dynamic and complex luminal environment, may have several advantages over conventional dissolution methodologies when assessing the performance of oral formulations in either the fasted or fed states.

14.4 Conclusion

In conclusion, the use of in vitro dissolution to select formulations for preclinical and clinical use remains constrained by the use of equipment and methodologies originally developed to support a quality control function. Integration of in vitro biorelevant dissolution testing with physiology-based pharmacokinetic modeling is an important consideration for the ultimate objective of accurately predicting in vivo performance and is a key objective of the Innovative Medicines Initiative OrBiTo program, which is seeking to develop the next generation of predictive tools for oral absorption through a pre-competitive collaboration between academic, industrial, and specialist technology providers.[157] There is a clear need to improve the current in vitro tools to more closely simulate key aspects of gastrointestinal physiology, so that formulation decisions can be made on robust and biorelevant data. Newer systems such as the DGM and the TIM-1 gastrointestinal model illustrate potential new design paths for the next generation of biorelevant dissolution systems and if current drawbacks such as sample throughput, in-line analysis, and more accurate replication of the membrane permeation step can be addressed, they may at last offer formulation scientists a realistic alternative to in vivo testing.

References

1. Edwards, L. J., The dissolution and diffusion of aspirin in aqueous media. *Trans. Faraday Soc.*, 1951. **47**, 1191–1210.

2. Dokoumetzidis, A., Macheras, P., A century of dissolution research: from Noyes and Whitney to the biopharmaceutics classification system. *Int. J. Pharm.*, 2006. **321**(1–2), 1–11.

3. McAllister, M., Dynamic dissolution: a step closer to predictive dissolution testing? *Mol. Pharm.*, 2010. **7**(5), 1374–1387.

4. Benet, L. Z., Wu, C.-Y., Custodio, J. M., Predicting drug absorption and the effects of food on oral bioavailability. *Bulletin Technique Gattefossé*, 2006. **99**, 9–16.

5. Sherry Ku, M., Dulin, W., A biopharmaceutical classification-based Right-First-Time formulation approach to reduce human pharmacokinetic variability and project cycle time from First-In-Human to clinical Proof-Of-Concept. *Pharm. Dev. Technol.*, 2012. **17**(3), 285–302.

6. Kawabata, Y., Wada, K., Nakatani, M., Yamada, S., Onoue, S., Formulation design for poorly water-soluble drugs based on biopharmaceutics classification system: basic approaches and practical applications. *Int. J. Pharm.*, 2011. **420**(1), 1–10.

7. Carino, S. R., Sperry, D. C., Hawley, M., Relative bioavailability estimation of carbamazepine crystal forms using an artificial stomach-duodenum model. *J. Pharm. Sci.*, 2006. **95**(1), 116–125.

8. Carino, S. R., Sperry, D. C., Hawley, M., Relative bioavailability of three different solid forms of PNU-141659 as determined with the artificial stomach-duodenum model. *J. Pharm. Sci.*, 2010. **99**(9), 3923–3930.

9. Kong, F., Singh, R. P., A human gastric simulator (HGS) to study food digestion in human stomach. *J. Food Sci.*, 2010. **75**(9), E627–E635.

10. Roman, M. J., Burri, B. J., Singh, R. P., Release and bioacessibility of β–carotene from fortified almond butter during in vitro digestion. *J. Agric. Food Chem.*, 2012. **60**, 9659–9666.

11. Kostewicz. E. S., Abrahamsson, B., Brewster, M., Brouwers, J., Butler, J., Carlert, S., Dickinson, P. A., Dressman, J., Holm, R., Klein, S., Mann, J., McAllister, M., Minekus, M., Muenster, U., Müllertz, A., Verwei, M., Vertzoni, M., Weitschies, W., Augustijns, P., In vitro models for the prediction of in vivo performance of oral dosage forms. *Eur. J. Pharm. Sci.*, 2014. **57**, 342–366.

12. Minekus, M., Marteau, P., Havenaar, R, Huis in 't Veld, J. H. J., A multi compartmental dynamic computer-controlled model simulating the

stomach and small intestine. *ATLA Altern. Lab. Anim.*, 1995. **23**, 197–209.

13. Wickham, M. J. S., Faulks, R. M., Mann, J., Mandalari, G., The design, operation, and application of a dynamic gastric model. *Dissolut. Technol.*, 2012. **19**(3), 15–22.

14. Lambert, R., Martin, F., Vagne, M., Relationship between hydrogen ion and pepsin concentration in human gastric secretion. *Digestion*, 1968. **1**(2), 65–77.

15. Schmidt, H. A., Fritzlar, G., Dölle, W., Goebell, H., [Comparative studies on the histamine and insulin stimulated acid pepsin secretion in patients suffering from ulcus duodeni and control persons]. *Dtsch Med Wochenschr*, 1970. **95**(40), 2011–2016.

16. Mudie, D. M., Amidon, G. L., Amidon, G. E., Physiological parameters for oral delivery and in vitro testing. *Mol. Pharm.*, 2010. **7**(5), 1388–1405.

17. Dressman, J. B., et al., Upper gastrointestinal (GI) pH in young, healthy men and women. *Pharm. Res.*, 1990. **7**(7), 756–761.

18. Efentakis, M., Dressman, J. B., Gastric juice as a dissolution medium: surface tension and pH. *Eur. J. Drug Metab. Pharmacokinet.*, 1998. **23**(2), 97–102.

19. Hoerter, D., Dressman, J. B., Influence of physicochemical properties on dissolution of drugs in the gastrointestinal tract. *Adv. Drug Deliv. Rev.*, 1997. **25**, 3–14.

20. Dressman, J. B., et al., Dissolution testing as a prognostic tool for oral drug absorption: immediate release dosage forms. *Pharm. Res.*, 1998. **15**(1), 11–22.

21. Vertzoni, M., et al., Simulation of fasting gastric conditions and its importance for the in vivo dissolution of lipophilic compounds. *Eur. J. Pharm. Biopharm.*, 2005. **60**(3), 413–417.

22. Ogata, H., et al., Development and evaluation of a new peroral test agent GA-test for assessment of gastric acidity. *J. Pharmacobiodyn.*, 1984. **7**(9), 656–664.

23. Kalantzi, L., et al., Characterization of the human upper gastrointestinal contents under conditions simulating bioavailability/bioequivalence studies. *Pharm. Res.*, 2006. **23**(1), 165–176.

24. Gibaldi, M., Feldman, S., Mechanisms of surfactant effects on drug absorption. *J. Pharm. Sci.*, 1970. **59**(5), 579–589.

25. Solvang, S., Finholt, P., Effect of tablet processing and formulation factors on dissolution rate of the active ingredient in human gastric juice. *J. Pharm. Sci.*, 1970. **59**(1), 49–52.

26. Ruby, M. V., et al., Estimation of lead and arsenic bioavailability using a physiologically based extraction test. *Environ. Sci. Technol.*, 1996. **30**(2), 422–430.

27. Anwar, S., Fell, J. T., Dickinson, P. A., An investigation of the disintegration of tablets in biorelevant media. *Int. J. Pharm.*, 2005. **290**(1–2), 121–127.

28. Galia, E., Horton, J., Dressman, J. B., Albendazole generics–a comparative in vitro study. *Pharm. Res.*, 1999. **16**(12), 1871–1875.

29. Zhao, F., et al., Effect of sodium lauryl sulfate in dissolution media on dissolution of hard gelatin capsule shells. *Pharm. Res.*, 2004. **21**(1), 144–148.

30. Vertzoni, M., et al., Simulation of fasting gastric conditions and its importance for the in vivo dissolution of lipophilic compounds. *Eur. J. Pharm. Biopharm.*, 2005. **60**(3), 413–417.

31. Vertzoni, M., et al., Estimation of intragastric solubility of drugs: in what medium? *Pharm. Res.*, 2007. **24**(5), 909–917.

32. Hörter, D., Dressman, J. B., Influence of physicochemical properties on dissolution of drugs in the gastrointestinal tract. *Adv. Drug Deliv. Rev.*, 2001. **46**(1–3), 75–87.

33. Pedersen, B. L., et al., A comparison of the solubility of danazol in human and simulated gastrointestinal fluids. *Pharm. Res.*, 2000. **17**(7), 891–894.

34. Lindahl, A., et al., Characterization of fluids from the stomach and proximal jejunum in men and women. *Pharm. Res.*, 1997. **14**(4), 497–502.

35. Macheras, P., Koupparis, M., Tsaprounis, C., Drug dissolution studies in milk using the automated flow injection serial dynamic dialysis technique. *Int. J. Pharm.*, 1986. **33**(1–3), 125–136.

36. Macheras, P. E., Koupparis, M. A., Antimisiaris, S. G., Drug binding and solubility in milk. *Pharm. Res.*, 1990. **7**(5), 537–541.

37. Jantratid, E., et al., Dissolution media simulating conditions in the proximal human gastrointestinal tract: an update. *Pharm. Res.*, 2008. **25**(7), 1663–1676.

38. Galia, E., et al., Evaluation of various dissolution media for predicting in vivo performance of class I and II drugs. *Pharm. Res.*, 1998. **15**(5), 698–705.

39. Nicolaides, E., et al., Forecasting the in vivo performance of four low solubility drugs from their in vitro dissolution data. *Pharm. Res.*, 1999. **16**(12), 1876–1882.

40. FDA, Guidance for industry: food-effect bioavailability and fed bioequivalence studies. Food and Drug Administration, Editor, 2002: Rockville, MD.

41. Macheras, P. E., Koupparis, M. A., Antimisiaris, S. G., Effect of temperature and fat content on the binding of hydrochlorothiazide and chlorothiazide to milk. *J. Pharm. Sci.,* 1988. **77**(4), 334–336.

42. Macheras, P., Koupparis, M., Apostolelli, E., Dissolution of 4 controlled-release theophylline formulations in milk. *Int. J. Pharm.,* 1987. **36**(1), 73–79.

43. Dressman, J. B., Thelen, K., Jantratid, E., Towards quantitative prediction of oral drug absorption. *Clin. Pharmacokinet.,* 2008. **47**(10), 655–667.

44. Fotaki, N., Symillides, M., Reppas, C., Canine versus in vitro data for predicting input profiles of L-sulpiride after oral administration. *Eur. J. Pharm. Sci.,* 2005. **26**(3–4), 324–333.

45. Jantratid, E., Dressman, J. B., Biorelevant dissolution media simulating the proximal human gastrointestinal tract: an update. *Dissolut. Technol.,* 2009. **16**(3), 21–25.

46. Clarysse, S., et al., Postprandial evolution in composition and characteristics of human duodenal fluids in different nutritional states. *J. Pharm. Sci.,* 2009. **98**(3), 1177–1192.

47. Persson, E., et al., The effects of food on the dissolution of poorly soluble drugs in human and in model small intestinal fluids. *Pharm. Res.,* 2005. **22**(12), 2141–2151.

48. Kalantzi, L., et al., Canine intestinal contents vs. simulated media for the assessment of solubility of two weak bases in the human small intestinal contents. *Pharm. Res.,* 2006. **23**(6), 1373–1381.

49. Davenport, H. W., Secretion of the Bile, in: *Physiology of the Gastrointestinal Tract.* 5th ed. Chapter 11. 1982, Chicago, USA: Year Book Medical Publishers.

50. Mithani, S. D., et al., Estimation of the increase in solubility of drugs as a function of bile salt concentration. *Pharm. Res.,* 1996. **13**(1), 163–167.

51. Takano, R., et al., Oral absorption of poorly water-soluble drugs: computer simulation of fraction absorbed in humans from a miniscale dissolution test. *Pharm. Res.,* 2006. **23**(6), 1144–1156.

52. Soderlind, E., et al., Simulating fasted human intestinal fluids: understanding the roles of lecithin and bile acids. *Mol. Pharm.,* 2010. **7**(5), 1498–1507.

53. Marques, M., Dissolution media simulating fasted and fed states. *Dissolut. Technol.,* 2004. **11**(2), 16.

54. Fagerberg, J. H., et al., Dissolution rate and apparent solubility of poorly soluble drugs in biorelevant dissolution media. *Mol. Pharm.,* 2010. **7**(5), 1419–1430.

55. Zoeller, T., Klein, S., Simplified biorelevant media for screening dissolution performance of poorly soluble drugs. *Dissolut. Technol.,* 2007. **14**(4), 8–13.

56. Lobenberg, R., et al., Dissolution testing as a prognostic tool for oral drug absorption: dissolution behavior of glibenclamide. *Pharm. Res.,* 2000. **17**(4), 439–444.

57. Khan, M. Z., et al., Classification of loratadine based on the biopharmaceutics drug classification concept and possible in vitro-in vivo correlation. *Biol. Pharm. Bull.,* 2004. **27**(10), 1630–1635.

58. Sunesen, V. H., et al., In vivo in vitro correlations for a poorly soluble drug, danazol, using the flow-through dissolution method with biorelevant dissolution media. *Eur. J. Pharm. Sci.,* 2005. **24**(4), 305–313.

59. Vertzoni, M., et al., Dissolution media simulating the intralumenal composition of the small intestine: physiological issues and practical aspects. *J. Pharm. Pharmacol.,* 2004. **56**(4), 453–462.

60. Wei, H., Lobenberg, R., Biorelevant dissolution media as a predictive tool for glyburide a class II drug. *Eur. J. Pharm. Sci.,* 2006. **29**(1), 45–52.

61. Kleberg, K., et al., Biorelevant media simulating fed state intestinal fluids: colloid phase characterization and impact on solubilization capacity. *J. Pharm. Sci.,* 2010. **99**(8), 3522–3532.

62. Markopoulos, C., et al., Biorelevant media for transport experiments in the Caco-2 model to evaluate drug absorption in the fasted and the fed state and their usefulness. *Eur. J. Pharm. Biopharm.,* 2014. **86**(3), 438–448.

63. Boni, J. E., et al., Instant FaSSIF and FeSSIF—biorelevance meets practicality. *Dissolut. Technol.,* 2009. 41–45.

64. Kloefer, B., et al., Study of a standardized taurocholate–lecithin powder for preparing the biorelevant media FeSSIF and FaSSIF. *Dissolut. Technol.,* 2010. 6–13.

65. Cummings, J. H., et al., Short chain fatty acids in human large intestine, portal, hepatic and venous blood. *Gut,* 1987. **28**(10), 1221–1227.

66. Fotaki, N., Symillides, M., Reppas, C., In vitro versus canine data for predicting input profiles of isosorbide-5-mononitrate from oral extended release products on a confidence interval basis. *Eur. J. Pharm. Sci.,* 2005. **24**(1), 115–122.

67. Vertzoni, M., et al., Biorelevant media to simulate fluids in the ascending colon of humans and their usefulness in predicting intracolonic drug solubility. *Pharm. Res.,* 2010. **27**(10), 2187–2196.

68. Diakidou, A., et al., Characterization of the contents of ascending colon to which drugs are exposed after oral administration to healthy adults. *Pharm. Res.,* 2009. **26**(9), 2141–2151.

69. Dressman, J. B., Reppas, C., In vitro-in vivo correlations for lipophilic, poorly water-soluble drugs. *Eur. J. Pharm. Sci.,* 2000. **11**(2), S73–80.

70. Charkoftaki, G., et al., Biopharmaceutical classification based on solubility and dissolution: a reappraisal of criteria for hypothesis models in the light of the experimental observations. *Basic Clin. Pharmacol. Toxicol.,* 2010. **106**(3), 168–172.

71. Gray, V., et al., The science of USP 1 and 2 dissolution: present challenges and future relevance. *Pharm. Res.,* 2009. **26**(6), 1289–1302.

72. Schiller, C., et al., Intestinal fluid volumes and transit of dosage forms as assessed by magnetic resonance imaging. *Aliment. Pharmacol. Ther.,* 2005. **22**(10), 971–979.

73. Memis, D., Turan, A., Karamanlioglu, B., Guler, T., Yurdakoc, A., Pamukçu, Z., Turan, N., Effect of preoperative oral use of erythromycin and nizatidine on gastric pH and volume. *Anaesth. Intensive Care,* 2002. **30**, 428–432.

74. Memis, D., Turan, A., Karamanlioglu, B., Saral, P., Türe, M., Pamukçu, Z., The effect of intravenous pantoprazole and ranitidine for improving preoperative gastric fluid properties in adults undergoing elective surgery. *Anesth. Analg.,* 2003. **97**, 1360–1363.

75. Dressman, J. B., Fleisher, D., Mixing-tank model for predicting dissolution rate control of oral absorption. *J. Pharm. Sci.,* 1986. **75**(2), 109–116.

76. Kostewicz, E. S., et al., Forecasting the oral absorption behavior of poorly soluble weak bases using solubility and dissolution studies in biorelevant media. *Pharm. Res.,* 2002. **19**(3), 345–349.

77. Gao, Z., In vitro dissolution testing with flow-through method: a technical note. *AAPS PharmSciTech,* 2009. **10**(4), 1401–1405.

78. Cohen, J. L., et al., The development of USP dissolution and drug release standards. *Pharm. Res.,* 1990. **7**(10), 983–987.

79. Sunesen, V. H., et al., Effect of liquid volume and food intake on the absolute bioavailability of danazol, a poorly soluble drug. *Eur. J. Pharm. Sci.*, 2005. **24**(4), 297–303.

80. Weitschies, W., Kosch, O., Mönnikes, H., Trahms, L., Magnetic Marker Monitoring: an application of biomagnetic measurement instrumentation and principles for the determination of the gastrointestinal behavior of magnetically marked solid dosage forms. *Adv. Drug Deliv. Rev.*, 2005. **57**(8), 1210–1222.

81. Weitschies, W., Blume, H., Monnikes, H., Magnetic marker monitoring: high resolution real-time tracking of oral solid dosage forms in the gastrointestinal tract. *Eur. J. Pharm. Biopharm.*, 2010. **74**, 93–101.

82. Laulicht, B., Tripathi, A., Schlageter, V., et al., Understanding gastric forces calculated from high-resolution pill tracking. *Proc. Natl. Acad. Sci. U. S. A.*, 2010. **107**, 8201–8206.

83. Burke, M., Maheshwari, C. R., Zimmerman, B. O., Pharmaceutical analysis apparatus and method. 2006: USA Pat.

84. Garbacz, G., Klein, S., Weitschies, W., A biorelevant dissolution stress test device - background and experiences. *Expert Opin. Drug Deliv.*, 2010. **7**(11), 1251–1261.

85. Garbacz, G., Kandzi, A., Koziolek, M., Mazgalski, J., Weitschies, W., Release characteristics of quetiapine fumarate extended release tablets under biorelevant stress test conditions. *AAPS PharmSciTech*, 2014. **15**(1), 230–236.

86. Garbacz, G., Golke, B., Wedemeyer, R. S., Axell, M., Söderlind, E., Abrahamsson, B., Weitschies, W., Comparison of dissolution profiles obtained from nifedipine extended release once a day products using different dissolution test apparatuses. *Eur. J. Pharm. Sci.*, 2009. **38**(2), 147–155.

87. Garbacz, G., Wedemeyer, R. S., Nagel, S., Giessmann, T., Mönnikes, H., Wilson, C. G., Siegmund, W., Weitschies, W., Irregular absorption profiles observed from diclofenac extended release tablets can be predicted using a dissolution test apparatus that mimics in vivo physical stresses. *Eur. J. Pharm. Biopharm.*, 2008. **70**(2), 421–428.

88. Garbacz, G., Weitschies, W., Investigation of dissolution behavior of diclofenac sodium extended release formulations under standard and biorelevant test conditions. *Drug Dev. Ind. Pharm.*, 2010. **36**(5), 518–530.

89. Garbacz, G., Cadé, D., Benameur, H., Weitschies, W., Bio-relevant dissolution testing of hard capsules prepared from different shell

materials using the dynamic open flow through test apparatus. *Eur. J. Pharm. Sci.*, 2014. **57**, 264–272.

90. Wilde, L., Bock, M., Glöckl, G., Garbacz, G., Weitschies, W., Development of a pressure-sensitive glyceryl tristearate capsule filled with a drug-containing hydrogel. *Int. J. Pharm.*, 2014. **461**(1–2), 296–300.

91. USP, The United States Pharmacopeia (USP 29). 2006, Rockville MD: Pharmacopeial Convention, Inc.

92. Jones, H. M., et al., Predicting pharmacokinetic food effects using biorelevant solubility media and physiologically based modelling. *Clin. Pharmacokinet.*, 2006. **45**(12), 1213–1226.

93. Fleisher, D., et al., Drug, Meal and Formulation Interactions Influencing Drug Absorption After Oral Administration. *Clin. Pharmacokinet.*, 1999. **36**(3), 233–254.

94. Yu, L. X., et al., The effect of food on the relative bioavailability of rapidly dissolving immediate-release solid oral products containing highly soluble drugs. *Mol. Pharm.*, 2004. **1**(5), 357–362.

95. Singh, B. N., A quantitative approach to probe the dependence and correlation of food-effect with aqueous solubility, dose/solubility ratio, and partition coefficient (Log P) for orally active drugs administered as immediate-release formulations. *Drug Dev. Res.*, 2005. **65**(2), 55–75.

96. Lentz, K., Current methods for predicting human food effect. *AAPS J.*, 2008. **10**(2), 282–288.

97. Ottaviani, G., et al., What is modulating solubility in simulated intestinal fluids? *Eur. J. Pharm. Sci.*, 2010. **41**(3–4), 452–457.

98. Wu, C. Y., Benet, L. Z., Predicting drug disposition via application of BCS: transport/absorption/elimination interplay and development of a biopharmaceutics drug disposition classification system. *Pharm. Res.*, 2005. **22**(1), 11–23.

99. Amidon, G., et al., A theoretical basis for a biopharmaceutic drug classification: the correlation of in vitro drug product dissolution and in vivo bioavailability. *Pharm. Res.*, 1995. **12**(3), 413–420.

100. Gu, C.-H., et al., Predicting effect of food on extent of drug absorption based on physicochemical properties. *Pharm. Res.*, 2007. **24**(6), 1118–1130.

101. Clarysse, S., et al., Postprandial changes in solubilizing capacity of human intestinal fluids for BCS class II drugs. *Pharm. Res.*, 2009. **26**(6), 1456–1466.

102. Ghazal, H. S., et al., In vitro evaluation of the dissolution behaviour of itraconazole in bio-relevant media. *Int. J. Pharm.*, 2009. **366**(1–2), 117–123.

103. Walravens, J., Brouwers, J., Spriet, I., Tack, J., Annaert, P., Augustijns, P., Effect of pH and comedication on gastrointestinal absorption of posaconazole: monitoring of intraluminal and plasma drug concentrations. *Clin. Pharmacokinet.*, 2011. **50**(11), 725–734.

104. Charman, W. N., et al., Absorption of danazol after administration to different sites of the gastrointestinal tract and the relationship to single- and double-peak phenomena in the plasma profiles. *J. Clin. Pharmacol.*, 1993. **33**(12), 1207–1213.

105. Klein, S., Stein, J., Dressman, J., Site-specific delivery of anti-inflammatory drugs in the gastrointestinal tract: an in-vitro release model. *J. Pharm. Pharmacol.*, 2005. **57**(6), 709–719.

106. Jantratid, E., et al., Application of biorelevant dissolution tests to the prediction of in vivo performance of diclofenac sodium from an oral modified-release pellet dosage form. *Eur. J. Pharm. Sci.*, 2009. **37**(3–4), 434–441.

107. Samaha, D., Shehayeb, R., Kyriacos, S., Modeling and comparison of dissolution profiles of Diltiazem modified-release formulations. *Dissolut. Technol.*, 2009. 41–46.

108. Parrott, N., et al., Predicting pharmacokinetics of drugs using physiologically based modeling–application to food effects. *AAPS J.*, 2009. **11**(1), 45–53.

109. Shono, Y., et al., Prediction of food effects on the absorption of celecoxib based on biorelevant dissolution testing coupled with physiologically based pharmacokinetic modeling. *Eur. J. Pharm. Biopharm.*, 2009. **73**(1), 107–114.

110. Wagner, C., et al., Utilizing in vitro and PBPK tools to link ADME characteristics to plasma profiles: case example nifedipine immediate release formulation. *J. Pharm. Sci.*, 2013. **102**(9), 3205–3219.

111. Okumu, A., DiMaso, M., Löbenberg, R., Dynamic dissolution testing to establish in vitro/in vivo correlations for montelukast sodium, a poorly soluble drug. *Pharm. Res.*, 2008. **25**(12), 2778–2785.

112. Perng, C.-Y., et al., Assessment of oral bioavailability enhancing approaches for SB-247083 using flow-through cell dissolution testing as one of the screens. *Int. J. Pharm.*, 2003. **250**(1), 147–156.

113. Tomaszewska, I., In vitro and Physiologically Based Pharmacokinetic models for pharmaceutical cocrystals. *Thesis*, Pharmacy and Pahrmacology Department. 2013, University of Bath.

114. Takano, R., et al., Rate-limiting steps of oral absorption for poorly water-soluble drugs in dogs; prediction from a miniscale dissolution

test and a physiologically-based computer simulation. *Pharm. Res.,* 2008. **25**(10), 2334–2344.

115. United States Pharmacopeia and National Formulary USP 29-NF 24. 2007, Rockville, MD: The United States Pharmacoeial Convention, Inc.

116. Grassi, M., Coceani, N., Magarotto, L., Modeling partitioning of sparingly soluble drugs in a two-phase liquid system. *Int. J. Pharm.,* 2002. **239**(1–2), 157–169.

117. Shi, Y., Gao, P., In vitro test method for poorly soluble drugs, in *AAPS Annual Symposium.* 2009: Los Angeles.

118. Vangani, S., et al., Dissolution of poorly water-soluble drugs inbiphasic media using USP 4 and fiber optic system. *Clin. Res. Regul. Aff.,* 2009. **26**(1–2), 8–19.

119. Gu, C.-H., et al., Using a novel multicompartment dissolution system to predict the effect of gastric pH on the oral absorption of weak bases with poor intrinsic solubility. *J. Pharm. Sci.,* 2005. **94**(1), 199–208.

120. Fallingborg, J., Intraluminal pH of the human gastrointestinal tract. *Dan. Med. Bull.,* 1999. **46**, 183–196.

121. Kararli, T., Composition of the gastrointestinal anatomy, physiology and biochemistry of humans and commonly used laboratory animals. *Biopharm. Drug Dispos.,* 1995. **16**, 351–380.

122. Frank, K. J., Locher, K., Zecevic, D. E., Fleth, J., Wagner, K. G., In vivo predictive mini-scale dissolution for weak bases: advantages of pH-shift in combination with an absorptive compartment. *Eur. J. Pharm. Sci.,* 2014. **61**, 32–39.

123. Lamberti, G., Cascone, S., Iannaccone, M., Titomanlio, G., In vitro simulation of drug intestinal absorption. *Int. J. Pharm.,* 2012. **439**(1–2), 165–168.

124. Kumar, S., Xu, X., Gokhale, R. Burgess, D. J., Formulation parameters of crystalline nanosuspensions on spray drying processing: a DoE approach. *Int. J. Pharm.,* 2014. **464**, 34–45.

125. Kataoka, M., Masaoka, Y., Yamazaki, Y., Sakane, T., Sezaki, H., Yamashita, S., In vitro system to evaluate oral absorption of poorly water-soluble drugs: simultaneous analysis on dissolution and permeation of drugs. *Pharm. Res.,* 2003. **20**(10), 1674–1680.

126. Kataoka, M., Yokoyama, T., Masaoka, Y., Sakuma, S., Yamashita, S., Estimation of P-glycoprotein-mediated efflux in the oral absorption of P-gp substrate drugs from simulataneous analysis of drug dissolution and permeation. *Eur. J. Pharm. Sci.,* 2011. **44**, 544–551.

127. Kataoka, M., Sugano, K., da Costa Mathews, C., Wong, J. W., Jones, K. L., Masaoka, Y., Sakuma, S., Yamashita, S., Application of dissolution/permeation system for evaluation of formulation effect on oral absorption of poorly water-soluble drugs in drug development. *Pharm. Res.,* 2012. **29**, 1485–1494.

128. Kataoka, M., Yano, K., Hamatsu, Y., Masaoka, Y., Sakuma, S., Yamashita, S., Assessment of absorption potential of poorly-soluble drugs by using the dissolution/permeation system. *Eur. J. Pharm. Biopharm.,* 2013. **85**, 1317–1324.

129. Buch, P., et al., IVIVC in oral absorption for fenofibrate immediate release tablets using a dissolution/permeation system. *J. Pharm. Sci.,* 2009. **98**(6), 2001–2009.

130. Wickham, M. J. S., Faulks, R. M., Mann, J., Mandalari, G., The design, operation, and application of a dynamic gastric model. *Dissolut. Technol.,* 2012. **19**(3), 15–22.

131. Marciani, L., et al., Assessment of antral grinding of a model solid meal with echo-planar imaging. *Am. J. Physiol.,* 2001. **280**(5, Pt. 1), G844–G849.

132. Marciani, L., et al., Gastric response to increased meal viscosity assessed by echo-planar magnetic resonance imaging in humans. *J. Nutr.,* 2000. **130**(1), 122–127.

133. Marciani, L., et al., Effect of meal viscosity and nutrients on satiety, intragastric dilution, and emptying assessed by MRI. *Am. J. Physiol.,* 2001. **280**(6, Pt. 1), G1227–G1233.

134. Marciani, L., et al., Intragastric oil-in-water emulsion fat fraction measured using inversion recovery echo-planar magnetic resonance imaging. *J. Food Sci.,* 2004. **69**(6), E290–E296.

135. Wickham, M., Faulks, R., Mills, C., In vitro digestion methods for assessing the effect of food structure on allergen breakdown. *Mol. Nutr. Food Res.,* 2009. **53**(8), 952–958.

136. Mercuri, A., Passalacqua, A., Wickham, M. S., Faulks, R. M., Craig, D. Q., Barker, S. A., The effect of composition and gastric conditions on the self-emulsification process of ibuprofen-loaded self-emulsifying drug delivery systems: a microscopic and dynamic gastric model study. *Pharm. Res.,* 2011. **28**(7), 1540–1551.

137. Vardakou, M., Mercuri, A., Naylor, T. A., Rizzo, D., Butler, J. M., Connolly, P. C., Wickham, M. S., Faulks, R. M., Predicting the human in vivo performance of different oral capsule shell types using a novel in vitro dynamic gastric model. *Int. J. Pharm.,* 2011. **419**(1–2), 192–199.

138. Mann, J. C. P., Samuel, R., A formulation case study comparing the dynamic gastric model with conventional dissolution methods. *Dissolut. Technol.*, 2012. **19**(4), 14–19.

139. Blanquet, S., Zeijdner, E., Beyssac, E., Meunier, J.-P., Denis, S., Havenaar, R., Alric, M., A dynamic artificial gastrointestinal system for studying the behavior of orally administered drug dosage forms under various physiological conditions. *Pharm. Res.*, 2004. **21**(4), 585–591.

140. David, S. E., Strozyk, M. M., Naylor, T. A., Using TNO gastro-intestinal model (TIM1) to screen potential formulations for a poorly soluble development compound. *J. Pharm. Pharmacol.*, 2010. **62**(10), 1236–1237.

141. Souliman, S., Beyssac, E., Cardot, J.-M., Denis, S., Alric, M., Investigation of the biopharmaceutical behavior of theophylline hydrophilic matrix tablets using USP methods and an artificial digestive system. *Drug Dev. Ind. Pharm.*, 2007. **33**(4), 475–483.

142. Souliman, S., et al., A level A in vitro/in vivo correlation in fasted and fed states using different methods: applied to solid immediate release oral dosage from. *Eur. J. Pharm. Sci.*, 2006. **27**(1), 72–79.

143. Chen, L., Hebrard, G., Beyssac, E., Denis, S., Subirade, M., In vitro study of the release properties of soy-zein protein microspheres with a dynamic artificial digestive system. *J. Agric. Food Chem.*, 2010. **58**, 9861–9867.

144. Krul, C. A. M., et al., Application of a dynamic in vitro gastrointestinal tract model to study the availability of food mutagens, using heterocyclic aromatic amines as model compounds. *Food Chem. Toxicol.*, 2000. **38**, 783–792.

145. Lila, M. A., Ribnicky, D. M., Rojo, L. E., Rojas-Silva, P., Oren, A., Havenaar, R., Janle, E. M., Raskin, I., Yousef, G. G., Grace, M. H., Complementary approaches to gauge the bioavailability and distribution of ingested berry polyphenolics. *J. Agric. Food Chem.*, 2012. **60**(23), 5763–5771.

146. Verwei, M., et al., Folic acid and 5-methyl-tetrahydrofolate in fortified milk are bioaccessible as determined in a dynamic in vitro gastrointestinal model. *J. Nutr.*, 2003. **133**, 2377–2383.

147. Verwei, M., et al., Predicted serum folate concentrations based on in vitro studies and kinetic modeling are consistent with measured folate concentrations in humans. *J. Nutr.*, 2006. **136**, 3074–3078.

148. Ameer, B., et al., Absolute and relative bioavailability of oral acetaminophen preparations. *J. Pharm. Sci.*, 1983. **72**(8), 955–958.

149. Divoll, M., et al., Effect of food on acetaminophen absorption in young and elderly subjects. *J. Clin. Pharmacol.*, 1982. **22**(11–12), 571–576.

150. Rostami-Hodjegan, A., et al., A new rapidly absorbed paracetamol tablet containing sodium bicarbonate. I. A four-way crossover study to compare the concentration-time profile of paracetamol from the new paracetamol/sodium bicarbonate tablet and a conventional paracetamol tablet in fed and fasted volunteers. *Drug Dev. Ind. Pharm.*, 2002. **28**(5), 523–531.

151. Rygnestad, T., Zahlsen, K., Samdal, F. A., Absorption of effervescent paracetamol tablets relative to ordinary paracetamol tablets in healthy volunteers. *Eur. J. Clin. Pharmacol.*, 2000. **56**(2), 141–143.

152. Naylor, T. A., Connolly, P. C., Martini, L. G., Elder, D. P., Minekus, M., Havenaar, R., Zeijdner, E., Use of a gastro-intestinal model and Gastroplus™ for the prediction of in vivo performance. *Appl. Ther. Res.*, 2006. **6**(1), 15–19.

153. Brouwers, J., Anneveld, B., Goudappel, G. J., Duchateau, G., Annaert, P., Augustijns, P., Zeijdner, E., Food-dependent disintegration of immediate release fosamprenavir tablets: in vitro evaluation using magnetic resonance imaging and a dynamic gastrointestinal system. *Eur. J. Pharm. Biopharm.*, 2011. **77**(2), 313–319.

154. Dickinson, P. A., Abu Rmaileh, R., Ashworth, L., Barker, R. A., Burke, W. M., Patterson, C. M., Stainforth, N., Yasin, M., An investigation into the utility of a multi-compartmental, dynamic, system of the upper gastrointestinal tract to support formulation development and establish bioequivalence of poorly soluble drugs. *AAPS J.*, 2012. **14**(2), 196–205.

155. Minekus, M., Jelier, M., Xiao, J.-Z., Kondo, S., Iwatsuki, K., Kokubo, S., Bos, M., Dunnewind, B., Havenaar, R., Effect of partially hydrolyzed guar gum (PHGG) on the bioaccessibility of fat and cholesterol. *Biosci. Biotechnol. Biochem.*, 2005. **69**, 932–938.

156. Reis, P. M., Raab, T. W., Chuat, J. Y., Leser, M. E., Miller, R., Watzke, H. J., Holmberg, K., Influence of surfactants on lipase fat digestion in a model gastrointestinal system. *Food Biophys.*, 2008. **3**, 370–381.

157. Lennernäs, H., Aarons, L., Augustijns, P., Beato, S., Bolger, M., Box, K., Brewster, M., Butler, J., Dressman, J., Holm, R., Julia Frank, K., Kendall, R., Langguth, P., Sydor, J., Lindahl, A., McAllister, M., Muenster, U., Müllertz, A., Ojala, K., Pepin, X., Reppas, C., Rostami-Hodjegan, A., Verwei, M., Weitschies, W., Wilson, C., Karlsson, C., Abrahamsson, B., Oral biopharmaceutics tools – time for a new initiative – an introduction to the IMI project OrBiTo. *Eur. J. Pharm. Sci.*, 2014. **57**, 292–299.

Chapter 15

Clinically Relevant Dissolution for Low-Solubility Immediate-Release Products

Paul A. Dickinson,[a] Talia Flanagan,[b] David Holt,[b] and Paul W. Stott[b]

[a] *Seda Pharmaceutical Development Services, The BioHub at Alderley Park, Alderley Edge, Cheshire SK10 4TG, UK*
[b] *Global Product Development, Pharmaceutical Development and Technology, AstraZeneca, Silk Road, Macclesfield, Cheshire SK10 2NA, UK*
paul.dickinson@sedapds.com, talia.flanagan@astrazeneca.com, david.holt@astrazeneca.com, paul.stott@astrazeneca.com

15.1 Introduction

The previous chapter discussed the developments in dissolution testing during the late 1990s and early 2000s when the focus was on biorelevant dissolution testing and more specifically the development of dissolution media that physicochemically mimic the fluid in the GI tract. These approaches have proved useful for the development and choice of formulations with improved in vivo release; however, generally these methods have only provided,

Poorly Soluble Drugs: Dissolution and Drug Release
Edited by Gregory K. Webster, J. Derek Jackson, and Robert G. Bell
Copyright © 2017 Pan Stanford Publishing Pte. Ltd.
ISBN 978-981-4745-45-1 (Hardcover), 978-981-4745-46-8 (eBook)
www.panstanford.com

a priori, a qualitative understanding of relative performance of different formulations. Based on this understanding, developments in dissolution in the late 2000s and 2010s have focused on two areas: advanced dissolution testing (also discussed in the previous chapter) and an area that has become known as "clinically relevant dissolution testing." Both of these areas have become the focus of dissolution testing research as there is a growing recognition for the need to quantitatively predict product performance in vivo. It is the latter area, clinically relevant dissolution testing, which is the focus of this chapter. In terms of the drug development process/timeline this area becomes most important when the late stage product has been largely developed/chosen and the sponsor is preparing for late-stage efficacy studies, submission of the marketing application (e.g., NDA), and commercial supply. This is in contrast to biorelevant testing (i.e., testing with media replicating the GI milieu such as FaSSIF), which has most value in the early stages of drug development (pre-phase 1 to phase 2) when sponsors are going through the iterative process of developing formulations good enough to meet the needs of the early clinical program.

During the 2000s, changes in regulatory thinking stimulated by the US FDA critical path initiative which identified a lack of innovation in pharmaceutical manufacturing[1] and led to the introduction of ICH Q8[2] and the concept of quality by design (QbD) resulted in a reconsideration of pharmaceutical quality. For instance, ICH Q8 introduced the concept of the quality target product profile, which is defined as "a prospective summary of the quality characteristics of a drug product that ideally will be achieved to ensure the desired quality, taking into account safety and efficacy of the drug product." As this specifically called out the clinical performance of the product (safety and efficacy), this led to a discussion on pharmaceutical tests that can control this. Against this backdrop, the role of dissolution testing to ensure pharmaceutical quality has been debated.[3] During these debates the tension between the dissolution tests to control in vivo performance and dissolution tests which control manufacturing consistency (the more common role of dissolution testing prior to the QbD initiatives) has been debated. This will be briefly discussed in this chapter in the control strategy section. At the 2013

AAPS Annual Meeting during a roundtable discussion[4] speakers from the US FDA Biopharmaceutics Team (Angelica Dorantes) and EMA (Evangelos Kotzagiorgis) both stated that in their opinion the primary role of dissolution testing in a regulatory setting is to ensure consistent clinical performance,[5,6] with Dr. Dorantes stating that "clinically relevant specifications are those that consider the clinical impact of variations in the critical quality attributes (CQA) and process parameters assuring a consistent drug product and therefore a consistent efficacy and safety profile." Additionally, for immediate-release products, Dr. Dorantes stated that there would generally be a single time point specification when $Q = 80\%$.

So the challenge faced by pharmaceutical scientists is how to develop a dissolution test that can both be capable of being validated to meet regulatory analytical expectations and thus is suitable as a quality test but is also able to demonstrate a link to clinical performance so that the specification can be set which ensures clinical performance. Generally demonstration of bioequivalence via a clinical bioequivalence study is the standard applied to ensure clinical performance of two products (in this sense product is taken as the formulation and manufacturing process used to form it). As discussed later, for Biopharmaceutics Classification System (BCS) class 1 drugs and BCS class 3 drugs,[7] in some territories, biowaivers are possible where comparative dissolution across the physiological pH range is considered sufficient to ensure bioequivalence.

For BCS class 2 and 4 drug products, a bioequivalence study is generally required as in vitro tests that, a priori, can be used to assess the impact of the changes on clinical safety and efficacy are not readily available.[a] Thus for BCS class 2 and 4 products, in order to establish that a dissolution method is capable of ensuring safety and efficacy there is need to confirm a link between dissolution performance and clinical pharmacokinetics. Prior to 2008, at least for BCS class 2 products, this was generally considered to be the need to develop an in vitro–in vivo correlation (IVIVC) and more specifically a level A IVIVC defined as "a point-to-point relationship

---

[a]It should be noted for some small changes, which have been shown to have a low probability of affecting drug dissolution release and thus safety and efficacy, can be assessed using dissolution testing.[8–10]

between in vitro dissolution and the in vivo input rate."[10] This understanding was based on the seminal BCS paper of Amidon and coworkers,[11] which stated that for a BCS 2 product "IVIVC expected if in vitro dissolution rate is similar to in vivo dissolution rate, unless dose is very high" for which the corollary is if an IVIVC is not demonstrated, then the dissolution method is not clinically relevant. Amidon and coworkers also stated that limited or no IVIVC was expected for a BCS class 4 product, implying that clinically relevant dissolution was not possible for a BCS class 4 product. These conclusions, however, preceded the definition of the values for the BCS boundaries in the FDA BCS guidance.[7] In one of the first papers to link the concepts of quality by design to clinical performance, "Clinical Relevance of Dissolution Testing in Quality by Design," Dickinson and coworkers demonstrated that only under some conditions would an IVIVC for BCS class 2 products be achieved.[12] That was only when gastric emptying rate was at the high end of the physiological range and permeability very high (much higher than necessary to be considered a high permeability compound) could a level A IVIVC be established. Under more physiological relevant simulations gastric emptying and or permeability had a dampening effect on absorption such that dissolution was not rate determining and an IVIVC was not produced. These observations were supported by the work of Polli and coworkers that pre-dated the BCS framework but demonstrated a similar conclusions if the compounds they had investigated were classified by BCS class.[13,14]

The potential rate determining/limiting steps are shown in a simplified schematic (Fig. 15.1). Four major kinetic processes can be considered: gastric emptying, dissolution, permeation (together dissolution and permeation constitute absorption), and intestinal transit. For most drugs, absorption does not occur in the stomach, so this is the first potential rate-limiting step (if dissolution and permeation are rapid in comparison). In the fasted state, gastric emptying is generally taken to occur as a first-order processes with a half-life of about 15 min, although in practice it is highly variable. After emptying from the stomach, drug needs to be in solution to allow permeation and absorption to occur. If dissolution is slow, then this can be rate-limiting, but if permeation is slow and dissolution rapid, then permeation is rate-limiting. Generally

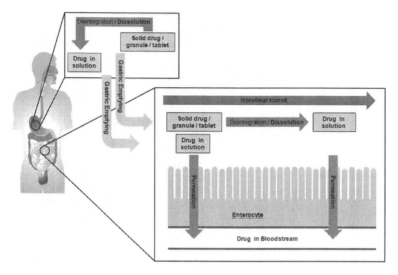

Figure 15.1 Simplified schematic of potential rate determining steps involved in drug absorption.

it is desirable for immediate-release products that dissolution and permeation are as rapid as possible (although the formulator has most control over dissolution) because if either of these processes are slow relative to intestinal transit (on average 3–4 hours),[15,16] then undissolved/unabsorbed drug will enter the colon where drug absorption is likely to be substantially reduced. Drug absorption can be much more complex with other processes such as precipitation, complexation, and drug efflux from enterocytes occurring simultaneously; however, it is useful for formulators and analysts to consider product performance in the context of this framework of potential rate-determining steps.

From their analysis Dickinson and coworkers identified three possible outcomes for BCS class 2 products when trying to relate in vitro dissolution performance to clinical pharmacokinetic performance:

(1) A classical level A IVIVC is established.
(2) A "safe space" is observed where no effect is seen in the clinical study but in vitro dissolution is sensitive to product and process changes.

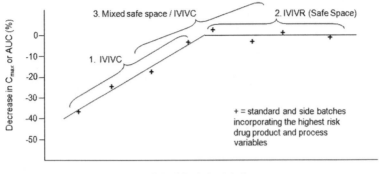

Figure 15.2 Plot of change in AUC and or $C_{\max}$ versus in vitro dissolution rate demonstrating safe space and IVIVC outcomes.

(3) A mixed safe space/IVIVC result in which clinical pharmacokinetics are only affected for a few of the variants tested clinically.

This is represented visually in Fig. 15.2.

The implicit assumption for these three outcomes is that in vitro dissolution is mechanistically similar to in vivo dissolution. A fourth potential outcome is that there are differences in in vivo pharmacokinetics that are not replicated in vitro (rather no differentiation in vitro or different rank order) proving that in vitro dissolution is not mechanistically similar to in vivo dissolution and the dissolution method is not clinically relevant.

From their analysis, Dickinson and coworkers concluded that, generally, for the magnitude of dissolution perturbations that could be expected to be introduced into formulations by competent formulators (by deliberate variations in input materials or processing conditions), "safe space" or "mixed safe space/IVIVC" is the most likely outcome. The FDA have recently recognized this potential for a "safe space," as will be discussed further in the section on control strategy.[5, 17]

The 2008 paper on clinical relevance of dissolution testing proposed a five-step process to develop a clinically relevant test and ensure the clinical quality of product. These were:

(1) Perform a quality risk assessment (QRA) to allow the most relevant risks to in vivo dissolution to be identified.
(2) Develop dissolution test(s) with physiological relevance that is most likely to identify changes in dissolution.
(3) Understand the importance of changes to these most relevant manufacturing variables on clinical quality (for BCS 2 and 4 compounds this would require some clinical pharmacokinetics data).
(4) Establish the dissolution limit which ensures clinical quality.
(5) Ensure dissolution within established limits to ensure clinical quality is used to define the product control strategy.[b]

This five-step process will be illustrated in detail throughout the rest of the chapter using case studies. Employees of the US FDA[18] have also subsequently postulated a similar stepwise process based on the above. Specifically:

(1) Manufacture tablet variants with different release characteristics (this is step 1 and part of step 2 from the 2008 paper) and select the optimal dissolution method with adequate discriminatory power.
(2) Determine bioavailability for tablet variants.
(3) Determine dissolution rates resulting in similar bioavailability.
(4) Choose design Space to ensure bioequivalent product performance.

Dr. Marroum, although no longer at the FDA, has subsequently written a detailed report which showed this approach as an expansion of the five steps described in 2008 by Dickinson et al.[19]

15.1.1 Benefits of Developing a Clinically Relevant Dissolution Method?

Developing a clinically relevant dissolution method for BCS 2 and 4 compounds requires a different approach to the development of the test and importantly the use of some clinical pharmacokinetic data.

---

[b]The 2008 paper [12] refers to defining design space but the authors acknowledge that since that time focus has moved to establishment of a robust and meaningful control strategy.

Thus it is likely to be more resource intensive that a more traditional QC approach to developing the dissolution test. Consequently, against this likely increased cost, it is important to understand the potential benefits of this approach.

The most obvious benefit is to the patient as there is improved assurance that future production lots will be bioequivalent to those used in clinical studies. The sponsor may also benefit by being able to make manufacturing changes without the need for bioequivalence studies or increased probability of passing any bioequivalence studies.

As described later in the chapter, the development of a clinically relevant dissolution test may result in a release specification that is "wider" than would have been accepted in the absence of proof of clinical relevance. This in turn should allow fewer constraints on input material properties and process parameter ranges. Overall, this should lead to a more robust supply line, reduced drug product shortages, and reduced manufacturing costs, thus meeting some of the objectives of the FDA critical path initiative.[1]

Speakers from FDA have described three approaches that support the benefits that may accrue from developing a clinically relevant dissolution test.[5,17] The first is the traditional QC approach when there is no data linking clinical and dissolution performance. In this approach, limited regulatory flexibility is expected with a tighter dissolution limit based on the pivotal batches and a tighter design space based on f2 testing. The second approach is a clinically relevant dissolution test based on safe space. This would result in a wider dissolution specification if supported by the data (compared to just using pivotal batches with no clinical link) and a more flexible design space, although in this approach the flexibility may be limited to CMC changes investigated in vivo. The final approach is one where an IVIVC has been developed which can result in wider specifications and an enhanced flexible design space. It is interesting to note not only that demonstration of clinical relevance of the dissolution test can lead to a wider dissolution specification but also impacts on the approved manufacturing process.

In the following sections of this chapter the five step process is discussed in more detail and illustrated using a case study (Example 1). Further case studies are briefly discussed at the end of this chapter to exemplify how different compounds properties and pharmaceutical development requirements impact the outcome from the five-step process.

15.2 Step 1: Perform a Quality Risk Assessment (QRA) to Allow the Most Relevant Risk to in vivo Dissolution to Be Identified

15.2.1 Assessment of Biopharmaceutics Risk

Before the likelihood of a given manufacturing process or formulation change affecting in vivo performance can be assessed, it is necessary to understand the driving forces for, and potential limitations to, absorption for the compound under development as these form inputs for the risk assessment. Development of a clinically relevant dissolution test therefore begins with careful consideration of the biopharmaceutics properties of the API and delivery system. The biopharmaceutics risk associated with a compound determines both the likelihood that a given process change will have a significant impact on in vivo performance, and our ability to detect this using in vitro tests. These considerations feed into the quality risk assessment, which forms step 1 of the 5-step process to establishing a clinically relevant dissolution test.

Pertinent physicochemical properties of the case study compound are shown in Table 15.1.

Table 15.1 Physicochemical and biopharmaceutical properties for an example BCS class 2 compound

Solubility at 37°C	pH 1.2	>300 mg/250 mL
	pH 4.5	>300 mg/250 mL
	pH 6.8	88 mg/250 mL
	FaSSIF	>300 mg/250 mL
Permeability		High
Dose		100–300 mg
Formulation		Immediate-release tablet

This compound falls into BCS class 2, as it has high permeability but the highest dose is not soluble in 250 mL or less across the physiological pH range. However, the high solubility in FaSSIF media indicates that the compound is likely to also be highly soluble in the intestinal environment in vivo.

The BCS is often used as a starting point to understand the likely drivers and limitations for absorption. The BCS provides a well-established framework for risk assessment of IR products, based on their biopharmaceutics properties. In developing a clinically relevant dissolution test, the BCS is effectively a body of prior knowledge, as it establishes boundaries within which drug products are known to be of very low biopharmaceutics risk, and defines specific in vitro tests as having relevance to assess and manage this risk, i.e., clinical relevance. For BCS class 1 and 3 compounds, the link between dissolution in simple aqueous buffers across the physiological pH range and in vivo performance is already well established. It is accepted for such compounds that if two formulations show rapid dissolution in simple aqueous buffers across the physiological pH range under mild agitation conditions as per the BCS guidance (i.e., pH 1.2, 4.5, and 6.8, USP1 100 rpm or USP2 50 rpm), their in vivo performance will be equivalent. If a drug product falls into one of these classes, the risk of a process or formulation change having a significant impact on clinical performance is generally accepted to be very low, as long as dissolution performance is maintained within the specified acceptance criteria and excipients are low risk; this is the basis on which BCS-based biowaivers are granted.

The BCS criteria for low-risk compounds (i.e., class 1 and 3) are by necessity set very tightly, as they are intended to be applicable to all compounds administered as immediate release products, from any manufacturing process and site. BCS classes 2 and 4, therefore, span a wide range of compound properties, and hence a wide spectrum of biopharmaceutics risk. To develop clinically relevant dissolution tests for these compounds, a more bespoke biopharmaceutics risk assessment than BCS is required, tailored to the specific attributes of the compound and formulation under consideration. This should make use of three key elements:

- Generic prior knowledge and literature precedent
- Existing clinical data for the compound and formulation under consideration
- Dissolution testing in biorelevant media and advanced apparatus

15.2.1.1 Generic prior knowledge and literature precedent

Given the conservative nature of the BCS classification boundaries, there will be instances where, despite being formally classified as BCS 2 or 4, a compound will in fact behave more like a BCS 1 or 3 compound in terms of product performance. For example, if a compound is near the boundary of the solubility or permeability criteria, and in silico modeling suggests that this will not limit absorption, or if a compound is not near the classification boundary but passes the very rapidly dissolving criteria in simple aqueous buffers across the physiological pH range (perhaps because the drug is presented as a salt), the performance profile is likely to be similar to that of a class 1 or 3 compound. In these circumstances, formulators and analysts may be confident to apply the principles of a BCS class 1 or 3 to the establishment of a control strategy, although the regulatory burden of proof is likely to be higher than if a compound fell clearly within these categories.

There is a large body of useful information in the literature which can act as prior knowledge around the likely level and nature of biopharmaceutics risk associated with compounds outside BCS class 1 and 3. The WHO series of biowaiver monographs discuss in detail the level and nature of bioequivalence risk for various marketed products, and suggest clinically relevant dissolution tests to assess these risks.[20] As such, they may offer useful insights, if a monograph is available for a product with similar physicochemical and clinical characteristics, and a useful starting point for a clinically relevant dissolution testing strategy. The continuum of risk that the biopharmaceutics properties of drugs represents has been discussed in the literature with numerous publications subdividing the BCS categories to more thoroughly describe biopharmaceutics risk, either from first principles or based on a review of existing data.[21–23] This prior knowledge of absorption risk and clinically relevant testing conditions for compounds of similar properties is a useful input into the risk assessment process, and forms a good starting point for developing the product-specific understanding needed to develop a clinically relevant dissolution test.

Based on biopharmaceutics properties of the case study compound described herein (Table 15.1), the development team considered the risk of process and formulation parameters impacting on in vivo performance to be relatively low, as in spite of the compounds formal BCS2 classification it was likely to act as a highly soluble drug in vivo. However, the initial risk assessment was performed in the very early days of the development of clinically relevant tests and the project team decided to take a conservative approach based on incomplete in vitro dissolution at intestinal pH and the ability to a priori define a dissolution specification that would ensure clinical quality, as described below.

15.2.1.2 Existing clinical data

Clinical data on the current and previous formulations is an important building block in the development of a clinically relevant dissolution test. Comparison of in vitro and in vivo data from formulation comparison studies can provide important information on the clinical relevance or otherwise of a dissolution test. For example, the ability to detect and exclude a formulation which did not show equivalent performance in the clinic would provide a high degree of confidence in the clinical relevance of a dissolution test. Conversely, data which show equivalence is maintained in spite of wide-ranging changes in process and formulation can provide assurance of product robustness, and are a useful starting point to understand the in vivo relevance of any discrimination seen in in vitro dissolution tests (i.e., is the dissolution test over or under discriminating?).

Existing clinical data also offer an opportunity to develop an in silico absorption model, which could be used to assess the robustness of the absorption process to changes in product dissolution. Cook[24] describes a Bayesian approach to IVIVC/R where a model is built and revised based on learning from previous studies, and can be used to design subsequent formulation studies by predicting the likely impact of a change in dissolution profile on clinical performance based on the clinical experience to date. Such models are useful to build internal confidence during development of a clinically relevant dissolution test. There is also increasing interest from health authorities in the use of in silico absorption modeling; potential advantages and barriers to wider use were recently described by Jian et al.[a, 25]

15.2.1.3 Dissolution testing in biorelevant media and apparatus

In assessing the level and nature of biopharmaceutics risk, two key questions must be considered: firstly, how likely is it that a reasonable change in process or formulation would have a significant impact on clinical performance; secondly, if this occurred, could it be detected in vitro? A priori biorelevant media and conditions are very useful in understanding the robustness of a product to process and formulation changes. As they are designed to mimic the intestinal environment, these media and conditions are likely to be biorelevant for many drug products.

Simple aqueous buffers of established clinical relevance (i.e., the BCS buffers) can be used both to generate product and process understanding, and as routine quality control media. However, for many BCS class 2 and 4 compounds, poor solubility may prevent the use of simple aqueous buffers as dissolution media as the dose will not be soluble in dissolution apparatus volumes (normally 900 mL). It is unlikely that a test in which only a small percentage of the dose dissolves will provide insight into the clinical performance of the product. This is because in vivo a much greater extent of dissolution would occur due to permeation removing dissolved drug from the GI lumen allowing further dissolution to occur. More complex

---

[a]A recent publication by Pepin et al. (2016) describes the use of in silico absorption modelling to inform the dissolution specification for an oral immediate release product. X. J. Pepin, T. R. Flanagan, D. J. Holt, A. Eidelman, D. Treacy, and C. E. Rowlings (2016). Justification of drug product dissolution rate and drug substance particle size specifications based on absorption PBPK modelling for lesinurad immediate release tablets. *Mol. Pharm.* **13**, 3256–3269.

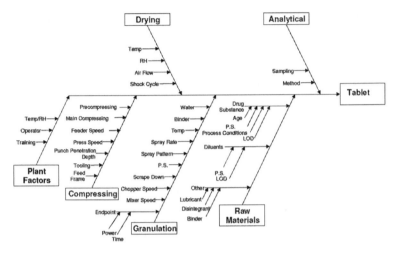

Figure 15.3 Example Ishikawa (fishbone) diagram taken from ICH Q8 (R2). Copyright © ICH.

biorelevant media designed to mimic the intestinal environment (e.g., FaSSIF) and/or complex apparatus which mimics GI conditions (e.g., TNO-TIM1[26]) can be useful for such compounds. While the complexity associated with these tests precludes their use in a routine QC setting, they can be used to develop product-specific knowledge and understanding of the factors which are important for the in vivo performance, and hence would need to be detected by a clinically relevant dissolution QC test which uses simpler media and conditions.

15.2.2 Assessment of Drug Product Risk

The first step in assessing drug product risks should utilize a cross-functional team of experts to identify potential variables from the formulation and manufacturing process, which could have an impact on the desired critical quality attributes. As presented in ICH Q8 (R2) (ICH Harmonised Tripartite Guideline, Pharmaceutical Development, 2009),[2] an Ishikawa (fishbone) diagram can be used to brainstorm and visualize the potential causes which could lead to product failure (see Fig. 15.3).

In the case study example presented herein for a low-solubility drug, it was especially relevant to identify and understand the raw materials and process variables that could impact upon clinical safety and efficacy through effects on in vivo dissolution using failure mode effects analysis (FMEA). The key question asked during this risk assessment process was "what is the overall risk of each potential failure mode affecting in vivo pharmacokinetics (and therefore patient safety and efficacy)?" As stated previously, a relatively conservative approach was adopted in the initial FMEA with potential minor changes in dissolution being scored quite highly.

This chapter and case study focuses on the drug product risk assessment associated with poor in vivo performance but in practice the risk assessment process should capture all risk that may affect the totality of drug product quality resulting in a comprehensive assessment that informs the entire development plan.

FMEA involves assessing and prioritizing formulation and process risks, through consideration of the probability of a failure (P), the severity of such a failure (S) and the detectability of the failure if it were to occur (D). The higher the likelihood of a failure occurring is, the higher the score for P. The more severe the consequence of a failure is, the higher the score for S. The greater the chance that the failure will not be detected until late in the manufacturing process then the higher the score for D.

The output of FMEA risk assessment for the BCS class 2 case study compound (Table 15.1) presented as an immediate-release tablet and manufactured by a conventional wet granulation process is shown in Table 15.2. The scores for P, S, and D are assigned values between 1 and 5, and the scores are then multiplied together to give the risk priority number (RPN) for each possible failure.

Based on the output from the risk assessment, the highest risk failure modes with the potential to retard dissolution rate and impact in vivo pharmacokinetics were identified. From Table 15.2 it can be seen that highest risks to in vivo product performance are changes in drug substance particle size, wet mass over-granulation, increased level of binder, and decreased level of disintegrant. The respective in vitro dissolution rate retardation mechanisms are impact of drug surface area on dissolution rate, impact of

Table 15.2 Summary of failure modes effects analysis

Risk area	Failure mode	Failure effect	P	S	D	RPN
					Risk analysis	
	Formulation					
Drug substance	Changes in particle size or properties, e.g., shape, surface energy, bulk properties	Changes in dissolution performance, thus impacting on clinical performance	4	5	5	100
Excipients	Increasing the level of binder	Impeded tablet disintegration leading to altered clinical performance	4	5	5	100
	Decreasing the level of disintegrant	Impeded tablet disintegration leading to altered clinical performance	4	5	5	100
	Lubricant variability, e.g., surface area, concentration, leading to impeded wetting of drug particle	Changes in dissolution behaviour leading to altered clinical performance	3	5	5	75
	Variability in filler level or properties within specified grade e.g., surface area, particle size, wettability	Changes in granule properties resulting in altered disintegration/ dissolution profile	2	5	5	50
Tablet coating process	Variability in coating thickness	Effect on dissolution and therefore clinical performance	2	5	5	50
	Process					
Dry mixing	Insufficient mixing; blend uniformity of drug substance	High and low doses across batch of tablets	2	3	4	24

Wet granulation	Failure to control granulation end-point; Over-granulation	Decreased granule porosity, decreased water ingress and decreased dissolution rate. Adverse effect on clinical quality	4	5	5	100
	Excessive water added or holding the wet mass for significant time before drying	Decrease in disintegrant performance; wicking and swelling during wet massing. Decreased dissolution rate and consequent adverse effect on clinical performance. Potential formation of alternative polymorphic form.	2	5	5	50
Milling of dried granules	Incorrect dry milling parameters, e.g., impeller speed, screen size	Effect on granule size leads to altered dissolution and adverse effect on clinical performance	2	5	5	50
Blending	Blending time too long, leading to hydrophobic coat of lubricant around granules	Decreased dissolution rate leads to adverse effect on clinical performance	3	5	5	75
Tablet compression	Increased compression force	Increased density of tablets causes decreased water ingress and decreased dissolution rate. Adverse effect on clinical performance	3	5	5	75

granule properties on water ingress, and impact of slowed tablet disintegration on dissolution.

15.3 Step 2: Develop Dissolution Test(s) with Physiological Relevance that is Most Likely to Identify Changes in Dissolution

Based on the output from the initial FMEA relating to in vivo performance, several tablet variants were manufactured incorporating the highest risk product and process variables. For simplicity, only the four tablet variants that were subsequently taken into a clinical study are described. These tablets were based on the proposed commercial tablet formulation, which is represented as the standard tablet (variant A) and a drug substance particle size variant (variant B), a process variant (variant C), and a formulation variant (variant D).

In vitro evaluation of these batches also supported the development of appropriate dissolution methods for use in ongoing manufacturing process development and routine quality control. Within this context, and by consideration of ICH Q6A guidance,[27] the objective was to provide dissolution methods that have a range of capabilities:

(1) ability to detect the impact of minor process and formulation changes
(2) ability to detect changes in performance of the product on storage (stability indicating)
(3) ability to achieve complete dissolution within an appropriate timescale for an immediate-release tablet (e.g., less than 60 min)
(4) ability to ensure in vivo performance, i.e., it can be used to set a specification which ensures that tablets will give equivalent clinical performance to those used in pivotal clinical studies

Dissolution testing initially focused on the evaluation of aqueous buffers across the physiological pH range 1.2 to 6.8, utilizing standard pharmacopeial apparatus 2 (see Fig. 15.4). In pH 1.2 media it was not possible to discriminate between tablet variants A, B,

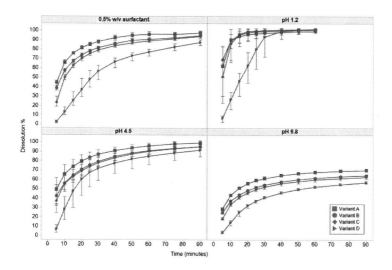

Figure 15.4 In vitro dissolution profiles for tablet variants.

and C, although some discrimination was demonstrated between tablet variant D and the other tablet variants. In pH 4.5 media, the observed within-batch variability was high, and consequently these conditions were not the optimum for differentiating between most relevant processing and formulation variables. In pH 6.8 media reasonable discrimination between tablet variants was achieved; however, incomplete drug release was observed (<70% in 60 min) because of the poor aqueous solubility of the drug at neutral pH.

From these studies, it was concluded that aqueous buffers did not provide the optimum conditions for a test capable of differentiation between processing and formulation variables for the tablets.

In accordance with dissolution testing guidance for industry,[28] the use of surfactants to increase solubility was evaluated. Dissolution in a surfactant media exhibited the potential for differentiation between processing and formulation variables (see Fig. 15.4). Surfactant-based media were considered to be valid scientifically as surfactants increase solubility via micellar solubilization, which, based on drug in FaSSIF solubility, also occurred in the small intestine as a consequence of the bile acid mixed micelles present. In some cases there may be concern that the increased solubility afforded by surfactants may mask important changes in dissolution

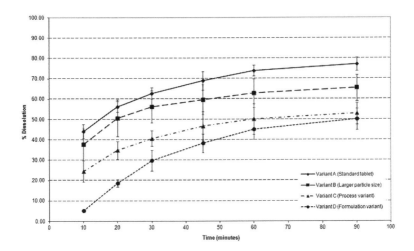

Figure 15.5   FaSSIF in vitro dissolution profiles for tablet variants.

rate (i.e., lead to an under-discriminatory dissolution test). The optimum surfactant concentration was identified at which the rate of tablet dissolution was sufficiently slow to provide the potential for discrimination between tablet variants while affording complete dissolution within a timescale appropriate for use as a routine control test.

The physiological relevance of the 0.5% w/v surfactant media was further confirmed by investigating the dissolution of tablet variants A to D in media thought to be more representative of the small intestine milieu (i.e., FaSSIF[29]). The dissolution profiles obtained for each of the tablets variants in FaSSIF are presented in Fig. 15.5. These data further support the physiological relevance of the 0.5% w/v surfactant media, with the nature of the discrimination observed for the dissolution in FaSSIF media being similar to that observed in 0.5% w/v surfactant media. In FaSSIF media dissolution was more variable and incomplete as a consequence of the coning that occurred at the 50 rpm stirring speed.

The in vitro performance (dissolution rate) of the three tablet variants, plus the standard tablet, was assessed using the preferred method, the dissolution test with surfactant. The tablet variant batches provided a good range of in vitro dissolution profiles. It can also be seen that discrimination between the variants follows the same rank order for each of the dissolution methods' conditions.

The dissolution profiles obtained demonstrated differentiation between the tablet variants, indicating that the methods can detect changes in the drug product caused by deliberately varying the highest risk process parameters and product attributes identified from the quality risk assessment.

Based on these in vitro dissolution profiles, and potential biopharmaceutical risks, these tablet variants were dosed in a human volunteer study, as this would allow the best opportunity to establish an in vitro–in vivo correlation (IVIVC) should these variables be important.

15.4 Step 3: Understand the Importance of Changes to these Most Relevant Manufacturing Variables on Clinical Quality

Immediate-release products of BCS class 2 and 4 drugs will likely require some sort of comparative clinical pharmacokinetics data to understand the impact of dissolution changes on in vivo performance. This is in contrast to BCS 1 and 3 immediate-release products where prior knowledge alone may be sufficient to understand the (lack of) impact of dissolution changes of in vivo performance, as discussed earlier. For BCS 2 and 4 immediate-release products the nature of clinical study required has not been extensively discussed in the literature. The broad options are

(1) a relative bioavailability study comparing one or more variants to the standard/reference variant
(2) a bioequivalence study comparing one or more variants to the standard/reference variant
(3) an IVIVC-type study where an "instant" releasing formulation is dosed as well as the tablet variants to allow deconvolution of the in vivo dissolution profile and thus allow an IVIVC to be established.
(4) A Bayesian approach in which different variants are dosed in different clinical studies (possibly where the primary objective

of the clinical study is not related to the dissolution test—for instance, a drug–drug interaction study) and the comparisons on pharmacokinetics performance are made across study.[24]

The basic design of the relative bioavailability and bioequivalence studies are similar. The standard design is a randomized cross-over design in which the dosing sequence for the tablets is randomized for each subject. A variant is dosed and plasma analyzed for drug concentration. Plasma samples are taken at suitable time points and for a suitable length of time to allow the maximum plasma concentration (known as $C_{\max}$) and total exposure to the drug (known as area under the plasma concentration time curve (AUC)) to be determined. The "washout period," the time between the single doses of the variants, depends on the plasma half-life of drug and is ideally long enough to ensure that any remaining concentration of drug in the plasma at the time of the next dosing is less than 5% of $C_{\max}$. The data is then interpreted in the context of the log transformed treatment (T)/reference (R) ratio and 90% confidence limit (CI) for $C_{\max}$ and AUC. The standard tablet would be the reference (R) treatment and the variants considered the test (T) treatment, with ratios <1 indicating lower exposure from the variant and greater than 1, higher exposure. For a relative bioavailability study there is no predefined criteria for these metrics, but for a product to be considered bioequivalent, which carries particular regulatory significance, the 90% confidence limit for $C_{\max}$ and AUC treatment/reference ratio must fall between 0.8 and 1.25.[30] Study size is largely determined by the within subject variability of drug pharmacokinetics and the precision of the T/R ratio required, with larger N required to reduce the width of the confidence interval for any given within subject variability. Thus bioequivalence studies usually contain more subjects than relative bioavailability studies to ensure that the derived CI is sufficiently small to allow BE to be concluded if there is no difference in the performance of T and R.

IVIVC studies are similar to relative bioavailability studies except that "instant releasing" formulation is also administered. This allows the pharmacokinetics of this formulation to be "subtracted" from the pharmacokinetics for the tablet variants (deconvolution). Generally, this "instant releasing" formulation would be a non-precipitating solution given orally or possibly as an intravenous injection. Use of an IV injection allows for assessment of systemic PK (post-absorption). These data can be used in the deconvolution of the remaining PK profile to reduce the oral PK data to the absorption profile. In contrast the oral solution pharmacokinetics includes both the post-absorption kinetics as well as the permeation/absorption step, so when used for deconvolution of the variant(s), the pharmacokinetic profiles which remain are reduced to the in vivo dissolution profile. This is only the case if in vivo dissolution is slower than gastric emptying or permeation; if not, then deconvolution will show apparently instant dissolution, which is likely an artifact of the mathematical process. This would be the situation for a safe space example and so an IVIVC-type approach where safe space is expected may be less valuable. However, relative bioavailability data showing that a solution produces similar pharmacokinetics to tablet variants does provide strong (and intuitively compelling) evidence that, within the range of in vitro dissolution studied, dissolution is not important.

Bayesian/across-study comparison is outside the scope of this chapter, and readers are directed to Cook's publication for an introduction to the subject.[24] The work of Samatani and coworkers, although focused on cross-study comparison of different long-acting intramuscular injections, and not specifically linking product performance to dissolution testing, does show how powerful cross-study comparisons can be made to understand product performance.[31-33]

In the case study described here, an IVIVC study was chosen with a non-precipitating solution being dosed in period 1 and the tablet variants randomized for dosing in periods 2 to 4 using an incomplete block design approach. This was due to the long half-life of the drug, which would have required subjects to stay on study for an unrealistically long time if they received all four variants.

Due to ethical and logistical reasons, in vivo testing of all potential variables and range of variables identified in the FMEA was impractical. Therefore, only tablet variants manufactured to incorporate relevant retardation mechanisms through changes of only the *highest*-risk raw materials and process variables were assessed in the IVIVC study.

The choice of variants for inclusion in a study is governed by two broad concerns: firstly the need to generate in vivo data to link to dissolution performance, which requires variants with different dissolution performance, and secondly what variations to product and manufacturing process are likely to be made after product approval.

Considering this second driver, there are two broad approaches. In the first approach, which has been described by some employees of the FDA, at least for a "safe space" outcome, only future changes to CMC variables that have been specifically assessed in a clinical study can be supported by dissolution testing.[5] So, for instance, if a change in particle size was required in the future, then a particle size variant would need to have been tested in the in vivo study to establish the clinical relevance of the dissolution test such that future changes could be justified by comparative dissolution testing only.

Another approach that may be justifiable by applying a risk-based approach is that the highest risks should be investigated in vivo. Then the most appropriate discriminatory in vitro validated dissolution test would be chosen after the results of the IVIVC study were known. By definition this test would be sensitive for the higher risks. If this is also able to detect changes in the lower risk variables, the test could then be used to understand the impact and to control the impact of any changes to lower risk variables (as these variables have a lower probability of impacting in vivo performance from the start).

In this case study the four variants were shown to be equivalent to oral solution and, although not designed as a bioequivalence study, a post hoc analysis showed similar performance of each variant to the oral solution (Table 15.3).

Table 15.3 Treatment to reference (oral solution) ratios (90% confidence interval) after dosing various tablet variants with different dissolution properties to healthy volunteer, corrected for potency

	Variant A	Variant B	Variant C	Variant D
AUC	1.07 (1.01–1.15)	1.09 (1.03–1.17)	1.05 (0.99–1.12)	1.04 (0.97–1.12)
C_{max}	1.02 (0.93–1.12)	0.96 (0.87–1.05)	0.96 (0.88–1.05)	0.95 (0.86–1.04)

15.5 Step 4: Establish the Dissolution Limit which Ensures Clinical Quality (i.e., no Effect by Changes)

From step 3 for a poorly soluble compound, two broad outcomes are possible when considering setting a dissolution limit that ensures clinical quality:

(1) safe space or
(2) IVIVC or mixed safe space/IVIVC being established

For the safe space outcome the slowest dissolution profile tested in the clinical study becomes the limit that ensures clinical quality. In this situation this limit is set at the boundary of knowledge rather than on a biological effect, and it is unknown how much slower dissolution would need to be to affect in vivo performance. If the outcome is the establishment of an IVIVC or mixed safe space/IVIVC, a dissolution limit that controls C_{max} and AUC to within the required extent should be chosen. The authors consider this would be to limit the variation in AUC and C_{max} by a maximum of $+/-$ 10%, as IVIVC predicts mean values. For most drugs, a mean product performance difference of 10% in AUC or C_{max} is likely to be maximum difference that could be tested in a clinical bioequivalence and result in the 80% CI still being within 0.8–1.25 with a practical number of subjects.

In the case study presented, a safe space was established and thus the dissolution profile/limit used in subsequent product and process establishment activities was based on the slowest dissolving variant dosed to volunteers: variant D. Thus, if a product was produced which had a dissolution profile faster than that of variant D, then in vivo performance would be comparable to pivotal clinical trials material. The clinical relevance of the dissolution test is confirmed by being able to define a limit which guarantees consistent bioavailability of the product. As the tablet variants dosed covered a broad range of different dissolution retardation mechanisms expected from an IR product manufactured by wet granulation, this provided additional confidence in both the discriminating nature of the dissolution method and the clinical boundary of performance.

15.6 Step 5: Ensure Dissolution within Established Limits to Ensure that Clinical Quality is used to Define the Product Control Strategy

As discussed in ICH Q8 (R2) (ICH Harmonised Tripartite Guideline. Pharmaceutical Development, 2009), a control strategy is a planned set of controls, derived from current product and process understanding that ensures process performance and product quality. The controls can include parameters and attributes related to drug substance and drug product materials and components, facility and equipment operating conditions, in-process controls, finished product specifications, and the associated methods and frequency of monitoring and control.

In developing the control strategy for in vivo product performance, consideration should be given to how in vitro measures of dissolution performance can be used to assess the impact of input material attributes and manufacturing process parameters on product quality.

Within the context of in vivo product performance, the case study example presented illustrates the approach to be used for a class 2 drug in connecting in vitro dissolution testing to bioavailability and thereby ensuring the clinical quality of a product. Multivariate manufacturing process investigations utilizing the surfactant-based clinically relevant dissolution test, and the limit provided by variant D, were used to establish the control strategy for the product.

15.6.1 The Role of Multivariate Manufacturing Process Investigations in Establishing a Control Strategy

In line with ICH Q8 (R2), multivariate manufacturing process investigations should be used to demonstrate an enhanced knowledge of product performance over a range of material attributes, manufacturing process options and process parameters. This understanding can be gained by application of, for example, modern statistical design of experiments (DoE), process analytical technology (PAT), and/or prior knowledge. As previously described, appropriate use of quality risk management principles can be helpful in prioritizing the additional pharmaceutical development studies to collect such knowledge. This knowledge can subsequently be used to define both critical and noncritical sources of variability in the product and process, and establish appropriate controls.

For the case study example BCS class 2 drug, a number of multivariate design of experiments were executed to evaluate the cause and effect relationships between manufacturing unit operations and intermediate product attributes, and the downstream effects on dissolution performance. Cause-and-effect relationships were established between intermediate granule surface area (GSA), tablet disintegration time, and dissolution rate in the 0.5% w/v surfactant media (see Fig. 15.6). To describe the combined rate limiting contribution of the physical properties of GSA and disintegration time, they were expressed as a ratio.

A large value of this ratio is indicative of either a high GSA or a low disintegration time, or both, leading to a fast rate of dissolution. Conversely, a low value of this ratio indicates either a low GSA or a high disintegration time, or both, leading to a slower rate of dissolution. It is clear that at each time point there is a stronger correlation at low ratios of GSA to disintegration time, but that as this ratio increases the extent of dissolution reaches a plateau, indicating that these physical mechanisms of dissolution retardation are no longer rate-limiting to in vitro dissolution.

The enhanced knowledge of the factors that impact dissolution rate, alongside other aspects of product quality (e.g., assay and content uniformity), were used to define a control strategy that would deliver future product with a dissolution profile faster than variant D, thus providing assurance of clinical quality (see Table 15.4).

As verification that the control strategy was justified, an experimental batch was produced that represented manufacture at the extremes of the established control strategy with respect to dissolution rate retardation (tablet variant X, see Fig. 15.7). These data demonstrate that tablets manufactured when operating at the extremes of the control strategy provide dissolution profiles faster than variant D, and therefore clinical quality will always be ensured.

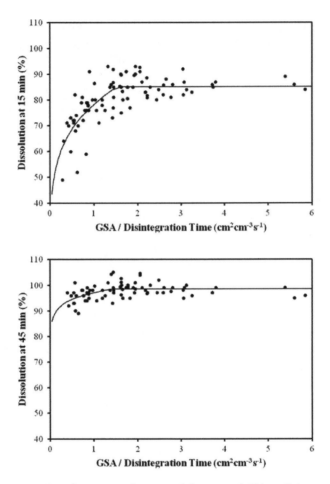

Figure 15.6 Dissolution as a function of the ratio of GSA to disintegration time expressed as dissolution at 15 and 45 min in 0.5% w/v surfactant media.

15.6.2 Setting an Appropriate Dissolution Specification as Part of the Control Strategy

In accordance with ICH Q6A guidance (ICH Harmonised Tripartite Guideline. Specifications: Test Procedures and Acceptance Criteria for New Drug Substances and New Drug Products: Chemical Structures. 1999), the dissolution specification (i.e., the test procedure

Table 15.4   Control strategy for dissolution

Control strategy element	Specific controls applied
Input material controls	Quantitative composition for all drug product components
	Specification limit for drug substance particle size
Manufacturing process parameter controls	Granulation water quantity
	Granulation mixing time
	Dry granule milling screen size
In-process testing	Unmilled granule moisture content
	Milled granule surface area (GSA)
	Tablet core weight
	Tablet core disintegration time
End-product testing	Dissolution method and specification

and acceptance criteria) plays a major role in ensuring a consistently high quality of the drug product at release and during shelf life. A single-point dissolution specification is normally considered suitable for immediate-release dosage forms.

A generic approach to the role of BCS class in the establishment of a dissolution specification that controls clinical quality is described in Table 15.5.

Table 15.5   The role of drug BCS class in defining the approach to setting clinically relevant dissolution specifications

BCS class	Approach to setting specification that ensures clinical performance
1	≥80% within 30 min in most discriminating aqueous media (pH 1.2–6.8). If slower, set limit based on clinical data
2	Specification set based on clinical "bioavailability" data
3	≥80% within 15 min in most discriminating aqueous media (pH 1.2–6.8). If slower, set limit based on clinical data
4	Limit set on case-by-case basis

After completing writing this chapter the USA FDA issued a draft guidance on specification setting for none narrow therapeutic index BCS 1 and 3 drugs which suggests the use of 500 mL of 0.01M HCl aqueous media.
FDA Draft Guidance: Dissolution Testing and Specification Criteria for Immediate-Release Solid Oral Dosage Forms Containing Biopharmaceutics Classification System Class 1 and 3 Drugs. http://www.fda.gov/downloads/Drugs/GuidanceComplianceRegulatory Information/Guidances/UCM456594.pdf

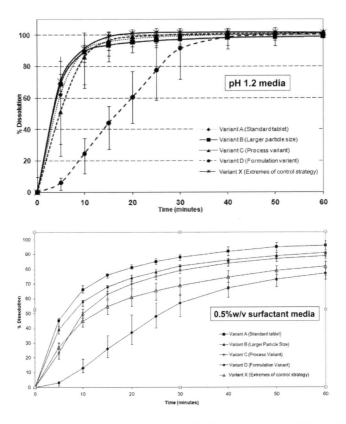

Figure 15.7   Dissolution performance of tablet variants A, B, C, D, and X.

In setting the dissolution specification, there are three general aspects to consider:

(i) the discriminatory nature of the dissolution test conditions
(ii) the amount of drug dissolved to ensure consistent batch-to-batch bioavailability (commonly referred to as the Q value)
(iii) an appropriate dissolution time point to measure the Q value

For immediate-release drug products where changes in dissolution rate have been demonstrated to significantly affect bioavailability, it is desirable to develop test conditions which can distinguish batches with unacceptable bioavailability. If changes in formulation or process variables significantly affect dissolution and such changes are not controlled by another aspect of the specification, it may also be appropriate to adopt dissolution test conditions which can distinguish between these changes.

Where dissolution significantly affects bioavailability, the acceptance criteria should be set to reject batches with unacceptable bioavailability. Otherwise, test conditions and acceptance criteria should be established which pass clinically acceptable batches.

In situations where an immediate-release dosage form provides close to 100% release in the selected dissolution test, a Q value in excess of 80% is generally not used, because allowance needs to be made for assay and content uniformity ranges, and for the inherent variability in the reproducibility of the dissolution test. It is a common regulatory agency expectation to provide Q values set at either 75% or 80% (80% particularly in USA). Setting Q values at these levels provides assurance of batch-to-batch consistency in dissolution performance, and therefore minimizes the probability of bioinequivalence.

With respect to the dissolution time point for immediate-release dosage forms, it is most common to see the specification set at either 15 or 30 min. The use of time points earlier than 15 min are not generally used as allowance needs to be made for disintegration time, and because dissolution differences in this early timeframe are unlikely to lead to in vivo differences, this would provide an unnecessarily stringent limit on the manufacturing process capability.

In situations where it has been demonstrated that dissolution rate does not impact in vivo pharmacokinetics within tested ranges, it may be more appropriate to select a dissolution time point later than 30 min, particularly if the dissolution test conditions are highly discriminatory.

For the BCS class 2 case study example a "safe space" specification can be set on the basis of no effect seen in the clinical study, and the slowest dissolution profile tested in the clinical study (i.e., variant D). Consideration is also given to the batch manufactured at the extremes of the established control strategy with respect to dissolution rate retardation (tablet variant X) and the standard tablets from the pivotal clinical batches. Two specification options are available (see Fig. 15.8):

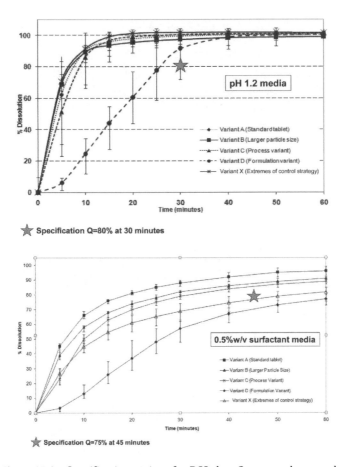

Figure 15.8   Specification options for BCS class 2 case study example.

(i) a specification of $Q = 80\%$ at 30 min in pH 1.2 media

(ii) a specification of $Q = 75\%$ at 45 min in 0.5% w/v surfactant media

Both of these options described would satisfy the requirements of ICH Q6A in providing a specification that will pass clinically acceptable batches, and reject batches with unacceptable bioavailability. Selection of the final dissolution test conditions and specification will be dependent upon an assessment of the overall residual risks associated with the control strategy, with due consideration of both biopharmaceutical and manufacturing process risks. This will inform the desired dissolution method detectability requirements for the quality control test, i.e., the level of discriminating capability required. It should be noted that FDA normally requires a Q value set at 80%, which could ultimately influence the choice of dissolution test employed for routine quality control. The specification for the case study and the different discussions that occurred with regulators in different countries been discussed previously.[34] The challenges in establishing global harmonization of the final dissolution specification for this product illustrate that this is an evolving area of regulatory and pharmaceutical science and warrants further investigation.

15.7 Further Examples of the Application of the 5-Step Approach

The sections above have used a detailed case study to exemplify the application of the 5-step process, to develop a clinically relevant dissolution test and acceptance criterion for a BCS class 2 compound. This approach can be applied across a broad spectrum of compound properties and drug products; the decisions taken at the various stages will depend on the properties of API and the drug product under consideration, the amount and type of clinical data already available, and the desired level of flexibility in the control strategy. Additional examples, based on the authors' experience, of the application of the 5-step process are briefly described below.

15.7.1 Example 2: A BCS Class 4 Drug with Reasonable Permeability and Poor Aqueous Solubility

The compound had moderate permeability and poor aqueous solubility; however, solubility in FASSIF was high, indicating that reasonable solubility in the small intestine was likely due to bile acid solubilization.

15.7.1.1 Step 1: Perform a quality risk assessment (QRA) allowing the most relevant risk to in vivo dissolution to be identified

A QRA indicated that disintegration time and granulation parameters posed the highest risk to in vivo dissolution (input API attributes had already been fixed in an area of low risk). Tablet variants were manufactured incorporating these highest-risk failure mechanisms: two granulation process variants, and a formulation variant with a lower level of disintegrant.

15.7.1.2 Step 2: Develop dissolution test(s) with physiological relevance that is most likely to identify changes in dissolution

Three surfactant containing media were investigated (poor solubility prevented the use of aqueous buffers). Performance was also assessed in FaSSIF as a known biorelevant media. Dissolution was impaired for the process and formulation variants in all media; the same rank order of profiles was seen in each condition, but with different degrees of discrimination.

15.7.1.3 Step 3: Understand the importance of changes to these most relevant manufacturing variables on clinical quality

A clinical relative bioavailability study was performed to assess the impact of the changes in in vitro dissolution on in vivo performance. All of the tablet variants were bioequivalent to the standard tablet.

15.7.1.4 Step 4: Establish the dissolution limit which ensures clinical quality

The clinical study had established a range of in vitro dissolution profiles within which bioequivalence was ensured; a specification anywhere within this region would be suitable to ensure clinical quality.

15.7.1.5 Step 5: Ensure dissolution within established limits to ensure that clinical quality is used to define the product control strategy

The in vivo data was used to underpin the control strategy, to ensure that every tablet produced during commercial manufacture will be bioequivalent to the pivotal safety and efficacy studies. The impact of lower-risk variables which had not been dosed in the clinical study was assessed through in vitro testing only.

One of the surfactant-based media (giving a similar degree of discrimination to FaSSIF) was selected for use as a clinically relevant QC release test.

15.7.2 Example 3: A BCS Class 4 Drug Near the Boundary of the High Solubility and Permeability Thresholds

The case study has been simplified. The compound was highly soluble across the pH range of 1.2 to 6.8 and had a very rapid dissolution rate being presented as a salt; however, above pH 6.8, solubility did not meet the requirements for high solubility according to the BCS. Permeability was moderate to high.[a]

Therefore, the risk of drug salt solubility and dissolution being rate-limiting to absorption was considered to be low. This conclusion was supported by the lack of differences in drug plasma exposure (AUC, C_{max}, and t_{max}) between a tablet and an oral solution. Tablets are rapidly dissolving in aqueous pharmacopeial buffers across the physiologically relevant pH range of 1.2 to 6.8 due to the high solubility and rapid dissolution rate of drug salt.

---

[a]Note since this chapter was completed the USA FDA have issued draft BCS guidance which would mean that this compound would now be considered to have high solubility. FDA Guidance for Industry: Waiver of In Vivo Bioavailability and Bioequivalence Studies for Immediate-Release Solid Oral Dosage Forms Based on a Biopharmaceutics Classification System. http://www.fda.gov/downloads/Drugs/GuidanceComplianceRegulatoryInformation/Guidances/UCM070246.pdf

15.7.2.1 Step 1: Perform a quality risk assessment (QRA) allowing the most relevant risk to in vivo dissolution to be identified

On the basis of the biopharmaceutics information described above, the risk of process and formulation parameters impacting in vitro dissolution and consequently in vivo exposures was considered to be low. Clinical data were available showing similar pharmacokinetics between a tablet and an oral solution and dose-proportional increases in exposure, supporting that dissolution was not rate-limiting for in vivo absorption. The level of risk was therefore considered to be commensurate with that of a BCS class 3 drug.

15.7.2.2 Step 2: Develop dissolution test(s) with physiological relevance that is most likely to identify changes in dissolution

Dissolution testing in simple aqueous buffers across the physiological pH range (1.2, 4.5, and 6.8) under mild agitation conditions was used, as per the BCS guidelines.

15.7.2.3 Step 3: Understand the importance of changes to these most relevant manufacturing variables on clinical quality

On the basis of the physicochemical properties of the compound and existing clinical data, it was considered appropriate to use the dissolution tests specified in the BCS guidance in conjunction with the predefined acceptance criteria for BCS class 3 drugs to assess the potential in vivo impact of process and formulation parameters.

Changes to input drug substance particle size and manufacturing parameters within a relevant range were shown not to impact in vitro dissolution performance (i.e., rapid dissolution [>85% in 15 min]) was maintained across the physiological pH range.

15.7.2.4 Step 4: Establish the dissolution limit which ensures clinical quality

The dissolution limit which ensures quality for BCS class 3 compounds was applied; tablet variants having >85% dissolution (i.e., $Q = 80$) in 15 min across the physiological pH range were considered to be bioequivalent.

15.7.2.5 Step 5: Ensure dissolution within established limits to ensure that clinical quality is used to define the product control strategy

The tests and conditions described in steps 3 and 4 were used to define a control strategy which ensured that equivalent tablet performance to pivotal safety and efficacy studies was maintained during routine manufacture.

15.8 Summary

The authors are firmly of the belief that the development of increased knowledge and understanding proposed under the auspices of quality by design are predicated on an insight into and control of in vivo performance.

A structured approach to development which endeavors to identify and evaluate risk to clinical performance and, where appropriate, test the impact of these risks in vivo has been presented. Each compound should be considered on its own merits, but the application of the proposed 5-step approach will ensure that all important factors are considered and will ultimately lead to the establishment of a robust control strategy.

The benefits of such an approach include enhanced security of product supply, the ability to optimize the manufacturing process and demonstrate the (lack of) impact of any proposed change, and an improved assurance of the clinical quality of product supplied to patients.

It is acknowledged that the dissolution test and its associated specification is expected to serve a number of purposes and at times the demands of discriminatory power and demonstration of complete release can be at conflict. Further work in our scientific understanding and regulatory harmonization are required if we are to realize the full benefit of a move toward more clinically relevant dissolution specifications.

References

1. U. S. Food and Drug Administration (2004). Innovation or Stagnation: Challenge and Opportunity on the Critical Path to New Medical Products, http://www.fda.gov/downloads/ScienceResearch/SpecialTopics/CriticalPathInitiative/CriticalPathOpportunitiesReports/ucm113411.pdf, accessed March 20, 2014.

2. Guidance for Industry Q8(R2) Pharmaceutical Development, 2009 (ICH Q8 (R2)). http://www.fda.gov/downloads/Drugs/GuidanceComplianceRegulatoryInformation/Guidances/ucm073507.pdf, accessed March 20, 2014.

3. A. Selen, M. T. Cruañes, A. Müllertz, P. A. Dickinson, J. A. Cook, J. E. Polli, F. Kesisoglou, J. Crison, K. C. Johnson, G. T. Muirhead, T. Schofield, and Y. Tsong (2010). Meeting report: applied biopharmaceutics and quality by design for dissolution/release specification setting: product quality for patient benefit. *AAPS J.* **12**, 465–472.

4. Dickinson and Cook (2013). Must We Celebrate all Diversity? Regulatory Drivers and Perspectives on Differences in Requirements for Dissolution Release Methods and Specifications that Lead to Global Inconsistencies. AAPS Annual Meeting and Exposition, Wed Nov 13, http://www.aaps.org/pastmeetings/, accessed 20 March 2014.

5. A. Dorantes (2013). FDA Regulatory Perspective. 2013 AAPS Annual Meeting and Exposition, Wed Nov 13, http://www.aaps.org/pastmeetings/, accessed 20 March 2014.

6. E. C. Kotzagiorgis (2013). EU Regulatory Requirements. 2013 AAPS Annual Meeting and Exposition, Wed Nov 13, http://www.aaps.org/pastmeetings/, accessed 20 March 2014.

7. U. S. Food and Drug Administration (2000). Guidance for Industry Waiver on In vivo Bioavailability and Bioequivalence Studies for Immediate-Release Solid Oral Dosage forms based on a Biopharmaceutics Classification System.

8. U. S. Food and Drug Administration (1995). Guidance for Industry Immediate Release Solid Oral Dosage Forms Scale-Up and Postapproval Changes: Chemistry, Manufacturing, and Controls, In Vitro Dissolution Testing, and In Vivo Bioequivalence Documentation.

9. The European Agency for the Evaluation of Medicinal Products (EMEA) (2000). CfPMPC. Note for Guidance on quality of modified release products: A: oral dosage forms, B: transdermal dosage forms (quality).

10. U. S. Food and Drug Administration (1997). Guidance for Industry Extended Release Oral Dosage Forms: Development, Evaluation, and Application of In Vitro/In Vivo Correlations.

11. G. L. Amidon, H. Lennernäs, V. P. Shah, and J. R. Crison (1995). A theoretical basis for a biopharmaceutic drug classification: the correlation of in vitro drug product dissolution and in vivo bioavailability. *Pharm. Res.* **12**, 413–420.

12. P. A. Dickinson, W. W. Lee, P. W. Stott, A. I. Townsend, J. P. Smart, P. Ghahramani, T. Hammett, L. Billett, S. Behn, R. C. Gibb, and B. Abrahamsson (2008). Clinical Relevance of Dissolution Testing in Quality by Design. *AAPS J.* **10**, 380–390.

13. J. E. Polli, J. R. Crison, and G. L. Amidon (1996). Novel approach to the analysis of in vitro–in vivo relationships. *J. Pharm. Sci.* **85**, 753–760.

14. J. E. Polli, and M. J. Ginski (1998). Human drug absorption kinetics and comparison to Caco-2 monolayer permeabilities. *Pharm. Res.* **15**, 47–52.

15. L. X. Yu, and G. L. Amidon (1999). A compartmental absorption and transit model for estimating oral drug absorption. *Int. J. Pharm.* **186**, 119–125.

16. E. L. McConnell, H. M. Fadda and A. W. Basit (2008). Gut instincts: Explorations in intestinal physiology and drug delivery. *Int. J. Pharm.* **364**, 213–226.

17. C. Moore (2011). Specifications for QbD Containing Applications. DIA CMC Workshop: Translating Science into Successful Submissions, Washington, D. C. February 9.

18. P. Marroum (2010). Dissolution and Its Relevance in Life Cycle of a Product an FDA Perspective PSWC2010, New Orleans, October.

19. P. J. Marroum (2012). Clinically Relevant Dissolution Methods and Specifications. American Pharmaceutical Review accessed at http://www.americanpharmaceuticalreview.com/Featured-Articles/38389-Clinically-Relevant-Dissolution-Methods-and-Specifications/accessed on 5th Jan 2014.

20. https://www.fip.org/bcs_monographs.

21. J. M. Butler and J. B. Dressman (2010). The developability classification system: application of biopharmaceutics concepts to formulation development. *J. Pharm. Sci.* **99**, 4940–4954.

22. L. X. Yu, G. L. Amidon, J. E. Polli, H. Zhao, M. U. Mehta, D. P. Conner, V. P Shah, L. J. Lesko, M. L. Chen, V. H. Lee, and A. S. Hussain (2002). Biopharmaceutics classification system: the scientific basis for biowaiver extensions. *Pharm. Res.* **19**, 921–925.

23. C. A. Bergström, S. B. Andersson, J. H. Fagerberg, G. Ragnarsson, and A. Lindahl (2013). Is the full potential of the biopharmaceutics classification system reached? *Eur. J. Pharm. Sci.* **57**, 224–231.

24. J. A. Cook (2012). Development strategies for IVIVC in an industrial environment. *Biopharm. Drug Dispos.* **33**, 349–353.

25. W. Jiang, S. Kim, X. Zhang, R. A. Lionberger, B. M. Davi, D. P. Conner D. P., and L. X. Yu (2011). The role of predictive biopharmaceutical modeling and simulation in drug development and regulatory evaluation. *Int. J. Pharm.* **418**, 151–160.

26. P. A. Dickinson, R. Abu Rmaileh, L. Ashworth, R. A. Barker, W. M. Burke W. M., C. M. Patterson N. Stainforth, and M. Yasin (2012). An investigation into the utility of a multi-compartmental, dynamic, system of the upper gastrointestinal tract to support formulation development and establish bioequivalence of poorly soluble drugs. *AAPS J.* **14**, 196–205.

27. I. C. H. Harmonised Tripartite Guideline Q6A (1999). Specifications: Test procedures and acceptance criteria for new drug substances and new drug products: Chemical structures.

28. U. S. Food and Drug Administration (1997). Guidance for Industry Dissolution Testing of Immediate Release Solid Oral Dosage Forms.

29. J. B. Dressman and C. Reppas (2000). In vitro-in vivo correlations for lipophilic, poorly water-soluble drugs. *Eur. J. Pharm. Sci.* **11**(Suppl 2), S73–S80.

30. U. S. Food and Drug Administration (2014) Guidance for Industry. Bioavailability and Bioequivalence Studies Submitted in NDAs or INDs—General Considerations.

31. M. N. Samtani, S. Gopal, C. Gassmann-Mayer, L. Alphs, and J. M. Palumbo (2011). Dosing and Switching Strategies for Paliperidone Palmitate Based on Population Pharmacokinetic Modelling and Clinical Trial Data. *CNS Drugs* **25**, 829–845.

32. S. Gopal, C. Gassmann-Mayer, J. Palumbo, M. N. Samtani, R. Shiwach and L. Alphs (2010). Practical guidance for dosing and switching paliperidone palmitate treatment in patients with schizophrenia. *CMR* **26**, 377–387.

33. M. N. Samtani, A. Vermeulen and K. Stuyckens (2009). Population PK of intramuscular paliperidone palmitate in patients with schizophrenia: a novel once-monthly, long-acting formulation of an atypical antipsychotic. *CPK* **48**, 585–600.

34. R. J. Timko, P. A. Dickinson, D. J. Holt, F. J. Montgomery, G. K. Reynolds, P. W. Stott and A. Watt (2013). Quality-by-Design: Differing Dissolution Philosophies. AAPS Annual Meeting. Henry B. Gonzalez Convention Center, San Antonio TX. T3376 http://abstracts.aaps.org/published/ContentInfo.aspx?conID=43664.

Chapter 16

The QbD Approach to Method Development and Validation for Dissolution Testing

Alger D. Salt

H&A Scientific, 105 A Regency Blvd, Greenville, NC 27834, USA
alger.salt@hascientific.com

16.1 Introduction

Quality by design (QbD) is a structured approach to the development and delivery of products, technologies, and processes.[1] Historically, QbD has focused on manufacturing processes and product attributes, but the underlying principles are now routinely applied to analytical methods. In this chapter we will define the four main components of QbD and how the QbD approach can be applied to developing and validating a dissolution test method that uses UV spectrophotometry as the analytical measurement.

Method transfers are part of a successful product life cycle because work initiated in R&D is ultimately implemented in the

Poorly Soluble Drugs: Dissolution and Drug Release
Edited by Gregory K. Webster, J. Derek Jackson, and Robert G. Bell
Copyright © 2017 Pan Stanford Publishing Pte. Ltd.
ISBN 978-981-4745-45-1 (Hardcover), 978-981-4745-46-8 (eBook)
www.panstanford.com

QC laboratories at manufacturing sites.[2,3] Prior to embracing QbD principles, companies often learned (and many times re-learned) that problems with analytical methods were often discovered during method transfers. This is not a good time to discover such problems, because it is difficult to make changes to the method and fixing the problems can cause costly delays and much frustration to those involved. Obviously, developing the method right the first time is best. In addition, as you will learn in this chapter, the QbD approach can provide some regulatory flexibility which better allows a pathway for continuous improvement as time rolls on and as the method evolves throughout the product life cycle.

16.2 Stages of QbD

The four stages of the QbD approach are

- Design Intent
- Design Selection
- Control Definition
- Control Verification

Definitions for each of the above follow in the subsequent sections. Note that the control definition stage is the focal point of this chapter and is considered by some as the primary component of the QbD approach.

16.3 Design Intent

Design intent is a formal step where the purpose for the dissolution method is stated. Note that the purpose depends on the phase of the life cycle within the drug development project. Less information (or at least different information) may be needed from dissolution test results in the early phases of product development than for a method intended to be used in a phase III stability program or as a QC release test. So, be aware that the design intent (the purpose of the method) often evolves over time.

For example, the design intent for the dissolution test of a simple immediate-release product in an early development phase might be to select components and processing parameters to develop the formulation. Biorelevant media are typically used for this screening. Once a formulation has been developed a dissolution test for late phase stability or for QC release will be needed. It must be discriminating to critical manufacturing processes. In other words, it must yield a passing result for good tablets and fail those that are produced outside of the known limits of the manufacturing parameters, such as particle size or compression force. Another consideration at this phase might be to use a simpler dissolution medium and/or apparatus that may not be biorelevant but may be suitable for a method that is discriminating with respect to the product's critical quality attributes.

It is also important to state the desired (or required) method performance criteria. Specifying the method in terms of the required performance may allow for some future regulatory flexibility and will better enable continuous improvement in the future. For example, instead of specifying HPLC as the analytical measurement, consider specifying the accuracy, precision, specificity, and other attributes of the analytical measurement. Doing this could allow measurement by direct UV spectrophotometry, which might be faster and more amenable to automation, assuming that it meets the agreed and approved method performance criteria.

Note that the performance of an analytical method is typically stated in terms of accuracy, precision, linearity, and specificity. The performance of a dissolution test method is often stated in its ability to discriminate certain attributes of a formulation such as tablet coating thickness, the compression force used to make the tablet, or particle size of the drug substance in the tablet. It is good to include all of this in the stated purpose of the dissolution test method. Again, be aware that the purpose may change and evolve over time.

The design intent of a modified or extended release formulation will likely be more complex because there may be multiple specifications at multiple sample intervals and will require some relation to in-vivo data. The release mechanisms are very likely to be more complex for a modified release formulation. The design intent of the dissolution test is tied to the design intent of the formulation itself.

16.4 Design Selection

Design selection is QbD terminology for "method development." The goal is to design a method that will be fit for the purposes that were identified in the design intent phase. Ideally, downstream customers and stakeholders should have some say into the method design. Although it sounds simple, there are a myriad of choices available when developing the method. Some are obvious and some require thought and more information. For example,

- Does a dissolution test method already exist? If so, are you constrained to it?
- Does it make sense to do this on apparatus 1 or 2 or should apparatus 4 be considered?
- Where is the most likely site of absorption?
- Will absorption be limited by permeability or by dissolution rate?
- Is a rapid or slow dissolution rate desired?

To select the composition of the medium it may be useful to review plots of solubility vs pH of the active. This information is usually available for any active pharmaceutical ingredient (API) that has been incorporated into a formulation. You might also ask: Is a surfactant needed? Will the method require a media addition step to effect a pH change? Is it better to measure with straight UV or by HPLC? If UV is feasible then can quantitation be done with an extinction coefficient or response value?

It is helpful to understand the mechanisms of drug release from the formulation. Any of the sub-processes shown in Fig. 16.1 could be rate limiting and could hold the key to a better understanding of the product. For example, if disintegration is the rate limiting step and truly determines bioavailability, then perhaps a dissolution test is not appropriate. A disintegration test may be more appropriate as a quality control measure and as a measure of biorelevance. If dissolution of drug particles is likely the rate-limiting step, then

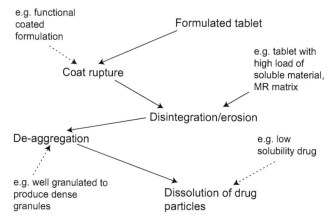

Breaking down the dissolution process - for a specific product, what are the rate-limiting/critical quality attributes?

Figure 16.1 Drug release mechanisms.

the intent should be to design a method that is discriminating with respect to the drug particle size distribution.

16.5 Control Definition

In this section we will follow a step-by-step process to establish the control definition. But first a couple of definitions and mnemonics will be helpful.

Robustness is the capacity of the analytical method performance characteristics to remain unaffected by small variations in the parameters that are likely to be encountered under normal conditions.

Ruggedness is the capacity of the analytical method performance characteristics to remain unaffected by variations in noise factors that are likely to be encountered under normal conditions.

The terms *ruggedness* and *robustness* are easily confused with each other. They are often incorrectly used interchangeably. The graphic and mnemonic in Fig. 16.2 can help you correctly associate

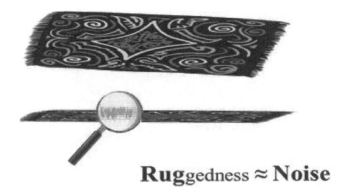

Ruggedness ≈ Noise

Figure 16.2 Robustness or ruggedness? Which is which?

ruggedness with noise. Think about a carpet, or rug. The carpet hairs will look like noise when viewed from the edge of the rug.

After having established that the method is fit for purpose, it may be a good time to gather the key stakeholders (current and future) of this method together for two events: a brainstorming exercise and a method walkthrough. In the brainstorming session participants list all known method parameters that could affect the variability or performance of the method. It might be useful to organize these in a table or set of tables (such as in Table 16.1). Sometimes it is helpful to consider the following major sources of variability: equipment, environment, method, measurement, materials, and people. These are often presented as the major branches of a fishbone cause and

Table 16.1 Variables list (IR formulation with UV measurement)

Variable	Class
UV wavelength	X
pH of medium (acid concentration)	X
Ambient laboratory conditions	N
Temperature of medium	C
Surfactant level	X
Surfactant supplier	N
UV integration time	X
Recirculation volume	C
Agitation rate (paddle speed)	X

effect diagram. The second event is a method walkthrough where the method is actually followed in detail on paper and/or in the lab to understand its operation and possibly to identify any additional parameters that could contribute to the method performance. Again, it is best to include current and future stakeholders in this exercise.

Classify each parameter as controlled (C), experimental (X), or noise (N). Controlled (C) parameters are usually fixed and are specified in the method. One example might include the temperature of the medium in the dissolution vessels (which is typically tightly controlled at 37°C). Experimental parameters are usually continuous variables that can be varied within a limited range. Examples include media composition, analytical wavelength, or column temperature. Noise factors are uncontrolled or those that are not controllable. They are usually discrete variables such as site, instrument, or surfactant supplier. Some advise initially classifying all parameters as controlled by default and then reclassifying some as experimental or noise. Table 16.1 is an example showing a partial list of parameters and how each is classified as X (experimental), C (controlled), or N (noise).

Note that the terms parameter, variable, and factor are often used interchangeably. Statisticians often prefer the term *factor* to mean a parameter or variable.

Experimental (X) parameters deemed critical will be the subject of a robustness assessment. Noise (N) parameters deemed critical will be the subject of a ruggedness assessment.

16.5.1 Robustness Assessment

Let's consider the robustness assessment first. Extract all of the experimental parameters from the list and rank them in terms of their potential impact on the method performance attributes such as linearity, accuracy, and precision. You may wish to consider other method attributes such as specificity, selectivity, peak shape, tailing, or resolution for a method involving HPLC measurement. Table 16.2 shows an example where the pH, surfactant levels, analytical wavelength, instrument integration time, and paddle speed were considered in terms of their possible effect on method performance. Higher numbers indicate more negative effect on the three method

Table 16.2   Example prioritization table

Experimental parameter	Linearity	Accuracy	Precision	Score (sum)	Rank
pH of medium	7	9	5	21	H
Surfactant level	3	7	3	13	M
UV wavelength	7	7	5	19	H
Integration time	1	1	1	3	L
Paddle speed	?	?	?	?	?

performance attributes: linearity, accuracy, and precision. In this case only odd numbers from 1 to 9 were allowed. The sum of these numbers represents the perceived risk that a specific parameter could negatively affect the performance of the method.

The objective is to design a set of experiments to assess the robustness of the method. The prioritization matrix is one of the intermediate outputs from this process. It represents a risk assessment of the experimental parameters and provides a ranking that allows us to determine which ones to use in a design of experiments (DOE) for the robustness assessment. In this example three of the five experimental parameters are ranked high, one as medium, and one as low. Question marks are listed for paddle speed and this is addressed shortly.

In this example, the pH of the medium and the UV analytical wavelength are likely to have the greatest effect on the performance of the method. Surfactant level is ranked medium while integration time is ranked low. So we will use the medium and high ranked factors to design and perform a series of experiments to demonstrate the robustness of the method.

Should paddle speed be considered an experimental factor? Perhaps not, but it should be classified as a controlled parameter. Typically one should expect a relationship between paddle speed and dissolution rate. Typically, we would not want a dissolution test method that is robust with respect to paddle speed. Such a method would not be discriminating for the critical quality attributes of the formulation. So, in some cases we have to balance our expectations between robustness and discrimination. That said, the relationship

Table 16.3 Example DOE design table

Experimental parameter	Nominal	Low	High
H of medium	1.5	1.2	1.8
Surfactant conc. (%)	1.0	0.5	1.5
Analytical wavelength (nm)	295	300	305

between paddle speed and dissolution rate should be studied and well understood.

So, getting back to the construction of our robustness assessment, typically we would retain the factors ranked high and perhaps those ranked as medium in the DOE to assess or to demonstrate robustness. Only three parameters are displayed in the example presented in Table 16.3. These could be evaluated one at a time in separate experiments; however, a factorial design would be a more efficient way to assess the effects of multiple variables.

Consider decoupling the dissolution-related parameters from the analytical measurement parameters. So, it may make sense to perform two separate robustness assessments.

16.5.2 Ruggedness Assessment

The first step in assessing the effect of uncontrolled variables, or noise, is to perform failure mode effect and analysis (FMEA). List all of the noise (N) parameters. Postulate or predict the effect of each factor on method performance. Rank each effect in terms of risk (high, medium, and low). List any controls that are in place or can be put into place to eliminate or reduce the risk of failures. Determine if the factor will be included in the ruggedness assessment.

Risk analysis is central to the QbD approach. It generally involves assessment within three dimensions: (i) the estimated frequency that a problem will occur, (ii) the likelihood of detecting the problem, and (iii) the impact that the problem will have on the product or process.

An example FMEA is illustrated in Table 16.4 for a dissolution test method with online UV measurement. It includes UV flowcells, standard preparation technique, lot-to-lot variability of components in the dissolution medium, measurement noise, and environmental

Table 16.4 Example FMEA and risk analysis table

Parameter	Failure mode	Potential effect	Frequency	Detection	Impact	Risk score (product)
UV flowcell	Path lengths may differ between flowcells	Cell path length has a direct proportional effect on response	1	1	7	7
Sonic bath	Inconsistent energy imparted between units	Undissolved material could result in lower than expected standard concentration	4	7	7	196
Surfactant source	Quality/purity many vary between suppliers	Purity of the surfactant could affect the amount in the medium	4	7	7	196
UV instrument	Instrument performance may vary	Could yield different results of same solution on different instruments	1	4	4	16
Site	Unknown differences between sites	Could yield inaccurate results	4	7	4	112
Analyst	Unknown differences between analysts	Could yield inaccurate results	1	4	4	16

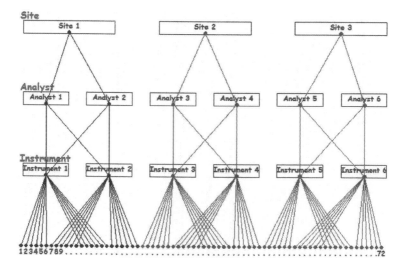

Figure 16.3   Schematic of MSA design.

effects. Sonic baths are notorious for delivering inconsistent energy, from one unit to the next. We really have no control over this. If they are used in the preparation of standard or sample solutions, then consider this as a possible noise factor. In this example, only values of 1, 4, and 7 are used to represent lowest to highest risk in terms the likely frequency of the failure, probability of detection, and impact or severity.

The FMEA represents the risk assessment of the method parameters that were classified as noise variables. The FMEA helps determine which parameters should be included in a measurement systems analysis or MSA, namely those with the highest risk scores.

The intent of the MSA is to characterize and better understand the sources of variability. The diagram in Fig. 16.3 shows the design of an example MSA intended to characterize variability from 3 different sites, 6 different analysts, and 6 different instruments. There is cross-over between the analysts and instruments. In this example, each analyst operates two different instruments. In this design the two analysts at each site will use two different instruments. A total of 72 test results will be produced. Note that an MSA can also serve as the method transfer.

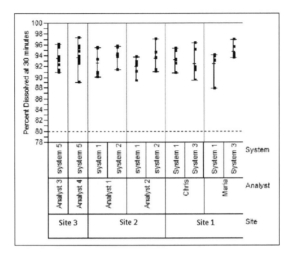

Figure 16.4 Schematic of MSA design.

Results (% dissolved after 30 min) from this example MSA form the ruggedness assessment. This box plot in Fig. 16.4 shows the results of this study in which two analysts worked with two instruments at three different sites. We see that the three sources all contribute about evenly to the variability of the results. This type of box plot can quickly provide a good idea of the relative variance from the different noise sources within the MSA design. Statistical processing tools such as ANOVA can extract and quantitate the contribution from each of these different sources to the total variability. Help from a statistician can prove very useful in the design and analysis of the ruggedness assessment.

16.5.3 Control Definition Table

The final step in the control definition stage is to build a control definition table that lists both the noise parameters and the experimental parameters. Some of this information comes from the results of the robustness assessment and some from the ruggedness assessment. A control definition table (such as the example in Table 16.5) should contain the following columns:

Table 16.5 Example control definition table

Parameter	Rationale for impact assessment	Impact classification	Design space	Action plan
pH of medium	Assessed in robustness study	Impact confirmed	0.008 N to 0.012 N	Specify 0.009 N to 0.011 N in the procedure
Surfactant level	Assessed in robustness study	Impact confirmed	0.3% to 0.7%	Specify 0.4% to 0.6% in the procedure
UV wavelength	Assessed in robustness study	Impact confirmed	295 nm to 305 nm	Specify that the wavelength accuracy of the UV instrument must be certified to within +/−2 nm
Sonic bath	Assessed in ruggedness study.	Impact not confirmed	Four different models were assessed	Ensure that standard solutions exhibit no cloudiness or un-dissolved material
Surfactant source	Assessed in ruggedness study	Impact not confirmed	Surfactant from three different suppliers were assessed.	Procure surfactant only from certified suppliers
UV flowcell	Assessed in ruggedness study	Impact confirmed	36 different flowcells at three different sites were assessed.	Certify flowcells via known UV standard solutions and procedures

- Rationale for the impact assessment on method performance
- Classification of the parameter indicating whether the parameter is known to have impact, confirmed no impact, or whether it is considered low risk but not confirmed.
- Design space, the known acceptable range for the parameter
- Action plan, a list action steps to be taken to ensure control and that the method performance is within the design space

A good control definition can possibly provide some flexibility when it becomes necessary to change an operating parameter within a method that has been filed and/or approved by a regulatory agency. It is easy to justify such a change if it is still within the known acceptable ranges or if it is known to not have an effect on the performance of the method.

16.6 Control Verification

The control verification phase is where it is confirmed that the method is fit for purpose. This may require additional validation experiments and it should also include metrics about the performance of the method once it has been implemented. This is also a good place to document the history of the method, including;ow it was developed and how the control strategy was established.

16.7 Continuous Improvement

Allow for continuous improvement. This may require you to demonstrate that the improved method meets or exceeds the registered performance criteria. Hence, this means register the method performance criteria rather than registering hard-coded method conditions that may be too specific or too detailed. Regulatory bodies are likely to agree with this approach because it encourages continuous improvement.

16.8 A Case Study

Consider this a negative example, because it was a case where QbD principles were not applied. It involves an IR tablet product that was in late-stage development. The dissolution test method had already been filed. There was considerable resistance to making any changes to the method before it was implemented as a release test in a quality control laboratory of a manufacturing site.

Direct UV measurement at 293 nm was used as the analytical response for percent dissolved as per Eq. 4.1. Some questionable results were obtained from dissolution tests. Values of around 110% dissolved were obtained on product for which the content was known to be around 98%. The dissolution medium is a simple 0.01N HCl solution. Investigation revealed that the UV response (absorbance at 293 nm) was dependent on the pH, so a procedural method control was established. The method states that standard solutions must be prepared from the same batch of medium used for the dissolution test. This was, in effect, a testimony that the method was not robust. This is not ideal and it is not conducive to a high-throughput laboratory environment.

The investigation showed that the analytical response depends on the acid concentration or pH of the dissolution medium. UV spectra in the range of 270 nm to 330 nm of the active pharmaceutical ingredient in different levels of acid are shown in Fig. 16.5. The analytical response (absorbance at 293 nm) is affected by changes in the acid concentration. Differences in pH or acid concentration between the standard solutions and the dissolution sample solutions could yield a bias in the analytical response and hence in the results. Figure 16.6 shows that the UV absorbance at 293 nm decreases with increasing acid concentrations. This could explain the high results observed prior to the procedural control. Changes in the analytical response of the standard have a direct and proportional effect on the analytical result (percent dissolved). Eq. 16.1 shows that the sample result is inversely proportional to the absorbance of the standard solution. So, if the standard solution was prepared with in a solution with higher acid concentration (lower pH) than the dissolution medium, the analytical results would be

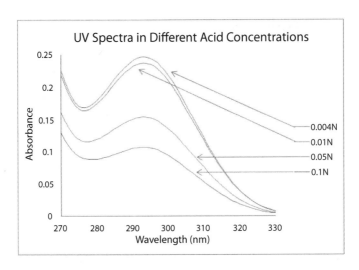

Figure 16.5   Effect of acid concentration on UV spectra.

biased high.

$$\% \text{ Dissolved } = 100 \times \frac{A_{\text{spl}} \times C_{\text{std}}}{A_{\text{std}} \times V_{\text{spl}} \times L} \qquad (16.1)$$

where

A_{spl} = absorbance of the sample solution,

C_{std} = concentration of the standard solution (in mg/mL),

A_{std} = absorbance of the standard solution,

V_{spl} = volume of the sample solution (in mL), and

L = the label claim or dosage strength (in mg).

Equation 16.1: Calculation of % dissolved

So what could or should have been done to remove or to at least minimize the dependency of analytical response to the acid concentration in the medium? Figure 16.7 shows a wider spectral view of the four UV spectra from 200 nm to 350 nm of the active pharmaceutical ingredient dissolved in the same four different levels of acid (0.004 N to 0.1 N), as shown in Fig. 16.5. There are three wavelengths where there is no pH dependency. These are called the isosbestic points and appear as the intersections of the spectra at or around 205 nm, 225 nm, and 239 nm. The response (UV absorbance) at these three wavelengths is immune to changes in pH.

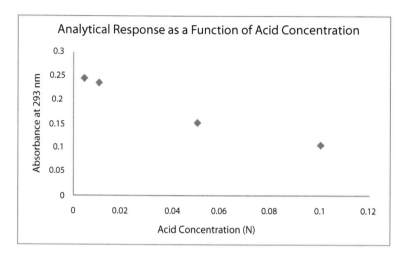

Figure 16.6 Effect of acid concentration on analytical response at 293 nm.

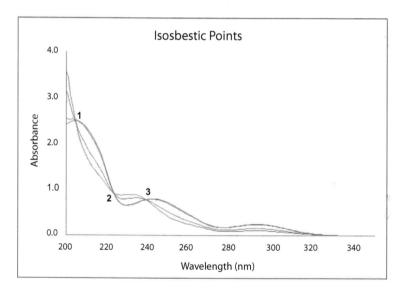

Figure 16.7 Isosbestic points within the UV spectrum.

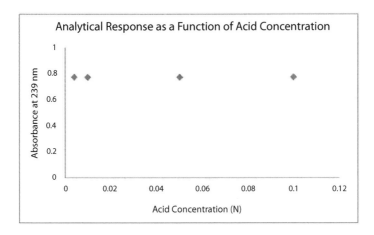

Figure 16.8   Effect of acid concentration on analytical response at 239 nm.

Absorbance measurements at wavelengths below 230 nm are often noisy because of the lower intensity of deuterium lamps that are the most common source for UV spectrophotometers. Dissolved oxygen can also affect the UV absorbance at these lower wavelengths. So, 239 nm would likely be the best of these three choices. The plot of the absorbance at 239 nm in Fig. 16.8 shows no appreciable dependence upon acid concentration.

So, it appears that using 239 nm as opposed to 293 nm as the analytical wavelength would solve the problem of pH dependency and possibly negate the need for the procedural control requiring that standard solutions be prepared from the exact same preparation as the dissolution medium. The main point of this negative case study is that all of this may have been discovered and determined had a quality by design approach been used initially to develop the dissolution method.

16.9 Why Implement QbD for Analytical Methods?

So we've just reviewed an example of what can happen when QbD is not properly employed. So, let's review some of the many reasons why one should implement QbD principles in the life cycle of analytical methods:

- to understand sources of variability
- to develop robust and rugged methods
- to set specifications that are aligned with dosage form's performance objective(s)
- to better understand critical attributes of manufacturing processes and of the dosage form itself
- to build knowledge about the product
- to allow regulatory flexibility for method changes
- to better enable continuous improvement

16.10 Summary

Traditional approaches to method development and validation focus on analytical measurements. Often, problems are associated with the preparation of the sample or standard solutions. There is a lot of knowledge associated with any given analytical method. All of this knowledge is often not conveyed in the written procedure. A structured approach that is guided by QbD principles provides a better chance of detecting and resolving problems that likely would be revealed during a transfer of a method developed by the traditional approach. The following bullets summarize the main points of this chapter:

- Traditional approaches do not always yield methods that are robust, rugged, and fit for purpose over the lifecycle.
- The QbD approach includes some different terminology such as deign intent, design selection, etc.
- Include stakeholders (current and future) in the development process. Leverage multiple brains and use brainstorming tools as early as you can in the product life cycle to identify and classify all factors and method parameters.
- Risk analysis is central to the QbD approach.
- Identify and classify method parameters as controlled, experimental, or noise.

- Assess risk of experimental parameters via prioritization matrix. Use DOEs to assess robustness.
- Assess risk of noise parameters via FMEA. Use MSAs to assess the effect of noise factors.
- Register method performance criteria as opposed to specific method conditions.
- Understand and formally declare the intent of your dissolution test method.
- Make allowance for that to change during the life cycle of the product.

It has been suggested that we should change the term *technology transfer* and call it *confirmation of manufacturability*. Rather than specifying specific hard-coded method conditions in your submissions, register the method performance criteria. This may allow some regulatory flexibility and allow for continuous improvement during the life cycle of the product and the method.

References

1. Puertollano, M. M., Cartwright, T., Aylott, M., and Kaye, N. (2009). Assessing an analytical method of the dissolution profile of an extended-release tablet in accordance with QbD, *Tablets and Capsules*, January 30.
2. Raska, C. S., Bennett T. S., and Goodberlet, S. A. (2010). Risk-based analytical method transfer: application to large multi-product transfers, *Anal. Chem.*, **82**, 5932–5936.
3. Nethercote, P., Borman, P., Bennett, T., Martin, G., and McGregor, P. (2010). QbD for better method validation and transfer, *Pharmaceutical Manufacturing*, April 13.

Chapter 17

Regulatory Considerations in Dissolution and Drug Release of BCS Class II and IV Compounds

Robert G. Bell[a] and Laila Kott[b]

[a] *Drug & Biotechnology Development, LLC, 406 South Arcturas Avenue, Suite 5, Clearwater, FL 33765, USA*
[b] *Takeda Pharmaceuticals International Co., 40 Landsdowne Street, Cambridge, MA 02139, USA*
rgb@drugbiodev.com, laila.kott@takeda.com

17.1 Introduction: What Are the Regulatory Implications That the Pharmaceutical Scientists Must Face When Developing Dissolution Methods for Poorly Soluble Drugs?

Regulatory aspects regarding the dissolution of poorly soluble oral solid drug compounds involve developing and documenting the appropriate dissolution procedures and specifications that ensure product quality over its life cycle as well as, when needed, the conducting of bioequivalence (BE) studies. For poorly soluble

Poorly Soluble Drugs: Dissolution and Drug Release
Edited by Gregory K. Webster, J. Derek Jackson, and Robert G. Bell
Copyright © 2017 Pan Stanford Publishing Pte. Ltd.
ISBN 978-981-4745-45-1 (Hardcover), 978-981-4745-46-8 (eBook)
www.panstanford.com

compounds, dissolution tests are often empirically derived and the specifications may not be reflective of product quality. Without the appropriate dissolution methods, formulation development is hampered and can result in product delays and regulatory failures. Regardless of the solubility of your dosage form, it is fundamental to understand the dissolution characteristics of the active pharmaceutical ingredient (API) and drug product. U.S. registrations require relevant dissolution specifications for release and testing of most oral solid dosage forms, and the pharmaceutical scientist must comply with the regulations as well as the intent of the associated guidelines. Dissolution is cited in 21 CFR §211, §314, §320, §343 in the Codes of the Federal Registrar (CFR):

- 21 CFR §211. 110 deals with sampling and testing of in-process materials and drug products to assure the sponsor examines, among other specifications, dissolution time and rate,[1]
- 21 CFR §314.53(b)(2)(5) addresses comparative in vitro dissolution testing on 12 dosage units each of the executed test batch and the new drug application product,[2] and
- 21 CFR §343.90 addresses aspirin dissolution and drug release testing.[3]

Sections 21 CFR §320.24(b)(5) and 21 CFR §320.22(d)(3) address product quality bioavailability and bioequivalence assessment using *in vitro* approaches.[4,5] The more relevant regulation in regards to poorly soluble drugs is 21 CFR §320.33(e), which deals with the physicochemical evidence and criteria to assess actual or potential bioequivalence problems.[6] The dissolution of poorly soluble compounds is specifically addressed in 21 CFR §320.33(e):

- 21 CFR §320.33(e)(1): The active drug ingredient has a low solubility in water, e.g., less than 5 mg per 1 mL.
- 21 CFR §320.33(e)(2): The dissolution rate of one or more such products is slow, e.g., less than 50% in 30 min when tested using either a general method specified in an official compendium or a paddle method at 50 revolutions per minute in 900 mL of distilled or deionized water at 37°C, or differs significantly from that of an appropriate reference

material such as an identical drug product that is the subject of an approved full new drug application.

- 21 CFR §320.33(e)(3): The particle size and/or surface area of the active drug ingredient is critical in determining its bioavailability.
- 21 CFR §320.33(e)(4): Certain physical structural characteristics of the active drug ingredient, e.g., polymorphic forms, conformers, solvates, complexes, and crystal modifications, dissolve poorly and this poor dissolution may affect absorption.
- 21 CFR §320.33(e)(5): Such drug products have a high ratio of excipients to active ingredients, e.g., greater than 5 to 1.
- 21 CFR §320.33(e)(6): Specific inactive ingredients, e.g., hydrophilic or hydrophobic excipients and lubricants, either may be required for absorption of the active drug ingredient or therapeutic moiety or, alternatively, if present, may interfere with such absorption.

In addition, there are numerous FDA and United States Pharmacopeia (USP) (Table 17.1) guidance documents regarding dissolution.[7,8]

Quality by design (QdD), as described in ICH Q8R2,[9] is a systematic approach to development that begins with predefined objectives and emphasizes product and process understanding and control that is based on sound science and quality risk management. Contained within ICH Q8R2 is reference to the quality target product profile (QTPP). The QTPP is a prospective summary of the quality characteristics of a drug product that ideally will be achieved to ensure the desired quality, taking into account safety and efficacy of the drug substance, excipients, intermediates (in-process materials), and drug product.[9] Considerations for the QTPP include

- intended use in clinical setting, route of administration, dosage form, delivery systems
- dosage strength(s)
- container closure system
- therapeutic moiety release or delivery and attributes affecting pharmacokinetic characteristics (e.g., dissolution, aerodynamic performance, etc.) appropriate to the drug product dosage form being developed

Table 17.1   FDA and USP guidances for dissolution testing

Date	Guidance
1950 USP	Disintegration testing (incomplete test) "disintegration does not imply complete solution of the tablet or even of its active ingredient"
1970 USP	USP Dissolution Apparatus 1, Basket, USP 18
1976 USP	USP Dissolution Apparatus 2 (paddle), USP 19
1978 USP	Reciprocating piston, Flow-through cell, Dissolution calibrators
1978 FDA	FDA/DPA Guidelines for dissolution testing
1984 FDA	FDA/DPA Guidelines for dissolution testing addendum
1990 USP	USP Apparatus 5 (Paddle over Disc), 6 (Cylinder), or 7 (Reciprocating Holder)
1991 USP	USP Dissolution Apparatus 3 (Reciprocating cylinder)
1995 USP	USP Dissolution Apparatus 4 (Flow-through cell)
1995 FDA	Guidance for Industry, SUPAC-IR: Immediate-Release Solid Oral Dosage Forms: Scale-Up and Post-Approval Changes: Chemistry, Manufacturing and Controls, In vitro Dissolution Testing, and in vivo Bioequivalence
1997 FDA	Guidance for Industry, SUPAC-MR: Modified Release Solid Oral Dosage Forms Scale-Up and Postapproval Changes: Chemistry, Manufacturing, and Controls; In vitro Dissolution Testing and in vivo Bioequivalence
1997 FDA	Guidance for Industry, Dissolution Testing of Immediate Release Solid Oral Dosage Forms
1997 FDA	Guidance for Industry, Extended Release Oral Dosage Forms: Development, Evaluation, and Application of in vitro/in vivo Correlations
2010 FDA	Guidance for Industry, The Use of Mechanical Calibration of Dissolution Apparatus 1 and 2 – Current Good Manufacturing Practice (CGMP)
2010 ICH	Guideline Q4B annex 7 (R2) to note for evaluation and recommendation of pharmacopoeial texts for use in the ICH regions on dissolution test-general chapter (this is basically the harmonization of EP/JP/USP disolution methods
2011 FDA	Guidance for Industry, Q4B Evaluation and Recommendation of Pharmacopoeial Texts for Use in the ICH Regions, Annex 7(R2) Dissolution Test General Chapter
2015 FDA	DRAFT Dissolution Testing and Specification Criteria for Immediate-Release Solid Oral Dosage Forms Containing Biopharmaceutics Classification System Class 1 and 3 Drugs

- drug product quality criteria (e.g., sterility, purity, stability, and drug release) appropriate for the intended marketed product.

These considerations supplement the Target Product Profile, as described in the FDA Guidance for Industry and Review Staff—Target Product Profile—A Strategic Development Process Tool, March 2007.[10]

These guidances encourage an "evolution of knowledge" to understand the CMC quality aspects and procedures to the relationship with the product and patient. Drug release, along with purity, strength and stability, are critical quality attributes of the drug substance and drug product to ensure the required product quality.

Perhaps the most relevant for dissolution and poorly soluble compounds is the Biopharmaceutics Classification System (BCS), as described in August 2000 U.S. FDA Guidance for Industry.[11] BCS is based on the drug's aqueous solubility and intestinal permeability, and has been widely used for years to predict drug absorption during the pharmaceutical development lifecycle. Under the BCS, it may be possible to receive in vivo bioequivalence waiver (IND, NDA, ANDA, post-approval changes) for immediate release oral solid dosage forms that are highly soluble and permeable, use inactive ingredient database (IID) and have a wide therapeutic window (BCS class I) (Table 17.2).[12] For BCS class II and class IV drugs, there are fewer options; however, researchers are considering further subdividing these classes into neutral molecules, weak acids, and weak bases. Working within these subdivisions, there is opportunity to further develop dissolution requirements for biowavers.[13,14]

Recent Generic Drug User Fee Act (GDUFA) initiatives involves examining the permeability and absorption data of BCS class III

Table 17.2 Biopharmaceutics Classification System as defined by Amidon et al.[12]

Biopharmaceutical classification	Solubility	Permeability
I	High	High
II	Low	High
III	High	Low
IV	Low	Low

drugs with the aim of possibly extending biowaivers to BCS class III drugs in the hopes of eliminating the need for in vivo bioequivalence studies.[15]

The FDA BCS Guidance[11] defines a high drug substance solubility, permeability and rapid dissolution as

- "highly soluble" when the highest dose strength is soluble in 250 ml or less of aqueous media over the pH range of 1–7.5 at 37°C as demonstrated by the appropriate method (shake-flask, titration method) and analyzed using a validated stability-indicating assay
- "highly permeable" when the extent of absorption in humans is determined to be >90% of an administered dose, based on mass-balance or in comparison to an intravenous reference dose or in vitro intestinal permeability methods
- "rapidly dissolving" when greater than 85% of the labeled amount of drug substance dissolves within 30 min using USP apparatus I or II in a volume of less than 900 mL buffer solutions with the appropriate f2 comparisons

A challenging aspect of developing poorly soluble drug formulations is the development of appropriate "biorelevant" dissolution methods for these products, which correlate with in vivo dissolution. Low solubility, according to the BCS, would be considered any drug requiring greater than 250 mL of aqueous media over the pH range of 1–7.5 at 37°C.[11] The BCS class II and IV compounds contain low-solubility drugs in combination with high and low permeability, respectively. In many instances, a solubilizing agent is used to improve the dissolution rate and overall solubility, which, depending on the additive, may impact the biorelevance. It may be possible for BCS class II drugs to have an in vitro–in vivo correlation but unlikely for BCS class IV drug product. Lack of in vitro–in vivo correlation for BCS class II drug product would necessitate bioequivalence studies. BCS class IV drugs would require bioequivalence (and possibly clinical) studies.[16]

Many therapeutic small molecules are lipophillic and have poor aqueous solubility[17] and thus fall into BCS II and IV classification. The pharmaceutical scientists should be aware of the quality and regulatory challenges of developing and validating dissolution methods for BCS class II and IV drugs.

17.2 Classification Systems: BCS versus BDDCS versus DCS

In 1995 Amidon et al.[12] introduced the now familiar biopharmaceutics classification system (BCS) for drug products. This is a scheme of correlating the in vitro dissolution behavior to in vivo bioavailability based on the solubility and permeability of the drug substance. This scheme was one of the first to correlate what is measured in the laboratory with that which is occurring in the body. This approach was widely accepted by the pharmaceutical industry and in 2000 was adopted by the FDA, as a science-based approach for cases requesting biowaivers for BCS class I drugs (see Table 17.2).[11]

Wu and Benet[18] suggested modifying the BCS, to better predict the drug disposition in the body, by incorporating such things as routes of drug elimination, adsorption, effects of transporters, and food effects. In their Biopharmaceutics Drug Disposition Classification System (BDDCS), they proposed that it would be simpler to assign classes based on metabolism instead of by permeability.

More recently, however, both the BCS and BDDCS classification systems have been challenged by the Developability Classification System (DCS), which takes the lessons learned from the previous two systems and focuses more on the drug developability.[19] This approach also includes contributions from the excipients and can be used as an aid for the formulator.

An overview of the three classification systems is shown in Table 17.3 and more in-depth descriptions of the approaches are described below.

17.2.1 Overview of the Three Classification Systems

17.2.1.1 Biopharmaceutics Classification System (BCS)[12]

The BCS was designed to correlate in vitro testing to in vivo bioavailabilty. It assumes that drug dissolution and gastrointestinal

Table 17.3 Comparison of the BCS vs BDDCS vs DCS

	Class I	Class II		Class III	Class IV
BCS	High solubility, high permeability	Low solubility, high permeability		High solubility, low permeability	Low solubility, low permeability
BDDCS	High solubility, extensive metabolism	Low solubility, extensive metabolism		High solubility, poor metabolism	Low solubility, poor metabolism
		Class II a	Class IIb		
DCS	Good solubility* and permeability	Good permeability, poor solubility*: dissolution rate limited	Good permeability, poor solubility*: solubility limited	Good solubility*, poor permeability	Poor solubility*, poor permeability

*Solubility based on a volume of 500 mL versus 250 mL for BCS/BDDCS.

permeability are the determining factors which influence the extent and rate of drug absorption; however, dose is also an important consideration and the plasma is assumed to be the physiological sink.

In the BCS (and BDDCS, see below) two definitions are key:

1. Highly soluble = when the highest dose strength is soluble in 250 mL or less of aqueous media over a pH range of 1 to 7.5.[11]
2. Highly permeable = when the administered dose is determined to be 90% or more absorbed in humans.[11]

This classification system considers that two drug products with the same API exhibiting the same concentration profile over time at the membrane surface will have the same extent and rate of absorption. From this it is inferred that if two drug products exhibit the same in vivo dissolution profile, then they will show the same extent and rate of drug absorption in the gut/intestine. What is assumed is that there are no other components in the drug product that affect intestinal transit or permeability. (Note: This is a point to which we will return when discussing other classification systems).

In vitro–in vivo correlation or IVIVC is a measure of how closely our in-laboratory dissolution measurements can describe what is happening in the body. The concept being that, by taking the data from a relatively simple in vitro test, one can understand how the drug will fare in patients. This type of testing can obviously simplify drug development, however a good correlation is not always possible. The IVIVC for the BCS classes are:

- Class I: Correlation only if dissolution is slower than the rate of gastric emptying, otherwise none.

  - If the above case is not true, where dissolution is *faster* than gastric emptying, then the highly soluble drug will be in solution as therefore, readily transported and consequently absorbed into the small intestine. These are the conditions (solubility, dissolution and intestinal permeability) under which the biowaivers are scientifically sound.[20]

  - From the USP <1090> (World Health Organization [WHO] approach): Dosage forms of drug substances that are highly soluble, highly permeable, and rapidly dissolving are eligible for biowaivers under the following conditions:

* 85% or more of the dosage form dissolves in 30 min or less and the dissolution profile of the generic product is similar to that of the reference product in pH 1.2, 4.5, and 6.8 buffer, using the basket method at 100 rpm or the paddle method at 50 rpm (FDA) or 75 rpm (WHO), and meets the criterion of dissolution profile similarity, $f_2 \geq 50$.
* If both the reference and the generic dosage forms are very rapidly dissolving (i.e., 85% dissolution in 15 min or less in all three media under the above test conditions), then profile determination is not necessary.

- Class II: Correlation if in vitro and in vivo dissolution rates are similar. If dose is high, IVIVC may break down.

  - From the USP <1090>(WHO approach): Dosage forms of drug substances with high solubility only in pH 6.8 and high permeability (low solubility by definition, BCS class II) are eligible for biowaivers, provided that

    * the dosage form is rapidly dissolving (85% or more in 30 min or less) in pH 6.8 buffer.
    * the generic product exhibits dissolution profiles similar to those of the comparator product in buffers at pH 1.2, 4.5, and 6.8.

- Class III: Limited to none, however, recent GDUFA initiatives involve examining the permeability and absorption data of BCS class III Drugs with the aim of possibly extending biowaivers to BCS class III Drugs in the hopes of eliminating the need for in vivo bioequivalence studies.[15]

  - From the USP <1090> (WHO approach): Dosage forms of drug substances that are highly soluble and have low permeability are eligible for biowaivers under the following conditions:

    * Both the reference and the generic dosage forms are very rapidly dissolving (85% dissolution in 15 min or less in all three media under the test conditions given

above), and they do not contain any excipients and/or inactive substances that are known to alter gastrointestinal motility and/or permeability or influence drug absorption.

* Firms should show that the quantity of excipients used is consistent with the intended use. When new excipients and/or atypically large amounts of commonly used excipients are included in the dosage form, additional information documenting the absence of any significant impact on bioavailability of the drug is required.

- Class IV: None.

For water-insoluble drugs, the dissolution media becomes more difficult to determine, but can be overcome with the addition of surfactants, with sodium dodecylsufate (SDS), often used in place of bile salts. In choice of media, sink conditions should be maintained, if possible. Sink conditions are defined as the drug being soluble in only 20% to 30% of the media used (roughly $3\times$ to $5\times$ conditions). Finally, the BCS highlights particle size as a critical parameter for dissolution.

As the BCS became more commonly applied, some questions arose, such as What is its applicability to controlled release drugs? Are the dissolution criteria too restrictive (85% by 30 min)? And what part, if any, do the excipients play in affecting the dissolution profile?[21] However, despite some of these drawbacks, by 2010, the BCS classification had been adopted by many regulatory authorities (European Medicines Agency [EMA], Center for Drug Evaluation and Research [CDER], World Health Organization [WHO]) and has become a key tool used for the granting of biowaivers.[22]

17.2.1.2 Biopharmaceutics Drug Disposition Classification System (BDDCS)[18]

Wu and Benet proposed a moderate change to the BCS classification criteria, with the intention of improving the in vivo predictability of the four classes. They found that class I and class II drugs were predominately eliminated by metabolism whereas, class III and class IV compounds were predominately eliminated unchanged in the urine. This led to a prediction that metabolism was a better predictor than permeability. Therefore, the major difference between the BCS and BDDCS is whether permeability is defined as the extent of drug transport across the cell membranes or if it is viewed as a rate.[23] The key assumptions of BDDCS were as follows:[18]

- Transporter effects will be minimal for class I compounds.
- Efflux transporter effects will predominate for class II compounds.
- Transporter–enzyme interplay in the intestines will be important for class II compounds that are substrates for CYP3A and phase 2 conjugation enzymes.
- Absorptive transporter effects will predominate for class III compounds.

They also hypothesized that drug–transporter interactions may be the primary mechanism for food effects as they suspected that high-fat meals would inhibit both in influx and efflux of drug transporters. Specifically for each class they describe the following:[18]

- High-fat meals will have no significant effect on increasing bioavailability for class I compounds.
- High-fat meals will increase bioavailability for class II compounds.
- Formulation changes that markedly increase the solubility of class II compounds will decrease or eliminate the high-fat meal effects of these drugs.
- High-fat meals will decrease the bioavailability for class III compounds.

Other parameters that have been considered while developing the BDDCS are post-absorption effects, intravenous dosing, and drug–drug interactions.

Permeability determinations have proven to be difficult to perform, yielding an amount of uncertainty with each measurement. With the BCS guidance's lack of a single unified permeability method,[24] coupled with the fact that permeability in humans is more of an absorption measurement rather than an indication of bioavailability, Wu and Benet propose:[18]

Designation of the major route of drug elimination as part or instead of the permeability criteria (see Table 17.3) would reduce the regulatory burden for many more class I compounds, would eliminate the ambiguity and difficulty in determining 90% (or 85%) absorption for Classes I and II compounds, and would allow predictability of absorption and disposition characteristics of drugs in all four Biopharmaceutics Drug Disposition Classification System classes.

17.2.1.3 Developability Classification System (DCS)[19]

As the BCS and, to some extent the BDDCS, classification systems were applied more universally, it was noted that the effects of the excipients were not considered. For example, Grattan et al. discovered that the dissolution rate/behavior of acetaminophen, can be altered by the addition of sodium bicarbonate.[25] In 2001, Dressman et al. reviewed some of the short falls/deficiencies in the BCS. Building upon that, Butler and Dressman have also looked at the drug classification systems and have suggested that there should be a way to better categorize drugs by the factors limiting their oral absorption. They have proposed their own developability classification system, which focuses less on bioequivalence and more on oral drug product development, taking QbD into consideration.

Their modified classification system includes the following concepts (illustrated in Fig. 17.1):[19]

- An estimate of human fasted intestinal solubility (e.g., by using FaSSIF) as the primary measure of in vivo solubility useful for the prediction of the extent of human absorption.

  – This includes the modification of the definition of solubility, by incorporating a volume of 500 mL (considered more representative of the fluid volume in the GI tract) over the standard 250 mL, used in the BCS and BDDCS. This increases the number of compounds that could be considered class I or class III.

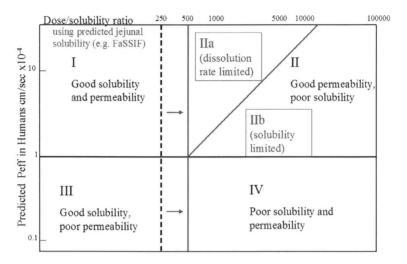

Figure 17.1 The Developabilty Classification System. Reproduced with permission from Wiley Butler and Dressman [19]. Copyright (2010) Wiley.

- A solubility-limited absorbable dose (SLAD) concept, based on the idea that for class II drugs, at least, permeability and solubility are compensatory.
- Dissolution rate, expressed as a target drug particle size rather than dose/solubility ratio, provides a better means of assessing the development risks and critical quality attributes (CQAs) for drugs with dissolution rate limited extent of absorption.

The authors state that the BCS is useful to the pharmaceutical industry because it is simple and, therefore, also keep their classification system simple, but more realistic for developability analysis. A key advantage of the system, as described by Butler and Dressman, is that it gives early information to the formulator "as to which poorly soluble drugs can be adequately formulated via simple size control versus which are likely to need specialist solubilisation techniques to obtain complete oral absorption and to avoid solubility related food effects."[19]

The DCS system is most useful for drugs which are not ionized under physiological pH's. If a drug is ionized, its permeability and solubility profile down the GI tract is more difficult to define as it can change as it travels through the system. In this case, a point in the GI tract, where the most significant absorption occurs, should be chosen as a point at which to examine the drug. The authors also advocate for estimating intestinal solubility using biorelavent fluids, arguing that this way the solubility of especially lipophilic drugs in the GI tract are more accurately estimated. Any error in the estimated solubility is largely due to the uncertainty of the true composition of the biological media, as the exact composition and capacity of biological buffer solutions, as well as the complex composition of bile salts and other solubilizing agents in the body, making it difficult to produce the perfect synthetic biological media.

Overall the authors of the DCS consider that understanding/measuring the factors that would put a drug's performance at risk, in vivo are the ones that are most helpful in deciding on formulation strategies and for highlighting CQAs.[19]

17.3 Biorelavent Dissolution Methods versus Standard Quality Dissolution Methods

For both the in vivo behavior and in vitro assessment of a drug formulation, the dissolution test is used, as it is useful for both quality control and development purposes.[26] The quality control dissolution test often employs simple aqueous buffers as they are very reproducible, however, when using dissolution to aid in formulation development, more complex, biorelevant media that simulate gastrointestinal (GI) fluids yield more relevant data, as they better suggest what the drug will encounter in the body. However, a precise definition of biorelevant dissolution is still a topic of discussion. The definition/description that currently seems to most aptly resonate within the pharmaceutical industry of biorelevant dissolution is dissolution that mimics in vivo dissolution.[27]

In 2008 the participants in the workshop "Bioequivalence, Biopharmaceutics Classification System, and Beyond" published a summary of what they considered the ideal attributes of in vitro dissolution testing:[24]

- Sensitivity to product changes such that in vitro dissolution testing ensures high quality and consistent product performance.
- Predictability of in vivo drug product performance such that human studies can be reduced and product development accelerated.
- Recognition by regulatory policies and procedure to support product applications.

This has been how the standard quality dissolution testing has been approached/viewed since the introduction of the BCS. However, during this workshop, some consideration was also given to biorelevant dissolution methodologies and how they could be treated the same or different than the quality control dissolution methods.[24]

While traditional quality control tests are suitable for passing or failing a batch of drug product, they often have limited relevance to physiological performance. As such, there is an ever growing trend to test drug products using more biorelevant media. From this trend, several considerations arise:

- Do we really need a quality control test, if a biorelevant test tells us all we need to know about the drug product? Or do we need both?
- Can we use the biorelevant test as the quality control test?
- Are they truly better predictors of performance?
- Are biorelevant dissolution methods as reproducible/rugged as the standard quality control tests used up until now?
- Are they telling us the whole story? Or just another part of the story?
- What is the industry trend?

Biorelevant dissolution media has been investigated since the 1990s. Galia et al.[28] studied some class I and II drugs in simulated gastric fluid (with and without pepsin), simulated intestinal fluid including fed and fasted equivalents (FeSSIF/FaSSIF), and milk. For the class I drugs, they found that choice of media (except for milk) had little influence on the dissolution profile. However, for class II drugs different media produced very different dissolution profiles, and as such the choice of dissolution media should be considered carefully.

On the basis of the classification systems discussed previously, one can conclude that biorelevant dissolution media would be unnecessary for class I drugs and by extension even class III drugs, as these are the drugs that readily dissolve, but are absorbed under a wide range of conditions. However, biorelevant dissolution media can be important in gaining knowledge of the fate of poorly soluble drugs,[26,27] which in turn can be a reflection of overall quality or future performance in the body.

A biorelevant medium that contains surfactants, will have a reduced surface tension in comparison to water, which aids in wetting the solids, increasing sink conditions and, if both fasted and fed dissolution media are used, can provide indication of potential food effects.[26] As always, which media to choose depends heavily on the type of formulations and the characteristics of the compound. Knowledge of the compound can also aid in deciding what biorelevant parameters are important. If pH does not affect solubility, then, perhaps, a neutral media (or one at the pH of the expected site of dissolution or absorption) exploring the amounts and/or types of surfactants would be an appropriate biorelevant dissolution media. Several papers and reviews on biorelevant dissolution media have compiled the compositions of fasted and fed stomach, upper small and lower small intestines.[26,29,30] A summary of the currently used compositions of fasted state simulated gastric fluid (FaSSGF), fed state simulated gastric fluid (FeSSGF), fasted state simulated intestinal fluid (FaSSIF), and fed state simulated intestinal fluid (FeSSIF) are summarized in Tables 17.4 to 17.6.

The USP describes a *biorelevant medium* in chapter <1092>, "The Dissolution Procedure: Development and Validation,"[8] as a medium that has some relevance to the in vivo performance of the dosage unit and is based on a mechanistic approach that considers the absorption site, if known, and whether the rate-limiting step to absorption is the dissolution or permeability of the compound. In some instances, the biorelevant medium and the time points will be different from the test conditions chosen for the regulatory test. Media that simulate the fed and fasted states reflect changes in pH, bile concentrations, and osmolarity after

Table 17.4 Compositions of FaSSGF and FeSSGF[29, 30]

	FaSSGF	FeSSGF
Sodium taurocholate (µM)	80	
Lecithin (µM)	20	
Pepsin (mg/mL)	0.1	
Acetic acid (mM)		17.12
Sodium acetate (mM)		29.75
Milk/acetate buffer (mM)		1:1
Water	qs 1L	
Sodium chloride (mM)	34.2	237.02
Hydrochloric acid to adjust to pH	1.6	5.0
Osmolality (mOsmol/kg)	127.05±2.5	400
Buffer capacity (mEq/pH/L)		25

Table 17.5 Compositions of FaSSIF

	FaSSIF[30, 31]	FaSSIF-V2[29, 30]
Sodium taurocholate (mM)	3	3
Lecithin (mM)	0.75	0.2
NaH_2PO_4 (mM)	28.7	
Maleic acid (mM)		19.12
Water (L)	1	1
Sodium chloride (mM)	105.7	68.62
Sodium hydroxide (mM)	Use to adjust pH	34.8
pH	6.5	6.5
Osmolality (mOsmol/kg)	~270	180±10
Buffer capacity (mEq/pH/L)	~12	10

meal intake and therefore have a composition different from that of typical USP compendial media. These fed and fasted simulated media are primarily used to establish in vitro–in vivo correlations during formulation development and to assess potential food effects and are not intended for quality control purposes. For quality control purposes, the substitution of natural bile surfactants with the appropriate synthetic surfactants is permitted.

Despite the fact that the use of biorelevant dissolution media appears to have many benefits, there is no current regulatory requirement and the FDA does not feel that there is a sufficient

Table 17.6 Compositions of FeSSIF

	FeSSIF[12,30]	FeSSIF-V2[29,30]
Sodium taurocholate (mM)	15	10
Lecithin (mM)	3.75	2
Glyceryl monooleate (mM)		5
Sodium oleate (mM)		0.8
Maleic acid (mM)		55.02
Acetic acid (mM)	144	
Water (L)	1	1
Sodium chloride (mM)	203	125.5
Sodium hydroxide (mM)	101	81.65
pH	5.0	5.8
Osmolality (mOsmol/kg)	~670	390±10
Buffer capacity (mEq/pH/L)	~72	25

body of evidence to recommend biowaivers for class II drugs based on the use of biorelevant dissolution media.[27] There is, however, support within the industry to have biorelevant dissolution media serve both as a quality control test and a possible surrogate of drug bioavailability and bioequivalence.[32]

17.3.1 Setting Dissolution Specifications

Since 1999 the following have been part of the ICH guidelines for setting the dissolution specifications for new drug products (from ICH Q6A):[33]

- "The specification for solid oral dosage forms normally includes a test to measure release of drug substance form the drug product."
- "Single point measurements are normally considered suitable for immediate-release (IR) dosage forms."
- "For modified-release dosage forms (MR), appropriate test conditions and sampling procedures should be established."
- "For immediate-release drug products where changes in dissolution rate have been demonstrated to significantly affect bioavailability, it is desirable to develop test conditions which can distinguish batches with unacceptable bioavailability."

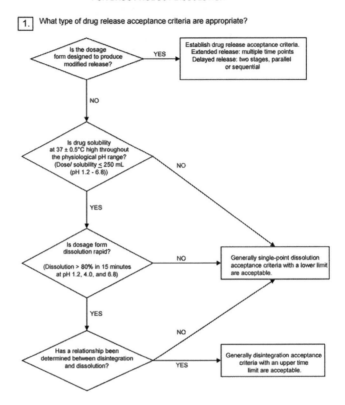

DECISION TREES #7: SETTING ACCEPTANCE CRITERIA
FOR DRUG PRODUCT DISSOLUTION

1. What type of drug release acceptance criteria are appropriate?

Is the dosage form designed to produce modified release? — YES → Establish drug release acceptance criteria. Extended release: multiple time points Delayed release: two stages, parallel or sequential

NO

Is drug solubility at 37 ± 0.5°C high throughout the physiological pH range? (Dose/ solubility ≤ 250 mL (pH 1.2 - 6.8)) — NO

YES

Is dosage form dissolution rapid? (Dissolution > 80% in 15 minutes at pH 1.2, 4.0, and 6.8) — NO → Generally single-point dissolution acceptance criteria with a lower limit are acceptable.

YES

NO

Has a relationship been determined between disintegration and dissolution? — YES → Generally disintegration acceptance criteria with an upper time limit are acceptable.

Figure 17.2 Part 1 of the ICH decision tree for dissolution testing specification setting.[33]

- "For extended-release drug products, in vitro/in vivo correlation may be used to establish acceptance criteria when human bioavailability data are available for formulations exhibiting different release rates." When this data is not available and release is not independent of the conditions of the in vitro test, then acceptance criteria should be based on available batch data.
- The ICH clearly states that the permitted variability should not exceed ±10% of the label claim. As an example, from the ICH guidance, the total allowable variability is 20%.

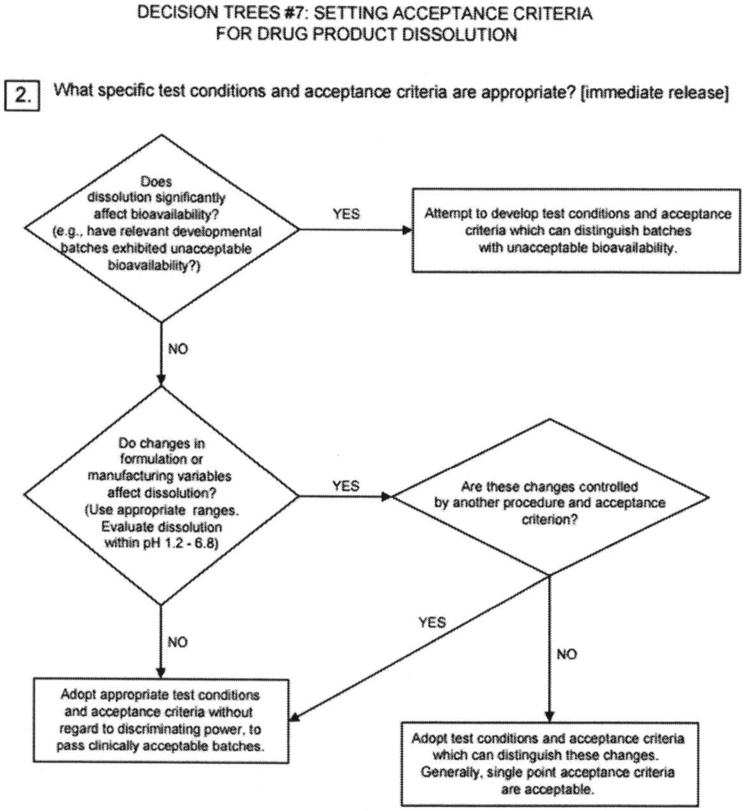

DECISION TREES #7: SETTING ACCEPTANCE CRITERIA
FOR DRUG PRODUCT DISSOLUTION

2. What specific test conditions and acceptance criteria are appropriate? [immediate release]

Figure 17.3 Part 2 of the ICH decision tree for dissolution testing specification setting.[33]

Therefore, if the requirements is a label claim of 50% ± 10%, then the acceptable label claim range would be 40% to 60%, unless a wider range is supported by a bioequivalency study (see Fig. 17.4).

The suggested requirements of dissolution testing based on BCS class and supported by the Center for Drug Evaluation and Research (CDER) are:[12,34]

- Class I: For immediate release (IR) dosage forms: 85% dissolved in less than 15 min

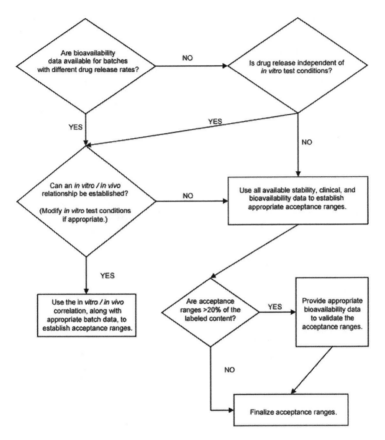

Figure 17.4 Part 3 of the ICH decision tree for dissolution testing specification setting.[33]

- Class II: At least 85% dissolution at several physiological pH's and dissolution profile containing 2 to 6 data points
- Class III: For IR dosage forms: 85% dissolved in less than 15 min
- Class IV: Case by case

For modified-release dosage forms, acceptance criteria are often based on three time points:[35]

1. An early time point to identify any occurrence of dose dumping
2. A time point that demonstrates the extended release and characterizes the release profile
3. A later time point to show that most of the dose was delivered, targeting at least 80% of label claim

17.3.2 Stage 1, Stage 2, Stage 3, or Level 1, Level 2, Level 3 Testing

Looking at specifications strictly from a quality control perspective, we note that according to the USP chapter on dissolution <711>,[8] there are three stages of testing that can be performed on a drug product, prior to rejecting a batch.[8] When developing a method and defining a specification, many industry researchers follow the practice of setting the dissolution specifications tight enough that batches will fail Stage 1/Level 1 testing a certain percentage of the times (i.e., 10% to 15% of the batches will go on to Stage 2 or Level 2 testing). There is, however, no written guidance to that explicitly states the Stage1/Level 1 failure rate. Additionally, the FDA has been suggesting, especially for generic drug products, but others as well, that testing start with $n = 12$ dosage forms, which essentially means that one should start with Stage 2/Level 2 testing.[36]

17.3.3 Considering Specifications Using Biorelevant Methods

At the 2013 AAPS workshop on biorelevant dissolution testing considerations, the FDA Office of Generic Drugs informed the group that the "FDA is asking for biorelevant data to set specifications, which in an ideal case will accompany the product throughout its entire lifecycle."[27] The FDA elaborated that method development should focus on setting "meaningful specifications" based on the drug product's in vivo performance and that such "meaningful specifications" could aid in the application of the biowaiver process.[27] Finally, the FDA also suggested that biorelevant dissolution data may also show that the product (and methods) are rugged.[27]

As previously discussed, the ICH and in the original CDER guidance, the possibility of using biorelevant dissolution media was considered.[34] Others have considered that dissolution testing can be a surrogate for bioavailability data and have considered that a discriminating dissolution method can work as a QC test and give clinically relevant information.[35,37]

Another consideration when setting dissolution specifications is, What do we really need dissolution specifications to tell us? The answers are fairly simple:

1. That the quality of a batch of drug product is acceptable.
2. That this batch will behave in the body the same way all other batches do and the drug will get to its target at the proper concentration.

The current standard QC tests are developed to ensure manufacturing quality, but are the tolerances imposed upon a drug product by the QC test appropriate to also show the same 'quality' or response in the body? How relevant is a QC dissolution test that fails a batch of drug product, when the response of that drug product is the same or better in the body?

For these questions, if we collect up the key points and suggestions for setting the QC specifications from the USP[8] and CDER,[34] we get:

a. "The specification for solid oral dosage forms normally includes a test to measure release of drug substance form the drug product." (USP)
b. "Single point measurements are normally considered suitable for immediate-release (IR) dosage forms." (USP)
c. Class II: At least 85% dissolution at several physiological pH's and dissolution profile containing 2 to 6 data points (CDER).
d. Class IV: Case by case (CDER).

When looking at class II and class IV drug products, to develop a method that addresses all the above points, one could be in a situation where the media that satisfies 85% dissolution across several physiological pH's is not physiological relevant, rendering this type of testing irrelevant.

Consider an IR dissolution test under a physiological pH and surfactant conditions, for where that drug is to be released in the body that yields 60% dissolved at 45 min and tablets with these results have been shown to be effective. Then consider a dissolution test for the same drug product that to gets to 85% dissolved, but you need to add 20% Tween and 10% isopropyl alcohol (IPA) to reach that level. What would be the response if

- both dissolution tests failed?
- the physiological one failed but the QC one passed?
- the QC test failed but the physiological one passed?

In the first case it is clear that there is something wrong with the batch. In the second case, are we really sure that this drug product will be effective in the body, if is it not releasing the active ingredient under the conditions it should? This result should raise some eyebrows. In the final case, if we show that it is releasing the active ingredient as it should, should we be concerned that it misses the mark under extreme solubilization conditions?

Are we looking at dissolution the wrong way? For years we have accepted that the dissolution specification of an IR tablet is 85% release in 15 min in 0.1 M HCl or even water, but is that at all reflective of how well that tablet will work for patients? Only a portion of the digestive tract is at low pH. What if as a tablet travels through a (compromised) digestive system, the conditions presented are insufficient to dissolve the active ingredient. How can we understand the absorption/PK of the tablet if we are performing QC testing under artificially optimal conditions, to which the drug product being tested may never be exposed? Would a dissolution specification of 60% release in 45 min in biorelevant dissolution media be considered unacceptable if this method very accurately predicted the drug product's efficacy and, from there, also quality?

There is a lot of discussion on this point in the industry and the current guidances do allow for the use of biorelavent media, but nothing concrete has been issued from the regulatory bodies allowing deviations from the "typical" specifications. The science, however, is emerging to support such arguments, so stay tuned!

17.4 Emerging Regulatory Topics

As novel dosage forms emerge, so will new technologies and equipment for in vitro release testing, and so will the regulations associated with the new technologies and equipment. Novel dosage forms include formulations other than oral solid dosage forms (pulmonary, etc.) that will require adaptation to existing dissolution technologies or the development of new dissolution technologies. The adapted or new dissolution technologies will require qualification, method and specification development, and validation and relevance of in vitro testing to predicting bioequivalence and subsequent BCS-based biowaivers. Modeling and statistical simulation related to in vitro release and dissolution testing and its relationship to clinical outcomes will provide a deeper understanding of the IVIV correlation of drug products. In addition, the principles of quality by design would encourage the use of BCS-based dissolution early in product development and utilize the data throughout the product's lifecycle to understand the IVIVC. Continued harmonization and communication of dissolution testing in the pharmacopeias and ICH will benefit the global community regarding regulatory expectations for current in vitro release testing and dissolution as well as lay the groundwork for appropriately developing technologies that would assist in dissolution testing of novel drugs and delivery technologies.

17.5 Relevant Guidelines

- 21 CFR §211
- 21 CFR §314
- 21 CFR §320
- 21 CFR §343
- Center for Drug Evaluation and Research (CDER) (1997). Guidance for industry: Dissolution testing of immediate release solid oral dosage forms. US Department of Health and Human Services, Food and Drug Administration.

- ICH Harmonized Tripatriot Guideline (1999). Specifications: Test procedures and acceptance criteria for new drug substances and a new drug products: Chemical substances, Q6A
- USP <711> Dissolution
- USP <1088> In vitro and in vivo Evaluation of Dosage Forms appears in 1995. This has the important Level A to C correlation information.
- USP <1090> Assessment of Drug Product Performance—Bioavailability, Bioequivalence, and Dissolution appears in 2009.
- USP <1092> The Dissolution Procedure: Development and Validation is from 2006.
- 2015 FDA - DRAFT Dissolution Testing and Specification Criteria for Immediate-Release Solid Oral Dosage Forms Containing Biopharmaceutics Classification System Class 1 and 3 Drugs
- 2010 ICH - Guideline Q4B annex 7 (R2) to note for evaluation and recommendation of pharmacopoeial texts for use in the ICH regions on dissolution test-general chapter.

Acknowledgments

Laila Kott would like to acknowledge Takeda Pharmaceuticals Inc. for allowing the time and resources for the writing of this chapter.

References

1. 21 CFR §211.110.
2. 21 CFR §314.53(b)(2)(5).
3. 21 CFR §343.90.
4. 21 CFR §320.24(b)(5).
5. 21 CFR §320.22(d)(3).
6. 21 CFR §320.33(e).
7. Buhse, C., and Gao, Z. (2012). Dissolution Testing: Evolving Dissolution Apparatus. Advisory Committee for Pharmaceutical Science and Clin-

ical Pharmacology. August 8, 2012. http://www.fda.gov/downloads/
AdvisoryCommittees/CommitteesMeetingMaterials/Drugs/Advisory
CommitteeforPharmaceuticalScienceandClinicalPharmacology/UCM31
5763.pdf.

8. United States Pharmacopeia and National Formulary (38/33), <711>,
   <724>, <1090>, <1092>. Rockville, MD: United States Pharmacopeia
   Convention. 2015.

9. International Conference on Harmonisation of Technical Requirements
   for Registration of Pharmaceuticals for Human Use (ICH), Q8(R2)
   Pharmaceutical Development.

10. FDA Guidance for Industry and Review Staff: Target Product Profile—A
    Strategic Development Process Tool. March 2007.

11. Food and Drug Administration. Waiver of in vivo Bioavailability and
    Bioequivalence Studies for Immediate Release Solid Oral Dosage Forms
    Based on a Biopharmaceutics Classification System: Guidance for
    Industry. Rockville, MD: US Department of Health and Human Services,
    FDA, Center for Drug Evaluation and Research. August 2000. Available
    at http://www.fda.gov/downloads/Drugs/GuidanceComplianceRegu-
    latoryInformation/Guidances/ucm070246.pdf.

12. Amidon, G. L., Lennernäs, H., Shah, V. P., and Crison, J. R. (1995).
    A theoretical basis for a biopharmaceutic drug classification: the
    correlation of the in vitro drug product dissolution and in vivo
    bioavailability, *Pharm. Res.*, **12**, 413–420.

13. Shah, V. P., and Amidon, G. L. (2014). Amidon, G. L., Lennernäs, H., Shah,
    V. P., and Crison, J. R. A theoretical basis for a biopharmaceutic drug
    classification: the correlation of the in vitro drug product dissolution
    and in vivo bioavailability, *Pharm. Res.*, **12**, 413–420; 1995—Backstory
    of BCS, *AAPS J.*, **16**, 894–898.

14. Tsume, Y., Mudie, D. M., Langguth, P., Amidon, G. E., and Amidon, G. L.
    (2014). The biopharmacuetics classification system: subclasses for in
    vivo predicitive dissolution (IPD) methodology and IVIVC, *Eur. J. Pharm.
    Sci.*, **57**, 152–163.

15. Food and Drug Administration. Generic Drug User Fee Act Program
    Performance Goals and Procedures. Rockville, MD: US Department of
    Health and Human Services, FDA, Center for Drug Evaluation and
    Research. Available at http://www.fda.gov/downloads/ForIndustry/
    UserFees/GenericDrugUserFees/UCM282505.pdf

16. Cook, J., Addicks, W., and Wu, Y. H. (2008). Application of the
    biopharmaceutical classification system in clinical drug development:
    an industrial view, *AAPS J.*, **10**, 306–310.

17. Rane, S. S., and Anderson, B. D. (2008). What determines drug solubility in lipid vehicles: is it predictable?, *Adv. Drug Deliv. Rev.*, **60**, 638–656.

18. Wu, C.-Y., and Benet, L. Z. (2005). Predicting drug disposition via application of BCS: transport/absorption/elimination interplay and development of a biopharmaceutics drug disposition classification system, *Pharm. Res.*, **22**, 11–23.

19. Butler, J. M., and Dressman, J. B. (2010). The developability classification system: Application of the biopharmaceutics concepts to formulation development, *J. Pharm. Sci.*, **99**, 4940–4954.

20. Zhang, X., Zheng, N., Lionberger, R. A., and Yu, L. X. (2013) Innovative approaches for the demonstration of bioequivalence: the US FDA perspective, *Ther. Deliv.*, **4**, 725–740.

21. Dressman, J., Butler, J., Hempenstall, J., and Reppas, C. (2001). Then BCS:Where do we go from here?, *Pharm. Tech.*, **25**, 68–76.

22. Karalis, V., Magklara, E., Shah, V. P., and Macheras, P. (2010). From drug delivery systems to drug release, dissolution, IVIVC, BCS, BDDS, bioequivalence and biowaivers, *Pharm. Res.*, **27**, 2018–2029.

23. Chen, M.-L., Amidon, G. L., Benet, L. Z., Lennernas, H., and Yu, L. X. (2011) The BCS, BDDCS, and regulatory guidances, *Pharm. Res.*, **28**, 1774–1778.

24. Polli, J. E., Abrahamsson, B. S. I., Yu, L. X., Amidon, G. L., Baldoni, J. M., Cook, J. A., Fackler, P., Hartauer, K., Johnston, G., Krill, S. L., Lippers, R. A., Malick, W. A., Shah, V. P., Sun, D., Winkle, H. N., Wu, Y., and Zhang, H. (2008). Summary workshop report: bioequivalence, biopharmaceutics classification system, and beyond, *AAPS J.*, **10**, 373–379.

25. Gratten, T., Hickman, R., Darby-Dowman, A., Hayward, M., Boyce, M., and Warrington, S. (2000). A five way crossover human volunteer study to compare the pharmacokinetics of paracetamol following oral administration of two commercially available paracetamol tablets and three development tablets containing paracetamol in combination with sodium bicarbonate or calcium carbonate, *Eur. J. Pharm. Biopharm.*, **49**, 225–229.

26. Müllertz, A. (2007). *Solvent Systems and Their Selection in Pharmaceutics and Biopharmaceutics*, ed. Augustijns, P., and Brewster, M. E. Chapter 6: Biorelevant dissolution media (Springer, New York), pp. 151–177.

27. Krämer, J. (2013). AAPS workshop report on biorelevant in vitro performance testing of orally administered dosage forms, *Dissolut. Technol.*, **20**, 45–48.

28. Galia, E., Nicolaides, E., Hörter, D., Löbenberg, R., Reppas, C., and Dressman, J. B. (1998). Evaluation of various dissolution media for

prediction in vivo performance of class I and class II drugs, *Pharm. Res.*, **15**, 698–705.

29. Jantratid, E., Janssen, N., Reppas, C., and Dressman, J. B. (2008). Dissolution media simulating conditions in the proximal human gastrointestinal tract: an update, *Pharm. Res.*, **25**, 1663–1676.

30. Klein, S. (2010). The use of biorelevant dissolution media to forecast the in vivo performance of a drug, *AAPS J.*, **12**, 397–406.

31. Marques, M. (2004). Dissolution media simulating fasted and fed states, *Dissolut. Technol.*, **11**, 16.

32. Gowthamarajan, K., and Singh, S. K. (2010). Dissolution testing for poorly soluble drugs: a continuing perspective, *Dissolut. Technol.*, **17**, 24–32.

33. ICH Harmonized Tripatriot Guideline (1999). Specifications: Test Procedures and Acceptance Criteria for New Drug Substances and a New Drug Products: Chemical Substances, Q6A.

34. Center for Drug Evaluation and Research (CDER) (1997). Guidance for Industry: Dissolution Testing of Immediate Release Solid Oral Dosage Forms. US Department of Health and Human Services, Food and Drug Administration.

35. Piscitelli, D. A., and Young, D. (1997) *In vitro—in vivo Correlations*, ed. Young, D. B., Devane, J. G., and Butler, J. Chapter 13: Setting dissolution specifications for modified-release dosage forms (Plenum Press, New York), pp. 159–166.

36. Anand, O., Yu, L. X., Conner, D. P., and Davit, B. M. (2011). Dissolution testing for generic drugs: an FDA perspective, *AAPS J.*, **13**, 328–335.

37. Marroum, P. J. (2012) Clinically relevant dissolution methods and specifications, *Am. Pharm. Rev.*, **15**, 36–41.

Chapter 18

Dissolution of Liquid-Filled Capsules Based Formulations

Rampurna Prasad Gullapalli

Dart NeuroScience LLC, 11278 Scripps Summit Drive, San Diego, CA 92131, USA
rampurnal@gmail.com

18.1 Introduction

Absorption of a compound from the gastrointestinal tract (GIT) requires the availability of the compound in a solution or solubilized form and its diffusion into and across the enterocytes lining the intestinal lumen (Fig. 18.1).[1,2] The advent of combinatorial chemistry and high-throughput screening (HTS) resulted in the identification of many highly potent compounds that usually have less than desirable biopharmaceutical properties, e.g., low aqueous solubility and high lipophilicity.[3,4] These compounds exhibit extremely low aqueous solubility throughout the physiological pH range, resulting in low and inconsistent oral bioavailability.[5,6] Poor aqueous solubility has been identified as the single largest physicochemical challenge for the oral absorption of compounds

Poorly Soluble Drugs: Dissolution and Drug Release
Edited by Gregory K. Webster, J. Derek Jackson, and Robert G. Bell
Copyright © 2017 Pan Stanford Publishing Pte. Ltd.
ISBN 978-981-4745-45-1 (Hardcover), 978-981-4745-46-8 (eBook)
www.panstanford.com

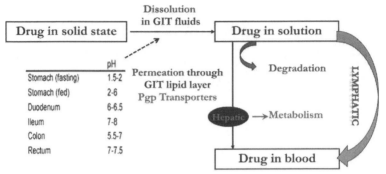

Figure 18.1 Schematic of factors influencing bioavailability of an orally administered compound.

and almost inevitably leads to their lower oral bioavailabilities from the conventional dosage forms (e.g., tablets, powder-filled capsules).[3]

The simplest way to present a compound to the GIT for absorption is to administer the compound as a solution or solubilized form, thereby removing any dissolution rate-limiting step in the absorption process.[2] As the compound is already in solution at the site of absorption, it could yield a faster, uniform, and enhanced absorption. As an alternative to a solution, some compounds with poor bioavailability were shown to provide exceptionally high bioavailability when dosed as non-aqueous suspensions.[7–10] For example, studies demonstrated that griseofulvin and phenytoin suspended in corn oil resulted in a significantly higher oral bioavailability compared to administration of an aqueous suspension in rodent model.[7,8] Larsen et al.[9] demonstrated danazol suspended in a lipid vehicle performed just as well as the lipid solutions. When a compound demonstrates sufficient solubility in a pharmaceutically acceptable non-aqueous vehicle and/or formulation of the compound in the non-aqueous vehicle is shown to be the practical option (a) to achieve its acceptable oral bioavailability, (b) dictated by its physical nature (e.g., low melting point, oily, waxy), or (c) dictated by its ultra-low to low dose requirements, liquid-filled capsules offer the advantage of conveniently delivering the non-aqueous formulation containing the compound as a unit dose solid dosage form.[11]

The availability of a compound formulated in a liquid-filled capsule for absorption depends on the initial dissolution, then rupture of the capsule shell and subsequent release and dissolution of its fill contents in the GIT fluids. These two processes need to be monitored at the time of release and during the shelf life of a capsule product. Liquid-filled capsules pose unique challenges during the development and application of dissolution methods due to the complex nature of the shell and fill materials. The shell material is prone to changes in its mechanical properties or cross-linking of gelatin, which results in changes in its solubility. The fill material, on the other hand, may exhibit changes in particle size distributions and/or polymorphic nature of the suspended material in a suspension fill or crystallization of a solubilized compound from a solution fill. In the latter case, the crystallization of the solubilized compound can occur either in the capsule dosage form or when the fill material encounters in vitro and in vivo aqueous fluids.[12–19]

Dissolution testing is a highly valuable tool to characterize liquid-filled capsule products in vitro and is used routinely (a) to assess batch-to-batch quality, (b) to monitor changes in the quality of a product during its shelf life, (c) to assess product sameness after scale-up and post approval changes (SUPAC),[20] (d) to comply with biowaiver requirement for a lower strength of a product,[21–23] and (e) to comply with biowaiver requirement for a product intended for local action in the GIT.[24] In addition, a dissolution method designed to produce in vitro–in vivo correlations (IVIVC) or in vitro–in vivo relationship (IVIVR) can be used to predict potential bioequivalency or bioinequivalency between products. The intent of this chapter is to provide an in-depth discussion on the factors affecting in vitro and in vivo dissolution of liquid-filled capsule products, development of dissolution methods for their routine quality control (QC) testing, and modifications to these methods to produce potential IVIVC and IVIVR.

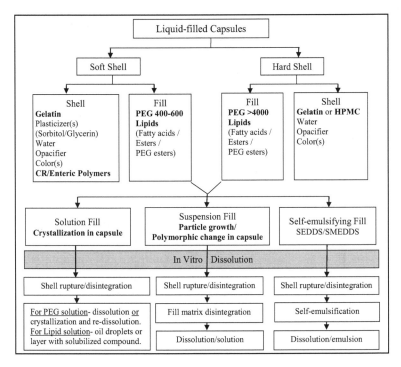

Figure 18.2 Schematic of typical composition of liquid-filled capsules and in vitro dissolution processes. Factors shown in bold can potentially effect dissolution.

18.1.1 Overview of Liquid-Filled Capsules

Liquid-filled capsules can be soft shell capsules (softgels) or hard shell capsules (gelatin or hydroxypropyl methylcellulose (HPMC)) (Fig. 18.2). Gelatin, the major component of gelatin capsule shells, is obtained by the partial hydrolysis of collagen derived from the skin, connective tissue, and bones of cattle, pigs, and fish. Gelatin contains a mixture of water-soluble proteins (84–90%), mineral salts (1–2%), and water (8–15%). The protein fraction consists mostly of amino acids (Table 18.1) linked by amide (peptide) bonds forming a linear polymer with a molecular weight ranging from 15,000 to 250,000 Daltons.[25–28] The high glass transition temperature of anhydrous gelatin (Tg > 100°C)[29–39] prevents it from forming a flexible and acceptable film readily during the manufacturing of

Table 18.1 Typical amino acid composition of gelatins from various sources (number of amino acid residues per 1000 total residues)[226]

Amino acid	Porcine skin (acid treated)	Bovine skin (alkali treated)	Bovine bone (alkali treated)
Alanine	111.7	112.0	116.6
Arginine	49.0	46.2	48.0
Aspartic acid	45.8	46.0	46.7
Glutamic acid	72.1	70.7	72.6
Glycine	330.0	333.0	335.0
Histidine	4.0	4.5	4.2
Hydroxylysine	6.4	5.5	4.3
Hydroxyproline	90.7	97.6	93.3
Isoleucine	9.5	12.0	10.8
Leucine	24.0	23.1	24.3
Lysine	26.6	27.8	27.6
Methionine	3.6	5.5	3.9
Phenylalanine	13.6	12.3	14.0
Proline	131.9	129.0	124.2
Serine	34.7	36.5	32.8
Threonine	17.9	16.9	18.3
Tyrosine	2.6	1.5	1.2
Valine	25.9	20.1	21.9

Arginine and lysine are involved in aldehyde induced gelatin cross-linking.

gelatin capsule shells. Water is an effective plasticizer for gelatin and reduces the Tg of gelatin proportionally to its water content.[29–32] However, due to its volatile nature, water will be lost during the drying process or on storage resulting in a brittle and fragile shell. Thus, non-volatile plasticizers (e.g., glycerin, sorbitol-sorbitan solution) are included in the production of gelatin ribbons for softgel capsules.

The shells of hard capsules are generally thinner and lighter than those of softgel capsules and are commonly manufactured from gelatin (or a non-gelatin material such as HPMC), water, and other minor additives such as opacifiers and colorants. Hard capsule shells, unlike those of softgel capsules, do not contain a non-volatile plasticizer and thus the presence of water (volatile plasticizer) in the shells is essential to maintain their integrity. Hard gelatin capsule shells must retain a moisture content of 10 to 18 percent to maintain their integrity and flexibility.[40,41] Below this range, the shells become brittle and are prone to breakage, while above this range, the shells may deform and become sticky. Thus, any alteration to this moisture range, due to migration of moisture between the shell and encapsulated fill or between the shell and external environment, could be detrimental to the integrity of the hard gelatin capsule shell. In contrast, the presence of a non-volatile plasticizer in the softgel capsule shell imparts elasticity to the shell that allows it to accommodate a wide range of fluctuations in its moisture content.

Types of fill formulations that can be encapsulated into capsules include solutions, suspensions, pastes, self-emulsifying drug delivery systems (SEDDS), and self-microemulsifying drug delivery systems (SMEDDS) (Fig. 18.2). Vehicles suitable for encapsulation into capsules can be broadly classified into (a) hydrophilic and (b) lipophilic (lipid based formulations).

Hydrophilic vehicles for softgel capsule fill formulations include lower molecular weight polyethylene glycols (e.g., PEG 400, PEG 600) and may also contain small amounts of propylene glycol, glycerin, ethyl alcohol, and water (Table 18.2). Polyethylene glycols of molecular weight higher than 4000 are suitable for encapsulating into hard gelatin capsules (Table 18.3).[11,42] Polyethylene glycols are ideal vehicles for solubilizing many poorly water-soluble compounds. However, the high affinity of these vehicles for water can potentially lead to the precipitation of the solubilized compounds when the formulations come into contact with an aqueous environment in vitro or in vivo.[12–19] This phenomenon may result in a higher variability in the dissolution profiles.

Lipophilic vehicles for softgel capsule and hard capsule fill formulations include free fatty acids (e.g., oleic acid), fatty acid esters of hydroxyl compounds such as ethyl alcohol, propylene glycol, glycerin, sorbitol, sucrose, polyethylene glycols, and polyethoxylated fatty acid esters (referred to as emulsifiers) (Tables 18.2 and 18.3). The fatty acid composition of these esters may vary from short chain (SC; <C8) to medium chain (MC; C8–C10) to long chain (LC; ≥C12). Lipid and other high molecular weight excipients are known to exist in several polymorphs. Polymorphism may not exist in excipients in liquid state. However, when a solid lipid matrix is designed to provide controlled release through

Table 18.2 Examples of liquid-filled softgel capsules requiring dissolution test as the performance test

Product	Fill excipients	Dissolution method
Agenerase® Amprenavir 50 mg & 150 mg GlaxoSmithKline	PEG 400 d-α-tocopheryl PEG 1000 succinate (TPGS) Propylene glycol	Apparatus 1 @ 50 rpm in 900 mL of 0.1N HCl @ 37 $\pm$ 0.5°C.
Targretin® Bexarotene 75 mg Eisai/Ligand	PEG 400 Polysorbate 20 Povidone K-90 BHA	Apparatus 2 @ 50 rpm in 900 mL of 0.05 M phosphate buffer, pH 7.5 containing 0.5% hexadecyltrimethylammonium bromide (HDTMA) @ 37 $\pm$ 0.5°C.
Zyrtec® Cetirizine 2HCl 10 mg McNeil Consumer	PEG 400 Sodium hydroxide Water	Apparatus 2 @ 50 rpm in 900 mL of water @ 37 $\pm$ 0.5°C.
Sandimmune® Cyclosporin A 25 mg, 50 mg & 100mg Novartis	Corn oil Ethyl alcohol Linoleoyl macrogolglycerides	Apparatus 2 @ 75 rpm in 500 mL (for 25 mg dose) or 1000 mL (for 100 mg dose) of 0.1 N HCl containing 2 mg/mL (for 25 mg dose) or 4 mg/mL (for 100 mg dose) of N, N-dimethydodecylamine N-oxide @ 37 $\pm$ 0.5°C.
Neoral® Cyclosporin A 25 mg & 100 mg Novartis	Corn oil mono-, di- and triglycerides Ethyl alcohol Cremophor RH 40 Propylene glycol dl-α-tocopherol	
Lanoxicap® Digoxin 0.05 mg, 0.1 mg & 0.2 mg GlaxoSmithKline	PEG 400 Ethyl alcohol Propylene glycol Water	Apparatus 1 @ 120 rpm in 600 mL of water @ 37 $\pm$ 0.5°C.
Marinol® Dronabinol 2.5 mg, 5 mg & 10 mg Unimed and Roxane	Sesame oil	Apparatus 2 @ 100/150 rpm in 500 mL of water, containing 10% Labrasol @ 37 $\pm$ 0.5°C.

(Contd.)

Table 18.2   (*Contd.*)

Product	Fill excipients	Dissolution method
Avodart® Dutasteride 0.5 mg GlaxoSmithKline	Medium chain mono- and diglycerides BHT	Apparatus 2 @ 50 rpm in 900 mL of 0.1 N HCl containing 2% sodium dodecyl sulfate @ 37 ± 0.5°C.
Zarontin® Ethosuximide 250 mg Pfizer	PEG 400	Apparatus 1 @ 50 rpm in 900 mL of phosphate buffer, pH 6.8 @ 37 ± 0.5°C.
VePesid® Etoposide 50 mg Bristol-Myers-Squibb	PEG 400 Glycerin Water Citric acid	Apparatus 2 @ 50 rpm in 900 mL of acetate buffer, pH 4.5 @ 37 ± 0.5°C.
Advil® Ibuprofen 200 mg Wyeth	PEG 600 Potassium hydroxide Water	Apparatus 1 @ 150 rpm in 900 mL of phosphate buffer, pH 7.2 @ 37 ± 0.5°C.
Advil PM Liqui-Gels® Ibuprofen 200 mg/ Diphenhydramine HCl 25 mg Wyeth	PEG 600 Potassium hydroxide Water	Apparatus 1 @ 100 rpm in 900 mL of phosphate buffer, pH 7.2 @ 37 ± 0.5°C.
Accutane® Isotretinoin 10 mg, 20 mg & 40 mg Roche	Soybean oil Beeswax Hydrogenated soybean oil Hydrogenated vegetable oil BHA Edetate disodium	Apparatus 1 @ 100 rpm in 900 mL of 0.05 M potassium phosphate dibasic buffer, pH 7.8 containing 0.5% lauryldimethylamine-oxide (LDAO) @ 37± 0.5°C.
Kaletra® Lopinavir 133.3 mg Ritonavir 33.3 mg Abbott	Oleic acid Cremophor EL Propylene glycol	Apparatus 2@ 50 rpm in 900 mL of 10 mM sodium phosphate monobasic solution containing 0.05 M polyoxyethylene 10 lauryl ether, pH 6.8 @ 37 ± 0.5°C.
Oxsoralen-Ultra® Methoxsalen 10 mg Valeant Pharmaceuticals	PEG 400	Apparatus 2 @ 50 rpm in 900 mL of Water @ 37 ± 0.5°C.

(*Contd.*)

Product	Fill excipients	Dissolution method
Aleve® Naproxen Sodium 220 mg Bayer HealthCare	PEG 400 Propylene glycol Povidone Lactic acid	Apparatus 2 @ 50 rpm in 900 mL of 0.1 M phosphate buffer (pH 7.4) @ 37 ± 0.5°C.
Procardia® Nifedipine 10 mg & 20 mg Pfizer	PEG 400 Glycerin Peppermint oil Sodium saccharin	Apparatus 2 @ 50 rpm in 900 mL of SGF @ 37 ± 0.5°C.
Nimotop® Nimodipine 30 mg Bayer	PEG 400 Glycerin Peppermint oil Water	Apparatus 2 @ 50 rpm in 900 mL of water, containing 0.5% sodium dodecyl sulfate @ 37 ± 0.5°C.
Zantac® Ranitidine HCl 150 mg & 300 mg GlaxoSmithKline	Medium chain triglycerides Gelucire® 33/01	Apparatus 2 @ 50 rpm in 900 mL of water @ 37 ± 0.5°C.
Norvir® Ritonavir 100 mg Abbott	Oleic acid Ethyl alcohol Cremophor EL BHT	Apparatus 2 @ 50 rpm in 900 mL of 0.1 N HCl, containing 0.025 M polyoxyethylene 10 lauryl ether @ 37 ± 0.5°C.
Fortovase® Saquinavir 200 mg Roche	Medium chain mono- and diglycerides Povidone dl-α-tocopherol	Apparatus 2 @ 50 rpm in 900 mL of citrate buffer containing 0.582% anhydrous dibasic sodium phosphate and 1.67g citric acid monohydrate @ 37± 0.5°C.
Aptivus® Tipranavir 250 mg Boehringer Ingelheim	Medium chain mono- and diglycerides Cremophor EL Propylene glycol Ethyl alcohol	Apparatus 2 @ 50 rpm in 900 mL of 0.05 M phosphate buffer pH 6.8 @ 37 ± 0.5°C.
Depakene® Valproic acid 250 mg Abbott	Neat active/no excipient.	Apparatus 2 @ 50 rpm in 900 mL of SIF TS without enzyme and with monobasic sodium phosphate (instead of monobasic potassium phosphate) and pH adjusted to 7.5 with 5M sodium hydroxide; containing 0.5% sodium dodecyl sulfate @ 37± 0.5°C.

Information was collected from Label, Prescription Information, FDA Approval Packages/ Dissolution Database and USP 36/NF 31 Monographs.

Table 18.3 Examples of liquid-filled hard capsules requiring dissolution test as the performance test

Product	Fill excipients	Dissolution method
Gengraf® Cyclosporine 25 mg & 100 mg AbbVie	Polyethylene glycol Cremophor EL Polysorbate 80 Propylene glycol Sorbitan monooleate Ethyl alcohol	Apparatus 2 @ 75 rpm in 500 mL (for 25 mg dose) or 1000 mL (for 100 mg dose) of 0.1 N HCl containing 4mg/mL of N, N-dimethydodecylamine N-oxide @ $37 \pm 0.5°C$.
Lipofen® Fenofibrate 50 mg & 150 mg Cipher Pharms	Gelucire 44/14 PEG 20000 PEG 8000 Hydroxypropyl cellulose Sodium starch glycolate	Apparatus 2 @ 75 rpm in 900 mL of pH 6.8 phosphate buffer containing 0.1% pancreatin and 2% Polysorbate 80 @ $37 \pm 0.5°C$.
Claravis® Isotretinoin 10 mg, 20 mg, 30 mg & 40 mg Teva	Hydrogenated vegetable oil Polysorbate 80 Soybean oil Beeswax Vitamin E BHA Edetate disodium	Apparatus 1 @ 100 rpm in 900 mL of 0.05 M potassium phosphate dibasic buffer, pH 7.8 containing 0.5% lauryldimethylamine-oxide (LDAO) @ $37 \pm 0.5°C$.
Hycamtin® Topotecan HCl 0.25 mg & 1 mg free base GlaxoSmithKline	Hydrogenated vegetable oil Glyceryl monostearate	Apparatus 2 @ 50 rpm in 500 mL of acetate buffer with 0.15% SDS, pH 4.5 @ $37 \pm 0.5°C$.
Vancocin® Vancomycin HCl 125 mg & 250 mg free base Viropharma	PEG 6000	Apparatus 1 @ 100 rpm in 900 mL of all three 0.1 N HCl (or 0.1 N HCl with NaCl at pH 1.2); pH 4.5 acetate buffer; and pH 6.8 phosphate buffer @ $37 \pm 0.5°C$.

Information was collected from Label, Prescription Information, FDA Approval Packages/ Dissolution Database and USP 36/NF 31 Monographs.

erosion mechanism or increased consistency (i.e., viscosifier) in a suspension fill formulation, any changes in the polymorphic nature of the matrix can have significant affect on the release properties of a compound.[43-48]

When a compound is soluble and has demonstrated stability in a non-aqueous vehicle, it can be encapsulated into capsules as a solution. Compounds that do not have sufficient solubility may require encapsulation as suspensions. In a suspension formulation, the dispersed material may undergo *Ostwald ripening* and/or *secondary nucleation*,[49] resulting in changes in the particle size distributions and/or polymorphic nature during the shelf life of a capsule product. These changes can potentially result in differences in the rate of dissolution and bioavailability with time.[50-54]

18.2 Dissolution of Liquid-Filled Capsules

18.2.1 Dissolution of Capsule Shell

The originally invented HPMC capsule shells contain a secondary gelling agent such as kappa-carrageenan (Quali-V;[®] Qualicaps) or gellan (Vcaps;[®] Capsugel) and potassium ions as gelling promoter, which are known to delay the shell dissolution in vitro and in vivo in some circumstances.[55,56] The dissolution of these HPMC capsule shells is dependent on both pH and composition of the dissolution media and usually longer than that of gelatin capsules. The delay in the dissolution of HPMC capsule shells is attributed to the interactions between the gelling agent and cations such as potassium and calcium present in the dissolution media. The newer HPMC capsule shells (Vcaps[®] Plus; Capsugel) without the secondary gelling agent and gelling promoter exhibit dissolution independent of both pH and ionic media.[57-59] The dissolution of the HPMC capsule shells is not influenced by temperatures between 10°C and 55°C in dissolution media with a pH of 5.8 or lower. In contrast, gelatin capsule shells are known to be soluble in aqueous media at body temperature, but insoluble at temperatures below 30°C.[60] The dissolution times of both gelatin and HPMC capsule shells are usually prolonged and more variable in phosphate buffer of pH 6.8.[60]

In vivo gamma scintigraphic studies in human subjects by Tuleu et al. (61) and Jones et al. (62) comparing HPMC capsules (with kappa-carrageenan as secondary gelling agent) and hard gelatin capsules suggested no significant differences in the disintegration times in either the fasting or fed state (mean $\pm$ SD): 8 ± 2 min and 7 ± 3 min for HPMC and gelatin capsules, respectively, under fasting condition and 16 ± 5 min and 12 ± 4 min for HPMC and gelatin capsules, respectively, under fed condition. In contrast, in vivo disintegration studies by Cole et al.[55] using gamma scintigraphic technique demonstrated delayed capsule disintegration (both initial and complete) for HPMC capsules (with gellan as secondary gelling agent) compared to that of gelatin capsules, i.e., initial disintegration at 0.47 ± 0.17 h vs. 0.13 ± 0.06 h and complete disintegration at 0.69 ± 0.29 h vs. 0.24 ± 0.14 h for HPMC and gelatin capsules, respectively, under fasting condition. Whereas, under fed condition, the initial disintegration times were 1.00 ± 0.37 h and 0.39 ± 0.36 h and the complete disintegration times were 1.61 ± 0.65 h and 1.22 ± 0.80 h for HPMC and gelatin capsules, respectively. However, these differences in the capsule disintegration times between HPMC and gelatin capsules were found to have no significant effect on the $AUC_{(0-\infty)}$ and C_{max} of the encapsulated compound in the study subjects.

Problems in the dissolution and rupture of gelatin capsule shells may become apparent (a) upon aging; (b) when exposed to physical conditions such as heat,[63,64] high temperature and humidity,[65] UV radiation,[66,67] γ-radiation,[68–72] light,[67,73,74] and rapid drying;[75] (c) when exposed to chemical substances such as aldehydes, ketones, imines, and carbodiimides;[76–88] and (d) when exposed to multivalent metal ions (e.g., Mg^{2+}, Fe^{3+}, Al^{3+}) present in the fill formulations or dyes used as colorants. These problems are attributed to cross-linking of gelatin (*pellicle formation*) that causes the gelatin shell to become swollen, tough, rubbery, and insoluble in water. Cross-linking of gelatin gives rise to the formation of a very thin film during the dissolution testing of a capsule product. The film is mechanically weak and can easily be punctured. However, the film does not disrupt easily with gentle agitation under normal dissolution conditions.[89]

18.2.1.1 Mechanism of gelatin cross-linking

The ultimate effects of exposure to elevated temperatures and relative humidities on the physical properties of the gelatin shell are known to be similar to those of exposure to aldehydes, though the mechanisms of chemical reactions occurring within the gelatin are distinctly different.[90, 91] Physically, exposure of gelatin shell to either elevated temperatures and relative humidities or aldehydes results in a decrease of its dissolution, swelling, and rate of water vapor transmission and an increase of its gel strength, a clear indication of the formation of three-dimensional networks within the gelatin.

Chemically, aldehydes are known to form methylene bonds between two amino groups on adjacent gelatin chains or within the same chain. The aldehyde induced cross-linking of gelatin is thought to involve the ε-amino functional groups present in the lysine moieties and the guanidino functional groups present in the arginine moieties of the gelatin chain.[77,78,92−97] In contrast, exposure of gelatin to elevated temperatures and relative humidities results in the formation of amide and ester bonds as illustrated below:[63,64]

$$Chain\ 1 - COOH + HO - Chain\ 2 \rightleftharpoons Chain\ 1 - CO - O - Chain\ 2 + H_2O$$

$$Chain\ 1 - COOH + H_2N - Chain\ 2 \rightleftharpoons Chain\ 1 - CO - NH - Chain\ 2 + H_2O$$

Whether it is thermally induced or chemically induced, cross-linking of gelatin results in the formation of three-dimensional molecular networks of a higher molecular weight with the loss of ionizable groups (i.e., R-NH$_2$, R-COOH and R-OH) than the original molecules, leading to the reduced aqueous solubility of gelatin.

The loss of ionizable groups in gelatin and the resulting decrease in its solubility may also arise from the ionic, hydrogen, and van der Waals interactions of these groups with other compounds used concurrently in the shell formulation, such as FD&C red # 3 and FD&C red # 40 dyes.[98,99] Exposure of gelatin to multivalent metal ions also results in the formation of weaker type cross-links involving the carboxylic acid groups of gelatin molecules and the metal ions.

18.2.1.2 Sources of aldehyde impurities

Aldehyde impurities in liquid-filled capsules may originate from the auto-oxidation of materials containing polyoxyethylene moieties in their structures (e.g., polyethylene glycols, methoxypolyethylene glycols, polyoxyethylene fatty acid esters).[100–106] Rayon coiler, included in the package presentations, is known to produce furfural (2-furaldehyde) under the accelerated stability conditions that could potentially cross-link gelatin.[107,108] Compounds containing carbonyl functional groups in their structures (e.g., nimesulide, rofecoxib, macrolide antibiotics) or compounds degrading into carbonyl impurities (e.g., hydrochlorothiazide) may also induce the cross-linking of gelatin.[109–111]

Polyethylene glycols and materials containing polyoxyethylene moieties in their structures are inherently susceptible to auto-oxidative degradation due to the presence of recurring ether groups in the polymer chains. These materials are known to undergo autooxidation in the presence of air (oxygen), moisture, heat, and light and produce reactive hydroperoxides (HPO) such as organic peroxides and hydrogen peroxide.[98–104,112–121] The organic peroxides further degrade to produce short chain aldehydes and carboxylic acids.[119–123] The amounts of these reactive peroxides, aldehydes, and carboxylic acids present in a polyethylene glycol or in a related material are dependent upon the molecular weight and age of the polymer and the extent of its exposure to air, temperature, humidity, and light during its storage and handling. The lower the molecular weight and the older the polymer, the greater the concentrations of these products. The relatively higher levels of these reactive products observed in a polyoxyethylene polymer of a lower molecular weight may be related to the increase in the mobility of the polymeric chains and increase in the diffusion rates of the reactants (i.e., radicals and oxygen) within the polymer due to its lower viscosity, thus allowing for the increased autooxidation reactions.[124] The peroxide, formic acid, and aldehyde levels reported in some commonly used pharmaceutical excipients are presented in Table 18.4.

Lipid excipients used in capsule fill formulations are also known to undergo autooxidative degradation to generate peroxides.[125–128]

Table 18.4 Impurity levels reported in some pharmaceutical excipients commonly used in liquid-filled capsules

Excipient	Level, ppm			
	Peroxides [227]	Formic acid [228]	Formaldehyde [124, 228]	Acetaldehyde [124]
Propylene glycol			0.3	0.5
PEG 200			107	12.5
PEG 400	1.7–59	469	102.5; 85.8	2.7
PEG 600			65.2	12.2
PEG 3500		1.7–2.3	0.3–1.2	
PEG 4000		1.7–14	0.4; 0.9–3.6	2.3
Diethyleneglycol monoethylether			1	2.8
Polysorbate 80			5.2	ND
Cremophor EL	0–2.9		3.5	ND
Cremophor RH 40	1.0			
Poloxamer 407			0.7	0.6
PEG 400 caprylate/ caprate glycerides			3.9–22.9	0.8–30.7
Vitamin E TPGS			0.2	ND
Medium chain mono- and di-glycerides			0.2	ND
Medium chain triglycerides			ND	ND
Corn oil			ND	ND
Povidone K-12	27			
Povidone K-17	65			
Povidone K-25/30	188	1990.5–3080.3	<0.2–0.4	
Povidone K-90		630.7	0.4	

ND: Not detected

The two common measures of peroxidation in lipids and oils are peroxide value and anisidine value. Peroxide values (primary oxidation products, e.g., hydroperoxides) of lipids and oils are known to reach a maximum and then decrease as the peroxides further degrade into secondary oxidation products (i.e., aldehydes), resulting in increased anisidine values.

Cross-linking caused by agents present in the fill results in the formation of a pellicle on the internal surface of the capsule shell, whereas, cross-linking caused by agents in contact with capsules, such as rayon coiler used as a packaging component, results in the formation of a pellicle on the external surface of the capsule shell. With increasing exposure time to the cross-linking agents, the cross-linking reaction is thought to proceed from the surface into the inside of the shell material, resulting in increased suppression of its aqueous solubility.[85,86,97] The extent of cross-linking may not be uniform within one capsule or among different capsules of the same batch, thus results in a higher variability in the dissolution results if the gelatin capsules are cross-linked.

18.2.1.3 Influence of proteolytic enzymes on cross-linked gelatin

The USP Chapters <711>[129] and <2040>[130] provide guidance and procedures for dissolution testing for dosage forms administered orally. For gelatin capsules that failed tier 1 dissolution testing (i.e., in a medium with no enzymes) due to *demonstrated* gelatin cross-linking, the USP recommends the use of proteolytic enzymes (i.e., pepsin or pancreatin) in the dissolution medium during tier 2 dissolution testing. USP Chapters <711> and <2040> and Gelatin Capsules Working Group[131] recommend use of the same dissolution medium during tier 2 testing as used in tier 1 dissolution testing, but with the addition of pepsin at 750,000 Units or less of activity per 1000 mL of an acidic medium or water or pancreatin at 1750 units or less of proteolytic activity per 1000 mL of a medium at or above pH 6.8. The use of proteolytic enzymes in the dissolution medium is known to offer an advantage in improving dissolution of only cross-linked gelatin capsules[129–132] but have no effect on the dissolution of uncross-linked gelatin capsules[133] or capsules made with non-gelatin material.[60] The use of these proteolytic enzymes in the dissolution medium is justified on the grounds that such enzymes are also present in the GIT.[65,84,134–136]

Gelatin capsules that failed the tier 1 dissolution testing but passed the tier 2 dissolution testing (Fail/Pass) were shown to be bioequivalent to those, which passed the tier 1 dissolution testing.[137–139] In a bioequivalence study using GI198745/dutasteride soft gelatin capsules in healthy male volunteers, capsules with reduced rate of drug release during in vitro testing due to gelatin cross-linking were found to be bioequivalent to fresh, non-cross-linked capsules.[137] Bexarotene (Targretin™) soft gelatin capsules that developed pellicles in dissolution testing were shown to display pharmacokinetics similar to those that did not form pellicles.[138] However, in few instances, capsules that failed both tiers of dissolution testing (Fail/Fail) were shown to be bioequivalent to those, which passed either one or both tiers of dissolution testing.[84, 140] In a bioequivalence study performed using acetaminophen soft and hard gelatin capsules, the moderately stressed capsules (Fail/Pass) were shown to be bioequivalent to the unstressed capsules, whereas the severely stressed capsules (Fail/Fail) were bioinequivalent to the unstressed capsules.[84] In some cases, though severe cross-linking (Fail/Fail) may not adversely affect the bioavailability (AUC and C_{max}) of a compound, but it may potentially delay the onset of its absorption (T_{max}) due to the delayed shell rupture in vivo.[141] Digenis et al.[141] studied the influence of extent of gelatin cross-linking on the in vivo rupture times of hard gelatin capsules containing amoxicillin in human subjects using gamma scintigraphy technique. The studies demonstrated delayed capsule rupture time in vivo with increase in the extent of gelatin cross-linking, i.e., at 7 ± 5 min, 22 ± 12 min, and 31 ± 15 min for unstressed (Pass/Pass), moderately stressed (Fail/Pass), and severely stressed (Fail/Fail) capsules, respectively, under fasting condition. Whereas, under fed condition, the rupture time delayed from 11 ± 7 min for the unstressed capsules to 23 ± 11 min and 71 ± 19 min for moderately stressed and severely stressed capsules, respectively. However, this delay in the capsule rupture times for both types of stressed capsules was shown to have no adverse effect on the $AUC_{(0-\infty)}$ and C_{max} of amoxicillin in human volunteers. The delayed shell rupture due to severe gelatin cross-linking could have bioequivalence implications for compounds with a narrow window for absorption in the GIT.[140]

The dissolution and bioavailability characteristics of softgels containing 0.05 mg of digoxin dissolved in PEG 400 were studied by Johnson et al.[142] Softgels stored at 37°C for 10 months showed slower dissolution in 0.6% HCl solution compared to that of initial samples, for example, time for 50% of labeled drug in solution delayed from 5 to 18 min, although complete dissolution obtained at 60 min in both cases. However, results from plasma and urinary studies in healthy volunteers showed that the aged softgels yielded similar AUC values with only a minor delay in T_{max} compared to fresh softgels.

Pepsin is shown to overcome the adverse effects of cross-linking of gelatin capsules on their dissolution in a media at and below pH 4.0.[132, 143] Pepsin begins to lose its ability to dissolve cross-linked gelatin at pH 6.0 and is completely unable to overcome the cross-linking at pH 6.8. This reduced effectiveness (proteolytic activity) of pepsin against cross-linked gelatin coincides with its reduced stability as a function of increasing pH.[135] Pepsin is reported to hydrolytically cleave the peptide bonds at the carboxylic acid ends of hydrophobic and aromatic amino acid residues (e.g., tyrosine, phenylalanine, methionine).[144] In contrast, pancreatin is known to hydrolytically cleave the peptide bonds at the carboxylic acid ends of lysine and arginine residues (side chains of the same amino acid residues are involved in aldehyde induced cross-linking) in the gelatin chains, thus permitting the dissolution and rupture of the cross-linked gelatin shell.[96] However, the lysine and arginine cross-links induced in gelatin by formaldehyde were reported to be resistant to hydrolysis by both of the proteolytic enzymes.[144] Severe cross-linking of gelatin (resulting in dissolution Fail/Fail outcome) was thought to result in the decreased availability of its peptide bonds towards the proteolytic enzymes, leading to decreased rate and extent of proteolysis of gelatin by these enzymes. The proteolytic enzymes may also cease to recognize the carboxyl ends of the aldehyde-derivatized lysine and arginine.[138]

Interestingly, dilute acid (low pH) environment can cleave the arginine-lysine cross-links, the extent of which increases with a decrease in the pH of the medium.[144] Thus, fed and fasted states can influence the in vivo dissolution and performance of cross-linked gelatin capsules. The mean gastric residence time (MGRT) in fasted state is usually lower than that in fed state. Consequently, a cross-linked gelatin capsule may pass through the stomach virtually intact in fasted state, releasing none or only a negligible portion of its contents. In contrast, in fed state, the same capsule may open in the stomach due to longer exposure time in the acidic environment and greater extent of hydrolysis of gelatin cross-links.

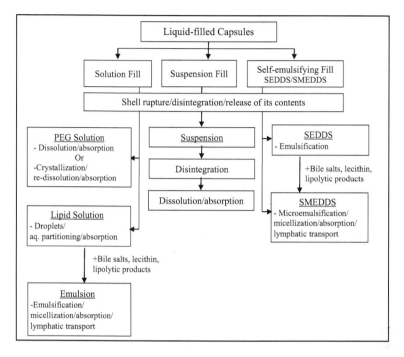

Figure 18.3 Schematic of probable in vivo dissolution pathways of liquid-filled capsules.

18.2.2 Dissolution of Capsule Fill

Once the capsule shell is ruptured, the release of the encapsulated compound for absorption depends upon the composition of the fill formulation (Figs. 18.2 and 18.3). For solution-filled capsules containing a highly soluble compound present in solution or solubilized form in a hydrophilic fill (e.g., polyethylene glycols), simple rupture of the capsule shell is sufficient to release the compound for absorption. On the other hand, for solution-filled capsules containing a highly water insoluble compound solubilized in a hydrophilic fill, upon contact with the dissolution medium, the solubilized compound may initially precipitate (*crash out*) and then slowly re-dissolve, resulting in erratic and inconsistent absorption.

For a simple lipid solution based fill formulation with no additional co-solvent or emulsifier, the fill, due to its lower density, rapidly floats to the surface of the dissolution medium in the form an immiscible layer or droplets. In presence of additional co-solvent and/or emulsifier, the lipid fill may self-disperse (self-emulsify) in the form of an emulsion or a microemulsion. In the case of self-microemulsifying and self-emulsifying drug delivery systems (SMEDDS and SEDDS, respectively), a compound dissolved in a lipophilic vehicle containing one or more emulsifiers and co-solvents, forms a microemulsion (droplet size ≤ 0.15 μm) when the ratio of emulsifier to lipid is high (>1) or a fine emulsion (droplet size >0.15 μm) when the ratio is <1, respectively, upon dilution with aqueous fluids in vitro or in vivo.[145]

The design of a lipid based formulation, in general, and SEDDS/SMEDDS, in particular, is usually based on the solubility of a compound in the formulation, the ease of dispersion of the formulation in an aqueous medium, and the particle size of the resulting emulsion droplets.[146–151] However, numerous studies suggested that the impact of lipid digestion products on the solubilization capacity of a lipid-based formulation is more critical to and more accurate representation of its in vivo performance.[147,148,152–154] Porter et al.[147] studied the in vitro and in vivo performance of SMEDDS formulations of long chain (LC-SMEDDS) and medium chain (MC-SMEDDS) lipids containing danazol, a poorly water soluble compound (aqueous solubility 0.42 μg/mL @ 37°C). These studies suggested, despite similar in vitro dispersion properties (i.e., particle size distributions) of both formulations, MC-SMEEDS resulted in a lower oral bioavailability in fasted beagle dogs. Interestingly, regardless of expected poor dispersibility of simple danazol in LC lipid solution formulation, the oral bioavailability results from the LC-SMEDDS and LC solution formulations were shown to be statistically indistinguishable. In vitro digestion studies demonstrated significant danazol precipitation from the MC-SMEDDS formulation when compared with the LC lipid formulations suggesting probable cause for the lower oral bioavailability of the former formulation. The subject of lipid digestion (lipolysis) process and its effects on the dissolution and absorption of compounds will be discussed in more detail in the Section 18.4.

For capsules filled with a suspension or a waxy matrix, a simple rupture of the shell may not be sufficient to deliver the compound for absorption. Measures of dispersion and subsequent solubilization of the compound in the dissolution medium are essential to evaluate the performance of the dosage form. Capsules containing a lipophilic matrix of a suspension, the dissolution profiles are dependent on the solubility of both the compound and the matrix in the dissolution medium and partitioning of the compound between the matrix and the medium.

Changes in the dissolution of a fill material are usually observed (a) when there is a change in particle size distributions (e.g., due to *Ostwald ripening*) and/or polymorphic nature (e.g., due to *secondary nucleation*) of the suspended material in a suspension fill formulation;[49] (b) when there is crystallization of a solubilized compound from a solution fill formulation;[155] or (c) when a poorly water soluble compound is dissolved in hydrophilic solvents and co-solvents. In the latter case, upon dilution with the GIT fluids in vivo or with the dissolution medium in vitro, the hydrophilic fill vehicle may dissolve or disperse in the aqueous fluids and thereby expose the solubilized water-insoluble compound to aqueous fluids, leading to erratic and inconsistent precipitating of the compound (*crashing out*).[12,14]

18.2.2.1 Use of surfactants in dissolution media

Surfactants are commonly used in the dissolution media during the dissolution testing of poorly soluble compounds and oily formulations.[1,136,156-164] The use of surfactants in the dissolution media has been proposed to be physiologically meaningful as these surfactants mimic those natural surfactants such as bile acids, bile salts, and lecithin present in the GIT.[163-165] The selection of a type and concentration of a surfactant and other solutes used in the medium for dissolution testing is based on a variety of factors such as (a) aqueous solubility, ionic nature (pK_a; pH-solubility profile) and dose of the compound; (b) compatibility of the surfactant with the compound; (c) compatibility of the surfactant with the capsule shell;[166-169] (d) critical micellar concentration (CMC) of the surfactant; and (e) degree of partitioning of the compound into the surfactant micelles (i.e., micellar loading).[159] The surfactant concentration must be kept at or above its CMC to obtain any substantial enhancement in the solubility of a compound. For an ionizable compound, pH of the medium and surfactant type and concentration may be varied simultaneously that can potentially offer substantial improvement in the solubilization of the compound. Some commonly used surfactants in the dissolution media include sodium lauryl sulfate (SLS), polyoxyethylene sorbitan monolaurate (polysorbate 20), polyoxyethylene sorbitan monooleate (polysorbate 80), polyoxyl castor oil (Cremophor EL), polyethylene glycol tert-octylphenyl ether (Triton), nonylphenol ethoxylate (Tergitol), cetyltrimethylammonium bromide (CTAB), hexadecyltrimethylammonium bromide (HTAB), lauryl dimethyl amine oxide (dodecyldimethylamine oxide, DDAO), lecithin, and cyclodextrins.[159]

Use of aqueous-organic solvent mixtures is strongly discouraged as the dissolution medium for gelatin capsules as these solvents could potentially prevent dissolution of capsule shells even when they are free from cross-linking, resulting in reaching misleading conclusions. In addition, the use of these aqueous-organic media has no relevance to the physiological environment and is not likely to generate meaningful data for in vivo interpretation.[136,163,164] If such a medium is used for dissolution testing of a poorly soluble compound, it is expected during regulatory filing to demonstrate that conventional tactics for getting adequate solubility and dissolution do not work.[170]

18.2.2.2 Challenges with use of surfactants in dissolution media

Surfactants are known for their denaturing effects on the proteolytic enzymes used in the dissolution media for cross-linked gelatin capsules.[171–174] The USP recommended levels of pepsin or pancreatin may be sufficient for the digestion of cross-linked gelatin capsules in the absence of a surfactant. The cross-linked gelatin capsules were shown to pass the tier 2 dissolution testing only when the enzyme was introduced into the medium initially followed by the surfactant few minutes afterward. These gelatin capsules failed to meet the dissolution specification when enzyme and surfactant were introduced into the medium together at the onset of dissolution testing. It is recommended to introduce the surfactant into the dissolution medium after the initial enzyme digestion of cross-linked gelatin capsules has occurred.[157, 162] The time of the digestion with proteolytic enzymes should be kept as short as possible and possibly not to exceed 15 min.[175] This digestion time should be included in the total time of the dissolution testing.

Sodium lauryl sulfate (SLS), an anionic surfactant, is known to interact with gelatin resulting in a complex formation at pH values equivalent to gastric pH.[167, 169, 174–178] Due to the high pK_a values of the amino functional groups in gelatin, these groups exist in their protonated form at gastric pH and thus are expected to undergo ionic interactions with the negative charges on an anionic surfactant such as sodium lauryl sulfate. The anionic charges of gelatin arising from the carboxylic side chains, on the other hand, may bind to cationic surfactants such as cetylpyridinium chloride (CPC), cetyltrimethylammonium bromide (CTAB), dodecylammonium chloride (DAC), and dodecyl trimethylammonium bromide (DTAB).[174, 179] Alternately, these surfactants may also bind to gelatin through hydrophobic interactions.[169, 180] These interactions could potentially influence the solubility and dissolution of the capsule shell adversely.[169, 177] Sodium lauryl sulfate is also known to interact with other ions present in the dissolution medium such as potassium, forming an insoluble complex.[181] In light of these interactions between gelatin and ionic surfactants, use of a non-ionic surfactant (e.g., polysorbate 20, polysorbate 80) may be considered in place of an ionic surfactant during the dissolution testing of poorly soluble compounds.[169, 182]

Purity and type of impurities present in a surfactant and other electrolytes used in the dissolution medium could influence the CMC of the surfactant and size and loading capacity of micelles for a compound and thus influence the ultimate solubility and dissolution rate of the compound in the dissolution medium.[159, 169, 183, 184] Any change in the quality profiles of a surfactant used in a routine quality control dissolution testing could potentially result in significant changes in the dissolution profiles of a capsule product that are unrelated to formulation or manufacturing process. When using a surfactant for in vitro dissolution testing of a dosage form, Crison et al.[184] recommended giving due consideration to the quality of the surfactant to achieve consistent and reliable dissolution results.

18.2.3 Factors Influencing Dissolution Stability of Liquid-filled Capsules

Several factors can influence the dissolution stability (change in dissolution profiles with time) of a liquid-filled capsule product. Major among them are (a) physicochemical properties of the compound and excipients, (b) impurity profiles of excipients, (c) nature of the shell material, (d) manufacturing process conditions, (e) packaging, and (f) storage conditions.

Compounds containing carbonyl functional groups in their structures or degrading into carbonyl impurities may cross-link gelatin shell and thereby reduce its dissolution.[109-111] On the other hand, compounds containing either a primary amine group (e.g., isoniazid, acyclovir) or a secondary amine group (e.g., ethambutol) may interact with any aldehyde impurities that may be present or form in the fill formulation, i.e., act as aldehyde scavengers, and thereby prevent the cross-linking of gelatin.[11] Crystallization of solubilized compounds in solution fills or particle growth due to *Ostwald ripening* or change of crystalline form into more thermodynamically stable polymorph (of lower solubility) in suspension fills are other potential sources for reduced dissolution.[155,185] In capsules encapsulated with caustic fill components such as acidic salts (e.g., ranitidine HCl, pseudoephedrine HCl, topotecan HCl), these components can potentially migrate into the capsule shells during manufacturing and shelf-life of the capsule products that results in hydrolysis of the polypeptide chains in gelatin. This, in turn, can weaken the shell and reduce its dissolution time with time.[186-189] Lipid and higher molecular weight excipients are known to exist in several polymorphs and changes in polymorphic form due to aging or changes in the manufacturing process can also affect the drug release properties.[43-47]

Polyethylene glycol related materials and lipids are known to undergo auto-oxidative reactions to produce aldehydes, which can cross-link gelatin and influence the dissolution stability of liquid-filled gelatin capsules adversely. These reactions can be minimized effectively by using an antioxidant (e.g., BHA, BHT, dl-α-tocopherol) in the fill formulation, purging the formulation with an inert gas such as nitrogen during its manufacturing, and thoroughly deaerating the formulation under vacuum to remove any dissolved oxygen before its encapsulation into capsules.[103, 106, 118, 190] In contrast to gelatin capsules, HPMC based capsule shells (e.g., Vcaps®, Vcaps® Plus) are inert and compatible with most compounds and excipients. These HPMC capsule shells are also not susceptible to cross-linking and thus the dissolution stability of liquid-filled HPMC capsules is not affected adversely in presence of aldehyde impurities.[57]

An aggressive rate and extent of drying process during softgel manufacturing can potentially lead to detrimental, often, irreparable changes in the structure and properties of the shell material.[75] These changes in the gelatin shell material may require a longer time for dissolution. Exposure of gelatin capsules to elevated temperatures and relative humidities (e.g., 40°C/75%RH) for prolonged periods has also been shown to prolong their dissolution times.

18.3 Development of Dissolution Methods

18.3.1 Development of Dissolution Medium

Based on the Biopharmaceutics Classification System (BCS) principles of solubility and permeability, a compound can be classified into one of four classes: class 1 (*highly soluble/highly permeable*), class 2 (*poorly soluble/highly permeable*), class 3 (*highly soluble/poorly permeable*), and class 4 (*poorly soluble/poorly permeable*) (Fig. 18.4).[5, 6] A compound is considered *highly soluble* when its highest dose strength is soluble in 250 mL or less of aqueous media over the pH range of 1–7.5. A compound is considered *highly permeable* when the extent of its absorption in humans is determined to be 90% or more of an administered dose based on a mass balance determination or in comparison to an intravenous reference dose. A dosage form is considered *rapidly dissolving* when no less than 85% of the labeled amount of the compound dissolves within 30 min, using USP apparatus 1 (basket method) at 100 rpm or apparatus 2 (paddle method) at 50 rpm in a volume of 900 mL or less in each

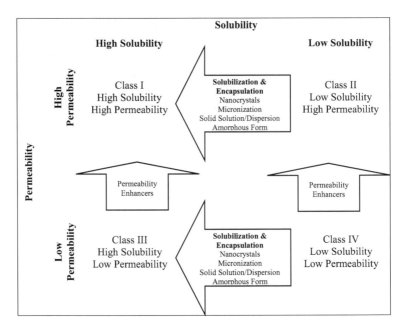

Figure 18.4 Role of Biopharmaceutics Classification System (BCS) in formulation development for poorly soluble compounds.

of the following media: (1) 0.1 N HCl or simulated gastric fluid USP without enzymes; (2) a pH 4.5 buffer; and (3) a pH 6.8 buffer or simulated intestinal fluid USP without enzymes.

Shah et al.[1] and Noory et al.,[164] based on the experience gained from the FDA field laboratories in the development of dissolution procedures for sparingly water soluble and water insoluble compounds, suggested stepwise evaluation of the effect of pH, the type of surfactant, and the concentration of the surfactant in the development of a suitable dissolution medium. This stepwise procedure involves evaluation of dissolution (a) in standard aqueous dissolution media as listed in the USP, including 0.1 N HCl (pH 1.0), pH 4.5 sodium acetate buffer, and pH 6.8 phosphate buffer to select a preferred pH medium; (b) in a 2% solution of a surfactant from cationic, anionic, and non-ionic classes (e.g., cetyltriammonium bromide, sodium lauryl sulfate and polysorbate, respectively) in the preferred pH medium to select a preferred class of surfactant; and (c) in media of increasing concentration of the preferred surfactant (e.g., 0.1, 0.25, 0.5, 0.75, 1.0, 2.0%) to maximize the sensitivity of the dissolution method. The aim of this stepwise procedure is to use the lowest amount of a surfactant to solubilize the compound in the selected dissolution medium that offers the maximum discriminatory power. Marques and Brown[191] cautioned on the use of surfactants in the dissolution media as surfactants can mask problems in the formulation or in the test. The authors suggested exploring other options such as (a) increasing the medium volume, agitation speed, flow rate or run time; (b) changing the composition of the dissolution medium (e.g., ionic strength, different counter-ion, different salts); or (c) using other type of USP dissolution apparatus (e.g., apparatus 3 or apparatus 4 in place of apparatus 1 and apparatus 2) before the use of a surfactant in the dissolution medium.

18.3.2 Selection of Dissolution Apparatus

The selection of a dissolution test apparatus is based on properties of the compound and capsule dosage form. USP basket method (apparatus 1) and USP paddle method (apparatus 2) are routinely used for dissolution testing of liquid-filled capsules at an agitation speed of 50 to 100 rpm and 50 or 75 rpm, respectively. Though both basket and paddle methods are typically used with media volumes from 500 to 1000 mL, they can also be adopted to 2000 to 4000 mL vessels for poorly soluble compounds and 100 to 250 mL vessels for highly potent, low dosage compounds. The operational minimum volume of the 1000 mL vessel is around 500 mL. The hydrodynamics below 500 mL become too unstable for routine dissolution testing. Smaller volume vessels with 100 mL and 200 mL capacities equipped with scaled down baskets and paddles have been developed to handle smaller volumes. The operational minimum for these smaller volume vessels is approximately 30 mL.[192] USP apparatus 1 and 2 have simple designs that can be used efficiently for routine QC purposes. However, these apparatus lack any resemblance to the GIT conditions and difficult to produce IVIVC and IVIVR.

Bottom et al.[80] studied the influence of the extent of cross-linking of nifedipine softgels by 10–20 ppm and 80 ppm formaldehyde in the fill on in vitro dissolution and bioavailability in human subjects. Softgels containing 0 ppm, 20 ppm, and 80 ppm formaldehyde in the fill were stored at 25°C for 1–1.5 years and dissolution test was performed as per the USP monograph (USP apparatus 2; 80% dissolved in 20 min). Softgels containing 10–20 ppm formaldehyde and stored at 25°C for 1.5 years were found to be fully bioavailable, whereas those containing 80 ppm formaldehyde and stored at 25°C for 1 year were found to be not bioavailable. However, the in vitro dissolution test in USP apparatus 2 showed that both batches of softgels failed to meet the dissolution criteria and thus found to be not discriminatory. When the dissolution testing was formed using USP apparatus 3 (reciprocating cylinder), softgels containing 80 ppm formaldehyde exhibited low dissolution, but those containing 10–20 ppm formaldehyde showed satisfactory dissolution. Based on these observations, the authors suggested that USP apparatus 3 might be more discriminatory than USP apparatus 2.

During dissolution testing of capsules filled with lipids and oils in USP apparatus 1 and 2, the fill components can form a layer above or droplets in the dissolution medium and/or adhere to the paddle or vessel walls, thereby, causing sampling challenges.[193] USP apparatus 4 (flow-through cell) has been re-configured for dissolution testing of capsules containing lipophilic fills.[130, 194] The standard flow-through cell (FTC) apparatus can operate in two different modes: (a) an open mode with fresh medium from the reservoir continuously passing through the cell and (b) a closed mode where a fixed volume of medium is recycled. The open mode is usually selected for a compound and dosage form that requires high volume of medium (e.g., poorly soluble compounds). The open mode allows fresh dissolution medium to be pumped through the flow cell continuously, thus maximizing the dissolution rate. The advantages of the FTC apparatus are it can be used with conventional dissolution media and biorelevant media and it allows a dosage form to be exposed continuously to various dissolution media during one experiment to simulate the changing conditions during its passage through the GIT.

The standard flow-through cell, typically used for conventional tablet or capsule formulations, is not suitable for lipid-filled capsule formulations. In the standard FTC, after rupture of the capsule

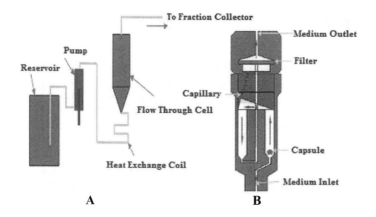

Figure 18.5 Schematics of USP dissolution apparatus 4 with open system (A) and modified flow through cell (B). Adapted from Hu et al. [194].

shell, the oil phase is quickly drawn into the filter, resulting in either clogging or passing through the filter. In the modified FTC design (Fig. 18.5), the lipid material from the ruptured capsule is trapped in a triangular area located at the top of cell, thereby, is prevented reaching the filter.[130,194] The dissolution medium that enters through the medium inlet continuously extracts the compound from the lipid layer as it flows through the cell. The dissolved compound is then collected and analyzed appropriately.

18.3.3 Alternate Dissolution Testing Procedures

USP General Chapter <701> Disintegration[195] suggests a disintegration test as the performance test for hard gelatin and softgel capsules. The disintegration test is performed in disintegration apparatus using water or a specified medium as the immersion fluid maintained at $37 \pm 2°C$. At the end of the time limit specified in the monograph, the basket is lifted from the fluid and observed the capsule. The test requirements are met if the capsule disintegrates completely. If one or two capsules fail to disintegrate completely, the test is repeated on 12 additional capsules. The requirements are met if not fewer than 16 of the total of 18 capsules tested disintegrate. The disintegration test does not imply complete dissolution of the capsule or its active content(s). USP General Chapter <701> defines complete disintegration as "that state in which any residue of the capsule, except fragments of insoluble capsule shell, remaining on the screen of the test apparatus or adhering to the lower surface of the disk, if used, is a soft mass having no palpably firm core."

USP General Chapter <2040> Disintegration and Dissolution of Dietary Supplements[129] suggests a disintegration test as the performance test for hard gelatin capsules. The disintegration test for hard gelatin capsules is performed in disintegration apparatus using a 0.05 M acetate buffer having a pH of 4.50 ± 0.05 and maintained at 37 ± 2°C. At the end of 30 min, the basket is lifted from the fluid and observed the capsules. The test requirements are met if all capsules disintegrate completely. If one or two capsules fail to disintegrate completely, the test is repeated on 12 additional capsules. The requirements are met if not fewer than 16 of the total of 18 capsules tested disintegrate. Similar to USP General Chapter <701> Disintegration, USP General Chapter <2040> defines complete disintegration as "that state in which any residue of the capsule, except fragments of insoluble capsule shell, remaining on the screen of the test apparatus or adhering to the lower surface of the disk, if used, is a soft mass having no palpably firm core."

For softgel capsules, USP General Chapter <2040> specifies the use of a rupture test as the performance test in Chapter <711> Dissolution[130] apparatus 2 (paddle) at 50 rpm with 500 mL of water as the immersion medium. The test requirements are met if all capsules rupture within 15 min or if not more than 2 of the total of 18 capsules tested rupture in more than 15 min but not more than 30 min. The difference is that for hard shell capsules, Chapter <2040> lists disintegration in pH 4.5 acetate buffer as the immersion medium with 30 minute time limit while rupture testing in water with 15 minute time limit for softgel capsules. In addition, USP General Chapter <2040> allows the use of purified pepsin at an activity of 750, 000 Units or less per 1000 mL of the medium for softgel capsules that do not conform to the above rupture test acceptance criteria. For capsules filled with solutions, suspensions or paste formulations, under the definition of USP General Chapters <701> and <2040>, rupture of capsule shells fulfills the endpoint

Table 18.5   Examples of liquid-filled capsules requiring rupture test as the performance test

Benzonatate capsules	Docusate sodium capsules
Calcifediol capsules	Doxercalciferol capsules
Calcitriol capsules	Dronabinol capsules
Chloral hydrate capsules	Ergocalciferol capsules
Chlorotrianisene Capsules	Ergoloid mesylate capsules
Clofazimine capsules	Ethchlorovynol capsules
Cod liver oil capsules	Progesterone capsules
Cyclosporine capsules	Pygeum capsules
Docusate calcium capsules	Saw palmetto capsules
Docusate potassium capsules	Schizochytrium oil capsules

criterion of the disintegration test and thus the endpoint is the same for both rupture and disintegration tests in this case.[196]

Examples of dissolution procedures used for some marketed liquid-filled softgel and hard shell capsule products are presented in Tables 18.2 and 18.3, respectively. Few liquid-filled capsule products requiring rupture test as the USP performance test are illustrated in Table 18.5.

18.4 Dissolution Procedures for IVIVC and IVIVR

Development of an appropriate dissolution procedure for generating successful IVIVC or IVIVR for liquid-filled capsules should be based on type of the fill formulation (lipid or non-lipid), dissolution media (composition and volume), and hydrodynamics (agitation intensity, fluid flow, and exposure of the formulation to various segments of the GIT). Dissolution procedures for generating IVIVC or IVIVR for non-lipid-filled capsules are similar to those used for conventional tablet and capsule dosage forms. The current discussion is limited to only for lipid-filled capsules, as they pose unique challenges during the development of dissolution procedures for IVIVC and IVIVR.

An understanding of how lipids of varying chain lengths of fatty acids could influence (a) the properties of their in vivo digestion (*lipolysis*) products; (b) the mode of their transportation and absorption; and (c) the solubilization and transportation of compounds across the GIT provides valuable guidance in the development of an appropriate in vitro dissolution method that can yield meaningful IVIVC and IVIVR. However, it is important to recognize that the in vitro digestion models and dissolution methods suffer from the lack absorption mechanisms to remove the digestion products and compounds solubilized within.

18.4.1 Lipolysis and Its Effects on Dissolution and Absorption of Compounds

The poor miscibility of the lipids (triglycerides) with the GIT aqueous environment may lead to highly variable gastric emptying and/or dispersion into an emulsion, which in turn, can result in variable absorption of the compound from the GIT.[152] The dispersibility of the lipids in vitro can be improved by including a surfactant either in the formulation or in the dissolution medium. In vivo, the digestion of the lipids within the GIT by the pancreatic lipase/co-lipase complex results into the generation of amphiphilic diglycerides, monoglycerides, and free fatty acids that can enhance the dispersion and dissolution of the lipids and the poorly soluble compound co-administered with the lipids. The release of biliary lipids from the gallbladder, primarily bile salts, phospholipids, and cholesterol, promotes the formation of a number of colloidal species in association with the digestion products within the small intestine that further enhances the dissolution of the compound.[197,198] Long chain triglycerides (LCT; $\geq$ C12) are thought to be capable of stimulating the gallbladder contraction and thereby elevating intestinal bile salts, phospholipids, and cholesterol levels. The extent of this stimulatory response appears to increase as the quantity of the LCT administered increases.[199] This effect is a likely contributor to the ability of the LCT-based formulations to mimic positive food effects and to enhance the oral bioavailability of some poorly soluble compounds. In contrast, administration of similar quantities of medium chain triglycerides (MCT; C6-C10) was shown to have relatively limited effects on the gallbladder contraction and would not stimulate appreciable increase in the intestinal concentration of biliary-derived lipids.[199]

Fatty acids with medium chain length directly enter the portal blood leading to the liver and then into the systemic circulation, thus circumventing the beneficial lymphatic transport mechanism (Fig. 18.1), whereas fatty acids with longer chain length are incorporated primarily into the chylomicrons and transported in the lymph.[200] Compounds transported from the intestinal lumen by the intestinal lymph gain direct access to the general circulation at the junction of the left internal jugular and left subclavian veins, thereby bypassing the hepatic system, a major site of metabolism for several compounds.[201] Lipophilic compounds with (a) a high logP (>5), (b) significant solubility in LCT (≥50 mg/mL), and (c) administered either in fed state or with an appropriate LCT source in fasted state are thought to potentially gain direct access to the systemic circulation through the intestinal lymphatic transport, resulting in the improved bioavailability of these compounds.[201–206]

In vitro digestion studies performed using sodium taurodeoxycholate/phosphatidyl choline at concentrations mimicking typical fasting[198,207,208] and fed state[197,198,208] suggested rapid and complete digestion of MCT compared to that of LCT.[209–217] In addition, the fatty acids and monoglycerides produced on digestion of MCT have greater aqueous solubility than those of LCT. The higher aqueous solubility of the digestion products of MCT also promotes their transport through the aqueous intestinal fluids and the static diffusion layer in vivo, resulting in a more rapid removal of these moieties from the surface of the digestion droplets due to their absorption. As a result of rapid removal and absorption, the quantity of MCT digestion products in the intestinal lumen decreases rapidly, resulting in reduced solubilization capacity and subsequent precipitation of lipophilic compounds solubilized within. The rapid absorption of MCT digestion products may occasionally lead to shorter T_{max} of lipophilic compounds formulated within, relative to the same formulated in LCT.[213] In vitro digestion studies have also suggested that MCT yield micelles of smaller radius (~3 nm) than of LCT (~8 nm), due to the relatively higher aqueous solubility of the digestion products of former compared to those of the latter.[210] Due to (a) more rapid absorption and resulting reduction in the MCT digestion products and associated reduction in the solubilization capacity for a lipophilic compound (i.e., precipitation) and (b) smaller micellar size and associated lower solubilization capacity within can potentially lead to reduced bioavailability (AUC and C_{max}).[206,214,218] By contrast, due to (a) slower absorption rate and (b) larger micellar size of the digestion products of LCT and associated higher solubility within, LCT are thought to increase the bioavailability of lipophilic compounds more effectively than MCT.[147,209,210,213,219,220]

18.4.2 Application of in vivo Processes to in vitro Dissolution Procedures

Dissolution testing in a biorelevant medium is a useful tool for qualitative prediction of formulation and food effects on the dissolution and availability of a compound for oral absorption. It can provide a more accurate simulation of in vivo pharmacokinetic profiles than that obtained in simulated gastric fluid (SGF) or simulated intestinal fluid (SIF). The use of biorelevant media such as fed state simulated intestinal fluid (FeSSIF) and fasted state simulated intestinal fluid (FaSSIF) is particularly important for poorly water-soluble compounds because they simulate the solubilizing environment of mixed micelles in the GIT. The biorelevant media comprise a bile salt and lecithin (Table 18.6), which are responsible for the emulsification and absorption of dietary fats in humans and animals. Lipid-based formulations that undergo lipolysis in vivo may require the addition of lipolytic products to the dissolution medium. The lipolytic products play an important role in the solubilization capacity in vitro and in vivo for poorly soluble compounds.

18.4.2.1 IVIVC and IVIVR using in vitro lipolysis models

Several in vitro lipolysis models that could mimic in vivo lipolytic processes have been evaluated to correlate the solubilization of compounds in the in vitro lipolytic media to the in vivo performance of lipid-based formulations.[147,153,154,220−224] The in vitro lipolysis models were successfully used to rank order the in vivo performance of several types of lipid-based formulations (e.g., LCT vs. MCT vs. SCT). Though the in vitro models may be able to predict solubilization process in vivo, they lack the in vivo pathways,

Table 18.6 Compositions of simulated intestinal fluids in fasted state (FaSSIF) and fed state (FeSSIF)[229]

Composition	FaSSIF-V2	FeSSIF-V2
Sodium taurocholate (mM)	3	10
Lecithin (mM)	0.2	2
Maleic acid (mM)	19.12	55.02
Sodium hydroxide (mM)	34.8	81.65
Sodium chloride (mM)	68.62	125.5
Glyceryl monooleate (mM)		5
Sodium oleate (mM)		0.8
Properties		
pH	6.50	5.8
Osmolality (mOsm/kg)	180 ± 10	390 ± 10
Buffer capacity (mmol/L/pH)	10	25
Recommended dissolution test volume	500 mL	1000 mL

the lipolytic products and compounds solubilized within would subsequently undergo during the absorption process (e.g., pre-systemic metabolism, transport mechanisms across GIT membranes, lymphatic transport). Another factor that can complicate IVIVC and IVIVR for lipid formulations is the inherent intra- and inter-subject variability of the lipid digestion process in vivo. Such variability is usually lower during in vitro digestion studies as these studies are performed under controlled conditions. As a result of these inherent differences, the in vitro models may not be ideal predictors of actual absorption profiles of the compounds in vivo.

Studies by Dahan and Hoffman[154, 223] demonstrated that in vitro lipolysis data were predictive of relative in vivo performance of several types of lipid formulations, except, in cases when lymphatic transport was a significant route of absorption of a compound. For example, the in vitro data indicated a rank order of MCT > LCT > SCT for progesterone and griseofulvin, which correlated with the in vivo performance rank order in rodents. On the other hand, for vitamin D3 (highly lipophilic compound with logP = 9.1 and known to be significantly transported via the mesenteric lymph), the in vitro and in vivo rank orders were MCT > LCT > SCT and LCT > MCT > SCT, respectively. However, in vivo performance rank order correlated with the in vitro rank order when the lymphatic transport mechanism was blocked.

Reymond et al.[153,225] investigated the distribution of cyclosporine in various phases during in vitro digestion of MCT and LCT (olive oil) and explored the relative in vivo performance of undigested and in vitro digested formulations administered intraduodenally to bile duct cannulated rats. These studies suggested presence of a significantly higher concentration of cyclosporine in the aqueous phase of in vitro lipolysis of MCT compared to that of LCT. However, in vivo performance was shown to be relatively higher from the LCT formulation. Increased permeability of the GIT membrane, induced by the long chain unsaturated fatty acids originating from LCT digestion, was attributed to the relatively higher in vivo performance from the LCT formulation.

The influence of intra- and intersubject variability of the lipid digestion process in vivo on the bioavailability is illustrated by the two types of commercially available lipidic formulations of cyclosporine. Cyclosporine is available as SEDDS (Sandimmune®; Novartis) and as SMEDDS (Neoral®; Novartis) (Table 18.2). SEDDS formulation comprises of cyclosporine dissolved in a blend of corn oil (LCT), linoleoyl macrogolglycerides, and ethyl alcohol, which forms a coarse emulsion on dispersion into an aqueous media.[230,231] The LCT excipients in SEEDS formulation require further lipolysis in vivo into diglycerides, monoglycerides, and free fatty acids, for efficient release and absorption of cyclosporine.[145,153,225] In contrast, the SEMEDDS formulation comprises of cyclosporine dissolved in a blend of corn oil mono- and di-glycerides, polyoxyl 40 hydrogenated castor oil, propylene glycol, ethyl alcohol, and dl-α-tocopherol, which spontaneously forms a microemulsion with a droplet size below 100 nm when introduced into an aqueous media.[230,231] The improved dispersion characteristics and presence of the rapidly absorbable mono- and diglycerides, which would not require further lipolysis in vivo (thus circumventing the lypolytic process) have been suggested to be responsible for the increased bioavailability and reduced inter- and intra- subject variability of cyclosporine from the SMEDDS formulation.[232,233] The bioavailability of cyclosporine from the SMEDDS formulation, for example, was shown to be significantly higher (174 to 239%), dose proportional, and free from food effects with reduced inter- and intra-subject variability compared to that from the SEDDS formulation.

18.5 Summary

Liquid-filled capsule dosage forms offer the unique advantage of delivering compounds solubilized or suspended in non-aqueous vehicles as a unit dose solid dosage form. The availability of a compound formulated in such a dosage form for absorption depends on the initial dissolution and rupture of the capsule shell and subsequent release and fate of its fill contents in the GIT fluids. A thorough understanding of (a) factors affecting dissolution and rupture of the capsule shell and release of its contents (e.g., gelatin cross-linking); (b) fate of the released contents (e.g., solubilization, recrystallization, emulsification, lipolysis); and (c) physiology of the GIT (e.g., fluid composition, residence time, fed/fasted condition, absorption mechanisms) is essential in the design of an optimal liquid-filled capsule product. Two types of in vitro procedures are currently being used to evaluate these processes, dissolution and lipolysis models.

In vitro dissolution methods are designed to differentiate any changes to a liquid-filled capsule product at the time of its manufacturing (e.g., composition, quality, process) and during its shelf life (e.g., gelatin shell cross-linking, recrystallization from solution fills, particle growth/polymorphic change in suspension fills). These changes may potentially result in differences in the rate of dissolution and bioavailability. In vitro lipolysis methods are designed to simulate in vivo lipolysis process and to correlate to the rank order in vivo performance of lipid-based formulations. These in vitro lipolysis models, however, may not be ideal predictors of actual absorption profiles of the compounds in vivo, as they lack the in vivo pathways for the lipolytic products and compounds solubilized within would subsequently undergo during the absorption process (e.g., pre-systemic metabolism, transport mechanisms across GIT membranes, lymphatic transport).

References

1. Shah, V. P., Noory, A., Noory, C., McCullough, B., Clarke, S., Everett, R., Naviasky, H., Srinivasan, B. N., Fortman, D., and Skelly, J. P. (1995). In vitro dissolution of sparingly water-soluble drug dosage forms. *Int. J. Pharm.*, **125**, 99–106.

2. Crowley, P. J., and Martini, L. G. (2004). Physicochemical approaches to enhancing oral absorption. *Pharm. Technol. Eur.*, **16**, 18–27.

3. Lipinski, C. A., Lombardo, F., Dominy, B. W., and Feeney, P. J. (1997). Experimental and computational approaches to estimate solubility and permeability in drug discovery and development settings. *Adv. Drug Deliv. Rev.*, **23**, 3–25.

4. Lipinski, C. (2003). Physicochemical properties in drug design/development. 2nd International Drug Discovery and Development Summit: Novel Concepts and Technologies to Accelerate Drug Development. Honolulu, HI.

5. Amidon, G. L., Lennernas, H., Shah, V. P., and Crison, J. R. (1995). A theoretical basis for a biopharmaceutic drug classification: the correlation of in vitro drug product dissolution and in vivo bioavailability. *Pharm. Res.*, **12**, 413–420.

6. FDA Guidance for Industry: Waiver of in vivo bioavailability and bioequivalence studies for immediate-release solid oral dosage forms based on a biopharmaceutics classification system. http://www.fda.gov/downloads/Drugs/GuidanceComplianceRegulatoryInformation/Guidances/ucm070246.pdf

7. Carrigan, P. J., and Bates, T. R. (1973). Biopharmaceutics of drugs administered in lipid-containing dosage forms I: GI absorption of griseofulvin from an oil-in-water emulsion in the rat. *J. Pharm. Sci.*, **62**, 1476–1479.

8. Chakrabarti, S., and Belpaire, F. M. (1978). Bioavailability of phenytoin in lipid containing dosage forms in rats. *J. Pharm. Pharmacol.*, **30**, 330–331.

9. Larsen, A., Holm, R., Pedersen, M. L., and Müllertz, A. (2008). Lipid-based formulations for danazol containing a digestible surfactant, Labrafil M2125CS: in vivo bioavailability and in vitro lipolysis in a dynamic lipolysis model. *Pharm. Res.*, **25**, 2769–2777.

10. Erlich, L., Yu, D., Pallister, D. A., Levinson, R. S., Gole, D. G., Wilkinson, P. A., Erlich, R. E., Reeve, L. E., and Viegas, T. X. (1999). Relative bioavailability of danazol in dogs from liquid-filled hard gelatin capsules. *Int. J. Pharm.*, **179**, 49–53.

11. Cole, E. T., Cade, D., and Benameur, H. (2008). Challenges and opportunities in the encapsulation of liquid and semi-solid formulations into capsules for oral administration. *Adv. Drug Deliv. Rev.*, **60**, 747–756.

12. Serajuddin, A. T. M., Sheen, P. C., and Augustine, M. A. (1986). Water migration from soft gelatin capsule shell to fill material and its effect on drug solubility. *J. Pharm. Sci.*, **75**, 62–64.

13. Groves, M. J., Bassett, B., and Sheth, V. (1984). The solubility of 17 b-oestradiol in aqueous polyethylene glycol 400. *J. Pharm. Pharmacol.*, **36**, 799–802.

14. Aungst, B. J., Nguyen, N. H., Rogers, N. J., Rowe, S. M., Hussain, M. A., White, S. J., and Shum, L. (1997). Amphiphilic vehicles improve the oral bioavailability of a poorly soluble HIV protease inhibitor at high doses. *Int. J. Pharm.*, **156**, 79–88.

15. Pouton, C. W. (2000). Lipid formulations for oral administration of drugs: non-emulsifying, self-emulsifying and 'self-microemulsifying' drug delivery systems. *Eur. J. Pharm. Sci.*, **11**(suppl. 2), S93–S98.

16. Roy, A., Winnike, R., Long, S., Rhodes, C., Bao, Z., and Robertson, G. (2001). Development of a soft gelatin capsule formulation of a poorly soluble compound X using a self emulsifying delivery system. In *Oral Drug Absorption: A Renaissance*, B. T. Gattefossé N°94. Gattefossé Corporation, Saint-Priest Cedex France, pp. 141–149.

17. Thelen, K., Jantratid, E., Dressman, J. B., Lippert, J., and Willmann, S. (2010). Analysis of nifedipine absorption from soft gelatin capsules using PBPK modeling and biorelevant dissolution testing. *J. Pharm. Sci.*, **99**, 2899–2904.

18. Quinn, K., Gullapalli, R. P., Merisko-liversidge, E., Goldbach, E., Wong, A., Liversidge, G. G., Hoffman, W., Sauer, J. M., Bullock, J., and Tonn, G. (2012). A formulation strategy for gamma secretase inhibitor ELND006, a BCS class II compound: development of a nanosuspension formulation with improved oral bioavailability and reduced food effects in dogs. *J. Pharm. Sci.*, **101**, 1462–1474.

19. Gullapalli, R., Wong, A., Brigham, E., Kwong, G., Wadsworth, A., Willits, C., Quinn, K., Goldbach, E., and Samant, B. (2012). Development of ALZET® osmotic pump compatible solvent compositions to solubilize poorly soluble compounds for preclinical studies. *Drug Deliv.*, **19**, 239–246.

20. FDA Guidance for Industry: SUPAC–IR. Immediate release solid oral dosage forms: scale-up and post approval changes: chemistry, manufacturing and controls, in vitro dissolution testing and in vivo bioequivalence documentation. Rockville, MD: FDA. 1995.

21. http://www.fda.gov/downloads/Drugs/GuidanceComplianceRegulatoryInformation/Guidances/UCM358187.pdf

22. http://www.fda.gov/downloads/Drugs/GuidanceComplianceRegulatoryInformation/Guidances/UCM209294.pdf

23. http://www.fda.gov/downloads/Drugs/GuidanceComplianceRegulatoryInformation/Guidances/ucm083267.pdf

24. http://www.fda.gov/downloads/Drugs/Guidances/UCM082278.pdf

25. Sobral, P. J. A., Menegalli, F. C., Hubinger, M. D., and Roques, M. A. (2001). Mechanical, water vapor barrier and thermal properties of gelatin based edible films. *Food Hydrocolloids*, **15**, 423–432.

26. Singh, S., Rama Rao, K. V., Venugopal, K., and Manikandan, R. (2002). Alteration in dissolution characteristics of gelatin-containing formulations. A review of the problem, test methods, and solutions. *Pharm. Technol.*, **26**, 36–58.

27. Chiou, B. S., Avena-Bustillos, R. J., Shey, J., Yee, E., Bechtel, P. G., Imam, S. H., Glenn, G. M., and Orts, W. J. (2006). Rheological and mechanical properties of cross-linked fish gelatins. *Polymer*, **47**, 6379–6386.

28. Price, J. C. (2006a). Gelatin. In *Handbook of Pharmaceutical Excipients*, 5th ed., Rowe, R. C., Sheskey, P. J., and Owen, S. C., eds. London: Pharmaceutical Press and Washington DC: American Pharmacists Association, pp. 295–298.

29. Yakimets, I., Wellner, N., Smith, A. C., Wilson, R. H., Farhat, I., and Mitchell, J. (2005). Mechanical properties with respect to water content of gelatin films in glassy state. *Polymer*, **46**, 12577–12585.

30. Marshall, A. S., and Petrie, S. E. B. (1980). Thermal transitions in gelatin and aqueous gelatin solutions. *J. Photogr. Sci.*, **28**, 128–134.

31. Pinhas, M. F., Blanshard, J. M. V., Derbyshire, W., and Mitchell, J. R. (1996). The effect of water on the physicochemical and mechanical properties of gelatin. *J. Therm. Anal.*, **47**, 1499–1511.

32. Coppola, M., Djabourov, M., and Ferrand, M. (2008). Phase diagram of gelatin plasticized by water and glycerol. *Macromol. Symp.*, **273**, 56–65.

33. Yannas, J. B., and Tobolsky, A. V. (1964). Viscoelastic properties of plasticized gelatin films. *J. Phys. Chem.*, **68**, 3880–3882.

34. Fraga, A. N., and Williams, R. J. J. (1985). Thermal properties of gelatin films. *Polymer*, **26**, 113–118.

35. Martucci, J. F., Ruseckaite, R. A., and Vazquez, A. (2006). Creep of glutaraldehyde-crosslinked gelatin films. *Mater. Sci. Eng. A*, **435–436** and 681–686.

36. Rivero, S., Garcia, M. A., and Pinotti, A. (2010). Correlation between structural, barrier, thermal and mechanical properties of plasticized gelatin films. *Innovative Food Science and Emerging Technologies*, **11**, 369–375.

37. Patil, R. D., Marka, J. E., Apostolovb, A., Vassilevab, E., and Fakirovb, S. (2000). Crystallization of water in some crosslinked gelatins. *Eur. Polym. J.*, **36**, 1055–1061.

38. Yannas, I. V., and Tobolsky, A. V. (1966). Transitions in gelatin-nonaqueous diluent systems. *J. Macromol. Chem.*, **1**, 723–737.

39. Finch, C. A., and Jobling, A. (1977). The physical properties of gelatin. In *The Science and Technology of Gelatin*, Ward, A. G., and Courts, A., eds. New York: Academic Press, pp. 285–286.

40. Kontny, M. J., and Mulski, C. A. (1989). Gelatin capsule brittleness as a function of relative humidity at room temperature. *Int. J. Pharm.*, **54**, 79–85.

41. Jones, B. E. (2004). Manufacture and properties of two-piece hard capsules. In *Pharmaceutical Capsules*, 2nd ed., Podczeck, F., and, Jones, B. E., eds. London: Pharmaceutical Press, pp. 93–94.

42. Bowtle, W. J., Barker, N. J., and Woodhams, J. (1988). A new approach to vancomycin formulation using filling technology for semi-solid matrix capsules. *Pharm. Technol.*, **12**, 86–97.

43. Khan, N., and Craig, Q. M. (2003). The influence of drug incorporation on the structure and release properties of solid dispersions in lipid matrices., *J. Control. Release*, **93**, 355–368.

44. Freitas, C., and Muller, R. H. (1999). Correlation between long-term stability of solid lipid nanoparticles (SLN) and crystallinity of the lipid phase. *Eur. J. Pharm. Biopharm.*, **47**, 125–132.

45. Brubach, J. B., Ollivon, M., Jannin, V., Mahler, B., Bougaux, C., Lesieur, C., and Roy, P. (2004). Structural and thermal characterization of mono- and diacyl polyoxyethylene glycol by infrared spectroscopy and X-ray diffraction coupled to differential calorimetry. *J. Phys. Chem. B*, **108**, 17721–17729.

46. Choy, Y. W., Khan, N., and Yuen, K. H. (2005). Significance of lipid matrix aging on in vitro release and in vivo bioavailability. *Int. J. Pharm.*, **299**, 55–64.

47. Brubach, J. B., Jannin, V., Mahler, B., Bourgaux, C., Lessieur, P., Roy, P., and Ollivon, M. (2007). Structural and thermal characterization of glyceryl behenate by X-ray diffraction coupled to differential calorimetry and infrared spectroscopy. *Int. J. Pharm.*, **336**, 248–256.

48. Chatham, S. M. (1987). The use of bases in semi-solid matrix formulations. *S. T. P. Pharma.*, **3**, 575–582.

49. Kipp, J. E. (2004). The role of solid nanoparticle technology in the parenteral delivery of poorly water-soluble drugs. *Int. J. Pharm.*, **284**, 109–122.

50. Singhal, D., and Curatolo, W. (2004). Drug polymorphism and dosage form design: a practical perspective. *Adv. Drug Deliv. Rev.*, **56**, 335–347.

51. Aguiar, A. J., Krc, J., Kinkel, A. W., and Samyn, J. C. (1967). Effect of polymorphism on the absorption of chloramphenicol from chloramphenicol palmitate. *J. Pharm. Sci.*, **56**, 847–853.

52. Aguiar, A. J., and Zelmer, J. E. (1969). Dissolution behavior of polymorphs of chloramphenicol palmitate and mefanamic acid. *J. Pharm. Sci.*, **58**, 983–987.

53. Brice, G. W., and Hammer, H. F. (1969). Therapeutic nonequivalence of oxytetracycline capsules. *J. Am. Med. Assoc.*, **208**, 1189–1190.

54. Liebenberg, W., de Villiers, M., Wurster, D. E., Swanepoel, E., Dekker, T. G., and Lotter, A. P. (1999). The effect of polymorphism on powder compaction and dissolution properties of chemically equivalent oxytetracycline hydrochloride powders. *Drug Dev. Ind. Pharm.*, **25**, 1027–1033.

55. Cole, E. T., Scott, R. A., Cade, D., Conner, A. L., and Wilding, I. R. (2004). In vitro and in vivo pharmacoscintigraphic evaluation of ibuprofen hypromellose and gelatin capsules. *Pharm. Res.*, **21**, 793–798.

56. El-Malah, Y., Nazzal, S., and Bottom, C. B. (2007). Hard gelatin and hypromellose (HPMC) capsules: estimation of rupture time by real-time dissolution spectroscopy. *Drug Dev. Ind. Pharm.*, **33**, 27–34.

57. Ku, M. S., Li, W., Dulin, W., Donahue, F., Cade, D., Benameur, H., and Hutchison, K. (2010). Performance qualification of a new hypromellose capsule: Part I. Comparative evaluation of physical, mechanical and processability quality attributes of VCaps Plus®, Quali-V® and gelatin capsules. *Int. J. Pharm.*, **386**, 30–41.

58. Ku, M. S., Lu, Q., Li, W., and Chen, Y. (2011). Performance qualification of a new hypromellose capsule: Part II. Disintegration and dissolution comparison between two types of hypromellose capsules. *Int. J. Pharm.*, **416**, 16–24.

59. Cadé, D. Vcaps® Plus Capsules: A new HPMC capsule for optimum formulation of pharmaceutical dosage forms. http://capsugel.com/media/library/WP-VcapsPlus_30270_FIN_10-8-12.pdf

60. Chiwele, I., Jones, B. E., and Podczeck, F. (2000). The shell dissolution of various empty hard capsules. *Chem. Pharm. Bull.*, **48**, 951–956.

61. Tuleu, C., Khela, M. K., Evans, D. F., Jones, B. E., Nagat, S., and Basit, A. W. (2007). A scintigraphic investigation of the disintegration behavior of capsules in fasting subjects: a comparison of hypromellose capsules containing carrageenan as a gelling agent and standard gelatin capsules. *Eur. J. Pharm. Sci.*, **30**, 251–255.

62. Jones, B. E., Basit, A. W., Tuleu, C., and Khela. (2012). The disintegration behaviour of capsules in fed subjects: a comparison of hypromellose (carrageenan) capsules and standard gelatin capsules. *Int. J. Pharm.*, **424**, 40–43.

63. Yannas, I. V., and Tobolsky, A. V. (1967). Cross-linking of gelatine by dehydration. *Nature*, **215**, 509–510.

64. Welz, M. M., and Ofner, C. M. (1992). Examination of self-crosslinked gelatin as a hydrogel for controlled release. *J. Pharm. Sci.*, **81**, 85–90.

65. Dey, M., Enever, R., Kraml, M., Prue, D. G., Smith, D., and Weierstall, R. (1993). The dissolution and bioavailability of etodolac from capsules exposed to conditions of high relative humidity and temperatures. *Pharm. Res.*, **10**, 1295–1300.

66. Matsuda, Y., Kouzuki, K., Tanaka, M., Tanaka, Y., and Tanigaki, J. (1979). Photostability of gelatin capsules: effect of ultraviolet irradiation on the water vapor transmission properties and dissolution rates of indomethacin. *Yakugaku Zasshi*, **99**, 907–913.

67. Singh, S., Manikandan, R., and Singh, S. (2000). Stability testing for gelatin-based formulations: rapidly evaluating the possibility of a reduction in dissolution rates. *Pharm. Technol.*, **24**(5), 58–72.

68. Bessho, M., Kojima, T., Okuda, S., and Hara, M. (2007). Radiation-induced cross-linking of gelatin by using γ-rays: insoluble gelatin hydrogel formation. *Bull. Chem. Soc. Jpn.*, **80**, 979–985.

69. Tomoda, Y., and Tsuda, M. (1961). The importance of hydroxyl radicals as intermediates in the cross-linking of high polymers by g-irradiation. *Nature*, 190, 905.

70. Tomoda, Y., and Tsuda, M. (1961). Some aspects of the crosslinking and degradation of gelatin molecules in aqueous solution irradiated by 60Co γ-rays. *J. Polym. Sci.*, **54**, 321–328.

71. Vieira, F. F., and Del Mastro, N. L. (2002). Comparison of γ-radiation and electron beam irradiation effects on gelatin. *Radiat. Phys. Chem.*, **63**, 331–332.

72. Cataldo, F., Ursini, O., Lilla, E., and Angelini, G. (2008). Radiation-induced crosslinking of collagen gelatin into a stable hydrogel. *J. Radioanal. Nucl. Chem.*, **275**, 125–131.

73. Rama Rao, K. V., Pakhale, S. P., and Singh, S. (2003). A Film Approach for the stabilization of gelatin preparations against cross-linking. *Pharm. Technol.*, **27**, 54–63.

74. Rama Rao, K. V., and Singh, S. (2002). Sensitivity of gelatin raw materials to cross-linking: the influence of bloom strength, type, and source. *Pharm. Tech.*, **26**(December), 42–46.

75. Reich, V. G. (1995). Effect of drying conditions on structure and properties of gelatin capsules: studies with gelatin films. *Pharm. Ind.*, **57**, 63–67.

76. Sheehan, J. C., and Hlavka, J. J. (1957). The cross-linking of gelatin using a water-soluble carbodiimide. *J. Am. Chem. Soc.*, **79**, 4528–4529.

77. Davis, P., and Tabor, B. E. (1963). Kinetic study of the crosslinking of gelatin by formaldehyde and glyoxal. *J. Polym. Sci. Part A,* **1**, 799–815.

78. Digenis, G. A., Gold, T. B., and Shah, V. P. (1994). Cross-linking of gelatin capsules and its relevance to their in vitro–in vivo performance. *J. Pharm. Sci.*, **83**, 915–921.

79. Tomihata, K., and Ikada, Y. (1996). Cross-linking of gelatin with carbodiimides. *Tissue Eng.*, **2**, 307–313.

80. Bottom, C. B., Clark, M., and Carstensen, J. T. (1997). Dissolution testing of soft shell capsules -acetaminophen and nifedipine. *J. Pharm. Sci.*, **86**, 1057–1061.

81. Gold, T. B., Buice, R. G., Lodder, R. A., and Digenis, G. A. (1998). Detection of formaldehyde-induced crosslinking in soft elastic gelatin capsules using near-infrared spectrophotometry. *Pharm. Dev. Technol.*, **3**, 209–214.

82. Fan, H., and Dash, A. K. (2001). Effect of cross-linking on the in vitro release kinetics of doxorubicin from gelatin implants. *Int. J. Pharm.*, **213**, 103–116.

83. Liang, H. C., Chang, W. H., Liang, H. F., Lee, M. H., and Sung, H. W. (2004). Crosslinking structures of gelatin hydrogels crosslinked with genipin or a water-soluble carbodiimide. *J. Appl. Polym. Sci.*, **91**, 4017–4026.

84. Meyer, M. C., Straughn, A. B., Mhatre, R. M., Hussain, A., Shah, V. P., Bottom, C. B., Cole, E. T., Lesko, L. L., Mallinowski, H., and Williams, R. L. (2000). The effect of gelatin cross-linking on the bioequivalence of

hard and soft gelatin acetaminophen capsules. *Pharm. Res.*, **17**, 962–966.

85. Maeda, T., and Motoyoshi, H. (1996). Formation and characterization of formaldehyde-crosslinked gelatin film. *Kobunshi Ronbunshu*, **53**, 155–157.

86. Maeda, T., and Motoyoshi, H. (1997). Effects of poly(ethylene glycol) addition on the properties of gelatin films crosslinked with formaldehyde vapor. *Kobunshi Ronbunshu*, **54**, 138–143.

87. Kaneko, S., Okitani, A., Hayase, F., and Kato, H. (1988). Chemical modification of gelatin with vaporized hexanal. *Nippon Shokuhin Kogyo Gakkaishi*, **35**, 271–277.

88. de Carvalho, R. A., and Grosso, C. R. F. (2004). Characterization of gelatin based films modified with transglutaminase, glyoxal and formaldehyde. *Food Hydrocolloids*, **18**, 717–726.

89. Carstensen, J. T., and Rhodes, C. T. (1993). Pellicule formation in gelatin capsules. *Drug Dev. Ind. Pharm.*, **19**, 2709–2712.

90. Hakata, T., Sato, H., Watanabe, Y., and Matsumoto, M. (1994). Effect of formaldehyde on the physicochemical properties of soft gelatin capsule shells. *Chem. Pharm. Bull.*, **42**, 1138–1142.

91. Hakata, T., Sato, H., Watanabe, Y., and Matsumoto, M. (1994). Effect of storage temperature on the physicochemical properties of soft gelatin capsule shells. *Chem. Pharm. Bull.*, **42**, 1496–1500.

92. Fraenkel-Conrat, H., Cooper, M., and Olcott, H. S. (1945). Reaction of formaldehyde with proteins. *J. Am. Chem. Soc.*, **67**, 950–954.

93. Taylor, S. K., Davidson, F., and Ovenall, D. W. (1978). Carbon-13 nuclear magnetic resonance studies on gelatin crosslinking by formaldehyde. *Photogr. Sci. Eng.*, **22**, 134–138.

94. Albert, K., Peters, B., Bayer, E., Treiber, U., and Zwilling, M. (1986). Crosslinking of gelatin with formaldehyde: a 13C NMR study. *Z. Naturforsch*, **41**, 351–358.

95. Albert, K., Bayer, E., Worsching, A., and Vogele, H. (1991). Investigation of the hardening reaction of gelatin with 13C labeled formaldehyde by solution and solid state 13C NMR spectroscopy. *Z. Naturforsch*, **46**, 385–389.

96. Gold, T. B., Smith, S. L., and Digenis, G. A. (1996). Studies on the influence of pH and pancreatin on 13C-formaldehyde induced gelatin crosslinks using nuclear magnetic resonance. *Pharm. Dev. Technol.*, **1**, 21–26.

97. Ofner, C. M., Zhang, Y., Jobeck, V. C., and Bowman, B. J. (2001). Crosslinking studies in gelatin capsules treated with formaldehyde and in capsules exposed to elevated temperature and humidity. *J. Pharm. Sci.*, **90**, 79–88.

98. Cooper, J. W., Ansel, H. C., and Cadwallader, D. E. (1973). Liquid and solid interactions of primary certified colorants with pharmaceutical gelatins. *J. Pharm. Sci.*, **62**, 1156–1164.

99. Gautam, J., and Schott, H. (1994). Interaction of anionic compounds with gelatin I: binding studies. *J. Pharm. Sci.*, **83**, 922–930.

100. Bindra, D. S., Williams, T. D., and Stella, V. J. (1994). Degradation of O6-Benzylguanine in aqueous Polyethylene glycol 400 (PEG 400) solutions: concerns with formaldehyde in PEG 400. *Pharm. Res.*, **11**, 1060–1064.

101. Azaz, E., Donbrow, M., and Hamburger, R. (1973). Incompatibility of non-ionic surfactants with oxidisable drugs. *Pharm. J.*, **211**, 15.

102. Hamburger, R., Azaz, E., and Donbrow, M. (1975). Autoxidation of polyoxyethylenic non-ionic surfactants and of polyethylene glycols. *Pharm. Acta. Helv.*, **50**, 10–17.

103. McGinity, J. W., Hill, J. A., and La Via, A. L. (1975). Influence of peroxide impurities in polyethylene glycols on drug stability. *J. Pharm. Sci.*, **64**, 356–357.

104. Donbrow, M., Azaz, E., and Pillersdorf, A. (1978). Autoxidation of polysorbates. *J. Pharm. Sci.*, **67**, 1676–1681.

105. Chafetz, L., Hong, W., Tsilifonis, D. C., Taylor, A. K., and Philip, J. (1984). Decrease in the rate of capsule dissolution due to formaldehyde from Polysorbate 80 autooxidation. *J. Pharm. Sci.*, **73**, 1186–1187.

106. Johnson, D. M., and Taylor, W. F. (1984). Degradation of fenprostalene in Polyethylene glycol 400 solution. *J. Pharm. Sci.*, **73**, 1414–1417.

107. Hartauer, K. J., Bucko, J. H., Cooke, G. G., Mayer, R. F., Schwier, J. R., and Sullivan, G. R. (1993). The effect of rayon coiler on the dissolution stability of hard-shell gelatin capsules. *Pharm. Technol.*, **17**, 76–83.

108. Schwier, J. R., Cooke, G. G., Hartauer, K. J., and Yu, L. (1993). Rayon: source of furfural-a reactive aldehyde capable of insolublizing gelatin capsules. *Pharm. Technol.*, **17**, 78–80.

109. Yamamoto, T., Kobayashi, M., and Matsuura, S. (1995). Gelatin coating composition and hard gelatin capsule. *US Patent* 5,419,916.

110. Gholap, D., and Singh, S. (2004). The influence of drugs on gelatin cross-linking. *Pharm. Technol.*, **28**, 94–102.

111. Adesunloye, T. A., and Stach. P. E. (1998). Effect of glycine/citric acid on the dissolution stability of hard gelatin capsules. *Drug Dev. Ind. Pharm.*, **24**, 493–500.

112. Bergh, M., Shao, L. P., Hagelthorn, G., Gafvert, E., Lars, J., Nilsson, G., and Karlberg, A. T. (1998). Contact allergens from surfactants. Atmospheric oxidation of polyethylene alcohols, formation of ethoxylated aldehydes, and their allergenic activity. *J. Pharm. Sci.*, **87**, 276–282.

113. Ha, E., Wang, W., and Wang, Y. J. (2002). Peroxide formation in polysorbate 80 and protein stability. *J. Pharm. Sci.*, **91**, 2252–2264.

114. Kumar, V., and Kalonia, D. S. (2006). Removal of peroxides in polyethylene glycols by vacuum drying: implications in the stability of biotech and pharmaceutical formulations. *AAPS PharmSciTech*, **7**, E1–E7.

115. Wasylaschuk, W. R., Harmon, P. A., Wagner, G., Harman, A. B., Templeton, A. C., Xu, H., and Reed, R. A. (2007). Evaluation of hydroperoxides in common pharmaceutical excipients. *J. Pharm. Sci.*, **96**, 106–116.

116. Kishore, R. S. K., Pappenberger, A., Dauphin, I. B., Ross, A., Buergi, B., Staempfli, A., and Mahler, H. C. (2011). Degradation of polysorbates 20 and 80: studies on thermal autoxidation and hydrolysis. *J. Pharm. Sci.*, **100**, 721–731.

117. Ray, W. J., and Puvathingal, J. M. (1985). A simple procedure for removing contaminating aldehydes and peroxides from aqueous solutions of polyethylene glycols and of nonionic detergents that are based on the polyoxyethylene linkage. *Anal. Biochem.*, **146**, 307–312.

118. Dyakonov, T., Muir, A., Nasri, H., Toops, D., and Fatmi, A. (2010). Isolation and characterization of cetirizine degradation product: mechanism of cetirizine oxidation. *Pharm. Res.*, **27**, 1318–1324.

119. Han, S., Kim, C., and Kwon, D. (1995). Thermal degradation of poly(ethyleneglycol). *Polym. Degrad. Stab.*, **47**, 203–208.

120. Han, S., Kim, C., and Kwon, D. (1997). Thermal/oxidative degradation and stabilization of polyethylene glycol. *Polymer*, **38**, 317–323.

121. Hovorka, S. W., and Schöneich, C. (2001). Oxidative degradation of pharmaceuticals: theory, mechanisms and inhibition. *J. Pharm. Sci.*, **90**, 253–269.

122. Donbrow, M., Hamburger, R., Azaz, E., and Pillersdorf, A. (1978). Development of acidity in non-ionic surfactants: formic and acetic acid. *Analyst*, **103**, 400–402.

123. Lloyd, W. G. (1956). The low temperature autoxidation of diethylene glycol. *J. Am. Chem. Soc.*, **78**, 72–75.

124. Li, Z., Kozlowski, B. M., and Chang, E. P. (2007). Analysis of aldehydes in excipients used in liquid/semi-solid formulations by gas chromatography–negative chemical ionization mass spectrometry. *J. Chromatogr. A*, **1160**, 299–305.

125. Kiritsakis, A. K., and Dugan, L. R. (1984). Effect of selected storage conditions and packaging materials on olive oil quality. *JAOCS*, **61**, 1868–1870.

126. Brodnitz, M. H. (1968). Autoxidation of saturated fatty acids. A review. *J. Agric. Food Chem.*, **16**, 994–999.

127. Stirton, A. J., Turer, H., and Riemenschneider, R. W. (1945). Oxygen absorption of methyl esters of fatty acids and the effect of antioxidants. *Oil Soap*, **22**, 81–83.

128. Shahidi, F., and Zhong, Y. (2005). Lipid oxidation: measurement methods. In *Edible Oil and Fat Products: Chemistry, Properties, and Health Effects*, 6th ed., Shahidi, F. (Bailey's Industrial Oil and Fat Products, Vol. 1), Hoboken, New Jersey: John Wiley & Sons, pp. 1–8.

129. Chapter <2040> Disintegration and Dissolution of Dietary Supplements. In United States Pharmacopeia and National Formulary USP 36-NF 31. The United States Pharmacopeial Convention, Inc., Rockville, MD, 2013.

130. Chapter <711> Dissolution. In United States Pharmacopeia and National Formulary USP 36-NF 31. The United States Pharmacopeial Convention, Inc., Rockville, MD, 2013.

131. Gelatin Capsule Working Group. (1998). Collaborative development of two-tiered dissolution testing for gelatin capsules and gelatin-coated tablets using enzyme-containing media. Pharmacopeial Forum, 24, 7045–7050.

132. Marques, M. R. C., Brown, W., Giancaspro, G., Davydova, N., Chang, E., Fringer, J., Hauck, W., and DeStefano, A. (2012). Challenges in dissolution testing. *Dissolut. Technol.*, **19**(3), 39–41.

133. Gallery, J., Han, J. H., and Abraham, C. (2004). Pepsin and pancreatin performance in the dissolution of crosslinked gelatin capsules from pH 1 to 8. *Pharm. Forum*, **30**, 1084–1089.

134. Murthy, K. S., Reisch, R. G. J., and Fawzi, M. B. (1989). Dissolution stability of hard-shell capsule products, Part II: the effect of dissolution test conditions on in vitro drug release. *Pharm. Technol.*, **13**(6), 53–58.

135. Piper, D. W., and Fenton, B. H. (1965). pH stability and activity curves of pepsin with special reference to their clinical importance. *Gut*, **6**, 506–508.

136. Shiu, G. K. (1996). Pharmaceutical product quality research: an overview for solid oral dosage forms. *J. Food Drug Anal.*, **4**, 293–300.

137. GlaxoSmithKline Clinical Study ID ARI10018: An Evaluation of the Bioequivalence of GI198745/dutasteride Soft Gelatin Capsules Compared to GI198745/dutasteride Cross-linked Gelatin Capsules in Healthy Male Volunteers. http://download.gsk-clinicalstudyregister.com/files/912.pdf

138. EMEA Scientific Discussion on Targretin. (2005). p 6. http://www.ema.europa.eu/docs/en_GB/document_library/EPAR_-_Scientific_Discussion/human/000326/WC500034204.pdf

139. Tian, H., Mathieu, M., Pan, W., Geng, W., Umagat, H., Tang, K., and Chokshi, H. (2003). Dissolution method development for a lipid based soft gelatin capsule formulation. AAPS Annual Meeting and Exposition, Salt Lake City, UT. http://www.aapsj.org/abstracts/Am_2003/AAPS2003-002565.PDF

140. Mhatre, R. M., Malinowski, H., Nguyen, H., Meyer, M. C., Straughn, A. B., Lesko, L. L., and Williams, R. L. (1997). The effects of cross linking in gelatin capsules on the bioequivalence of acetaminophen. *Pharm. Res.*, **14**, S-251.

141. Digenis, G. A., Sandefer, E. P., Page, R. C., Doll, W. J., Gold, T. B., and Darwazeh, N. B. (2000). Bioequivalence study of stressed and nonstressed hard gelatin capsules using amoxicillin as a drug marker and gamma scintigraphy to confirm time and GI location of in vivo capsules rupture. *Pharm. Res.*, **17**, 572–582.

142. Johnson, B. F., Mcauley, P. V., Smith, P. M., and French, J. A. G. (1977). The effects of storage upon in vitro and in vivo characteristics of soft gelatin capsules containing digoxin. *J. Pharm. Pharmacol.*, **29**, 576–578.

143. Marques, M. R. C., and Brown, W. (2013). Updates on USP activities related to dissolution, disintegration, and drug release. *Dissolut. Technol.*, **20**(3), 54–55.

144. Gold, T. B., and Digenis, G. A. (2012). Influence of acid and pepsin on 13C-formaldehyde induced gelatin crosslinks using carbon-13 NMR. http://www.metricsinc.com/wp-content/uploads/2012/07/Brad_Gold-2004.pdf

145. Gao, P., Charton, M., and Morzowich, W. (2006). Speeding development of poorly soluble/poorly permeable drugs by SEDDS/S-SEDDS formulations and prodrugs (Part II). *Am. Pharm. Rev.*, **9**, 16–23.

146. Charman, S. A., Charman, W. N., Rogge, M. C., Wilson, T. D., Dutko, F. J., and Pouton, C. W. (1992). Self-emulsifying drug delivery systems: formulation and biopharmaceutic evaluation of an investigational lipophilic compound. *Pharm. Res.*, **9**, 87–93.

147. Porter, C. J. H., Kaukonen, A. M., Boyd, B. J., Edwards, G. A., and Charman, W. N. (2004). Susceptibility to lipase-mediated digestion reduces the oral bioavailability of danazol after administration as a medium-chain lipid-based microemulsion formulation. *Pharm. Res.*, **21**, 1405–1412.

148. Cuine, J. F., Charman, W. N., Pouton, C. W., Edwards, G. A., and Porter, C. J. H. (2007). Increasing the proportional content of surfactant (Cremophor EL) relative to lipid in self-emulsifying lipid-based formulations of danazol reduces oral bioavailability in beagle dogs. *Pharm. Res.*, **24**, 748–757.

149. Shah, N. H., Carjaval, M. T., Patel, C. I., Infeld, M. H., and Malick, A. W. (1994). Self-emulsifying drug delivery systems (SEDDS) with polyglycolyzed glycerides for improving in vitro dissolution and oral absorption of lipophilic drugs. *Int. J. Pharm.*, **106**, 15–23.

150. Li, P., Ghosh, A., Wagner, R. F., Krill, S., Joshi, Y. M., and Serajuddin, A. T. M. (2005). Effect of combined use of nonionic surfactant on formation of oil-in-water microemulsions. *Int. J. Pharm.*, **288**, 27–34.

151. Kommuru, T. R., Gurley, B., Khan, M. A., and Reddy, I. K. (2001). Selfemulsifying drug delivery systems (SEDDS) of coenzyme Q10: formulation development and bioavailability assessment. *Int. J. Pharm.*, **212**, 233–246.

152. MacGregor, K. J., Embleton, J. K., Lacy, J. E., Perry, E. A., Solomon, L. J., Seager, H., and Pouton, C. W. (1997). Influence of lipolysis on drug absorption from the gastro-intestinal tract. *Adv. Drug Deliv. Rev.*, **25**, 33–46.

153. Reymond, J., and Sucker, H. (1988). In vitro model for ciclosporin intestinal absorption in lipid vehicles. *Pharm. Res.*, **5**, 673–676.

154. Dahan, A., and Hoffman, A. (2006). Use of a dynamic in vitro lipolysis model to rationalize oral formulation development for poor water soluble drugs: correlation with in vivo data and the relationship to intra-enterocyte processes in rats. *Pharm. Res.*, **23**, 2165–2174.

155. Bauer, J., Spanton, S., Henry, R., Quick, J., Dziki, W., Porter, W., and Morris, J. (2001). Ritonavir: an extraordinary example of conformational polymorphism. *Pharm. Res.*, **18**, 859–866.

156. Anderson, N. R., and Gullapalli, R. P. (2005). b-Carboline pharmaceutical compositions. *US Patent* 6,841,167.

157. Speer, J., Flynn, J., Gullapalli, R., Agarwal, R., and Muffuid, P. (2005). Influence of digestive enzymes on dissolution of a poorly water soluble compound from cross-linked capsules in sodium lauryl sulfate medium. AAPS Annual Meeting and Exposition, Nashville, TN. www.aapsj.org/abstracts/AM_2005/AAPS2005-000998.pdf

158. Hu, J., Kyad, A., Ku, V., Zhou, P., and Cauchon, N. (2005). A comparison of dissolution testing on lipid soft gelatin capsules using USP apparatus 2 and apparatus 4. *Dissolut. Technol.*, **12**(2), 6–9.

159. Brown, C. K., Chokshi, H. P., Nickerson, B., Reed, R. A., Rohrs, B. R., and Shah, P. A. (2004). Acceptable analytical practices for dissolution testing of poorly soluble compounds. *Pharm. Technol.* 56–65.

160. Nishimura, H., Hayashi, C., Aiba, T., Okamoto, I., Miyamoto, Y., Nakade, S., Takeda, K., and Kurosaki, Y. (2007). Application of the correlation of in vitro dissolution behavior and in vivo plasma concentration profile (IVIVC) for soft-gel capsules-a pointless pursuit? *Biol. Pharm. Bull.*, **30**, 2221–2225.

161. Rossi, R. C., Dias, C. L., Donato, E. M., Martins, L. A., Bergold, A. M., and Froehlich, P. E. (2007). Development and validation of dissolution test for ritonavir soft gelatin capsules based on in vivo data. *Int. J. Pharm.*, **338**, 119–124.

162. Joseph, M., Mayberry, C., and Schiermeyer, C. (2005). A 2-step dissolution method for the elimination of crosslinking in nimodipine soft gelatin capsules while providing solubility of a water-insoluble drug. AAPS Annual Meeting and Exposition, Nashville, TN. www.aapsj.org/abstracts/AM_2005/AAPS2005-001000.pdf

163. Shah, V. P., Konecny, J. J., Everett, R. L., McCullough, B., Noorizadeh, A. C., and Skelly, J. P. (1989). In vitro dissolution profile of water-insoluble drug dosage forms in the presence of surfactants. *Pharm. Res.*, **6**, 612–618.

164. Noory, C., Tran, N., Ouderkirk, L., and Shah, V. (2002). Steps for development of a dissolution test for sparingly water-soluble drug products. *Am. Pharm. Rev.*, **5**, 16–21.

165. Buri, P., and Humbert-Droz, J. (1983). Solubilisation de principes actifs insolubles par des constituants des sucs digestifs. *3rd Int. Cong. Pharm. Technol. Paris*, **4**, 136–143.

166. Larson, C. E., and Greenberg, D. M. (1933). A paradoxical solubility phenomenon with gelatin. *J. Am. Chem. Soc.*, **55**, 2798–2799.

167. Pillay, V., and Fassihi, R. (1999). Unconventional dissolution methodologies. *J. Pharm. Sci.*, **88**, 843–851.

168. Pillay, V., and Fassihi, R. (1999). A new method for dissolution studies of liquid-filled capsules employing nifedipine as a model drug. *Pharm. Res.*, **16**, 333–337.

169. Zhao, F., Malayev, V., Rao, V., and Hussain, M. (2004). Effect of sodium lauryl sulfate in dissolution media on dissolution of hard gelatin capsule shells. *Pharm. Res.*, **21**, 144–148.

170. Brian, R. R. (2001). Dissolution method development for poorly soluble compounds. *Dissolut. Technol.*, **8**, 1–5.

171. Bartnik, F. G. (1992). Interaction of anionic surfactants with proteins, enzymes and membranes. In: *Anionic Surfactants: Biochemistry, Toxicology, Dermatology*, 2nd ed., Gloxhuber, C., and Kunstler, K., eds. (Surfactant Science Series, Vol. 43), New York: Marcel Dekker, Inc., pp. 1–42.

172. Marchais, H., Cayzeele, G., Legendre, J. V., Skiba, M., and Arnaud, P. (2003). Cross-linking of hard gelatin carbamazepine capsules: effect of dissolution conditions on in vitro drug release. *Eur. J. Pharm. Sci.*, **19**, 129–132.

173. Savelli, G., Spreti, N., and Di Profio, P. (2005). Enzyme activity and stability control by amphiphilic self-organizing systems in aqueous solutions. *Curr. Opin. Colloid Interface Sci.*, **5**, 111–117.

174. Pennings, F. H., Kwee, B. L. S., and Vromans, H. (2006). Influence of enzymes and surfactants on the disintegration behavior of cross-linked hard gelatin capsules during dissolution. *Drug Dev. Ind. Pharm.*, **32**, 33–37.

175. General Chapter <1094> Capsules - Dissolution Testing and Related Quality Attributes. Pharmacopeial Forum 39(3): In-Process Revision. 2013.

176. Onesippe, C., and Lagerge, S. (2009). Study of the complex formation between sodium dodecyl sulphate and gelatin. *Colloids Surf. A*, **337**, 61–66.

177. Wüstneck, R., Wetzel, R., Buder, E., and Hermel, H. (1988). The modification of the triple helical structure of gelatin in aqueous solution I. the influence of anionic surfactants, pH-value, and temperature. *Colloid Polym. Sci.*, **266**, 1061–1067.

178. Best, R., and Zhao, F. (2009). Effect of sodium lauryl sulfate on dissolution of hard gelatin capsule formulations. AAPS Annual Meeting and Exposition, Los Angeles, CA. http://www.aapsj.org/abstracts/AM_2009/AAPS2009-002662.PDF

179. Wüstneck, R., Buder, E., Wetzel, R., and Hermel, H. (1989). The modification of the triple helical structure of gelatin in aqueous solution 3. the influence of cationic surfactants. *Colloid Polym. Sci.*, **267**, 429–433.

180. Fruhner, H., and Kretzschmar, G. (1989). Effect of pH on the binding of alkyl sulfates to gelatin. *Colloid Polym. Sci.*, **267**, 839–843.

181. Ropers, M. H., Czichocki, G., and Brezesinski, G. (2003). Counterion effect on the thermodynamics of micellization of alkyl sulfates. *J. Phys. Chem. B*, **107**, 5281–5288.

182. Tamaki, K., and Tamamushi, B. (1955). The interactions of gelatin molecule with surface active ions. *Bull. Chem. Soc. Jpn.*, **28**, 555–559.

183. Woolfrey, S. G., Banzon, G. M., and Groves, M. J. (1986). The effect of sodium chloride on the dynamic surface tension of sodium dodecyl sulfate solutions. *J. Colloid Interface Sci.*, **112**, 583–587.

184. Crison, J. R., Weiner, N. D., and Amidon, G. L. (1997). Dissolution media for in vitro testing of water-insoluble drugs: effect of surfactant purity and electrolyte on in vitro dissolution of carbamazepine in aqueous solutions of sodium lauryl sulfate. *J. Pharm. Sci.*, **86**, 384–388.

185. Chemburkar, S. R., Bauer, J., Deming, K., Spiwek, H., Patel, K., Morris, J., Henry, R., Spanton, S., Dziki, W., Porter, W., Quick, J., Bauer, P., Donaubauer, J., Narayanan, B. A., Soldani, M., Riley, D., and McFarland, K. (2000). Dealing with the impact of ritonavir polymorphs on the late stages of bulk drug process development. *Org. Process Res. Dev.*, **4**, 413–417.

186. Chopra, S. K., and Makadia, T. T. (1991). Pharmaceutical capsules containing panetidine. *US Patent* 5,028,432.

187. Hom, F. S., Veresh, S. A., and Miskel, J. J. (1973). Soft gelatin capsules I: factors affecting capsule-shell dissolution rate. *J. Pharm. Sci.*, **62**, 1001–1006.

188. Herzyk, T., Christensen, G. A., Patel, K., and Palepu, N. R. (1999). Pharmaceutical formulation for camptothecin analogues in gelatin capsules. WIPO International Publication Number WO/1999/06031.

189. Marriott, C., and Kellaway, I. W. (1978). The release of antibacterial agents from glycerolgelatin gels. *Drug Dev. Ind. Pharm.*, **4**, 195–208.

190. Stein, D., and Bindra, D. S. (2007). Stabilization of hard gelatin capsule shells filled with polyethylene glycol matrices. *Pharm. Dev. Tech.*, **12**, 71–77.

191. Marques, M. R. C., and Brown, W. (2013). Question and answer section. *Dissolut. Technol.*, **20**(3), 56.

192. Crist, G. B. (2009). Trends in small-volume dissolution apparatus for low-dose compounds. *Dissolut. Technol.*, **16**(1), 19–22.

193. Marques, M. R. C., Cole, E., Kruep, D., Gray, V., Murachanian, D., Brown, W. E., and Giancaspro, G. I. (2009). Liquid-filled gelatin capsules. *Pharmacopeial Forum*, **35**, 1029–1041.

194. Hu, J., Kyad, A., Ku, V., Zhou, P., Cauchon, N. (2005). A comparison of dissolution testing on lipid soft gelatin capsules using USP apparatus 2 and apparatus 4. *Dissolut. Technol.*, **12**(2), 6–9.

195. Chapter <701> Disintegration. In United States Pharmacopeia and National Formulary USP 36-NF 31. The United States Pharmacopeial Convention, Inc., Rockville, MD, 2013.

196. Almukainzi, M., Salehi, M., Chacra, N. A. B., and Löbenberg, R. (2011). Comparison of the rupture and disintegration tests for soft-shell capsules. *Dissolut. Technol.*, **18**(1), 21–25.

197. Hernell, O., Staggers, J. E., and Carey, M. C. (1990). Physical-chemical behavior of dietary and biliary lipids during intestinal digestion and absorption. 2. Phase analysis and aggregation states of luminal lipids during duodenal fat digestion in healthy adult human beings. *Biochemistry*, **29**, 2041–2056.

198. Staggers, J. E., Hernell, O., Stafford, R. J., and Carey, M. C. (1990). Physical-chemical behavior of dietary and biliary lipids during intestinal digestion and absorption. 1. Phase behavior and aggregation states of model lipid systems patterned after aqueous duodenal contents of healthy adult human beings. *Biochemistry*, **29**, 2028–2040.

199. Kossena, G. A., Charman, W. N., Wilson, C. G., O'Mahony, B., Lindsay, B., Hempenstall, J. M., Davison, C. L., Crowley, P. J., and Porter, C. J. H. (2007). Low dose lipid formulations: effects on gastric emptying and biliary Secretion. *Pharm. Res.*, **24**, 2084–2096.

200. Clement, J. (1980). Intestinal absorption of triglycerols. *Reprod. Nutr. Dev.*, **20**, 1285–1307.

201. Charman, W. N., and Stella, V. J. (1986). Estimating the maximal potential for intestinal lymphatic transport of lipophilic drug molecules. *Int. J. Pharm.*, **34**, 175–178.

202. Nankervis, R., Davis, S. S., Day, N. H., and Shaw, P. N. (1995). Effect of lipid vehicle on the intestinal lymphatic transport of isotretinoin in the rat. *Int. J. Pharm.*, **119**, 173–181.

203. Nankervis, R., Davis, S. S., Day, N. H., and Shaw, P. N. (1996). Intestinal lymphatic transport of three retinoids in the rat after oral

administration: effect of lipophilicity and lipid vehicle. *Int. J. Pharm.*, **130**, 57–64.

204. Porter, C. J. H. (1997). Drug delivery to the lymphatic system. *Crit. Rev. Ther. Drug Carrier Syst.*, **14**, 333–393.

205. Porter, C. J. H., and Charman, W. N. (1997). Uptake of drugs into the intestinal lymphatics after oral administration. *Adv. Drug Deliv. Rev.*, **25**, 71–89.

206. Caliph, S. M., Charman, W. N., and Porter, C. J. H. (2000). Effect of short-, medium-, and long-chain fatty acid-based vehicles on the absolute oral bioavailability and intestinal lymphatic transport of halofantrine and assessment of mass balance in lymph-cannulated and non-cannulated rats. *J. Pharm. Sci.*, **89**, 1073–1084.

207. Armand, M., Borel, P., Pasquier, B., Dubois, C., Senft, M., Andre, M., Peyrot, J., Salducci, J., and Lairon, D. (1996). Physicochemical characteristics of emulsions during fat digestion in human stomach and duodenum. *Am. J. Physiol.*, **271**, G172–G183.

208. Ladas, S. D., Isaacs, P. E. T., Murphy, G. M., and Sladen, G. E. (1984). Comparison of the effects of medium and long chain triglyceride containing liquid meals on gall bladder and small intestinal function in normal man. *Gut*, **25**, 405–411.

209. Sek, L., Porter, C. J. H., Kaukonen, A. M., and Charman, W. N. (2002). Evaluation of the in-vitro digestion profiles of long and medium chain glycerides and the phase behavior of their lipolytic products. *J. Pharm. Pharmacol.*, **54**, 29–41.

210. Christensen, J. O., Schultz, K., Mollgaard, B., Kristensen, H. G., and Mullertz, A. (2004). Solubilisation of poorly water-soluble drugs during in vitro lipolysis of medium- and long-chain triacylglycerols. *Eur. J. Pharm. Sci.*, **23**, 287–296.

211. Deckelbaum, R. J., Hamilton, J. A., Moser, A., Bengtsson-Olivecrona, G., Butbul, E., Carpentier, Y. A., Gutman, A., and Olivecrona, T. (1990). Medium-chain versus long-chain triacylglycerols emulsion hydrolysis by lipoprotein lipase and hepatic lipase: implications for the mechanisms of lipase action. *Biochemistry*, **29**, 1136–1142.

212. Tsuzuki, W., Ue, A., Nagao, A., and Akasaka, K. (2002). Fluorimetric analysis of lipase hydrolysis of intermediate- and long-chain glycerides. *Analyst*, **127**, 669–673.

213. Lee, K. W., Porter, C. J. H., and Boyd, B. J. (2013). The effect of administered dose of lipid-based formulations on the in vitro and in vivo performance of cinnarizine as a model poorly water-soluble drug. *J. Pharm. Sci.*, **102**, 565–578.

214. Bach, A. C., and Babayan, V. K. (1982). Medium-chain triglycerides: an update. *Am. J. Clin. Nutr.*, **36**, 950–962.

215. Hunt, J. N., and Knox, M. T. (1968). A relation between chain length of fatty acids and slowing of gastric emptying. *J. Physiol.*, **194**, 327–336.

216. Fernando-Warnakulasuriya, G., Staggers, J., Frost, S. C., and Ma, W. (1981). Studies on fat digestion, absorption, and transport in the suckling rat. I. Fatty acid composition and concentrations of major lipid components. *J. Lipid Res.*, **22**, 668–674.

217. Mascioli, E. A., Lopes, S., Randall, S., Porter, K. A., Kater, G., Hirschberg, Y., Babayan, V. K., Bistrian, B. R., and Blackburn, G. L. (1989). Serum fatty acid profiles after intravenous medium chain triglyceride administration. *Lipids*, **24**, 793–798.

218. Caspary, W. F. (1992). Physiology and pathophysiology of intestinal absorption. *Am. J. Clin. Nutr.*, **55**, 299S–308S.

219. Kossena, G. A., Boyd, B. J., Porter, C. J. H., and Charman, W. N. (2003). Separation and characterization of the colloidal phases produced on digestion of common formulation lipids and assessment of their impact on the apparent solubility of selected poorly water soluble drugs. *J. Pharm. Sci.*, **92**, 634–648.

220. Porter, C. J. H., Kaukonen, A. M., Taillardat-Bertschinger, A., Boyd, B. J., O'Connor, J. M., Edwards, G. A., and Charman, W. N. (2004). Use of in vitro lipid digestion data to explain the in vivo performance of triglyceride based lipid formulations for the oral administration of poorly water-soluble drugs: studies with halofantrine. *J. Pharm. Sci.*, **93**, 1110–1121.

221. Sek, L., Porter, C. J. H., and Charman, W. N. (2001). Characterisation and quantification of medium chain and long chain triglycerides and their in vitro digestion products, by HPTLC coupled with in situ densitometric analysis. *J. Pharm. Biomed. Anal.*, **25**, 651–661.

222. Zangenberg, N. H., Müllertz, A., Kristensen, H. G., and Hovgaard, L. (2001). A dynamic in vitro lipolysis model II: evaluation of the model. *Eur. J. Pharm. Sci.*, **14**, 237–244.

223. Dahan, A., and Hoffman, A. (2007). The effect of different lipid based formulations on the oral absorption of lipophilic drugs: the ability of in vitro lipolysis and consecutive ex vivo intestinal permeability data to predict in vivo bioavailability in rats. *Eur. J. Pharm. Biopharm.*, **67**, 96–105.

224. Thomas, N., Holm, R., Rades, T., and Müllertz, A. (2012). Characterising lipid lipolysis and its implication in lipid-based formulation development. *The AAPS J.*, **14**(4), 860–871.

225. Reymond, J., Sucker, H., and Vonderscher, J. (1988). In vivo model for ciclosporin intestinal absorption in lipid vehicles. *Pharm. Res.*, **5**, 677–679.

226. Eastoe, J. E., and Leach, A. A. (1977). Chemical constitution of gelatin. In *The Science and Technology of Gelatin*, Ward, A. G., and Courts, A., eds. New York: Academic Press, pp. 77–80.

227. Huang, T., Garceau, M. E., and Gao, P. (2003). Liquid chromatographic determination of residual hydrogen peroxide in pharmaceutical excipients using platinum and wired enzyme electrodes. *J. Pharm. Bio. Anal.*, **31**, 1203–1210.

228. Barrio, M. D., Hu, J., Zhou, P., and Cauchon, N. (2006). Simultaneous determination of formic acid and formaldehyde in pharmaceutical excipients using headspace GC/MS. *J. Pharm. Biomed. Anal.*, **41**, 738–743.

229. Jantratid, E., Janssen, N., Reppas, C., and Dressman, J. B. (2008). Dissolution media simulating conditions in the proximal human gastrointestinal tract: an update. *Pharm. Res.*, **25**, 1663–1676.

230. Vonderscher, J., and Meinzer, A. (1994). Rationale for the development of Sandimmune Neoral. *Transplant. Proc.*, **26**, 2925–2927.

231. Andrysek, T. (2003). Impact of physical properties of formulations on bioavailability of active substance: current and novel drugs with cyclosporine. *Mol. Immunol.* **39**, 1061–1065.

232. Kovarik, J. M., Mueller, E. A., Van Bree, J. B., Tetzloff, W., and Kutz, K. (1994). Reduced inter- and intraindividual variability in cyclosporine pharmacokinetics from a microemulsion formulation. *J. Pharm. Sci.*, **83**, 444–446.

233. Mueller, E. A., Kovarik, J. M., Van Bree, J. B., Tetzloff, W., Grevel, J., and Kutz, K. (1994). Improved dose linearity of cyclosporine pharmacokinetics from a microemulsion formulation. *Pharm. Res.*, **11**, 301–304.

Chapter 19

Current and Emerging Non-compendial Methods for Dissolution Testing

Namita Tipnis and Diane J. Burgess

Department of Pharmaceutical Sciences, University of Connecticut, Storrs, CT 06269, USA
d.burgess@uconn.edu

Dissolution testing of conventional and non-conventional dosage forms was introduced in the 1960s. Dissolution testing is a compendial test that is used by all pharmacopeias to evaluate drug release. Apart from the standardized tests, there are several non-compendial methods which are qualified and validated for use in drug release testing. This chapter is divided into five sections. The first section provides an introduction stating the importance of dissolution testing. The second section briefly discusses existing compendial methods along with any non-compendial modifications. The third section describes non-compendial apparatus for different types of dosage forms. The fourth section explains the various detection techniques such as UV imaging with Raman spectroscopy, FTIR-ATR, and fiber optics. The final section provides a summary of all findings.

Poorly Soluble Drugs: Dissolution and Drug Release
Edited by Gregory K. Webster, J. Derek Jackson, and Robert G. Bell
Copyright © 2017 Pan Stanford Publishing Pte. Ltd.
ISBN 978-981-4745-45-1 (Hardcover), 978-981-4745-46-8 (eBook)
www.panstanford.com

19.1 Introduction

Dissolution testing is an important quality control test in pharmaceutical development. It plays a major role in monitoring product quality and process performance in vitro as well establishing relationships with in vivo product performance.[1,2]

Initially, dissolution testing was developed only for oral immediate release formulations, but later was also made a requirement for modified release and other special dosage forms.[3] For immediate-release formulations, the term "dissolution test" is used, whereas for sustained release formulations, "drug release test" or "in vitro release test" is used.[3–5]

A single test cannot always be used to study various drug release properties and hence different apparatuses and procedures have been employed depending on the type of dosage form. Compendial methods should be the first choice for dissolution testing. For any dosage form, modifications of existing tests or alternative equipment may be used only when compendial tests have been proven unsuccessful. Non-compendial methods are any methods which are not included in official pharmacopeia. When a non-compendial method is used for drug release testing, the experimental test conditions, qualifications and validation steps should be in accordance with guidance set by the International Federation of Pharmaceutical Sciences (FIP) and U.S. Food and Drug Administration (FDA).[6–8]

During preclinical testing, phase I, and early phase II trials, it is acceptable to use a qualified method instead of a fully validated method.[9] A qualified method is one for which there exists sufficient data to determine reliability of the method and controls variability. The data generated during qualification studies can eventually be applied towards validation studies. When the product reaches phase III clinical trials, the authorities expect to conduct large-scale studies on products which are truly representative of the proposed marketed products. Accordingly, the release testing method should be fully validated prior to phase III trials. Method validation means documenting the performance characteristics of the method (reliability, sensitivity, specificity, precision, accuracy, quantification range, linearity of standard curve, and robustness) for the intended analytical application.[10-12] The compendial tests that are listed in the official pharmacopeia are qualified and validated methods. Qualification and validation must also be conducted for non-compendial tests.[13-15] Non-compendial tests can be used to establish standardized protocols and even in vitro–in vivo correlation can be achieved. Kumar et al.[16] have developed a Level A IVIVC for indomethacin gelatin nanoparticles. The in vitro release testing was conducted using a dialysis sac membrane of appropriate molecular weight cut-off. Cao et al.[17] have developed a Level A IVIVC for silybin meglumine hollow-sphere mesoporous silica nanoparticles. A dialysis bag in combination with USP apparatus 1 was used for in vitro release testing.

19.1.1 Need for Non-compendial Tests

During early product development stages, often-limited amounts of product are available for dissolution testing, in which case USP apparatus 1 and 2 are not suitable for use, as they require large quantities of product which maybe a concern for powdered dosage forms such as microspheres, liposomes, etc. Poor analytical sensitivity in some cases and lack of biorelevance may require use of non-compendial methods. Simulation of gastrointestinal transit conditions is not easily possible in apparatus 1 and 2. Media evaporation is a problem in apparatus 3 during longer test times.[18] For low-dose drugs, it might not be possible to detect drug concentration in media and thus conventional high-volume dissolution tests may fail.[19-22]

19.2 Overview of Compendial Methods for in vitro Release and Their Non-compendial Modifications

19.2.1 USP Apparatus 1, Rotating Basket, and USP Apparatus 2, Rotating Paddle

USP apparatus 1 is used for dissolution testing of various dosage forms such as single units (tablets, capsules, suppositories) and multiple units (encapsulated beads). USP apparatus 2 is used for dissolution testing of tablets, capsules, and particulates such as suspensions and powders. The conventional basket and paddle apparatus uses a 1000 mL hemispherical bottom vessel that contains dissolution media. The dosage form is placed in the basket or at the bottom of the vessel for USP apparatus 1 and 2, respectively. The stirring speed is maintained at 50–100 rpm for the basket apparatus and 25–75 rpm for the paddle apparatus.[23] Typically, stirring speeds outside these ranges are unacceptable since poor hydrodynamic conditions are associated with speeds below these ranges and high turbulence can occur at speeds above these ranges.[23] In order to maintain sink conditions for poorly soluble and bolus dosage forms, vessels up to 4000 mL have been incorporated into USP apparatus 1 and 2. The operational minimum volume for 1000 mL vessels is approximately 500 mL since the hydrodynamic conditions are unstable below 500 mL.[24] Smaller volumes have been used. For example, dissolution testing of proton pump inhibitors (omeprazole magnesium and esomeprazole magnesium) requires only 300 mL of 10 mM HCl as the release media.[24] A stainless steel wire helical sinker can be used with dosage forms that would otherwise float in the paddle apparatus. Commercially available sinkers need to be validated prior to use.[24–28]

Recently, small-volume dissolution vessels with scaled-down baskets and paddles have been developed (Fig. 19.1). These small-volume vessels can easily handle nano- and picogram levels of drug. The operational minimum volume of 100 mL and 200 mL vessels is approximately 30 mL.[23,29,30]

19.2.2 USP Apparatus 3: Reciprocating Cylinder

USP apparatus 3 is used for dissolution testing of encapsulated beads, tablets, and capsules. USP apparatus 3 consists of a set each of cylindrical flat-bottomed glass vessels and glass reciprocating cylinders. The glass-reciprocating cylinders are covered by mesh at the top and bottom. A motor driven assembly is needed to reciprocate the glass cylinders inside the vessels. The vessels are immersed in suitable aqueous media, which is typically kept at 37°C

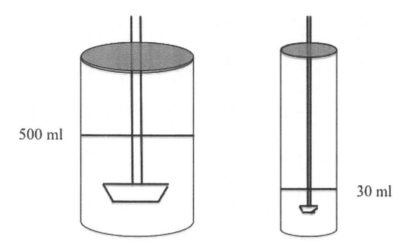

500 ml

30 ml

Figure 19.1 Representative diagram of a 1000 mL and a 100 mL (small volume) dissolution vessel with respective operational minimums.

$\pm$ 0.5°C during the test. The traditional vessel is 300 mL, which requires an operational minimum glass-reciprocating cylinder of approximately 150 mL.[24,31]

A non-compendial version that has been developed utilizes a 100 mL vessel with a scaled-down glass cylinder of approximately 50 mL.[24,32]

19.2.3 USP Apparatus 4: Flow-Through Cell

USP apparatus 4 is used for dissolution testing of tablets, capsules, implants, powders, granules, suppositories, soft gelatin capsules, ointments, suspensions, microspheres, and nanoparticles. The USP apparatus 4 was developed to simulate gastrointestinal conditions by exposing extended release and poorly soluble dosage forms to release media of differing pH.[33–35] The flow through cell assembly consists of a reservoir, pump, a flow-through cell, and a water bath. USP apparatus 4 can be used in both open- and closed-loop configurations. In the open-loop configuration, media passes once through the cell and then passes to a detector or is collected/discharged. The open-loop configuration therefore requires high media volume. In the closed-loop configuration, the media circulates through the cell during the entire testing period and samples are withdrawn from the media reservoir at periodic time intervals and replaced with fresh media. Zolnik et al.[36] have developed a modified USP 4 method for microspheres. Glass beads and microspheres are mixed to prevent microsphere aggregation. This system was run in the closed-loop configuration. The advantages of the closed-loop system include no media evaporation, easy sampling with media replacement, and flexibility of media volume.[37,38]

Non-compendial cells have been developed for USP apparatus 4 that allow dissolution testing of various different dosage forms (Fig. 19.2).[39] Studies are being carried out to reduce the media volume of the closed system to about 15 mL to facilitate dissolution testing of implants and other low-dose products.[40]

Bhardwaj et al.[41] have developed a novel dialysis adapter for USP apparatus 4 for nanoparticulate dosage forms (such as liposomes and nanoparticles) (Fig. 19.3). They compared this novel dialysis adaptor method with existing dialysis and reverse dialysis method. The results for three different formulations (solution, suspension, and liposomes) demonstrated that the dialysis adaptor method was not limited to flow rate and was independent of the volume present in the dialysis adaptor. The dialysis adaptor method was also able to discriminate dexamethasone release from various liposome formulations.

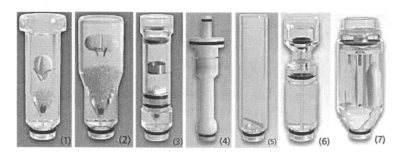

Figure 19.2 Different cell designs based on types of dosage form: (1) cell for tablets (12 mm), (2) cell for tablets (22.6 mm), (3) cell for powders and granulates, (4) cell for drug eluting stents (DES), (5) cell for large medical devices, (6) cell for implants, and (7) cell for suppositories and soft gelatin capsules. Adapted from Ag.[39]

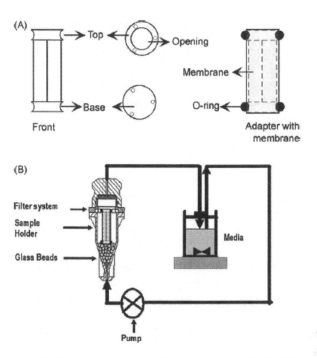

Figure 19.3 (a) Dialysis sac adaptor for apparatus 4. (b) Modified USP 4 with the dialysis sac adaptor placed in a flow-through cell. Figure shows a closed-loop system. Adapted from Bhardwaj and Burgess.[41]

19.2.4 USP Apparatus 5, Paddle over Disk, and USP Apparatus 6, Rotating Cylinder

USP apparatus 5 and 6 are used for dissolution testing of transdermal patches. Apparatus 5 uses the same paddle and hemispheric bottom vessel assembly as apparatus 2. In addition, it has a stainless steel disk placed at bottom of the vessel onto which the patch is placed. Apparatus 6 uses a stainless steel stirring element and a hemispheric bottom vessel. The transdermal patch is attached to the cylinder walls. Apparatus 5 is used for matrix-type transdermal patches, whereas apparatus 6 is used for reservoir-type transdermal patches.[24]

There have been no non-compendial modifications reported in the literature for apparatus 5 and 6.

19.2.5 USP Apparatus 7: Reciprocating Holder

USP apparatus 7 is used for dissolution testing of transdermal patches, tablets, capsules, implants, microspheres, and drug eluting stents. Apparatus 7 was introduced into the USP as a small-volume option for small transdermal patches. It was earlier known as the reciprocating disk apparatus and later renamed as the reciprocating holder apparatus. The rate of agitation is typically 30 cycles/min at an amplitude of 2 cm. The conventional vessel volume is 50–400 mL.[24]

19.2.6 Other Non-compendial Tests

19.2.6.1 Rotating bottle apparatus

The rotating bottle apparatus was introduced by Professor Beckett in 1970 for in vitro release testing of controlled-release formulations. The dosage form is placed in tubes or bottles, which are rotated using a revolving drum. The entire setup is placed in a water bath, which controls the temperature. The rotating bottle apparatus provides robust hydrodynamic conditions, i.e., the dosage form is able to rotate freely throughout the dissolution medium as the bottle rotates, which can be compared to USP apparatus 1 and 2, where different portions of the bulk medium move at different rates. The rotating bottle apparatus is mostly used for tablets and capsules. There are several shortcomings with this method. The system must be stopped during sampling and media replacement (which may affect the release profile), thus making it labor intensive. Dissolution testing in this apparatus may lead to inconsistencies in dissolution rate due to rough handling of dosage forms during the test. Due to frequent media replacements, dissolution testing in this apparatus is difficult to be automated.[24] For these reasons, the rotating bottle apparatus is not an official USP method.[42–45]

19.2.6.2 Biphasic dissolution method

The biphasic dissolution method is used for dissolution testing of controlled-release formulations and poorly soluble compounds.[46,47] For poorly soluble compounds, a non-sink biorelevant aqueous

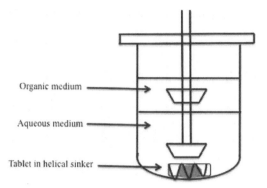

Figure 19.4 A typical representation of a double paddle biphasic dissolution apparatus.

medium is required, while the organic phase acts as the sink. In addition, this method may be used to model drug partitioning into fatty tissue at the subcutaneous site as well as across the intestinal lumen. The biphasic dissolution method involves both an aqueous phase and an organic phase. The dosage form is placed in the aqueous phase where dissolution of drug occurs, the drug then partitions into the organic phase. As initially dissolved drug partitions into the organic phase, more drug can dissolve into the aqueous phase and the cycle continues. The biphasic dissolution apparatus shown in Fig. 19.4 can also be connected to USP apparatus 4, in which the dosage form is place inside the USP apparatus 4 cell and a pump forces the aqueous medium through the cell. This biphasic dissolution method results in a non-saturable process and therefore sink conditions can be maintained throughout the testing period. This method can be used to investigate the impact of formulation changes on partitioning as well as dissolution and therefore may provide insight into drug absorption. The method can be scaled to reflect in vivo conditions.[48,49]

There are studies ongoing on selecting appropriate organic solvents as well as investigating the compatibility of this method with various BCS class II drugs. There are several limitations for this dissolution method. Solvent volatility is an issue, thus limiting the number of applicable organic media. In case of viscous organic solvents such as octanol, sample removal can be tedious.

19.2.6.3 Dissolution models simulating GI physical stress forces

Real-time imaging of transit of dosage form through gastrointestinal tract has shown that movement of dosage form is erratic with periods of rest and periods of high velocity. Garbacz et al.[50] developed a novel dissolution stress test device, which exposes the dosage form to an arbitrary sequence of movements, pressure waves, and periods of rest that may occur in the GI tract. The dissolution stress test device is capable of mimicking the irregular in vivo GI transit pattern and physical forces applied on dosage forms during gastric emptying and intestinal transit. The dosage form does not remain in constant contact with the media, which reproduces the effect of transit through an intestinal compartment with segmented fluid distribution. This device was used to obtain diclofenac release from extended-release formulations. Multiple peaks were observed in individual diclofenac plasma concentration profiles after dosing an extended release formulation, in contrast to a smooth continuous release profile observed with USP II apparatus.[51] Burke et al.[52] developed a simpler modification of USP II apparatus for replicating in vivo GI transit conditions. The dosage form is enclosed in a housing, which is subjected to forces of varying time intervals and intensities. Both of these apparatus can assess mechanical robustness of extended-release formulations. They can also predict the ability of these formulations to withstand gastric forces and passage through the pylorus or ileocecal junction.[51,53,54]

19.3 Current Non-compendial Tests for Special Dosage Forms

Due to the complexities in dissolution testing of dosage forms such as chewing gums/tablets, soft gelatin capsules, sublingual, buccal and floating tablets, as well as poorly soluble compounds, the current compendial tests may not be suitable for dissolution testing of these formulations. Accordingly, non-compendial apparatus have been developed for these special dosage forms.

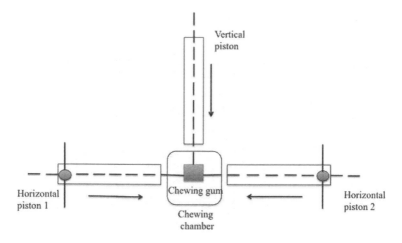

Figure 19.5 Representative diagram of chewing gum tester.

19.3.1 Chewing Gums and Tablets

Medicated chewing gum is a solid, single-dose preparation intended to be chewed and not swallowed. Chewing tablets are chewed and swallowed. A three-piston apparatus has been designed for dissolution testing of chewing gums and tablets. It consists of a chewing chamber (Fig. 19.5), which is maintained at 37°C ± 0.5°C. The dosage form is placed in the chewing chamber along with the release medium. Two electronically controlled horizontal pistons mechanically "chew" the dosage form. Compressed air is the driving force for the pistons. A third vertical piston alternates with the horizontal pistons and ensures proper positioning of the dosage form. The mechanical chewing rate can be varied. The volume of medium, the twisting movement, and the distance between the two horizontal pistons can be varied.[55–57]

19.3.2 Soft Gelatin Capsules

Soft gelatin capsules are single-unit dosage forms containing drug in either a hydrophilic or lipophilic liquid that is enclosed in a gelatin shell. USP apparatus 1 and 2 can be used for dissolution testing of soft gelatin capsules.[55] Dissolution testing using USP apparatus 2 can be problematic if the soft gelatin capsules float, in which

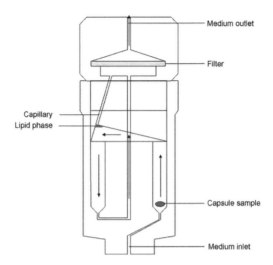

Figure 19.6 Representative diagram of soft gelatin capsule dissolution apparatus. Adapted from Azarmi et al.[56]

case the helical sinker should be used. When using USP apparatus 1, the soft gelatin capsules remain immersed within the basket. However, blockage of the basket mesh may occur, affecting the release profile.[58] A modified dual-chamber USP apparatus 4 cell has also been developed for in vitro release testing of soft gelatin capsules containing lipophilic liquids. A schematic diagram of this apparatus is shown in the Fig. 19.6. The lipoidal content rises after the capsule ruptures and reaches the triangular region on the top left side of the cell. The continuous flow of media through the lipoidal content results in drug extraction. The media exits the cell at the top and can then be collected to analyze the drug content.[58, 59]

19.3.3 Buccal and Sublingual Tablets

The buccal cavity has lower fluid volume, shorter residence time, and different composition compared to the stomach and intestine. Accordingly, there is a significant difference in buccal dissolution when compared to gastrointestinal dissolution. An in vitro release testing method designed by Rohm and Hass Laboratories-Springhouse consists of a continuous flow-through cell with a

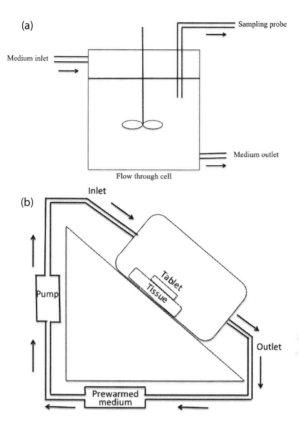

Figure 19.7 Representative diagram of (a) Rohm and Haas model and (b) Mumtaz and Ching model.

central shaft with propeller (Fig. 19.7a). The flow-through cell has a dip tube inlet with a filter for sampling and media. A pump drives the media through the cell.[60]

Another method designed by Mumtaz and Ching in 1995[61] is capable of evaluating drug release and the bioadhesive properties of buccal tablets. The apparatus developed by this group consists of a dissolution cell that holds the buccal tablet on a chicken pouch membrane (Fig. 19.7b). A chicken buccal pouch, freshly excised from the cheek area, was used in this model because of ease of availability and tissue uniformity. Pre-warmed dissolution medium circulates

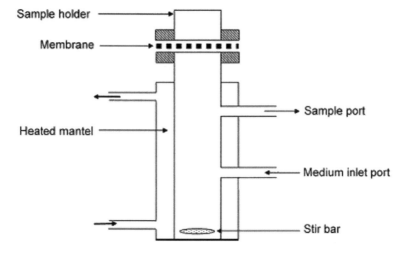

Figure 19.8 Representative diagram of Franz diffusion cell. Adapted from Azarmi et al.[56]

through cell. The outer assembly provides an angle which allows for gravitational flow of the dissolution medium.[55, 60, 62]

19.3.4 Semisolid Dosage Forms

Currently, there are no compendial drug release tests for semisolid dosage forms such as creams, ointments, and gels. The Franz diffusion cell (Fig. 19.8) is typically used for these dosage forms. The critical parameters in the Franz diffusion cell include diffusion cell volume and surface area, appropriate membrane, receiving medium, stirring speed, and temperature. The membrane should be nonreactive with the drug and should not contain leachables that interfere with the assay. To achieve sink conditions, receiver medium should have a high capability to dissolve the drug. There exists a lot of flexibility in terms of equipment design and experimental protocol. As a result of this, each laboratory can have its own set of Franz diffusion cells with varying physical dimensions and design characteristics.[63] Selection of pH of aqueous components depends on the formulation pH, the membrane, and the pH solubility of the drug.[64]

An enhancer cell designed by the Vankel Technology Group has also been used for dissolution testing of semisolids. It consists of a Teflon cell, the volume of which is adjustable and a screw cap to retain the artificial membrane. The gel or cream is introduced into the cell and the membrane provides a defined surface area to determine drug release.[56]

19.3.5 Poorly Soluble Compounds

Parikh et al.[65] developed the multi-compartment dissolution model to study dissolution of a weakly basic drug from a gastro-retentive dosage form. This method includes a gastric reservoir and an intestinal reservoir. The gastric reservoir contains 5 L of 0.1 N HCl to simulate the acidic conditions of the stomach, whereas the intestinal reservoir contains 5 L of 1.2 M alkaline borate buffers to mimic alkaline conditions. The gastric compartment (approximately 70 mL volume) has a side opening, which mimics the pylorus opening. The intestinal compartment has a volume of approximately 400 mL. There is an absorption compartment that receives fluid from the gastric and intestinal compartments at the rate of 4 mL/min. The absorption compartment is maintained at a constant agitation of 75 rpm. Samples are withdrawn at different time points from the absorption chamber and analyzed.[55]

19.3.6 Controlled-Release Parenterals

Controlled-release parenteral preparations are administered by non-oral routes. These dosage forms include microspheres, liposomes, nanoparticles, and implants. Modified USP apparatus 4, as described above, has been used for in vitro release testing of these dosage forms and is recommended for microsphere formulations.[37] Non-compendial "sample-and-separate" methods are also used for controlled release parenterals. In these methods, the formulation is dispersed in limited volumes of release media in vials or bottles. The vials or bottles are placed into water shaker baths and sampling is performed at specific time intervals depending on the drug and formulation. When sampling, the dosage forms such as microspheres, liposomes, and nanoparticles have to be

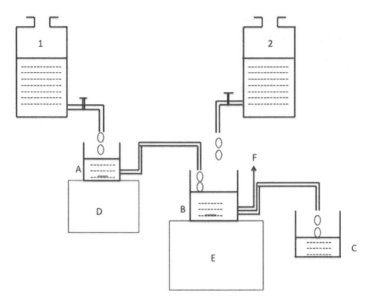

Figure 19.9 Representative diagram of a multi-compartment dissolution model. 1, Gastric reservoir; 2, intestinal reservoir; A, gastric compartment; B, intestinal compartment; C, absorption compartment; D and E, magnetic stirrers with heaters; F, filter.

separated from release media using centrifugation or filtration. These separation steps may lead to loss of the dosage form during sampling and the effect is cumulative with greater variability at later sampling time points. Other limitations of this method include potential violation of sink conditions, aggregation, and breakage of the dosage form due to forces involved in centrifugation and filtration. Also, the use of different sized vials or bottles makes inter-laboratory comparison difficult.[66]

The membrane diffusion method has also been used for in vitro release testing of liposomes, nanoparticles, and microspheres. This test is performed similarly to the sample and separate methods. The only difference is that a dialysis membrane of appropriate molecular weight cut-off is used to separate the formulation and the release media. The major limitation of this technique is that inadequate agitation inside the dialysis sac may cause aggregation. Another limitation of this technique is that appropriate selection of dialysis membrane must be made to avoid interactions between the membrane and drug as well as to allow rapid diffusion of the drug while retaining the dosage form.[67]

19.4 Detection Techniques for Dissolution Testing

There exists a need for development of new analytical techniques to understand the behavior of complex dissolution processes. Many imaging techniques have recently been studied for use in dissolution testing, including UV imaging with Raman spectroscopy, magnetic resonance imaging (MRI), fourier transform infrared spectroscopy (FT-IR) and near infrared (NIR). It is difficult to use FT-IR and NIR with an aqueous solvent due to the interfering absorption from aqueous solvent.

19.4.1 UV Imaging and Raman Spectroscopy

Traditional dissolution testing involves measurement of drug release into the bulk solution as a function of time. It provides no information about the sequence of events that occur close to the surface of the solid/tablet. UV imaging along with Raman spectroscopy have enabled real-time visualization of solution concentrations around a given solid sample. The real-time visualization of solution concentration around the solid samples provides new insights into the early stages of dissolution. During dissolution, nucleation may result because of local supersaturation and a new solid may be formed which may affect the dissolution rate.[68–70]

The UV imaging technique generates spatially and temporarily resolved solution phase information, called absorption maps. This technique, when combined with the channel flow cell method (Fig. 19.10), can provide information about the dissolution rate. A channel flow cell has a certain light path and cell volume in which the sample is placed and can be chosen as per experimental needs. The channel flow cell method is known to have well defined hydrodynamics. Raman spectroscopy plays a role in understanding the solid-state changes during dissolution testing.[71–73]

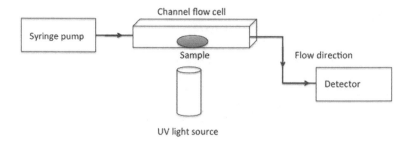

Figure 19.10 Schematic representation of UV imaging.

Boetker et al. used UV imaging and Raman spectroscopy to monitor drug dissolution of amlodipine and amlodipine besylate in real time.[74]

19.4.2 FTIR-ATR Spectroscopy

Fourier transform infrared-attenuated total reflection (FTIR-ATR) spectroscopy provides insight into various processes within a tablet and thus has been widely studied for use in tablet dissolution. The API and excipients (such as diluents, solvents, and binders) can be readily discriminated and quantified by the high chemical specificity of FTIR spectroscopy. This technique reveals the spatial distribution of these materials within the tablet as a function of time, thus facilitating the building of new mathematical models for the optimization of controlled drug delivery and thus allowing better understanding of the mechanism of drug release.[75,76] FTIR imaging allows for simultaneous acquisition of IR spectra in a relatively short amount of time (minutes) from different locations within the tablet. FTIR is dependent on the path length and can only analyze samples over short path lengths. Thus, tablets that are thicker than 10 μm cannot be analyzed using FTIR as it results in complete absorption of IR beam. However, FTIR combined with ATR is independent of sample thickness and provides high spatial resolution as well as high chemical specificity. All ATR crystals have different spectral ranges. Some of the commonly used ATR crystals are zinc selenide (ZnSe), germanium, thallium bromoiodide (KRS-5), and zinc sulfide

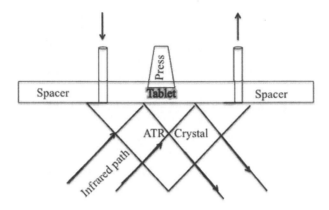

Figure 19.11 Schematic representation of FTIR-ATR representation.

(ZnS).[77] Figure 19.11 shows a typical representation of an FTIR-ATR experiment.[68,78]

19.4.3 UV Fiber Optics

UV fiber optics allows direct measurement of dissolved drug in the dissolution vessel. Individual probes are used for each dosage unit tested. One of the advantages of this technique is that tubing is not required and therefore analyte adsorption, leaching of tubing components, and effectiveness of cleaning procedures between dissolution runs are avoided. Another advantage is that sample separation is not required as sample analysis takes place in-line.[79–81] This cuts down greatly on the amount of time and effort needed in sample analysis. In situ measurements are possible, which means no samples have to be withdrawn during the test. In addition, real-time profiles can be obtained, providing instantaneous results on product performance.[82–84]

19.5 Conclusions

Compendial tests were designed for dissolution testing of conventional dosage forms. In recent years, modifications have been made to existing compendial tests and new tests have been developed to overcome the challenges that novel dosage forms such as liposomes, microspheres, and nanoparticles present. Non-compendial tests can provide reliable information on in vitro drug release, thus making qualification and validation of such methods possible. For many dosage forms, slight modifications of the existing methods are sufficient to achieve reproducible data. In addition, in vitro–in vivo correlations have been developed using non-compendial tests.

References

1. Bettini R. Pharmaceutical dissolution testing. *J. Controlled Release.* 1994;32:204–205. doi:10.1016/0168-3659(94)90064-7.

2. Dressman JB, Amidon GL, Reppas C, Shah VP. Dissolution testing as a prognostic tool for oral drug absorption: immediate release dosage forms. *Pharm. Res.* 1998;15:11–22. doi:10.1023/A:1011984216775.

3. Scheubel E, Lindenberg M, Beyssac E, Cardot J-M. Small volume dissolution testing as a powerful method during pharmaceutical development. *Pharmaceutics.* 2010;2(4):351–363. doi:10.3390/pharmaceutics2040351.

4. Uppoor VRS. Regulatory perspectives on in vitro (dissolution)/in vivo (bioavailability) correlations. *J. Controlled Release.* 2001;72:127–132. doi:10.1016/S0168-3659(01)00268-1.

5. FDA. Guidance for Industry Dissolution Testing of Immediate Release Solid Oral Dosage Forms. *Evaluation.* 1997.

6. Kauffman JS. Qualification and validation of USP Apparatus 4. *Dissolut. Technol.* 2005;12:41–43.

7. Taverniers I, De Loose M, Van Bockstaele E. Trends in quality in the analytical laboratory. II. Analytical method validation and quality assurance. *TrAC - Trends Anal. Chem.* 2004;23:535–552. doi:10.1016/j.trac.2004.04.001.

8. Swarbrick J. Drug dissolution testing: today and tomorrow. *Drug Dev. Commun.* 1976;2:429–438. doi:10.3109/03639047609051909.

9. Pharmaceutical Validation: Method Validation Guidelines. Available at: http://pharmaceuticalvalidation.blogspot.com/2009/12/method-validation-guidelines.html. Accessed July 27, 2015.

10. Rozet E, Ziemons E, Marini RD, Boulanger B, Hubert P. Validation of analytical methods involved in dissolution assays: acceptance

limits and decision methodologies. *Anal. Chim. Acta.* 2012;751:44–51. doi:10.1016/j.aca.2012.09.017.

11. Donato EM, Martins LA, Fröehlich PE, Bergold AM. Development and validation of dissolution test for lopinavir, a poorly water-soluble drug, in soft gel capsules, based on in vivo data. *J. Pharm. Biomed. Anal.* 2008;47:547–552. doi:10.1016/j.jpba.2008.02.014.

12. General Chapters: <1225> Validation of Compendial Methods. Available at: http://www.pharmacopeia.cn/v29240/usp29nf24s0_c1225. html. Accessed July 27, 2015.

13. Anand O, Yu LX, Conner DP, Davit BM. Dissolution testing for generic drugs: an FDA perspective. *AAPS J.* 2011;13:328–335. doi:10.1208/ s12248-011-9272-y.

14. Skelly JP, Gonzalez MA. FDA update–dissolution testing: simple tool– important contribution. *Eur. J. Pharm. Biopharm.* 1993;39:222–223.

15. Pharmacopoeia E. <711> Dissolution. 2011;1.

16. Kumar R, Nagarwal RC, Dhanawat M, Pandit JK. In-vitro and in-vivo study of indomethacin loaded gelatin nanoparticles. *J. Biomed. Nanotechnol.* 2011;7:325–333. doi:10.1166/jbn.2011.1290.

17. Cao X, Deng WW, Fu M, et al. In vitro release and in vitro-in vivo correlation for silybin meglumine incorporated into Hollow-type mesoporous silica nanoparticles. *Int. J. Nanomed.* 2012;7:753–762. doi:10.2147/IJN.S28348.

18. Stippler E. Compendial Dissolution: Theory and Practice. Available at: http://www2.aaps.org/uploadedFiles/Content/Sections_and_ Groups/Focus_Groups/In_Vitro_Release_and_Dissolution_Testing/Re-sources/IVRDTFGStippler2011.pdf. Accessed August 17, 2015.

19. Brown CK, Friedel HD, Barker AR, et al. FIP/AAPS joint workshop report: dissolution/in vitro release testing of novel/special dosage forms. *AAPS PharmSciTech.* 2011;12(2):782–794. doi:10.1208/s12249-011-9634-x.

20. Phillips DJ, Pygall SR, Cooper VB, Mann JC. Overcoming sink limitations in dissolution testing: a review of traditional methods and the potential utility of biphasic systems. *J. Pharm. Pharmacol.* 2012;64:1549–1559. doi:10.1111/j.2042-7158.2012.01523.x.

21. Pillay V, Fassihi R. Unconventional dissolution methodologies. *J. Pharm. Sci.* 1999;88:843–851. doi:10.1021/js990139b.

22. Dorozhkin SV. A review on the dissolution models of calcium apatites. *Prog. Cryst. Growth Charact. Mater.* 2002;44:45–61. doi:10.1016/ S0960-8974(02)00004-9.

23. Equipment D. *Dissolution Theory, Methodology.*

24. Crist GB. Trends in small-volume dissolution apparatus for low-dose compounds. *Dissolut. Technol.* 2009;28–31.

25. Qureshi SA, McGilveray IJ. Typical variability in drug dissolution testing: Study with USP and FDA calibrator tablets and a marketed drug (glibenclamide) product. *Eur. J. Pharm. Sci.* 1999;7:249–258. doi:10.1016/S0928-0987(98)00034-7.

26. Scholz A, Kostewicz E, Abrahamsson B, Dressman JB. Can the USP paddle method be used to represent in-vivo hydrodynamics? *J. Pharm. Pharmacol.* 2003;55:443–451. doi:10.1211/002235702946.

27. Long M, Chen Y. Dissolution testing of solid products. In: *Developing Solid Oral Dosage Forms.* 2009:319–340. doi:10.1016/B978-0-444-53242-8.00014-X.

28. Lee SL, Raw AS, Yu L. Dissolution testing. In: *Biopharmaceutics Applications in Drug Development.* 2008:47–74. doi:10.1007/978-0-387-72379-2.

29. US Department of Health and Human Services F and DAC for D, CDER E and R. *Guidance for Industry Dissolution Testing of Immediate Release Solid Oral Dosage Forms.* 1997. Available at: http://www.fda.gov/downloads/Drugs/.../Guidances/ucm070246.pdf\nhttp://scholar.google.com/scholar?hl=en&btnG=Search&q=intitle:Guidance+for+Industry:+Dissolution+Testing+of+Immediate+Release+Solid+Oral+Dosage+Forms#0\nhttp://scholar.google.com/scholar?hl=en&btnG=Search&q=intitle:FDA+guidance+for+industry+dissolution+testing+of+immediate+release+solid+oral+dosage+forms#0.

30. Bai G, Armenante PM, Plank R V., Gentzler M, Ford K, Harmon P. Hydrodynamic investigation of USP dissolution test apparatus II. *J. Pharm. Sci.* 2007;96:2327–2349. doi:10.1002/jps.20818.

31. Joshi A, Pund S, Nivsarkar M, Vasu K, Shishoo C. Dissolution test for site-specific release isoniazid pellets in USP apparatus 3 (reciprocating cylinder): Optimization using response surface methodology. *Eur. J. Pharm. Biopharm.* 2008;69:769–775. doi:10.1016/j.ejpb.2007.11.020.

32. Yu LX, Wang JT, Hussain AS. Evaluation of USP apparatus 3 for dissolution testing of immediate-release products. *AAPS PharmSci.* 2002;4:E1. doi:10.1208/ps040101.

33. Eaton JW, Tran D, Hauck WW, Stippler ES. Development of a performance verification test for USP apparatus 4. *Pharm. Res.* 2012;29:345–351. doi:10.1007/s11095-011-0559-6.

34. Looney TJ. USP apparatus 4 - applying the technology. *Dissolut. Technol.* 1997;4.

35. Jünemann D, Dressman J. Analytical methods for dissolution testing of nanosized drugs. *J. Pharm. Pharmacol.* 2012;64:931–943. doi:10.1111/j.2042-7158.2012.01520.x.

36. Zolnik BS, Leary PE, Burgess DJ. Elevated temperature accelerated release testing of PLGA microspheres. *J Controlled Release.* 2006;112:293–300. doi:10.1016/j.jconrel.2006.02.015.

37. Rawat A, Burgess DJ. USP apparatus 4 method for in vitro release testing of protein loaded microspheres. *Int. J. Pharm.* 2011;409:178–184. doi:10.1016/j.ijpharm.2011.02.057.

38. Rawat A, Bhardwaj U, Burgess DJ. Comparison of in vitro-in vivo release of Risperdal® Consta® microspheres. *Int. J. Pharm.* 2012;434:115–121. doi:10.1016/j.ijpharm.2012.05.006.

39. Ag S. USP 4 Flow-Through Dissolution Systems: 1–12.

40. Rawat A, Stippler E, Shah VP, Burgess DJ. Validation of USP apparatus 4 method for microsphere in vitro release testing using Risperdal Consta. *Int. J. Pharm.* 2011;420(2):198–205. doi:10.1016/j.ijpharm.2011.08.035.

41. Bhardwaj U, Burgess DJ. A novel USP apparatus 4 based release testing method for dispersed systems. *Int. J. Pharm.* 2010;388:287–294. doi:10.1016/j.ijpharm.2010.01.009.

42. Technologies A. Rotating Bottle Apparatus Operator's Manual.

43. Esbelin B, Beyssac E, Aiache JM, Shiu GK, Skelly JP. *A new method of dissolution in vitro, the "Bio-Dis" apparatus: comparison with the rotating bottle method and in vitro: in vivo correlations.* 1991.

44. Bernardez LA. Dissolution of polycyclic aromatic hydrocarbons from a non-aqueous phase liquid into a surfactant solution using a rotating disk apparatus. *Colloids Surfaces A Physicochem. Eng. Asp.* 2008;320:175–182. doi:10.1016/j.colsurfa.2008.01.044.

45. Thakker KD, Naik NC, Gray VA, Sun S. Fine tuning of dissolution apparatus for the apparatus suitability test using the USP dissolution calibrators. *Pharmacopeial Forum.* 1980;6:177–185.

46. Chiu R, Vangani S, Delbarrio M-A, et al. Dissolution of a poorly-water soluble drug - AMG 517 in biphasic media using USP apparatus 2 and 4. In: *AAPS Annual Meeting.* Vol 1000; 2007:1000–1000.

47. Shi Y, Gao P, Gong Y, Ping H. Application of a biphasic test for characterization of in vitro drug release of immediate release formulations of celecoxib and its relevance to in vivo absorption. *Mol. Pharm.* 2010;7:1458–1465. doi:10.1021/mp100114a.

48. Phillips DJ, Pygall SR, Cooper VB, Mann JC. Toward biorelevant dissolution: application of a biphasic dissolution model as a discriminating tool for HPMC matrices containing a model BCS class II drug. *Dissolut. Technol.* 2012;19:25–34.

49. Chaudhary RS, Gangwal SS, Gupta VK, Shah YN, Kindal KC, Khanna S. Dissolution system for nifedipine sustained-release formulations. *Drug Dev. Ind. Pharm.* 1994;20:1267–1274.

50. Garbacz G, Blume H, Weitschies W. Investigation of the dissolution characteristics of nifedipine extended-release formulations using USP apparatus 2 and a novel dissolution apparatus. *Dissolut. Technol.* 2009;16:7–13.

51. Mcallister M. Dynamic dissolution: a step closer to predictive dissolution testing? 2010;60(6):30–42.

52. Burke M, Kalantzi L PA. Pharmaceutical analysis apparatus and method. 2010. Available at: https://www.google.com/patents/US20100126287. Accessed August 16, 2015.

53. Koziolek M, Görke K, Neumann M, Garbacz G, Weitschies W. Development of a bio-relevant dissolution test device simulating mechanical aspects present in the fed stomach. *Eur. J. Pharm. Sci.* 2014;57:250–256. doi:10.1016/j.ejps.2013.09.004.

54. Jantratid E, Janssen N, Reppas C, Dressman JB. Dissolution media simulating conditions in the proximal human gastrointestinal tract: An update. *Pharm. Res.* 2008;25:1663–1676. doi:10.1007/s11095-008-9569-4.

55. Review Article. An updated review of dissolution apparatus for conventional and novel dosage forms. 2013;2(July):42–53.

56. Azarmi S, Roa W, Löbenberg R. Current perspectives in dissolution testing of conventional and novel dosage forms. *Int. J. Pharm.* 2007;328(1):12–21. doi:10.1016/j.ijpharm.2006.10.001.

57. Shah VP, Siewert M, Dressman J, Moeller H, Brown CK. Dissolution/in vitro release testing of special dosage forms. *In Vitro.* 2002;9:1–5.

58. Gullapalli RP. Soft gelatin capsules (softgels). *J. Pharm. Sci.* 2010;99: 4107–4148. doi:10.1002/jps.22151.

59. Hom FS. Soft gelatin capsules. Part 3. Accelerated method for evaluating the dissolution stability of various gel formulations. *Drug. Dev. Ind. Pharm.* 1984;10:275–287. doi:10.3109/03639048409064650.

60. Rachid O, Rawas-Qalaji M, Simons FER, Simons KJ. Dissolution testing of sublingual tablets: a novel in vitro method. *AAPS PharmSciTech.* 2011;12(2):544–52. doi:10.1208/s12249-011-9615-0.

61. Mumtaz AM, Ch'ng HS. Design of a dissolution apparatus suitable for in situ release study of triamcinolone acetonide from bioadhesive buccal tablets. *Int. J. Pharm.* 1995;121(2):129–139. doi:10.1016/0378-5173(94)00406-U.

62. Rachid O, Rawas-Qalaji M, Simons FER, Simons KJ. Rapidly-disintegrating sublingual tablets of epinephrine: Role of non-medicinal ingredients in formulation development. *Eur. J. Pharm. Biopharm.* 2012; 82:598–604. doi:10.1016/j.ejpb.2012.05.020.

63. Ng S-F, Rouse JJ, Sanderson FD, Meidan V, Eccleston GM. Validation of a static Franz diffusion cell system for in vitro permeation studies. *AAPS PharmSciTech.* 2010;11(3):1432–41. doi:10.1208/s12249-010-9522-9.

64. Bonferoni MC, Rossi S, Ferrari F, Caramella C. A modified Franz diffusion cell for simultaneous assessment of drug release and washability of mucoadhesive gels. *Pharm. Dev. Technol.* 1999;4:45–53. doi:10.1080/10837459908984223.

65. Parikh RK, Parikh DC, Delvadia RR, Patel SM. A novel multicompartment dissolution apparatus for evaluation of floating dosage form containing poorly soluble weakly basic drug. *Dissolut. Technol.* 2006;13(1): 14–19.

66. B. S. Zolnik, J-L. Raton and DJB. Application of USP apparatus 4 and in situ fiber optic analysis to microsphere release testing. *Dissolut. Technol.* 2005:11–14.

67. D'Souza SS, Faraj JA, DeLuca PP. A model-dependent approach to correlate accelerated with real-time release from biodegradable microspheres. *AAPS PharmSciTech.* 2005;6:E553–E564. doi:10.1208/pt060470.

68. Van der Weerd J, Kazarian SG. Combined approach of FTIR imaging and conventional dissolution tests applied to drug release. *J. Controlled Release.* 2004;98(2):295–305. doi:10.1016/j.jconrel.2004.05.007.

69. Niederquell A, Kuentz M. Biorelevant dissolution of poorly soluble weak acids studied by UV imaging reveals ranges of fractal-like kinetics. *Int. J. Pharm.* 2014;463:38–49. doi:10.1016/j.ijpharm.2013.12.049.

70. Savolainen M, Kogermann K, Heinz A, et al. Better understanding of dissolution behaviour of amorphous drugs by in situ solid-state analysis using Raman spectroscopy. *Eur. J. Pharm. Biopharm.* 2009;71:71–79. doi:10.1016/j.ejpb.2008.06.001.

71. Ozaki Y, Šašić S. Introduction to Raman Spectroscopy. *Pharm. Appl. Raman Spectrosc.* 2007:1–28. doi:10.1002/9780470225882.ch1.

72. Müller J, Brock D, Knop K, Axel Zeitler J, Kleinebudde P. Prediction of dissolution time and coating thickness of sustained release formulations using Raman spectroscopy and terahertz pulsed imaging. *Eur. J. Pharm. Biopharm.* 2012;80:690–697. doi:10.1016/j.ejpb.2011.12.003.

73. Østergaard J, Wu JX, Naelapää K, Boetker JP, Jensen H, Rantanen J. Simultaneous UV imaging and Raman spectroscopy for the measurement of solvent-mediated phase transformations during dissolution testing. *J. Pharm. Sci.* 2014;103:1149–1156. doi:10.1002/jps.23883.

74. Boetker JP, Savolainen M, Koradia V, et al. Insights into the early dissolution events of amlodipine using UV imaging and Raman spectroscopy. *Mol. Pharm.* 2011;8(4):1372–80. doi:10.1021/mp200205z.

75. Kazarian SG, Chan KLA. Applications of ATR-FTIR spectroscopic imaging to biomedical samples. *Biochim. Biophys. Acta, Biomembr.* 2006;1758:858–867. doi:10.1016/j.bbamem.2006.02.011.

76. Hind AR, Bhargava SK, McKinnon A. At the solid/liquid interface: FTIR/ATR–the tool of choice. *Adv. Colloid Interface Sci.* 2001;93:91–114. doi:10.1016/S0001-8686(00)00079-8.

77. FTIR and UV-Vis Accessories, ATR Specular and Diffuse Reflectance, Integrating Spheres, Microscopes and Transmission. Available at: http://www.piketech.com/ATR-Crystal-Selection.html. Accessed August 17, 2015.

78. Wray PS, Clarke GS, Kazarian SG. Dissolution of tablet-in-tablet formulations studied with ATR-FTIR spectroscopic imaging. *Eur. J. Pharm. Sci.* 2013;48:748–757. doi:10.1016/j.ejps.2012.12.022.

79. Crisp J. Introduction to fiber optics. In: *Introduction to Fiber Optics.* 2001:51–53. doi:10.1017/CBO9781139174770.

80. Mirza T, Liu Q, Vivilecchia R, Joshi Y. Comprehensive validation scheme for in situ fiber optics dissolution method for pharmaceutical drug product testing. *J. Pharm. Sci.* 2009;98:1086–1094. doi:10.1002/jps.21481.

81. Johansson J, Cauchi M, Sundgren M. Multiple fiber-optic dual-beam UV/Vis system with application to dissolution testing. *J. Pharm. Biomed. Anal.* 2002;29:469–476. doi:10.1016/S0731-7085(02)00091-2.

82. Martin CA. Evaluating the utility of fiber optic analysis for dissolution testing of drug products. *Dissolut. Technol.* 2003;10:37–39.

83. Glass AM, DiGiovanni DJ, Strasser TA, et al. Advances in fiber optics. *Bell Labs Tech. J.* 2000;5:168–187. doi:10.1002/bltj.2213.

84. Lu X, Xiao B, Lo L, Bolgar MS, Lloyd DK. Development of a two-step tier-2 dissolution method for blinded overencapsulated erlotinib tablets using UV fiber optic detection. *J. Pharm. Biomed. Anal.* 2011;56:23–29. doi:10.1016/j.jpba.2011.04.026.

Index

octanol-in-water 215
oil-in-water 162–163
enzymes 89, 134, 243, 315, 320,
    360–361, 364, 366–372, 379,
    440–441, 611, 618, 620, 624,
    628
  gut 127
  liver 127
  proteolytic 101, 366, 618, 620,
    624–625
ER, *see* extended release
ethanol 94, 98, 146, 165, 375–376
ethyl alcohol 608, 611–612, 638
eutectic mixtures 156
excipients 143–144, 163, 323,
    329, 355, 357, 374–375,
    400–401, 433, 436, 575, 579,
    583, 585, 626–627
  functional 12, 373
  high molecular weight 608
  hydrophilic 161, 163
  hydrophobic 575
  insoluble 338
  lipid-based 358
  lipidic 167
  polymeric 219, 231
  release-controlling 201
extended release (ER) 10, 96, 167,
    170, 216, 225, 252, 268, 470,
    474, 555, 576, 595, 665, 670

failure mode effects analysis
    (FMEA) 525, 533, 561, 563,
    572
fatty acids 167, 366, 465–466,
    469, 481, 606, 608, 633–635,
    638
felodipine 115, 219, 334,
    409–411, 460, 464
fenofibrate 115, 320, 339, 489,
    612
fiber optics 17, 661
fiberglass 288–290

fingolimod 95, 97
first-order absorption kinetics
    230
flow cells 109, 255–256, 259,
    261–262, 265, 630
flow-through cell (FTC) 255,
    259–260, 268–269, 362–363,
    397, 465, 482–483, 576, 630,
    665, 667, 672–673
FMEA, *see* failure mode effects
    analysis
foaming 245
food effects 126, 131, 141, 210,
    479–483, 485, 487, 489,
    491–493, 495, 579, 584, 586,
    636, 639
formulation approaches 135,
    140–141, 155, 218
  stabilized amorphous drug 396
formulation design 130, 194, 358
formulation development,
    toxicology 187
formulations
  amorphous-based drug 155
  bead 241
  bio-enhanced 490
  budesonide 482
  capsule fill 616
  chewable 10, 252
  controlled-release 168, 668
  controlled release nanoparticle
    333
  conventional micrometer-sized
    325
  extended-release 225, 555, 670
  fast-releasing 335
  functional-coated 557
  immediate-release 464, 662
  lipid-filled 362, 364
  lipid microparticle 339
  microsphere 675
  modified-release 492
  montelukast sodium 483
  nanoparticulate 315